Lecture Notes in Mechanical Engineering

Lecture Notes in Mechanical Engineering (LNME) publishes the latest developments in Mechanical Engineering—quickly, informally and with high quality. Original research reported in proceedings and post-proceedings represents the core of LNME. Volumes published in LNME embrace all aspects, subfields and new challenges of mechanical engineering.

To submit a proposal or request further information, please contact the Springer Editor of your location:

Europe, USA, Africa: Leontina Di Cecco at Leontina.dicecco@springer.com
China: Ella Zhang at ella.zhang@springer.com
India: Priya Vyas at priya.vyas@springer.com
Rest of Asia, Australia, New Zealand: Swati Meherishi at swati.meherishi@springer.com

Topics in the series include:

- Engineering Design
- Machinery and Machine Elements
- Automotive Engineering
- Engine Technology
- Aerospace Technology and Astronautics
- Nanotechnology and Microengineering
- MEMS
- Theoretical and Applied Mechanics
- Dynamical Systems, Control
- Fluid Mechanics
- Engineering Thermodynamics, Heat and Mass Transfer
- Manufacturing
- Precision Engineering, Instrumentation, Measurement
- Tribology and Surface Technology

Indexed by SCOPUS and EI Compendex. All books published in the series are submitted for consideration in Web of Science.

To submit a proposal for a monograph, please check our Springer Tracts in Mechanical Engineering at https://link.springer.com/bookseries/11693

Lahcen Azrar · Abdelilah Jalid · Samir Lamouri ·
Ali Siadat · Mourad Taha Janan · Fakher Chaari ·
Mohamed Haddar
Editors

Advances in Integrated Design and Production II

Proceedings of the 12th International Conference on Integrated Design and Production, CPI 2022, May 10–12, 2022, ENSAM, Rabat, Morocco

Springer

Editors
Lahcen Azrar
École Nationale Supérieure d'Arts et Métiers
Mohammed V University
Rabat, Morocco

Abdelilah Jalid
École Nationale Supérieure d'Arts et Métiers
Mohammed V University
Rabat, Morocco

Samir Lamouri
Arts et Metiers Institute of Technology
Paris, France

Ali Siadat
Arts et Metiers Institute of Technology
Paris, France

Mourad Taha Janan
École Nationale Supérieure d'Arts et Métiers
Mohammed V University
Rabat, Morocco

Fakher Chaari
National School of Engineers of Sfax
Sfax, Tunisia

Mohamed Haddar
National School of Engineers of Sfax
Sfax, Tunisia

ISSN 2195-4356 ISSN 2195-4364 (electronic)
Lecture Notes in Mechanical Engineering
ISBN 978-3-031-23614-3 ISBN 978-3-031-23615-0 (eBook)
https://doi.org/10.1007/978-3-031-23615-0

This Springer imprint is published by the registered company Springer Nature Switzerland AG
The registered company address is: Gewerbestrasse 11, 6330 Cham, Switzerland

Preface

The objectives of the CPI conference are to promote research, technology transfer and the creation of a space for constructive dialogue between different actors of industrial innovation. It allows highlighting current research work aimed at the control and better integration of activities related to the product life cycle. Design and Integrated Production represent the main technical activities that have played a fundamental role in creating and shaping products and systems.

The present edition of the conference has been held in Rabat between May 10 and 12, 2022, and comes to follow the past editions of 1995 and 1997 at the EST of Casablanca, 1999 at the FST of Tangier, 2001 at the EST of Fez, 2003 at the ENSAM of Meknes, 2005 at the EST of Casablanca, 2007 at the EMI of Rabat, 2009 at EST of FES, 2011 at the ENSA of Oujda, 2013 at the University of Tlemcen, 2015 in Tangier and 2019 in ENSA of FES.

This three-day event from May 10 to 12, organized by the Higher Engineering School ENSAM of Rabat in partnership with the Arts et Métiers Sciences et Technologies-Paris and the National School of Engineering Sfax (ENIS)-Tunisia, aims to create a platform for meetings and exchange of information between university researchers and industry in order to consolidate the links between research and industrial activity thus promoting collaboration between various research actors.

CPI-2022 has been an opportunity to share thoughts on the impact of Industrial Engineering and Industry 4.0 for the industrial performance of SMEs, modeling and simulation of safety-critical mechatronic systems, robotics and artificial intelligence, now a combination at the service of humans, and thus, identify future research avenues in the field of Design and Integrated Production.

CPI-2022 offered again an exciting technical program as well as networking opportunities. Outstanding scientists and industry leaders accepted the invitation for keynote speeches:

- Professor Jean Yves Choley, Laboratory Quartz, SUPMECA, Paris, France.
- Professor Mohamed Najib Ichchou, Laboratory of Tribology and System Dynamic, Ecole Centrale de Lyon, France.
- Professor Samir Lamouri, Arts et Métiers Sciences et Technologies-Paris, France.

- Professor Gérard Poulachon, ENSAM-Paris, France.
- Professor Nazih Mechbal, ENSAM-Paris, France.
- Professor Maxime Raison, Polytechnique Montreal, Canada.
- Professor Mounir Ben Amar, LSPM, CNRS, Paris, France.

The CPI-2022 conference takes up the following topics in its variety and discusses the state of the art and future trends:

Part One: Instrumentation, Measure and Metrology
Part Two: Industrial Engineering and Digital Factory
Part Three: Computing and Data-Driven Digital Industry
Part Four: Mechanical Engineering and Mechatronics
Part Five: Process and Integrated Manufacturing
Part Six: Integrated Design
Part Seven: Health, Bioengineering
Part Eight: Additive Manufacturing
Part Nine: Embedded Systems, Control and Modeling
Part Ten: Integrated Energy Production and Management

All contributions were subject to a double or triple review, and the review process was very competitive. Due to the time and conference schedule restrictions, we could finally accept only the best 61 papers. The conference had participants from different countries: Algeria, Cameroon, Canada, China, France, India, Russia, Tunisia, Viet Nam and Morocco.

This book is a collection of the best papers presented at the CPI-2022. The proceedings essentially consist of full paper submissions proposed by the session's chairs after oral presentation of the paper.

We would like to thank the organizing committee, scientific committee and all participants. Particular thanks are addressed to Springer for supporting the CPI-2022.

This conference would not have appeared without the support of several institutions and companies. We are mainly grateful to Mohammed V University and to the Hassan II Academy of Sciences and Techniques for their supports. Special thanks also go to the director of the ENSAM-Rabat as well as to SERDILAB and GASUP.

Rabat, Morocco — Lahcen Azrar
Rabat, Morocco — Abdelilah Jalid
Paris, France — Samir Lamouri
Paris, France — Ali Siadat
Rabat, Morocco — Mourad Taha Janan
Sfax, Tunisia — Fakher Chaari
Sfax, Tunisia — Mohamed Haddar

Contents

Instrumentation, Measure and Metrology

Estimation of the Cylindricity Defect on CMM by Coupling the Genetic Algorithm and the Interior Points Method

Jamal Gsim[1(✉)], Abdelilah Jalid[2], Mohamed Zeriab Es-sadek[1], and Malika Zazi[3]

[1] M2CS Laboratory, ENSAM, Mohammed V University, Rabat, Morocco
gsimjamale@gmail.com, m.essadek@um5r.ac.ma
[2] PCMT Laboratory, ENSAM, Mohammed V University, Rabat, Morocco
a.jalid@um5r.ac.ma
[3] LM2PI Laboratory, ENSAM, Mohammed V University, Rabat, Morocco
m.zazi@um5r.ac.ma

Abstract. In order to perform its function correctly in a mechanism, the specification of mechanical part must be estimated with sufficient precision to confirm its validity. The cylindricity error is estimated from a cloud of points taken from a CMM, and several methods to evaluate cylindricity have been proposed by different researchers. These adjustment methods are generally based on the least squares (LS) criteria or the minimum zone criteria, but these methods are not efficient in the case of the non convex and non linear problems. For that, authors use evolutionary algorithms like genetic algorithm or particle swarm optimization (PSO) algorithm, but also this type of algorithms don't guarantee the convergence. To remedy these problems the researchers use a coupling between these two types of problems. In this paper, an algorithm based on the coupling of the genetic algorithm and the interior point method is presented, then the algorithm is validated by comparing the obtained results with those of NIST.

Keywords: Cylindricity error · CMM cylindricity · Genetic algorithm · Hybrid algorithm

1 Introduction

The cylindricity error impacts the functionality of assemblies in mechanisms, which is why their determination is of great importance in the industrial field. On CMM a cloud of points are taken from a cylindrical surface, the operator estimates the cylindricity defect, the value of the estimated defect depends on the number of points. The association criterion and the algorithm of adjustment implemented in the machine software. The frequently mathematical optimization criteria used in the field are the least squares cylinder (LSC), the minimum zone cylinder (MZC), the minimum circumscribed cylinder (MCC) and the maximum inscribed cylinder (MIC). Several research works have treated the subject

L. Azrar et al. (Eds.): CIP 2022, LNME, pp. 3–11, 2023.
https://doi.org/10.1007/978-3-031-23615-0_1

such as Carr and Ferreira [2], they presented nonlinear optimization models for calculating cylindricity and straightness error of a median line which they solved using successive linear programs. A strategy based on geometric representation for minimum zone, evaluation of circles and cylinders is proposed by Lai and Chen [5]. This is an approximation strategy to the minimum zone circles or cylinders. Jalid et al. [4] worked on the estimation of parameters to determine the form defect of a surface based on the orthogonal distance regression ODR and provides the uncertainty associated. Generally, the modelling of the shape defect, in particular for the case of cylindricity is a non-linear and non-convex optimization problem, of which the standard optimization algorithms are likely to converge if the starting vector (an initial estimates) not correctly chosen and sometimes converges to a local optimum. In this article, we present a hybrid algorithm based on the coupling between the genetic algorithm which is an evolutionary algorithm and the interior points method which is a deterministic algorithm.

2 Mathematical Modelling of Cylindricity

Consider a cylinder of radius R, direction $d = \begin{pmatrix} x_d \\ y_d \\ z_d \end{pmatrix}$ and let A be a point of it axis (Fig. 1).

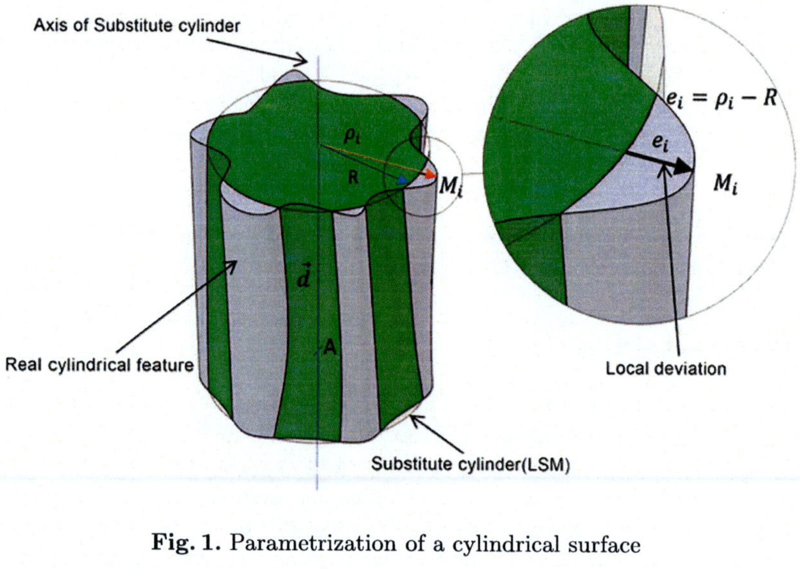

Fig. 1. Parametrization of a cylindrical surface

To determine cylindricity error by the least squares method (LSM), we must minimize the quantity:

$$W = \sum_{i=1}^{n} (\rho_i - R)^2 \tag{1}$$

where R is the radius of substitute cylinder and n the number of points collected. ρ is the radius measured from the axis to the point M_i:

$$\rho_i =< \vec{AM_i}, \vec{d} >= \left\| \begin{pmatrix} x_i - x_A \\ y_i - y_A \\ z_i - z_A \end{pmatrix} \wedge \begin{pmatrix} x_d \\ y_d \\ z_d \end{pmatrix} \right\| \tag{2}$$

Then, the point sought can be written $X = (x_1, x_2, x_3, x_4, x_5, x_6, x_7)^T$, where (x_1, x_2, x_3) are the coordinates of the point A which minimize the form defect, x_4 is the radius of the cylinder and (x_5, x_6, x_7) the coordinates of the optimal direction d of the cylinder.
Let $dat_{i,j},\ j = 1, 2, 3$ be the coordinates of the points M_i collected by the CMM, the problem is written:

$$\begin{cases} \min f(X) \\ a_1 \leq x_1 \leq b_1 \\ a_2 \leq x_2 \leq b_2 \\ a_3 \leq x_3 \leq b_3 \\ R_{min} \leq x_4 \leq R_{max} \\ x_5^2 + x_6^2 + x_7^2 = 1 \end{cases} \tag{3}$$

where the explicit definition of f is:

$$f(X) = \sum_{i=1}^{n} (\sqrt{((dat_{i,2} - x_2)x_7 - (dat_{i,3} - x_3)x_6)^2 + ((dat_{i,3} - x_3)x_5} \\ \sqrt{-(dat_{i,1} - x_1)x_7)^2 + ((dat_{i,1} - x_1)x_6 - (dat_{i,2} - x_2)x_5)^2} - x_4)^2$$

a_j and b_j are defined for $j = 1, 2, 3$ by:

$$a_j = \{\min dat_{i,j};\ i = 1, \ldots, n\} \quad \text{and} \quad b_j = \{\max dat_{i,j};\ i = 1, \ldots, n\}$$

and R_{min} and R_{max} are respectively the minimal and maximal radius. The cylindricity error (CE) is then estimated, after convergence of the algorithm, by:

$$CE = \max(\rho_i) - \min(\rho_i)$$

3 Resolution Method

The problem (3) is an optimization problem non-linear and non-convex, then the deterministic algorithms of optimisation don't give the global optimum. For that, the minimization algorithm applied to this problem is a hybrid algorithm based on the coupling of the genetic algorithm and the interior points method. It has been used in the case of flatness defect and has shown its efficiency [3].

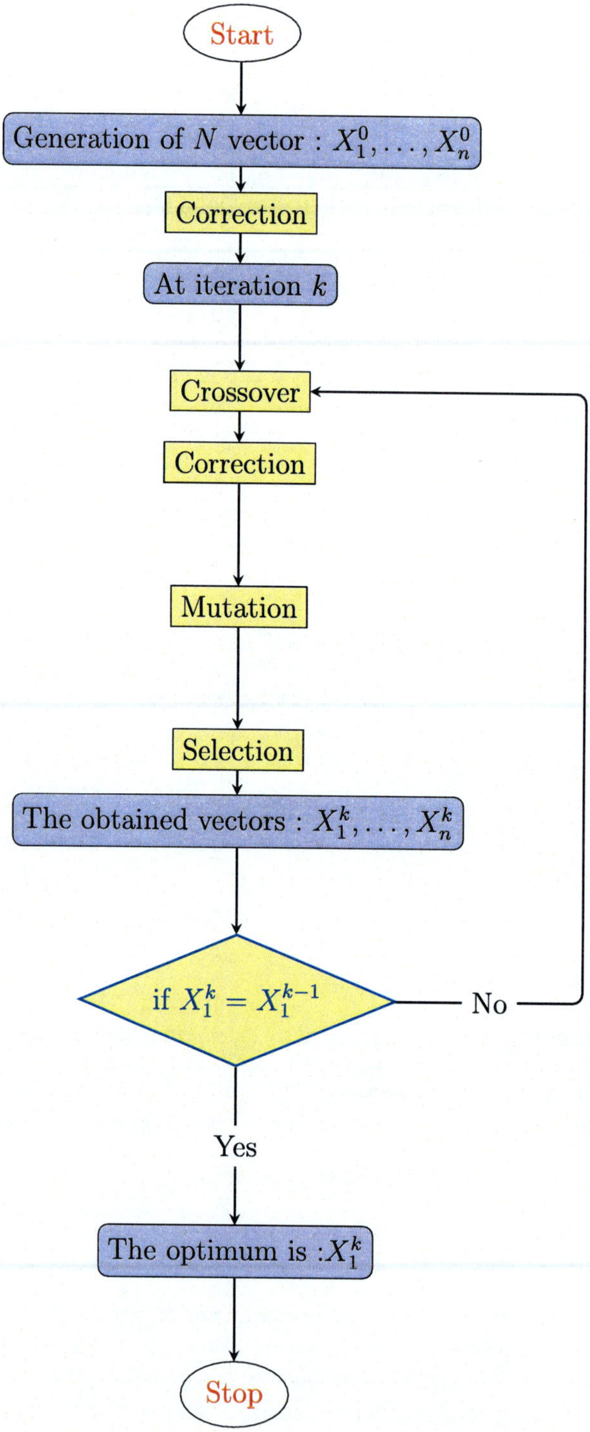

Fig. 2. The flowchart of the hybrid algorithm

3.1 Interior Points Method

Let the general optimization problem:

$$\begin{cases} \min f(x) \\ g_i(x) \leq 0 \;\; i = 1, \ldots, n \\ h_j(x) = 0 \;\; j = 1, \ldots, m \end{cases} \tag{4}$$

The difficulty in solving the problem (4) consists in finding a point which satisfies the constraints. The principle of the interior points method is to transform the problem into a problem without constraints, by using penalty functions. Then, the problem becomes:

$$\min f(x) - r \sum_{i=1}^{n} g_i(x) + \frac{1}{\sqrt{r}} \sum_{j=1}^{m} h_j^2(x) \tag{5}$$

where $r > 0$ is the penalty coefficient.
The problem (5) can be solved using any algorithm of optimization without constraints. The advantage of this method is that the optimum point obtained satisfies the constraints because the point cannot cross the boundary. But the disadvantage of this method is that it is local in the case where the function f is non-convex and non-linear.

3.2 Hybrid Algorithm

The genetic algorithm consists in applying three operators: Crossover, mutation and selection, to a population of N elements generated randomly, but the genetic doesn't converge to a global optimum when the domain is too large. For that a hybrid algorithm is proposed to solve the problem (3).
It consists in coupling the genetic algorithm to the interior points method by applying the interior point methods as the mutation operator of the genetic algorithm.
After the generation of N random vectors, and to be sure that these vectors satisfy the constraints, they are corrected towards the boundary using a correction procedure [1].
The crossover operator is applied to each two elements X and Y taken randomly from the population, and generating two new elements $\frac{X-Y}{2}$ and $\frac{X-Y}{3}$, then this points are corrected towards the boundary.
The selection operator consists in sorting the best N elements from the best to the worst.
The flowchart of the hybrid algorithm is illustrated in Fig. 2.

4 Results

To validate the method and the algorithm proposed, three data sets, provided by the National Institute of Standards and Technology (NIST) named cyl20,

cyl1 and cyl2, has been processed. The choice of these data sets highlights the problems most often encountered in optimization (convergence, local optimum, computing time ...).
By applying the proposed model, the parameters of the substitute cylinder estimated are compared to the reference values provided by NIST as shown in following subsections.

4.1 Data Set cyl20

This data set contains 48 points. The comparison between the results of the proposed method and NIST results is shown in Table 1.

Table 1. Comparison between the results found and those of NIST for data set cyl20

Parameters	NIST results	Proposed algorithm results	Gap
Point A	68.591540428292395007	67.9226563839905566	0.6688840443018336
	445.08766083052590703	445.0876353342608809	0.0000254962650388
	−346.62839561049204637	−346.6283853841056271	−0.0000102263863937
Direction d	−0.9999999995800821943	0.9999999993824427	−1.9999999989625250
	−0.00002742350347148223	0.0000332565312843	−0.0000606800347558
	$9.3694753653542153209.10^{-6}$	−0.0000113630052247	0.0000207324805900
Radius R	21.0454952157137	21.0454950694215590	0.0000001462922192
Cylindricity error	0.0374221461592938	0.0372806447385479	0.000141501420745899

Figure 3 shows the points collected by the CMM and the obtained cylinder by using the proposed method.

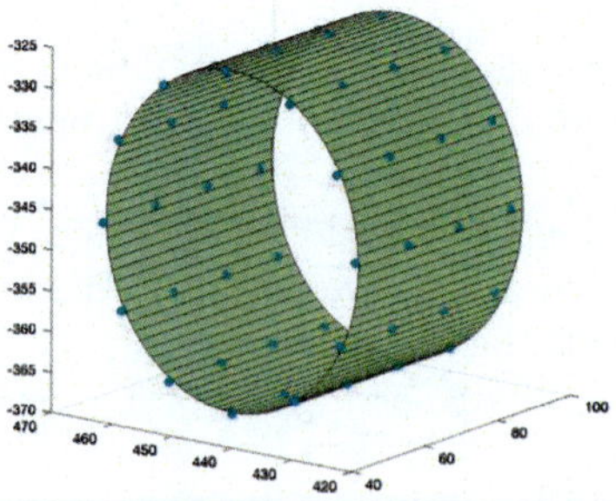

Fig. 3. The obtained cylinder for data set cyl20

4.2 Data Set cyl1

This data set contains 500 points. The comparison between the results of the proposed method and NIST results is shown in Table 2. Figure 4 shows the points collected by the CMM and the obtained cylinder by using the proposed method.

Table 2. Comparison between the results found and those of NIST for data set cyl1

Parameters	NIST results	Proposed algorithm results	Gap
Point A	28.135357897475979085	28.5835620604495837	−0.4482041629736000
	34.897139432596571311	34.9076330036649694	−0.0104935710684018
	−9.6627711654819512599	−10.0726934245531599	0.4099222590711500
Direction d	−0.73780670236054858417	−0.7378066452149147	−0.0000000571456340
	−0.017274414758069737819	−0.0172743611753182	−0.0000000535827515
	0.6747909784122925385	0.6747910422661949	−0.0000000638539021
Radius R	28.6446147537755	28.135357897475979085	0.5092568562996010
Cylindricity error	0.1932851823651980	0.1932851667346410	$1.56306.10^{-8}$

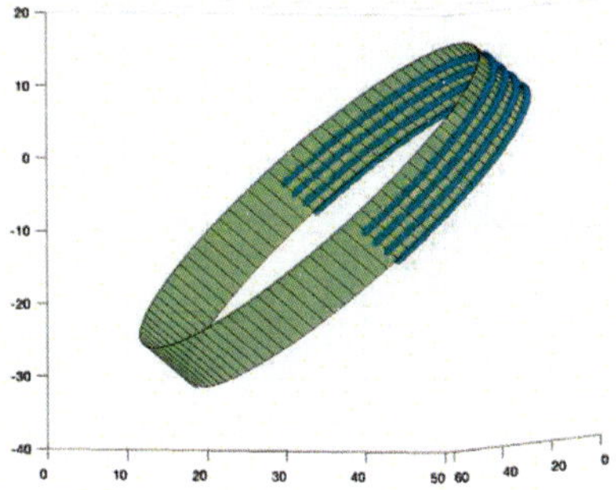

Fig. 4. The obtained cylinder for data set cyl1

4.3 Data Set cyl2

This data set contains 40 points. The comparison between the results of the proposed method and NIST results is shown in Table 3. Figure 5 shows the points collected by the CMM and the obtained cylinder by using the proposed method.

Table 3. Comparison between the results found and those of NIST for data set cyl2

Parameters	NIST results	Proposed algorithm results	Gap
Point A	581.32525620979540896	581.2078573998979891	0.1173988098973950
	650.66893190891241292	650.2528692999629811	0.4160626089494599
	−474.55000463702699317	−474.9229604659145707	0.3729558288875978
Direction d	0.20561439296272909606	0.2056142667386883	0.0000001262240408
	0.72871716896276960398	0.7287169948048681	0.0000001741579015
	0.65321819407105824746	0.6532184280897912	−0.0000002340187329
Radius R	0.391872983967585	0.3918705937057295	0.0000023902618550
Cylindricity error	0.0010929188251922	0.0010928583811002	$6.04441.10^{-08}$

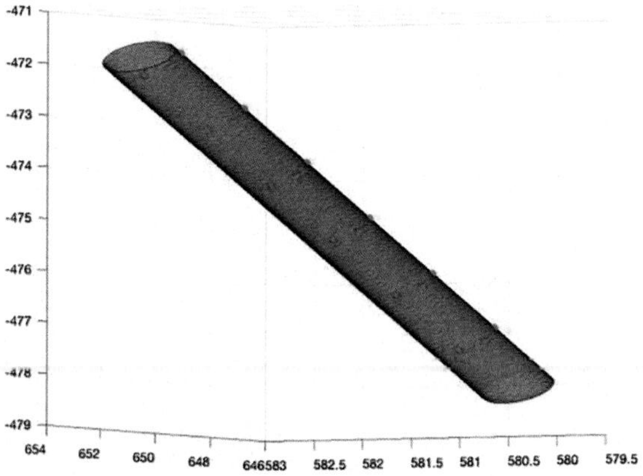

Fig. 5. The obtained cylinder for data set cyl2

5 Conclusion

The proposed study consists in estimating the cylindricity defect from a cloud of points taken from a CMM. The deterministic optimization methods require a starting point to converge.

This approach led us to build a robust algorithm, since it does not need a initial estimate, and gives the global optimum even if the problem is non-linear and non-convex. To validate our algorithm we compared the results found with those of the NIST organization. We note that the results obtained are good and the differences observed with those of NIST are very small.

References

1. Belkourchia, Y., Azrar, L., Es-sadek, M.Z.: Hybrid optimization procedure applied to optimal location finding for piezoelectric actuators and sensors for active vibration control. Appl. Math. Model. (2018)

2. Carr, K., Ferreira, P.: Verification of form tolerances. Part II: cylindricity and straightness of a median line. Precis. Eng. 144–156 (1995)
3. Es-sadek, M.Z., Jalid, A.: New method for estimating form defect of a feature by coupling the genetic algorithm and the interior points method: case of flatness in advances in integrated design and production. In: CPI 2019. Lecture Notes in Mechanical Engineering, pp. 286–291. Springer (2019)
4. Jalid, A., Hariri, S., Senelaer, J.P.: Estimation of form deviation and the associated uncertainty in coordinate metrology. Int. J. Qual. Reliab. Manag. (2015)
5. Lai, J., Chen, I.: Minimum zone evaluation of circles and cylinders. Int. J. Mach. Tools Manuf. **36**, 435–451 (1995)

Influence of the Adjustment Criterion and the Probing Strategy on the Measurement of the Cylindricity Defect

Djezouli Moulai-Khatir[1,2](✉), Abdelilah Jalid[3], and Nabil Habibi[3]

[1] LaRTFM Laboratory, National Polytechnic School, B.P. 1523 El M'naouer, Oran, Algeria
moulai_khatir@yahoo.fr, djezouli.moulai-khatir@univ-tlemcen.dz
[2] Faculty of Technology, Abou Bekr Belkaid University, B.P. 230, 13000 Tlemcen, Algeria
[3] ENSAM of Rabat (Ex ENSET), LaMIPI Laboratory, Mohammed V University, Street Royal Army, Madinat Al Irfane, 10100 Rabat, Morocco
{a.jalid,nabil.habibi}@um5r.ac.ma

Abstract. In precision mechanics, the control of mechanical parts is done on Coordinates Measuring Machines (CMM). The identification of surfaces in dimensional metrology from a cloud of points collected on the surface is a very important phase in subsequently determining shape defects. The least squares criterion has been the most widely used since the 1970s because it can be treated by a simple matrix calculation. Nowadays the minimum zone criterion "Chebyshev" is available on the majority of the software of the Coordinate Measuring Machines (CMM). Different strategies for measuring cylindricity are presented, birdcage strategy, circularity profile, according to the generatrix or by measuring distributed points. Subsequently the study focused on the circularity profile strategy, we showed the influence of the number of points on the cylindricity defect. This study also shows the impact of the adjustment criterion on this defect. An optimal number of points is determined approximately 16 points; this number must make it possible to find the compromise between desired precision and time of measurement. An application on an industrial part is presented.

Keywords: CMM · Adjustments criteria · Probing strategy · Cylindricity defect

1 Introduction

The verification and control of form defects are currently very important in the industrial field, because in many cases, the functionality of assemblies or mechanisms may depend on it (Moulai-Khatir 2017). The form defect is the maximum distance between the real surface and a theoretical surface in contact with it on the outer side of the material. According to the standard, the theoretical surface must be chosen to minimize this distance. Some standards define how to control shape, orientation, etc. However, the standards do not define how to actually measure these characteristics. The National Physics Laboratory (NPL) has published the Best Practice Guide (Flack 2014) for measuring the form defects. This guide is based on the collected experience of CMM manufacturers

L. Azrar et al. (Eds.): CIP 2022, LNME, pp. 12–18, 2023.
https://doi.org/10.1007/978-3-031-23615-0_2

and end-user. He studied some, but not all, of the deviations that could arise from the wrong selection of the measurement strategy. Vrba et al. (Vrba et al. 2013) studied the influence of the measurement strategy on the cylindricity error measured on a CMM. They demonstrated that the sampling strategy is significantly more influential than the evaluation method. Radouani et al. (Radouani and Anselmetti 2003) proposes methods for identification of geometrical surfaces based on a solver without small displacements hypothesis and calculates the form deviation. Dhanish (Dhanish 2002) has studied most of the current standards. He showed that they are based on the concept of minimum area. He also showed that none of them prescribes the method for measuring this minimum area. He developed a new algorithm for calculating the minimum deviation. Gapinski et al. (Gapinski et al. 2006) studied how the roundness deviation is measured on a CMM. They presented the influence of the number of probing points and the type of deviation on the result. They have demonstrated that the minimum number of probing points is never sufficient for the measurement, whereas too many points do not improve the measurement. They finally made recommendations on the number of probing points. Jalid et al. presented in (Jalid et al. 2015) the influence of the number of points on the defect of flatness and on the associated uncertainty, subsequently an optimal number of points was proposed.

In this context, this article discusses the influence of the fit criterion and the probing strategy on the measurement of the cylindricity defect. A comparative study between the two adjustment criteria: least squares and minimum zone "Chebyshev" will be presented. A program under Matlab code has been developed in this sense to determine the cylindricity defect according to the two criteria (least squares and minimum zone). This, by studying the effect of the probing strategy (number and position of points) on the measured shape defect (cylindricity).

2 Adjustment Criteria in Coordinate Metrology

2.1 Definitions

- **Fitting a perfect cylinder to a point cloud**

The metrologist identifies the real surface by a limited number of points M_i which must be representative of the surface. To perform the calculations necessary for verifying the specifications, it is often necessary to associate with this set of points a perfect theoretical surface, of the same nature as the nominal surface, which best represents the real surface according to the chosen association criterion.

- **Parameterization of a cylindrical surface and deviation e_i**

A cylindrical surface of radius R, longitudinal direction $\vec{d}\,(d_x, d_y, d_z)^T$, and A $(x_A, y_A, z_A)^T$ a point belonging to its axis, see Fig. 1. The setting chosen for the associated geometric elements is the same as that described in the standard ISO 10360-6 (ISO 10360-6 2002).

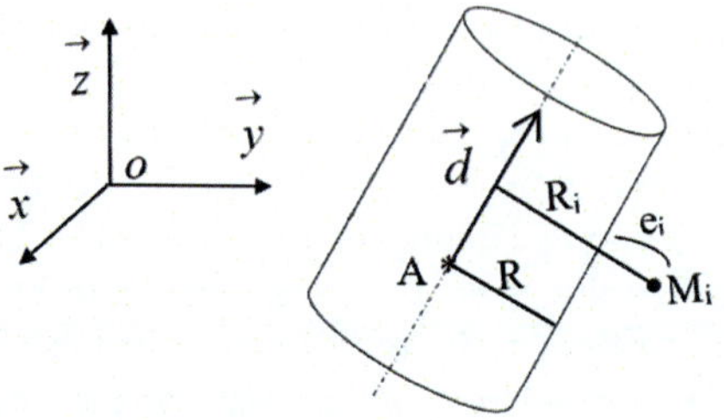

Fig. 1. A parameterization of a cylindrical surface

The radius R_i of the measured points M_i $(x_{M_i}, y_{M_i}, z_{M_i})^T$ is given by:

$$R_i = \frac{\left\| \vec{AM_i} \wedge \vec{d} \right\|}{\left\| \vec{d} \right\|} \quad (1)$$

Which gives the gap $e_i = R_i - R$.

The least squares criterion consists in minimizing the sum of the squares of the deviations e_i.

The problem results in:

$$\left\{ \begin{array}{l} \min \sum_{i=1}^{n} e_i^2 \\ with\ d_x^2 + d_y^2 + d_z^2 = 1 \end{array} \right\} \quad (2)$$

2.2 Least Squares Criterion

The least squares criterion is the most used by metrology software (Moulai-Khatir et al. 2018). The least squares surface is not tangent to the real surface. The associated least squares cylinder passes through the middle of the points cloud (Fig. 2).

2.3 Minimum Zone Criterion

This criterion is also called Chebyshev or tangent surface which minimizes the maximum distance (Moulai-Khatir et al. 2018). The cylindricity defect in this case represents the minimum area between two cylindrical envelopes, which encompass all the measured points. The standard defines for the deviation of cylindricity, the cylinder tangent to the surface of the free side of the material such that the maximum distance from the points of the real surface to this cylinder is minimum (Shunmugam 2001).

Based on the same parameterization as in Fig. 1, the vector of parameters is $a(x_a, y_a, z_a, d_x, d_y, d_z, R)^T$, we denoted by di(a) the distance from the measured point to the sought model. The minimum zone criteria is then expressed by:

$$\min_{a} \max_{i=1->n} |d_i(a)| \quad (3)$$

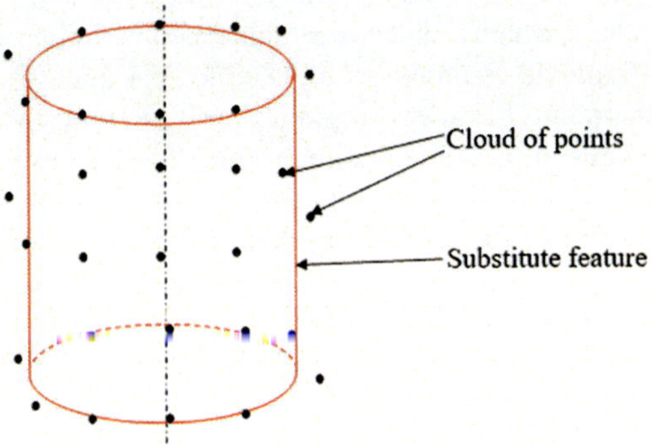

Fig. 2. A cloud of points and fitted least squares cylinder

The problem then consists in finding the vector of parameters, which minimizes the objective function. The search for a solution is made based on iterative numerical methods and requires a starting vector (initial estimates) in order to converge towards a global optimum.

3 Strategies for Measuring Cylindricity

The minimum number of points for measuring cylindricity is defined in ISO 12180-2:2011. In practice, it is often difficult to completely cover the cylindrical element given by the theoretical minimum density of points. The standard describes four measurement strategies: the bird-cage strategy, the roundness profiles strategy, the generatrix lines strategy, and the points strategy. These cylindricity measurement strategies are illustrated in Table 1 (ISO 12180-2:2011).

Table 1. Strategies for measuring cylindricity described by ISO 12180-2

Bird-cage strategy	Roundness profiles strategy	Generatrix lines strategy	Points strategy
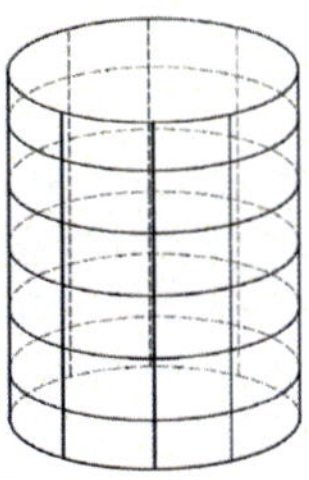	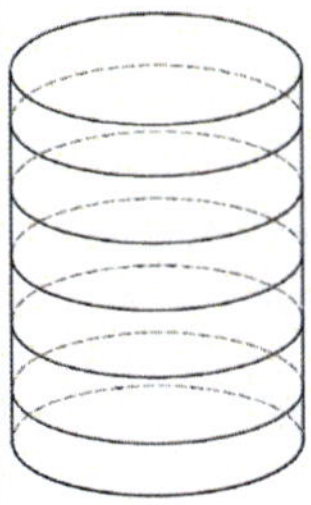	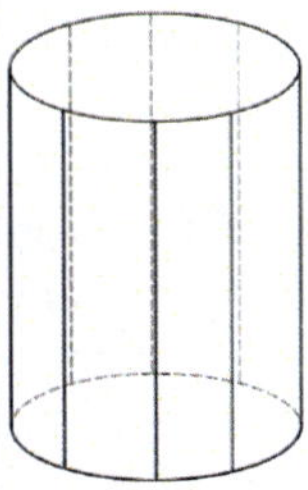	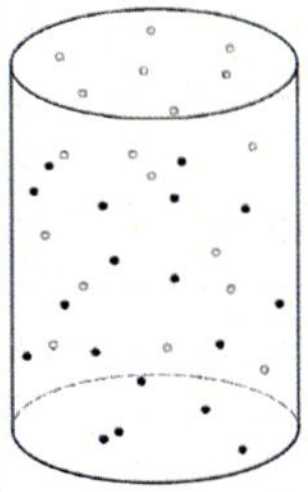

4 Methodology

In order to present our method, we will use as data the cloud of probed points of an industrial cylindrical part measured at 96 points on a CMM. When probing we used the roundness profiles strategy, which is the most suitable method for measuring cylindricity, as demonstrated by Zaimovic-Uzunovic et al. (Zaimovic-Uzunovic and Lemes 2018). Then we will export this points file in our program, developed under Matlab and including the criteria of least squares (Gauss) and minimum zone (Chebyshev). From the collected point cloud, we will measure the cylindricity defect of the industrial part according to the two criteria (least squares Sei2 and Chebyshev eiMax), for 96 measurement points (24 points per roundness profile). Then, during the various adjustments, we will degrade to 6, then 10 and finally 16 measurement points, as shown in Table 2.

Table 2. Probing strategy used (6, 10, 16 and 24 measuring points)

6 measuring points	10 measuring points	16 measuring points	24 measuring points
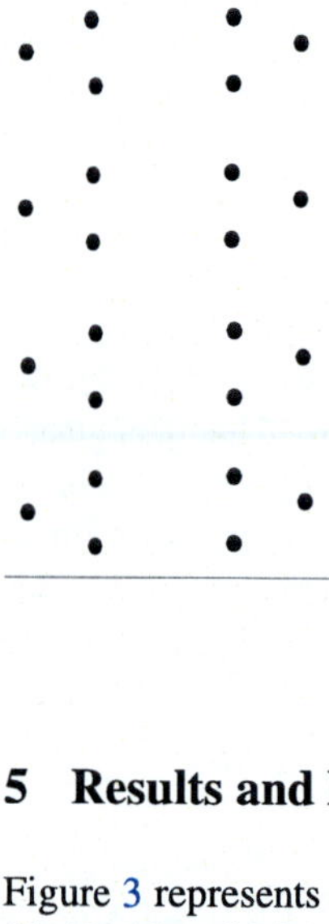	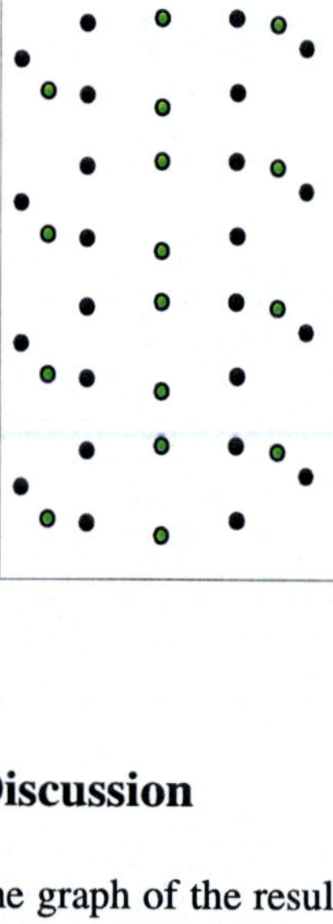	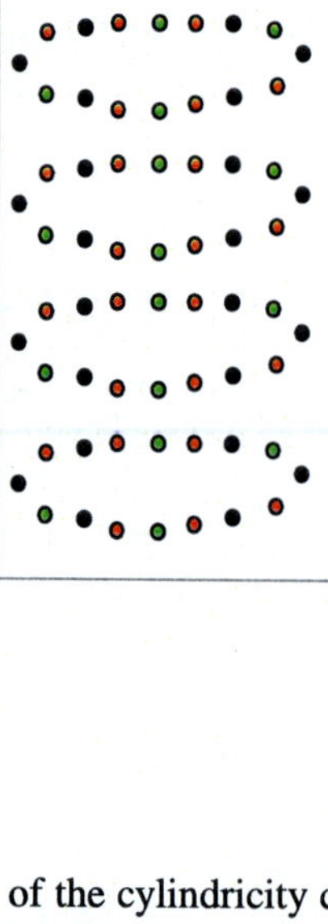	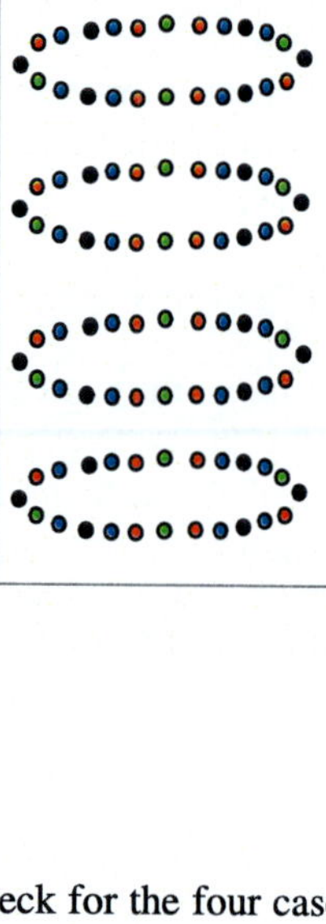

5 Results and Discussion

Figure 3 represents the graph of the results of the cylindricity check for the four cases (6, 10, 16 and 24 measurement points per roundness profile) and according to the two criteria: least squares (Sei2) and Chebyshev (eiMax).

The Gaussian least squares criterion (Sei2) is the fastest criterion to converge towards the shape defect. Chebyshev's criterion (eiMax) is the slowest, especially on low numbers of probing points (6 and 10 points), but it gives the lowest value of the defect. We also note that the number of points influences the value of the cylindricity defect, but from a number of points (16 in our case), this defect stabilizes and tends towards a value considered most probable (0.0375 mm).

6 Conclusion

In this article, we have shown that the measured cylindricity defect increases with the number of probing points. Chebyshev's criterion (eiMax) is the slowest to converge, with 10 points it is only 67% measured defect. The least squares criterion (Sei2) is the fastest

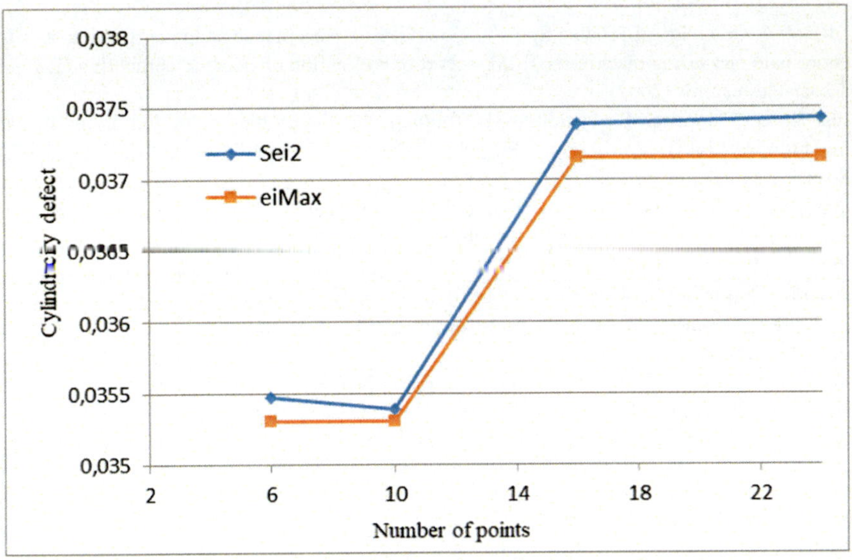

Fig. 3. Graph of the results obtained by measuring the cylindricity of industrial part according to both least squares (Sei2) and Chebyshev (eiMax) criteria

criterion. The minimum zone criterion (eiMax) requires many more probing points than the least squares criterion (Sei2). To measure a cylindricity defect, we recommend the use of the Gauss criterion (Sei2) with roundness profile strategy. Our contribution also proposes an optimal number of points, which is 16 points per roundness profile. This will reduce the measurement time and therefore the cost of control will be minimal. In perspective, we plan to extend our study to other flaws (orientation and position).

References

Moulai-Khatir, D.: Comparative study between two three-dimensional metrology software. Mech. Compendium **N°002**, 215–222 (2017). Tissemsilt University Center. Doi: https://doi.org/10.5281/zenodo.1175026

Flack, D.: CMM measurement strategies, measurement good practice guide 41. National Physical Laboratory (2014)

Vrba, I., Palencar, R., Hadzistevic, M., Strbac, B., Hodolic, J.: The influence of the sampling strategy and the evaluation method on the cylindricity error on a coordinate measurement machine. J. Prod. Eng. **16**(2), 53–56 (2013)

Dhanish, P.B.: A simple algorithm for evaluation of minimum zone circularity error from coordinate data. Int. J. Mach. Tools Manuf. **42**, 1589–1594 (2002). https://doi.org/10.1016/S0890-6955(02)00136-0

Gapinski, B., Grzelka, M., Rucki, M.: The roundness deviation measurement with coordinate measuring machines. Eng. Rev. **26**, 1–6 (2006). Rijeka, Croatia. UDK 531.7:53.083:528.7

Jalid, A., Hariri, S., Laghzale, N.E.: Influence of sample size on flatness estimation and uncertainty in three-dimensional measurement. Int. J. Metrol. Qual. Eng. **6**(1), 102 (2015). https://doi.org/10.1051/ijmqe/2015002

ISO 10360-6: Geometrical Product Specification (GPS)—Acceptance and reverification tests for coordinate measuring machines (CMM)—Part 6: Estimation of errors in computing Gaussian associated features (2002)

Moulai-Khatir, D., Pairel, E., Favreliere, H.: Influence of the probing definition on the flatness measurement. Int. J. Metrol. Qual. Eng. **9**, 15 (2018). Doi:https://doi.org/10.1051/ijmqe/2018011

Shunmugam, M.S.: Important aspects of form measurement and assessment of engineering surfaces. J. Inst. Eng. India Part Mc Mech. Eng. Div. (2001)

Radouani, M., Anselmetti, B.: Identification of real surfaces and inspection of the ISO specifications using a solver. Méc. Ind. **4**(3), 249–258 (2003). Elsevier

ISO 12180-2: Geometrical product specifications (GPS)—Cylindricity—Part 2: Specification operators (2011)

Zaimovic-Uzunovic, N., Lemes, S.: Cylindricity measurement on a coordinate measuring machine. In: Hamrol, A., Ciszak, O., Legutko, S., Jurczyk, M. (eds.) Advances in Manufacturing. LNME, pp. 825–835. Springer, Cham (2018). https://doi.org/10.1007/978-3-319-68619-6_80

Evaluation of the Performance of the Calibration Method by an Inter-laboratory Comparison

Kaoutar Bahassou(✉), Abdelouahhab Salih, and Abdelilah Jalid

Engineering and Health Sciences and Techniques, Research Team of Mechanical and Thermal Processes and Controls, ENSAM RABAT, Rabat, Morocco
el.k.bahassou@gmail.com, a.jalid@um5r.ac.ma

Abstract. This paper presents in first a method to adjust the coordinate measuring machines (CMMs) in the event of non-conformity, the process followed for the validation of our model and the results obtained. Then, secondly, an inter-laboratory comparison will be carried out to confirm the performance of the proposed method. The results obtained will be compared with those of a laboratory accredited by COFRAC in terms of dimensions. This is done by applying the standard deviation criterion developed in ISO 17043.

Keywords: Matrix correction · ISO 17043 · Inter-laboratory comparison · ISO 17025 · Declaration of conformity

1 Introduction

The calibration of CMM's has been the subject of multiple research studies. Indeed, different techniques are used to determine measurement errors either by laser or by using artefacts. The study proposed by G. Zhang [1] is to use rigid objects to describe the three-dimensional measuring machine, the author proposes 22 displacement lines in the machine's measuring volume and to validate the result, he measured the large diagonals of the machine. The path errors are obtained using the laser interferometer. While J. Chen [2] has developed a self-aligned laser interferometer system to reduce machine calibration time. G. Chen [3] proposes an alternative method to G. Zhang's method. This method consists in measuring the positioning errors of machine 3 along the 15 directions axes in order to quickly determine the 21 geometric deviations. Of these 15 directions, 7 are required by the ANSI/ASME standard for evaluating the performance of CNC machining centers. Therefore, only 8 additional positioning error measurements are required to evaluate the modelled machine. CK. Lim and M. Burdekin [4] designed a steel hole bar in the form of a 540 mm I-profile, characterized by good dimensional stability, good wear resistance and very good surface finish. The principle of this method is to measure this bar in 17 different orientations to determine the 21 geometric errors. Then, these 21 geometric errors were transformed into volumetric errors using the differential equations. E. Curran [5] proposed a quick method for verifying the CMM. This technique requires a telescopic ball bar that we can measure in plan, the first ball is

L. Azrar et al. (Eds.): CIP 2022, LNME, pp. 19–27, 2023.
https://doi.org/10.1007/978-3-031-23615-0_3

fixed with a magnetic holder, the second ball rotates around the first ball and is measured at several places by palpating 4 points at each place. This method allowed the authors to calculate the perpendicularity deviation, between the axes, in the measurement plane. Thus, a residual analysis is needed to explain the measurement errors caused by the perpendicularity deviations. S. Phillips [6] has developed a method for calibrating large three-dimensional measuring machines with spindles exceeding 4 m in length, this method uses a calibre equipped with a laser interferometer. A retro-reflector is attached to a sphere that is probed with the machine's stylus button. The retro-reflector and the sphere, ideally concentric, move along the gauge, the laser source is attached to the end of the gauge. The displacement of the sphere along the gauge is equal to the displacement of the retro-reflector which is measured by the laser.

Nowadays, the measurement techniques differ for each laboratory, there are those that use traditional measurement where the metrologist is required to have a good command of the quality of his measurements. Indeed, the distances are all measured in one direction perpendicular to the marble and referenced to a stack of wedges. Others uses the three-dimensional measurements where the metrologist has no longer this intuitive notion of the quality of the measured distances but he must carry out of his measurements in all confidence in the measuring machine and then periodically check the geometric quality.

In this paper, we will focus on a new approach that consists of global modeling the adjusted value, without specifying the geometric errors, by a mathematical model in matrix form following the measurements previously made according to ISO 10360-2 [7]. It should be remembered that measurements are taken by placing the wedge in 7 different orientations in the three axes, the three references plane and the diagonal of the measurement space. The measuring range used consists of probing a plane on one side of the wedge and a point in the centre of the opposite side [8–11]. The measurement of the wedge is then obtained by calculating the plane-point distance. These results were used to develop the correction matrix.

2 Identification of the Correction Matrix for Calibration of the Coordinate Measuring Machines

In a first approach, the adjusted values are linearly modelized by the equation of a line whose coefficients (the slope and the ordinate at the origin) are determined according to the measurements made.

$$V_{c/i} = A_i \times V_{lue} + B_i \tag{1}$$

where:

- V_{ci} and V_{luc} are the adjusted values respectively read on MMT according to the direction (i).
- A_i and B_i are the guiding coefficients of the correction line of the least-squares.

Then, we write this adjusted value in the ith position of the space. Indeed, the ith position of the wedge in the CMM measurement volume is defined by the two angles α_i and β_i and the reading value of the wedge as shown in Fig. 1.

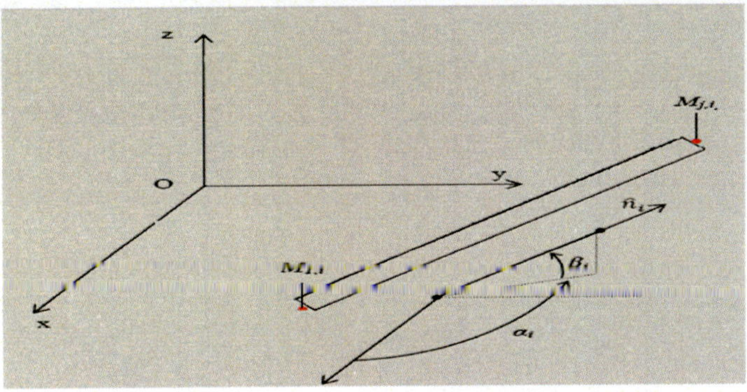

Fig. 1. The ith Position of the wedge in the space (xyz)

Where: $M_{1,i}M_{j,i} = \overrightarrow{\|M_{1,i}M_{j,i}\|} \times \overrightarrow{n_i}$ is the conventionally real distance between the two points $M_{1,i}$, $M_{j,i}$ and the normal vector $\overrightarrow{n_i}$.

The Jacobian matrix is defined by:

$$J = \begin{bmatrix} A_x \cos\alpha \cos\beta & 0 & 0 & B_x \cos\alpha \cos\beta \\ 0 & A_y \sin\alpha \cos\beta & 0 & B_y \sin\alpha \cos\beta \\ 0 & 0 & A_z \sin\beta & B_z \sin\beta \end{bmatrix} \tag{2}$$

A_x, A_y, A_z, B_x, B_y, and B_z are the guiding coefficients of the calibration line.

The normal vector is defined by:

$$\vec{n}_i = \begin{pmatrix} \cos\alpha_i \cos\beta_i \\ \sin\alpha_i \cos\beta_i \\ \sin\beta_i \end{pmatrix} \tag{3}$$

Experimentally, the read value between point-plane on the wedge is given by:

$$V_{lue/i} = \overrightarrow{V_{lue/i}} \times \overrightarrow{n_i} \tag{4}$$

The read value vector is composed of 3 values read along the axes of the machine (x, y, and z) and written as:

$$\vec{V}_{lue} = \begin{pmatrix} V_{lue/x} \\ V_{lue/y} \\ V_{lue/z} \\ 1 \end{pmatrix} \tag{5}$$

For the ith position of the wedge, the adjusted value of the measured value of the wedge is the projection of the measured value experimentally. For this distance, the value is:

$$V_c = (J^t \times n)^t \times V_{lue} \tag{6}$$

2.1 Calculation of the Uncertainty of the Correction Matrix

The uncertainty associated with this matrix depends on several factors (the angle of inclination of the wedge, the coefficients A_i and B_i of the calibration lines, as well as the reading value and the conventional value. This uncertainty is calculated by the GUM analytical method as follows:

$$U = \sqrt{\mathbf{U}_x^2 + \mathbf{U}_y^2 + \mathbf{U}_z^2} \tag{7}$$

Knowing:

$$U_x = \left[U_{A_x}^2 V_{lue_x}^2 + U_{B_x}^2 + U_{Vlue_x}^2 A_x^2\right]\left[\cos^2\alpha . \cos^2\beta\right] + \left[A_x V_{lue_x} + B_x\right]^2 \left[\sin^2\alpha . \cos_\beta^2 U_{\cos\alpha}^2 + \sin_\beta^2 . \cos_\alpha^2 U_{\cos\beta}^2\right]$$

$$U_y = \left[U_{A_y}^2 V_{lue_y}^2 + U_{B_y}^2 + U_{Vlue_y}^2 A_y^2\right]\left[\sin^2\alpha . \cos^2\beta\right] + \left[A_y V_{lue_y} + B_y\right]^2 \left[\cos^2\alpha . \cos_\beta^2 U_{\sin\alpha}^2 + \sin_\beta^2 . \sin_\alpha^2 U_{\cos\beta}^2\right]$$

$$U_z = \left[U_{A_z}^2 V_{lue_z}^2 + U_{B_z}^2 + U_{Vlue_z}^2 A_z^2\right]\left[\sin^2\beta\right] + \left[A_z V_{lue_z} + B_z\right]^2 \left[\cos_\beta^2 U_{\sin\beta}^2\right]$$

The wedges used are steel and nominal lengths of **125, 150, 175, 200, 250mm.** For **l = 125mm;** We obtain the standard uncertainty composed $\mathbf{uc_{/125(xyz)}} = \mathbf{1.16\mu m}$, In the case of a normal distribution, the expansion factor **k = 2** gives the expanded uncertainty with a **95%** confidence level: **U = k * uc = 2.33 µm.**

2.2 Measurements and Results of Identification

To extend the validity of the correction method, we used all the measurements made on the basis of standard 10360-2 and following the 7 directions. Using MATLAB, we were able to create a program to calculate the values adjusted by the proposed correction matrix. Then we started the comparison of the adjusted values with the conventional values of the wedges with an associated uncertainty (Table 1).

If we plotted the graph of the adjusted values adding the associated uncertainty and the two limits Vref + EMT and Vref -EMT:

In the Fig. 2, we try to demonstrate all the values adjusted by the correction matrix presented following the measurement directions. Once calculated, the results adding the associated uncertainties are within the interval [Vref-EMT, Vref + EMT], so the matrix becomes systematically valid, and could then be applied to other coordinate measuring machines with the same functionality, thus achieving the desired objective of declaring conformity to specifications.

We modelled the matrix to be able to determine the model errors between the reference value and the value adjusted by the model while following the measurement direction. Then, we corrected the standard values by the opposite process. The results obtained were compared and validated against the reference values (Table 2).

If we include the notion of capability, this difference between the conventional value of the wedge and the value adjusted by the model remains less than the maximum permissible error of the machine divided by 3 [9].

Table 1. The adjusted values by the correction matrix

	Adjusted values by matrix correction				
	125.00004	150.0004	175.00006	200.0006	250.0006
X	124.9998	150.0006	175.0003	200.0005	250.0005
Y	125.0001	150.0003	175.0013	200.0015	250.0019
Z	125.0011	150.0000	175.0009	200.0011	250.0022
XY	124.9998	150.0006	175.0003	200.0005	250.0005
XZ	125.0011	150.0000	175.0009	200.0011	250.0022
YZ	124.9998	150.0006	175.0003	200.0005	250.0005
XYZ	124.9998	150.0019	175.0013	200.0018	250.0015

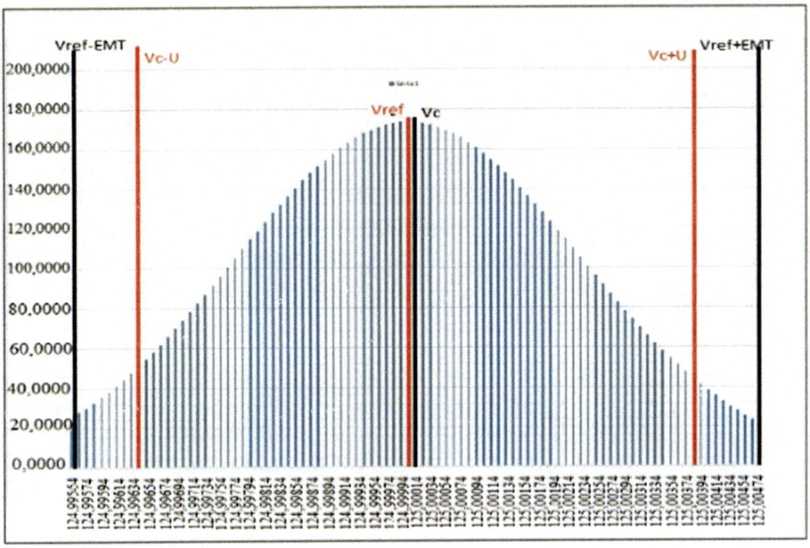

Fig. 2. Comparison of the adjusted value associated with uncertainties and the reference value

3 Evaluation of Performance by Inter-laboratories Comparison

The concept of interlaboratory comparison [10] is the most general one and covers several practices that must be identified. The definition of interlaboratory comparison is the organization, execution, and operation of measurements, calibrations on similar objects (samples, standards, etc.) by at least two different laboratories (participants) under predetermined conditions.

Our objective is to evaluate the performance of our proposed method. This evaluation consists of measuring a distance between two circle centres on a gauge block on our coordinate measuring machine and the results obtained by our software will be adjusted

Table 2. Error model between the conventional value and adjusted value

Error model E_m = Abs (Conventional value-adjusted value)					
	125.00004	150.0004	175.0001	200.0006	250.0006
E_{mx}(μm)	0.24	0.20	0.24	0.10	0.10
E_{my}(μm)	0.06	0.10	1.24	0.90	1.30
E_{mz}(μm)	1.06	0.40	0.84	0.50	1.60
E_{mxy}(μm)	0.24	0.20	0.24	0.10	0.10
E_{mzx}(μm)	1.06	0.40	0.84	0.50	1.60
E_{myz}(μm)	0.24	0.20	0.24	0.10	0.10
E_{mxyz}(μm)	0.24	1.50	1.24	1.20	0.90
EMT/3 (μm)	1.54	1.58	1.63	1.67	1.75

by the proposed correction matrix. This same distance is measured in the Air Metrology laboratory (accredited COFRAC) by an optical measuring machine.

To evaluate the performance of the correction method applied, the notion of the normalized variance will be used. This difference is most relevant in the case of a metrology inter-comparison, so it is important to be very precise in its calculation, which depends on the values of the participants (CMM-ENSET RABAT and Air Metrology) and the uncertainty associated with each value. If this difference is less than 1, then the method is validated. Otherwise, the measurement must be repeated and the correction better defined.

CMM ENSET		Air Metrology
		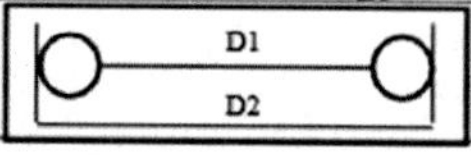
C_1	X=-25.017	D_1=140.599mm
	Y=-17.473	
	Z=0	
C_2	X=-175.588	D_2=160.539mm
	Y=-17.368	
	Z=0	
C_1C_2=150.571mm		$(D_1+D_2)/2$=150.569mm
$C_1C_{2adjusted}$=150.572mm		(D1+D2)/2=150.569mm
U_e=2.4 μm		U_e=4.44μm

The calculation of the normalized variation defined by:

$$E_n = \frac{|C_1C_2 - (D_1 + D_2)/2|}{\sqrt{U^2_{e(MMT)} + U^2_{e(Airmetro)}}} \tag{8}$$

This difference gives $\mathbf{E_n = 0.59 < 1}$, then we can conclude that the method is validated, the aptitude is sufficient and then our correction model is valid.

4 Declaration of Conformity with Risk

The objective of any calibration is to provide information with a level of reliability sufficient to be operated with low and acceptable risk. This usually raises the question of client and supplier risks, when the measurement result, affected by the uncertainty of his measurement, lead to potential doubt. This situation occurs when the result is within the interval defined by the specifications but for which due to the measurement uncertainty, there is a potential risk of having declared conformity by error, this is the case for customer risk. Similarly, a result outside the specification limits may lead to a declaration of non-conformity by error, we will be in supplier risk. Therefore, for optimal risk control, it is essential to know the dispersion of the measurement process. Indeed, the concept of "risk" was limited to determining the probability that a real value would be outside the limits of the specification, taking into account a measurement result and the associated uncertainty.

This is done by looking for the differences between each conventional value and its corresponding limit value reported on the standard deviation of the proposed model and then deducting the percentage of the loss of accuracy of the method. This percentage must not exceed 5% of the maximum tolerated risk to conclude on conformity. The experimental tests were carried out on a Mitutoyo C544 model machine installed in the metrology room in ENSET (Rabat, Morocco), the Error Maximum Tolerated is given by $\mathbf{EMT = 4\ \mu m + 5 * L/1000}$.

Table 3 presents the results of the declaration of conformity following the measurement volume xyz:

The maximum value of the method's accuracy degradation is given for the 150 mm conventional value wedge and it is 2.21% which is very low compared to the acceptable risk of 5%. It can, therefore, be concluded that the proposed method is at most accurate to 2.21%.

5 Conclusion

During the measurement, we did several tests and a calculation while ensuring that the way we will report would not give an incorrect result of the conventional value. Through to the correction matrix obtained previously, we will be able to recover the initial values of the standards while following the opposite way with a difference that does not exceed the maximum uncertainty of the correction matrix. Thus, any calculation of the measured value on the machine would be corrected by the correction matrix obtained by:

Therefore, in a way that has been demonstrated, we can say the following:

As long as the accuracy of the measuring instruments, especially the coordinate measuring machines, is fundamental in the industry, the model developed is validated and the CMM can be used reliably to obtain accurate results for both the supplier's and the customer's final inspection and therefore the risk of declaring conformity will be reduced.

Table 3. Declaration of conformity with risk for the different wedges following xyz

Declaration of conformity with risk (xyz)					
Matrix uncertainty U(mm)	0.00233	0.00246	0.00261	0.00277	0.00312
Reference value (mm)	125.00004	150.0004	175.00006	200.0006	250.0006
Adjusted value (xyz)(mm)	124.9998	150.0019	175.0013	200.0018	250.0015
$V\max = V_{adjusted} + U$ (mm)	125.00213	150.00436	175.00391	200.00457	250.00462
Reference value + U (mm)	125.002372	150.002863	175.00267	200.00337	250.003721
Zgauss Vmax	1.79414	3.21793	2.95029	2.86635	2.57680
Zgauss ($V_{ref} + U$)	2	2	2	2	2
Probability of Zgauss V_{max}	0.963605	0.999354	0.998413	0.997924	0.995014
Probability of Zgauss (V_{ref} + U)	0.977250	0.977250	0.977250	0.977250	0.977250
The risk limited	0.013645	0.022105	0.021163	0.020674	0.017764
Accuracy of the method (%)	1.36	2.21	2.12	2.07	1.78

References

1. Zhang, G.: Error compensation of coordinate measuring machines. CIRP Ann. **34**(1), 445–448 (1985)
2. Chen, J.S., Kou, T.W., Chiou, S.H.: Geometric calibration error of multi-axis machines using an auto-alignment laser interferometer. Precis. Eng. **23**(4), 243–252 (1999)
3. Chen, G., Yuan, J., Ni, J., Wu, S.M.: A displacement measurement approach for machine geometric error assessment. Int. J. Mach. Tools Manuf. **41**(1), 149–161 (2001)
4. Lim, C.K., Burdekin, M.: Rapid volumetric calibration of coordinate measuring machines using a hole bar artefact. Proc. Inst. Mech. Eng. Part B J. Eng. Manuf. **216**(8), 1083–1093 (2002)
5. Curran, E., Phelan, P.: Quick check error verification of coordinate measuring machine. J. Mater. Process. Technol. **155–156**, 1207–1213 (2004)
6. Phillips, S., Sawyer, D.: Laser trackers for large-scale dimensional metrology. Precis. Eng. **44**, 13–28 (2016)
7. ISO 10360-2: Spécification géométrique des produits gps essais de réception et de vérification périodique des machines a mesurer tridimensionnelles., partie 2 : MMT utilisés pour les mesures de dimensions linéaires
8. Bahassou, K., Salih, A., Jalid, A., Oubrek, M.: Modeling of the correction matrix for the calibration of measuring machines. Int. J. Mech. Eng. Technol. (IJMET) **8**(10), 862–870 (2017)
9. ISO 17043-2010: Évaluation de la conformité—Exigences générales concernant les essais d'aptitude

10. ISO 17025: Exigences générales concernant la compétence des laboratoires d'étalonnages et d'essais (2017)
11. Bahassou, K., Salih, A., Oubrek, M., Jalid, A.: Measurement uncertainty on the correction matrix of the coordinate measuring machine. Int. J. Adv. Res. Eng. Technol. (IJARET) **10**(2), 669–676 (2019)

Development of PPP Virtual Machine for the Study of the CMM Performance

Hasna Elbaggar(✉), Abdelhak Nafi, Mohammed Radouani, and Benaissa Elfahime

ENSAM, Moulay Ismail University ofMeknes, Meknes, Morocco
h.elbaggar@edu.umi.ac.ma, a.nafi@umi.ac.ma,
m.radouani@ensam-umi.ac.ma, b.elfahime@ensam-uci.ac.ma

Abstract. The growing use of PPP machines such as coordinate measuring machines (CMMs) and 3D printers or robots requires users to ensure the accuracy of the prismatic joint corresponding to the three axes. In order to model these kind of machines and identify the geometric errors related to the axes, it is very important to carry out tests on several machines whose errors are well known. However, in reality that will be impossible. In this paper, the study will focus on CMMs. The goal of this work is to develop a virtual machine (digital machine for tests) whose axis errors are well controlled and which are the sources of stylus tip position. The means and procedures for controlling the accuracy of the machine according to the ISO 10360 standard are used to produce measurement errors on gauge blocks (or by laser) as a function of the geometric errors related to the machine axes. This virtual machine will be the basis for the development of several machines with different topologies and will be used in future work to validate the results of real tests.

Keywords: Virtual CMM · Gauge block · Machine error · Volumetric error

1 Introduction

Coordinate measuring machine CMM is unavoidable in industry which requires high precision such as automotive and aerospace industry. It serves to check the dimensional and geometric conformity of mechanical parts. This machine consist of three prismatic axes (X - Y- Z) and a probing system, which ends with a contact sphere (the stylus tip). The principle of operation of a CMM is based on the reading of the displacements of the carriages for the three prismatic joints using the high precision rules to record the coordinates of the stylus tip each time it is in contact with the surface of the part to be measured. However, the accuracy of the CMM is strongly dependent on geometric errors related to prismatic joints of the machine. These affect the position of the stylus tip and generate a field of errors in the workspace of the machine. The points probed on the part to be measured do not correspond to the nominal values but to the real values stained with volumetric errors. In order to determine this field of errors and make a compensation, measurements must be made to quantify measurement errors and identify the sources of error through an analytical model. Various works have been

L. Azrar et al. (Eds.): CIP 2022, LNME, pp. 28–34, 2023.
https://doi.org/10.1007/978-3-031-23615-0_4

carried out on the calculations of measurement uncertainties and the performance of the machine either by laser or by using artefacts of (one, two and three dimensions) (Lee and Burdekin 2001) (Kruth et al 1994) (Bringmann et al 2005) (Kunzmann et al 1995). One-dimensional mechanical artifacts are the easiest to handle which are: step gauge, ball bar, hole bar and gauge blocks (Cauchick et al. 1996) (Miguel and King 1994) (Osawa et al 2002) (Ouyang and Jawahir 1995).

(Zhang et al 1988) used the rigid body hypothesis to model the coordinate measuring machine and they suggested 22 measurement positions to identify the 21 machine deviations using 1-D ball array. (Chen et al 2001) propose a method, which consists in quantifying the errors of measurement by laser in 15 directions. (Lim and Burdekin 2002) designed a hole bar in the shape of I-profile, characterized by good dimensional stability, good wear resistance and very good surface finish. They used 17 specific measurement positions. In order to carry out controlled tests and properly study the relationship between measurement errors and geometric errors, related to the axes of the machine, it is very interesting to develop a (virtual) digital machine thus making it possible to simulate several cases of tests and to study the limits of the analytical model and make modifications to it based on a mathematical analysis. The machine is represented by three prismatic joints. Each joint is attached to six kinematic errors. That gives rise to volumetric errors. Thus expressing the measurement errors in all directions regardless of the artefact used.

2 Position Error of the Stylus Tip

The coordinates X, Y and Z which define the positions of the carriages, are said nominally to define the position of Oz in the machine frame {F} (Fig. 1). The coordinates of the stylus tip t (xt, yt, zt), in the reference frame {F}, are then defined by axes coordinates X, Y and Z, by the length Ls (distance from the pivot of the articulated system to the centre of the stylus tip (t) and by the orientations angles of the articulated system, commonly referred to as (angles A and B).

The position of the stylus tip does not exactly correspond to its nominal position. Each joints X, Y and Z is associated respectively to the linear and angular deviation $(\vec{\delta}_X, \vec{\varepsilon}_X)$, $(\vec{\delta}_Y, \vec{\varepsilon}_Y)$ and $(\vec{\delta}_Z, \vec{\varepsilon}_Z)$. Where $\vec{\delta}$ and $\vec{\varepsilon}$ are the linear and the angular deviation vectors respectively.

The positions and rotations of the coordinates frames linked to the machine carriages are affected by geometric errors, which modeled as a small rotations (angular deviations) and small displacements (linear deviations) (Fig. 1).

$$\text{For x - axis}: \vec{\delta}_X = [\delta_X(X), \delta_Y(X), \delta_Z(X)]^T \quad ; \quad \vec{\varepsilon}_X = [\varepsilon_X(X), \varepsilon_Y(X), \varepsilon_Z(X)]^T$$
$$\text{For y - axis}: \vec{\delta}_Y = [\delta_X(Y), \delta_Y(Y), \delta_Z(Y)]^T \quad ; \quad \vec{\varepsilon}_X = [\varepsilon_X(Y), \varepsilon_Y(Y), \varepsilon_Z(Y)]^T$$
$$\text{For z - axis}: \vec{\delta}_Z = [\delta_X(Z), \delta_Y(Z), \delta_Z(Z)]^T \quad ; \quad \vec{\varepsilon}_X = [\varepsilon_X(Z), \varepsilon_Y(Z), \varepsilon_Z(Z)]^T$$

The real position of the stylus tip in the frame {F} is defined by the vector:

$$\overrightarrow{O_F T_R} = \vec{X} + \vec{\delta}_X + {}_{X'}^{X}R\left(\vec{Y} + \vec{\delta}_Y + {}_{Y'}^{Y}R\left(\vec{Z} + \vec{\delta}_Z + {}_{Z'}^{Z}R\vec{L}_S\right)\right)$$

The nominal position of the stylus tip in the frame {F} is defined by the vector:

$$\overrightarrow{O_F T_N} = \vec{X} + \vec{Y} + \vec{Z} + \vec{L_S}$$

where ${}^{F}_{F'}R$ stands for the rotation matrix of the frame *F'* relative to the frame *F*. this matrix is formed by the component of the angular deviation vectors

$$ {}^{X}_{X'}R = \begin{pmatrix} 1 & -\varepsilon_Z(X) & \varepsilon_Y(X) \\ \varepsilon_Z(X) & 1 & -\varepsilon_X(X) \\ -\varepsilon_Y(X) & \varepsilon_X(X) & 1 \end{pmatrix} ;$$

$$ {}^{X}_{X'}R^{Y}_{Y'}R = \begin{pmatrix} 1 & -\varepsilon_Z(X) - \varepsilon_Z(Y) & \varepsilon_Y(X) + \varepsilon_Y(Y) \\ \varepsilon_Z(X) + \varepsilon_Z(Y) & 1 & -\varepsilon_X(X) - \varepsilon_X(Y) \\ -\varepsilon_Y(X) - \varepsilon_Y(Y) & \varepsilon_X(X) + \varepsilon_X(Y) & 1 \end{pmatrix}$$

$$ {}^{X}_{X'}R^{Y}_{Y'}R^{Z}_{Z'}R = \begin{pmatrix} 1 & -\varepsilon_Z(X) - \varepsilon_Z(Y) - \varepsilon_Z(Z) & \varepsilon_Y(X) + \varepsilon_Y(Y) + \varepsilon_Y(Z) \\ \varepsilon_Z(X) + \varepsilon_Z(Y) + \varepsilon_Z(Z) & 1 & -\varepsilon_X(X) - \varepsilon_X(Y) - \varepsilon_X(Z) \\ -\varepsilon_Y(X) - \varepsilon_Y(Y) - \varepsilon_Y(Z) & \varepsilon_X(X) + \varepsilon_X(Y) + \varepsilon_X(Z) & 1 \end{pmatrix}$$

For each point M in the workspace, the volumetric error $\vec{e}_M$ is defined by:

$$\vec{e}_M = \overrightarrow{O_F T_R} - \overrightarrow{O_F T_N}$$

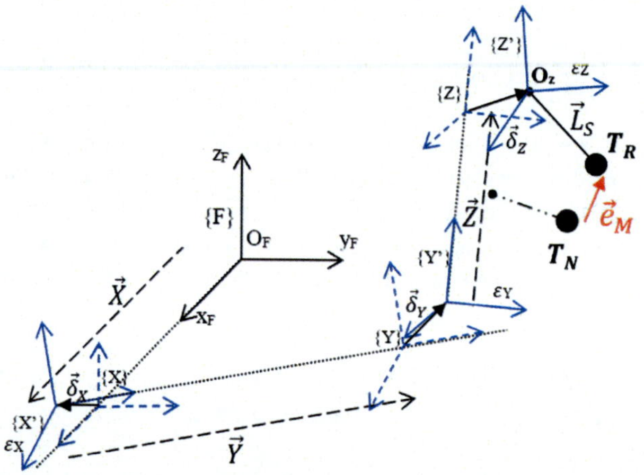

Fig. 1. Real and nominal position of the stylus tip in the coordinate system

3 Measurement Error in the Workspace

The measurement error $E_{i,j,k}$ between the point M_i and M_j, in the direction $\vec{u}_k$ is the difference between the value of the distance measured by the CMM and the distance's value obtained from the artefact (gauge block, step gauge,..., or using the laser). This error can be estimated using the volumetric error model (Fig. 2):

$$E_{1,j,k} = \left(\vec{e}_{M_j} - \vec{e}_{M_i}\right)^T \vec{u}_k$$

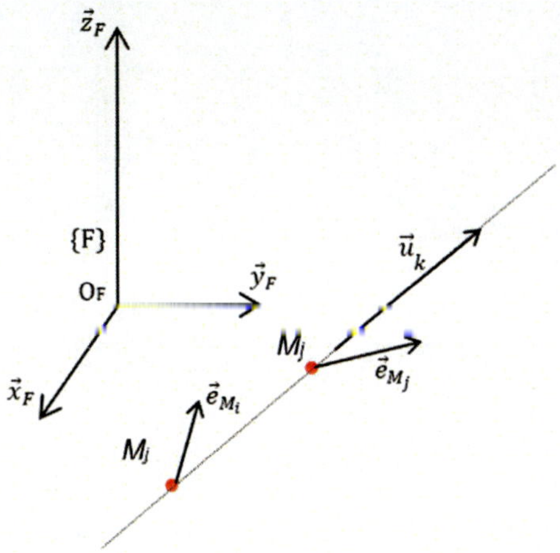

Fig. 2. Volumetric errors for points in the workspace and measurement direction

4 Virtual Machine and Application

Each deviation is modeled by a polynomial of degree 3 represented by four constant coefficients. The choice of order 3 is to simulate a shape close to reality in a simple way by avoiding the coupling between the coefficients during the system inversion and to avoid the numerical problems that appear with high powers such as the condition number of the matrix for example.

The machine has three axes. Then, we have 18 deviations, six for each axis. The machine is therefore defined by 72 coefficients. These coefficients are randomly generated. At each point M of the workspace, the volumetric error is calculated.

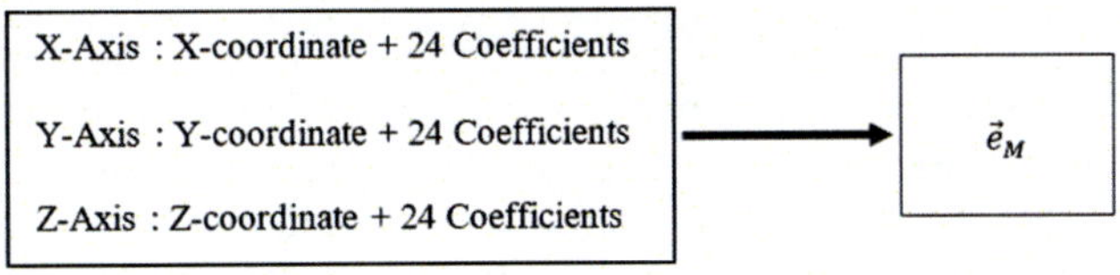

In order to practice and simulate the measurement errors, we choose the measurement method according to the standard ISO 10360.

We use five gauge blocks measured in seven positions as represented in the (Fig. 3). For each position the measurement are repeated tree times. For this simulation, five gauge blocks of lengths 2(in), 6(in), 10(in), 16(in) and 20(in) are used. In this simulation, we consider unidirectional measurements whose probe does not have too much impact. During the measurements, the probe remains vertical. For future work, we will take into account the probe error.

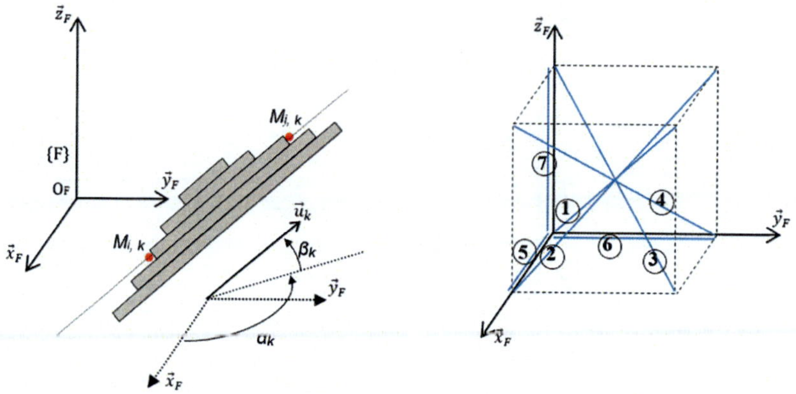

Fig. 3. Seven measurement positions according to the standard ISO 10360

5 Example of Measurement Result

The machine is randomly generated based on the 72 coefficients. Table 1 shows an example of this simulation.

The mean and standard deviation of the random function used are respectively: 5.0e−09 and 2.9e−09, which is applied for all the coefficients except for the out of squareness errors and scale errors that we propose to be able to identify them later with a reduced model.

For this example of a Virtual machine, and by performing tests according to the ISO 10360 standard, we obtain typical measurement results such that the measurement errors increase with the length of the wedges measured (Fig. 4).

6 Conclusion

A virtual CMM machine was developed and modeled by three axes interconnected by prismatic joints geometrically non-perfect. Geometric errors related to axes are among the main sources of systematic errors that affect the position of the stylus tip in the machine workspace. This virtual machine allowed us to reproduce the typical measurement errors on gauge blocks measured according to the recommendations of the ISO 10360 standard. This virtual machine will be the subject of development of other machine topologies in order to make a direct and inverse relationship between the geometric errors of the machine and the errors of observed measurements. In the perspective of this work, probe errors and other dynamic and thermal parameters will be taken into account in the model.

Table 1. Example for 72 coefficients defined a virtual machine (1.0e−03 *)

	$\delta_{i,j,0}$	$\delta_{i,j,1}$	$\delta_{i,j,2}$	$\delta_{i,j,3}$
$\delta_x(X)$; i = x et j = X	0.00000406	−0.90000000	0.00000929	0.00000346
$\delta_y(X)$; i = y et j = X	0.00000386	0.00000556	0.00000409	0.0000062
$\delta_z(X)$; i = z et j = X	0.00000609	0.0000026	0.000000084	0.0000079
$\varepsilon_x(X)$; i = x et j = X	0.00000166	0.000006805	0.00000541	0.00000746
$\varepsilon_y(X)$; i = y et j = X	0.29000000	0.00000233	0.00000207	0.00000125
$\varepsilon_z(X)$; i = z et j = X	0.60000000	0.00000456	0.00000219	0.0000082
$\delta_x(Y)$; i = x et j = Y	0.00000323	0.00000384	0.0000032	0.00000025
$\delta_y(Y)$; i = y et j = Y	0.00000769	−0.6000000	0.00000095	0.00000414
$\delta_z(Y)$; i = z et j = Y	0.00000234	0.00000991	0.00000747	0.00000731
$\varepsilon_x(Y)$; i = x et j = Y	0.7700000	0.00000755	0.00000748	0.0000078
$\varepsilon_y(Y)$; i = y et j = Y	0.00000692	0.00000980	0.00000543	0.0000036
$\varepsilon_z(Y)$; i = z et j = Y	0.0000082	0.00000234	0.00000338	0.0000074
$\delta_x(Z)$; i = x et j = Z	0.000008279	0.00000528	0.00000832	0.0000089
$\delta_y(Z)$; i = y et j = Z	0.00000293	0.000000514	0.0000055	0.00000242
$\delta_z(Z)$; i = z et j = Z	0.00000309	0.41000000	0.0000095	0.00000129
$\varepsilon_x(Z)$; i = x et j = Z	0.00000523	0.0000060	0.00000892	0.0000022
$\varepsilon_y(Z)$; i = y et j = Z	0.00000325	0.0000085	0.0000035	0.0000035
$\varepsilon_z(Z)$; i = z et j = Z	0.00000831	0.00000988	0.00000546	0.00000287

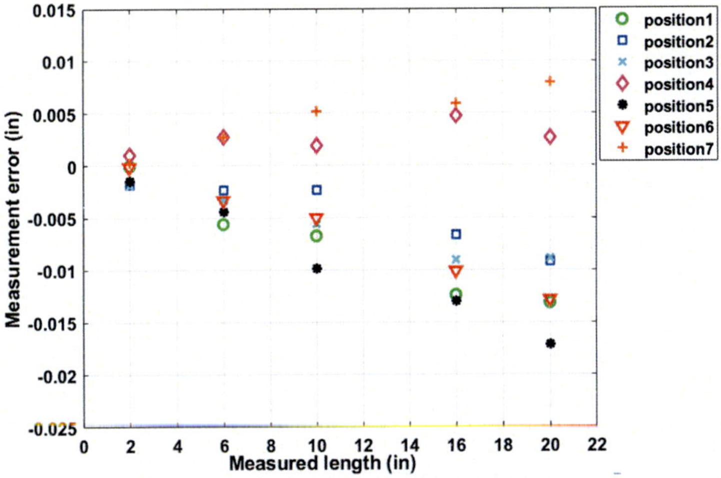

Fig. 4. Measurement result in the seven position according to the ISO 10360 standard

References

Cauchick-Miguel, P., King, T., Davis, J.: CMM verification: a survey. Meas.: J. Int. Meas. Confederation **17**, 1–16 (1996)

Miguel, P.A.C., King, T.G.: Co-ordinate measuring machines. Concept, classification and comparison of performance tests. Int. J. Qual. Reliab. Manag. **12**, 48–63 (1994)

Lee, E.S., Burdekin, M.: A hole-plate artifact design for the volumetric error calibration of CMM. Int. J. Adv. Manuf. Technol. **17**, 508–515 (2001)

Kruth, J.P., Vanherck, P., De Jonge, L.: Self-calibration method and software error correction for three-dimensional coordinate measuring machines using artefact measurements. Meas.: J. Int. Meas. Confederation **14**, 157–167 (1994)

Bringmann, B., Kung, A., Knapp, W.: A measuring artefact for true 3D machine testing and calibration. CIRP Ann. Manuf. Technol. **54**, 471–474 (2005)

Kunzmann, H., Trapet, E., Waldele, F.: Results of the international comparison of ball plate measurements in CIRP and WECC. CIRP Ann. Manuf. Technol. **44**, 479–482 (1995)

Osawa, S., Takatsuji, T., Noguchi, H., Kurosawa, T.: Development of a ball step-gauge and an interferometric stepper used for ball-plate calibration. Precis. Eng. **26**, 214–221 (2002)

Ouyang, J.F., Jawahir, I.S.: Ball array calibration on a coordinate measuring machine using a gage block. Meas.: J. Int. Meas. Confederation **16**, 219–229 (1995)

Zhang, G., Ouyang, R., Lu, B.: A displacement method for machine geometry calibration. Ann. CIRP **37** (1988)

Chen, G., Yuan, J., Ni, J.: Displacement measurement approach for machine geometric error assessment. Int. J. Mach. Tools Manuf **41**, 149–161 (2001)

Lim, C.K., Burdekin, M.: Rapid volumetric calibration of coordinate measuring machines using a hole bar artefact. Proc. Inst. Mech. Eng. Part B: J. Eng. Manuf. **216**, 1083–1093 (2002)

Statistical Analysis of Three-Dimensional Tolerances by Integrating Form Defects Using the Jacobian Torsor Model

Mustapha El Mouden(✉), Mouhssine Chahbouni, Driss Amegouz, and Said Boutahari

Université Sidi Mohammed Ben Abdellah/Ecole Supérieure de Technologie, Fez, Morocco
mm.mouden@gmail.com, mouhssine.chahbouni@usmba.ac.ma

Abstract. The Jacobian torsor model combines the Jacobian matrix for tolerance propagation and the torsor model describing tolerances. There are three translation vectors and three rotation vectors in the torsor model; this work introduces a constraint between the components and considers flatness defect. We'll combine the Jacobian torsor model and Monte Carlo simulation to statistically investigate parallelism tolerances with flatness fault. In this model, the assembly is segmented into surfaces, and their relative position is determined by the small displacement torsor parameters (SDT). We started with statistical tolerance analysis, and then included a flatness defect.

Keywords: Geometric tolerancing · Tolerance analysis · Jacobean torsor · Monte Carlo simulation · Constraint · Form defects

1 Introduction

The process of evaluating the total amount of manufacturing tolerances and the degree to which a mechanism can be assembled is referred to as a tolerance analysis. A preliminary examination of tolerances is required in order to estimate cumulative tolerances, eliminate failures that are a result of tolerance allocation, and predict whether or not components will be feasible. Because the requirements for tolerance in the industrial sector are becoming increasingly stringent and complex, tolerance analysis went from unidirectional 1D to three-directional 3D. Three-directional tolerance, unlike the unidirectional tolerance can define and measure the various categories of component tolerances and their interactions. These benefits are significant for specifying and allocating tolerances in complex mechanisms. Three-directional tolerance analysis approaches are new for expressing and transmitting tolerance in three-directional Euclidean space (Shen et al. 2015). These methods are able to take into consideration all possible forms of tolerances, including geometrical and dimensional ones, while also taking into account the interaction that exists between the two (Chen et al. 2015).

L. Azrar et al. (Eds.): CIP 2022, LNME, pp. 35–46, 2023.
https://doi.org/10.1007/978-3-031-23615-0_5

2 Model of the Jacobian Torsor

Assembly tolerances should quantify all variations; this comprises the dimensions and geometrical specifications for each component of the assembly, as well as the kinematic variations in any contacts or joints (Ghie et al. 1999) and (Chahbouni 2016). The small displacement torsor SDT may be used to compute the relative location of any two surfaces once a joint has been partitioned into surfaces (Zeng et al. 2017) and (Fréderic 2007). Tolerance zones specify the boundaries of small displacements among an associated surface and a nominal surface (El Feki 2011). Lafond and Laperrière (1999) proposed the Jacobian Torsor model, which is shown below:

$$\begin{bmatrix} \underline{u} & \overline{u} \\ \underline{v} & \overline{v} \\ \underline{w} & \overline{w} \\ \underline{\alpha} & \overline{\alpha} \\ \underline{\beta} & \overline{\beta} \\ \underline{\gamma} & \overline{\gamma} \end{bmatrix}_{FR} = \left[J1\ J2\ J3\ J4\ J5\ J6 \right]_{EF1} \cdots \left[J1\ J2\ J3\ J4\ J5\ J6 \right]_{EFN} \bullet \begin{bmatrix} \begin{bmatrix} \overline{u} & \underline{u} \\ \overline{v} & \underline{v} \\ \overline{w} & \underline{w} \\ \overline{\alpha} & \underline{\alpha} \\ \overline{\beta} & \underline{\beta} \\ \overline{\gamma} & \underline{\gamma} \end{bmatrix}_{EF1} \\ \cdots \\ \begin{bmatrix} \overline{u} & \underline{u} \\ \overline{v} & \underline{v} \\ \overline{w} & \underline{w} \\ \overline{\alpha} & \underline{\alpha} \\ \overline{\beta} & \underline{\beta} \\ \overline{\gamma} & \underline{\gamma} \end{bmatrix}_{EFN} \end{bmatrix} \tag{1}$$

$$And: \begin{bmatrix} \underline{u} & \overline{u} \\ \underline{v} & \overline{v} \\ \underline{w} & \overline{w} \\ \underline{\alpha} & \overline{\alpha} \\ \underline{\beta} & \overline{\beta} \\ \underline{\gamma} & \overline{\gamma} \end{bmatrix}_{FR} \textit{is the small displacement torsor}$$

$$Where: \begin{bmatrix} \underline{u} & \overline{u} \\ \underline{v} & \overline{v} \\ \underline{w} & \overline{w} \\ \underline{\alpha} & \overline{\alpha} \\ \underline{\beta} & \overline{\beta} \\ \underline{\gamma} & \overline{\gamma} \end{bmatrix}_{FR} \textit{is the functional requirement torsor}$$

3 Numerical Example

In this part, a tolerance study is completed by including a flatness defect to compare the statistical approach with and without a flatness defect, and calculating the non-conformity rate. This comparison will show how the non-conformity rate differs depending on whether or not flatness defect is taken into account. The unit under investigation is a stacking of three components (Fig. 1). The functional requirement that need to be checked first are the parallelism to S1,1 of the surface of the part 1 in relation to the reference surface S3,2 that belongs to the part 3, and this was done without taking into account the influence of flatness defect. After this, we analyzed tolerances while considering flatness defect, Fig. 1 presents the overall and elementary drawings.

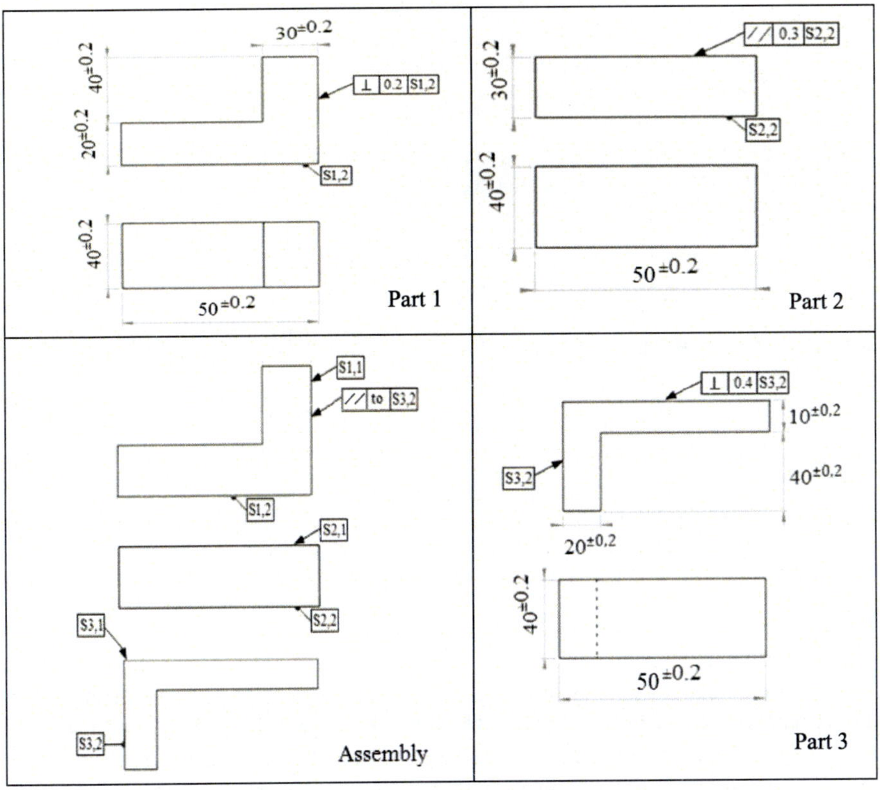

Fig. 1. Assembly and definition drawings

The features of the individual pieces serve as the basis for the establishment of the local coordinate systems, as shown in Fig. 2.

When all of the features and coordinate parameters of the assembly have been located, the surface graph of the assembly can be created as follows (Fig. 3):

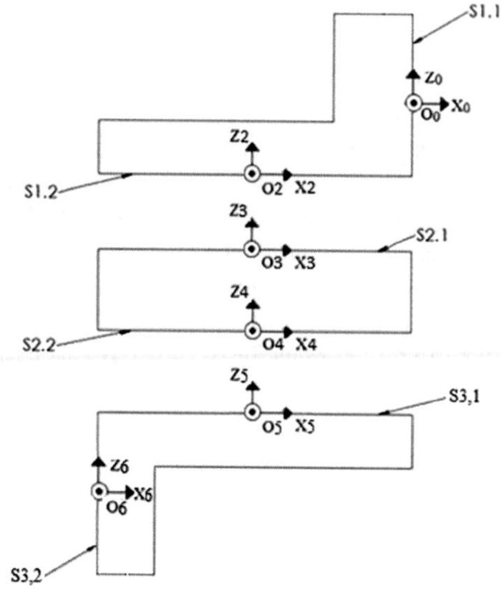

Fig. 2. Local coordinate systems

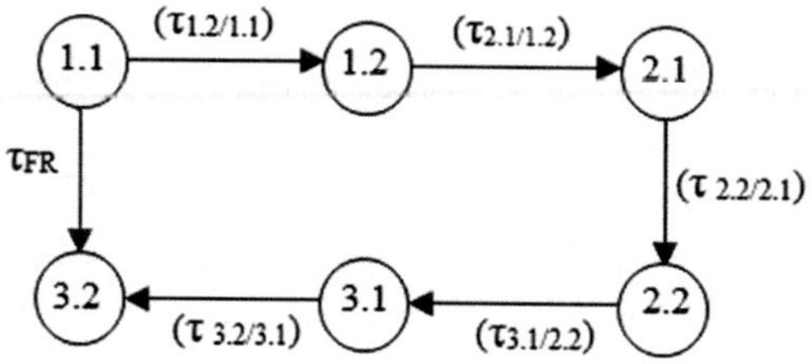

Fig. 3. Identification of the kinematic chain.

3.1 Statistical Tolerancing Without Taking into Account Flatness Defect

Without taking into consideration any flatness defect, we conducted an analysis of the tolerances of parallelism to use in this section. Therefore, the functional condition is made up of the torsors τ 2.2/2.1 and τ 1.2/1.1 and τ 3.2/3.1. Because it is presumed that there is perfect contact between the surfaces S2.2/S3.1 and S1.2/S2.1. The kinematic torsors τ 2.1/1.2 and τ 3.1/2.2 are both believed to have a value of zero. As a result, we are able to acquire the Jacobian torsor model, which can be expressed in the following way:

$$\begin{bmatrix} \underline{u} & \overline{u} \\ \underline{v} & \overline{v} \\ \underline{w} & \overline{w} \\ \underline{\alpha} & \overline{\alpha} \\ \underline{\beta} & \overline{\beta} \\ \underline{\gamma} & \overline{\gamma} \end{bmatrix}_{FR} = \left[\begin{bmatrix} 1&0&0&0&70&0 \\ 0&1&0&-70&0&25 \\ 0&0&1&0&-25&0 \\ 0&0&0&1&0&0 \\ 0&0&0&0&1&0 \\ 0&0&0&0&0&1 \end{bmatrix}_{FE1} \begin{bmatrix} 1&0&0&0&40&0 \\ 0&1&0&-40&0&0 \\ 0&0&1&0&0&0 \\ 0&0&0&1&0&0 \\ 0&0&0&0&1&0 \\ 0&0&0&0&0&1 \end{bmatrix}_{FE2} \begin{bmatrix} 1&0&0&0&0&0 \\ 0&1&0&0&0&0 \\ 0&0&1&0&0&0 \\ 0&0&0&1&0&0 \\ 0&0&0&0&1&0 \\ 0&0&0&0&0&1 \end{bmatrix}_{FE3} \right] \bullet \left[\begin{bmatrix} 0,0 \\ 0,0 \\ 0,0 \\ 0,0 \\ \pm 0.0033 \\ \pm 0.005 \end{bmatrix}_{FE1} \begin{bmatrix} 0,0 \\ 0,0 \\ 0,0 \\ \pm 0.0075 \\ \pm 0.006 \\ 0,0 \end{bmatrix}_{FE2} \begin{bmatrix} 0,0 \\ 0,0 \\ 0,0 \\ \pm 0.01 \\ \pm 0.008 \\ 0,0 \end{bmatrix}_{FE3} \right]^T \quad (2)$$

We may randomize each small displacement torsor component. By taking Eq. (2) as our starting point and then extracting the limits of each parameter to use. Table 1 shows simulation details.

The parameters of the functional requirements of parallelism, represented by β_{FR} and γ_{FR}, which were found using a statistical approach are provided on the last line of Table 1, and the picture that follows illustrates this simulation on Matlab (Fig. 4):

3.2 Tolerancing by Taking into Account Flatness Defect

This section introduces flatness defect analysis using the Jacobian matrix approach. We included flatness defect effects to show their impact on the study. To analyze tolerances while taking form defects DF into account, the following assumptions are made:

Each functional surface contains a flatness defect of 0.04 mm. For each of parts 1 and 3, the flatness defect is reported on one of the two surfaces and the other is considered perfect. In our case, S1.1 and S3.2 are considered perfect, while S1,2 and S3,1 possess a double flatness defect: DF*2 = 0.08 (Fig. 5).

During the assembly of the parts, the contact between surfaces S2.2/S3.1 and S1.2/S2.1 is not ideal; these zones accumulate all assembly flatness flaws (Fig. 5), now the torsors (τ 2.1/1.2) and (τ 3.1/2.2) computed with a form defect for each of them equal to: DF = 3*0.04 = 0.12. Table 2 illustrates torsors (τ 2.1/1.2) and (τ 3.1/2.2) representing the flatness defect:

Table 1. Parameter simulation

Torsor	Value	Distribution specified	Percentage of rejects %R	Statistical parameters	
				Mean	Standard deviation
$\tau_{1.2/1.1}$	$\begin{bmatrix} [0,0] \\ [0,0] \\ [0,0] \\ [0,0] \\ [\pm 0.0033] \\ [\pm 0.005] \end{bmatrix}_{O_0}$	Normale	5%	$u_u = 0$ $u_v = 0$ $u_w = 0$ $u_\alpha = 0$ $u_\beta = 0$ $u_\gamma = 0$	$\sigma_u = 0$ $\sigma_v = 0$ $\sigma_w = 0$ $\sigma_\alpha = 0$ $\sigma_\beta = 0.0019$ $\sigma_\gamma = 0.0029$
$\tau_{2.2/2.1}$	$\begin{bmatrix} [0,0] \\ [0,0] \\ [0,0] \\ [\pm 0.0075] \\ [\pm 0.006] \\ [0,0] \end{bmatrix}_{O_3}$	Normale	5%	$u_u = 0$ $u_v = 0$ $u_w = 0$ $u_\alpha = 0$ $u_\beta = 0$ $u_\gamma = 0$	$\sigma_u = 0$ $\sigma_v = 0$ $\sigma_w = 0$ $\sigma_\alpha = 0.0043$ $\sigma_\beta = 0.0014$ $\sigma_\gamma = 0$
$\tau_{3.2/3.1}$	$\begin{bmatrix} [0,0] \\ [0,0] \\ [0,0] \\ [\pm 0.01] \\ [\pm 0.008] \\ [0,0] \end{bmatrix}_{O_5}$	Normale	5%	$u_u = 0$ $u_v = 0$ $u_w = 0$ $u_\alpha = 0$ $u_\beta = 0$ $u_\gamma = 0$	$\sigma_u = 0$ $\sigma_v = 0$ $\sigma_w = 0$ $\sigma_\alpha = 0.0059$ $\sigma_\beta = 0.0045$ $\sigma_\gamma = 0$

(*continued*)

Table 1. (*continued*)

Torsor	Value	Distribution specified	Percentage of rejects %R	Statistical parameters	
				Mean	Standard deviation
τ_{FR}	$\begin{bmatrix} [0,0] \\ [0,0] \\ [0,0] \\ [0,0] \\ [\pm 0.012] \\ [\pm 0.005] \end{bmatrix}_{O_0}$	Normale	5%	$u_u = 0$ $u_v = 0$ $u_w = 0$ $u_\alpha = 0$ $u_\beta = 0$ $u_\gamma = 0$	$\sigma_u = 0$ $\sigma_v = 0$ $\sigma_w = 0$ $\sigma_\alpha = 0$ $\sigma_\beta =$ 0.0062 $\sigma_\gamma =$ 0.0029

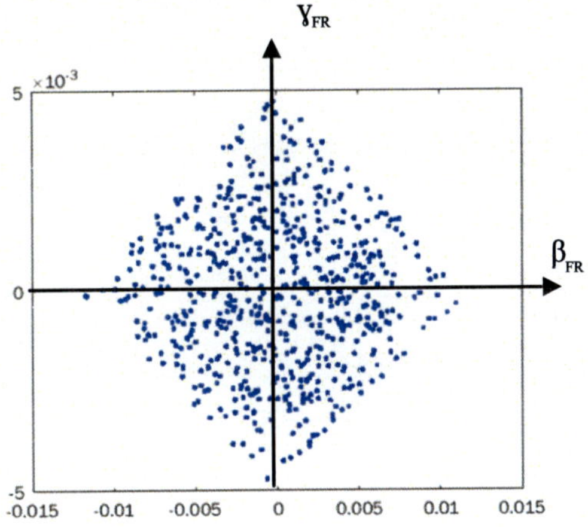

Fig. 4. Statistical tolerancing without taking flatness defect into account

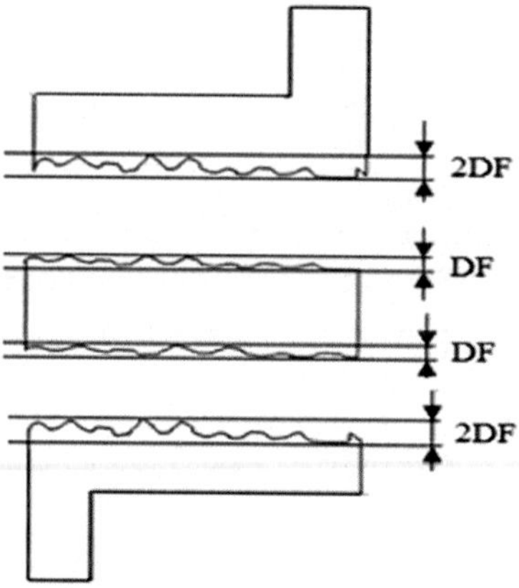

Fig. 5. Form defect of the assembly

Table 2. Parameter simulation

Torsor	Value	Distribution specified	Percentage of rejects %R	Statistical parameters	
				Mean	Standard deviation
$\tau_{2.1/1.2}$	$\begin{bmatrix} [0,0] \\ [0,0] \\ [0,0] \\ [\pm 0.003] \\ [\pm 0.0024] \\ [0,0] \end{bmatrix}$	Normale	5%	$u_u = 0$ $u_v = 0$ $u_w = 0$ $u_\alpha = 0$ $u_\beta = 0$ $u_\gamma = 0$	$\sigma_u = 0$ $\sigma_v = 0$ $\sigma_w = 0$ $\sigma_\alpha =$ 0.0017 $\sigma_\beta = 0.0014$ $\sigma_\gamma = 0$
$\tau_{3.1/2.2}$	$\begin{bmatrix} [0,0] \\ [0,0] \\ [0,0] \\ [\pm 0.003] \\ [\pm 0.0024] \\ [0,0] \end{bmatrix}$	Normale	5%	$u_u = 0$ $u_v = 0$ $u_w = 0$ $u_\alpha = 0$ $u_\beta = 0$ $u_\gamma = 0$	$\sigma_u = 0$ $\sigma_v = 0$ $\sigma_w = 0$ $\sigma_\alpha =$ 0.0017 $\sigma_\beta = 0.0014$ $\sigma_\gamma = 0$

(continued)

Table 2. (*continued*)

Torsor	Value	Distribution specified	Percentage of rejects %R	Statistical parameters	
				Mean	Standard deviation
τ_{FR}	$\begin{bmatrix} [0,0] \\ [0,0] \\ [0,0] \\ [0,0] \\ [\pm 0.00137] \\ [\pm 0.005] \end{bmatrix}$	Normale	5%	$u_u = 0$ $u_v = 0$ $u_w = 0$ $u_\alpha = 0$ $u_\beta = 0$ $u_\gamma = 0$	$\sigma_u = 0$ $\sigma_v = 0$ $\sigma_w = 0$ $\sigma_\alpha = 0$ $\sigma_\beta = 0.007$ $\sigma_\gamma = 0.0029$

As a result, the Jacobian torsor model with flatness defect may be expressed as follows:

$$\begin{bmatrix} \underline{u} & \overline{u} \\ \underline{v} & \overline{v} \\ \underline{w} & \overline{w} \\ \underline{\alpha} & \overline{\alpha} \\ \underline{\beta} & \overline{\beta} \\ \underline{\gamma} & \overline{\gamma} \end{bmatrix}_{FR} = \left[\begin{bmatrix} 1 & 0 & 0 & 0 & 70 & 0 \\ 0 & 1 & 0 & -70 & 0 & 25 \\ 0 & 0 & 1 & 0 & -25 & 0 \\ 0 & 0 & 0 & 1 & 0 & 0 \\ 0 & 0 & 0 & 0 & 1 & 0 \\ 0 & 0 & 0 & 0 & 0 & 1 \end{bmatrix}_{FE1} \begin{bmatrix} 1 & 0 & 0 & 0 & 40 & 0 \\ 0 & 1 & 0 & -40 & 0 & 0 \\ 0 & 0 & 1 & 0 & 0 & 0 \\ 0 & 0 & 0 & 1 & 0 & 0 \\ 0 & 0 & 0 & 0 & 1 & 0 \\ 0 & 0 & 0 & 0 & 0 & 1 \end{bmatrix}_{FE2/FE1} \right.$$

$$\left. \begin{bmatrix} 1 & 0 & 0 & 0 & 40 & 0 \\ 0 & 1 & 0 & -40 & 0 & 0 \\ 0 & 0 & 1 & 0 & 0 & 0 \\ 0 & 0 & 0 & 1 & 0 & 0 \\ 0 & 0 & 0 & 0 & 1 & 0 \\ 0 & 0 & 0 & 0 & 0 & 1 \end{bmatrix}_{FE2} \begin{bmatrix} 1 & 0 & 0 & 0 & 0 & 0 \\ 0 & 1 & 0 & 0 & 0 & 0 \\ 0 & 0 & 1 & 0 & 0 & 0 \\ 0 & 0 & 0 & 1 & 0 & 0 \\ 0 & 0 & 0 & 0 & 1 & 0 \\ 0 & 0 & 0 & 0 & 0 & 1 \end{bmatrix}_{FE3/FE2} \begin{bmatrix} 1 & 0 & 0 & 0 & 0 & 0 \\ 0 & 1 & 0 & 0 & 0 & 0 \\ 0 & 0 & 1 & 0 & 0 & 0 \\ 0 & 0 & 0 & 1 & 0 & 0 \\ 0 & 0 & 0 & 0 & 1 & 0 \\ 0 & 0 & 0 & 0 & 0 & 1 \end{bmatrix}_{FE3} \right]$$

$$\bullet \left[\begin{bmatrix} 0,0 \\ 0,0 \\ 0,0 \\ 0,0 \\ \pm 0.0033 \\ \pm 0.005 \end{bmatrix}_{FE1} \begin{bmatrix} 0,0 \\ 0,0 \\ 0,0 \\ \pm 0.0033 \\ \pm 0.0024 \\ 0,0 \end{bmatrix}_{FE2/FE1} \begin{bmatrix} 0,0 \\ 0,0 \\ 0,0 \\ \pm 0.0075 \\ \pm 0.006 \\ 0,0 \end{bmatrix}_{FE2} \right.$$

$$\left[\begin{bmatrix} 0,0 \\ 0,0 \\ 0,0 \\ \pm 0.0033 \\ \pm 0.0024 \\ 0,0 \end{bmatrix}_{FE3/FE2} \begin{bmatrix} 0,0 \\ 0,0 \\ 0,0 \\ \pm 0.01 \\ \pm 0.008 \\ 0,0 \end{bmatrix}_{FE3} \right]^T \quad (3)$$

Here the statistical study is repeated following the same steps described before, but the functional requirements determined in this part will give us the exact effect of the flatness defects. The statistical parameters of functional requirements are listed in last line of Table 2; the simulation is presented as follows (Fig. 6):

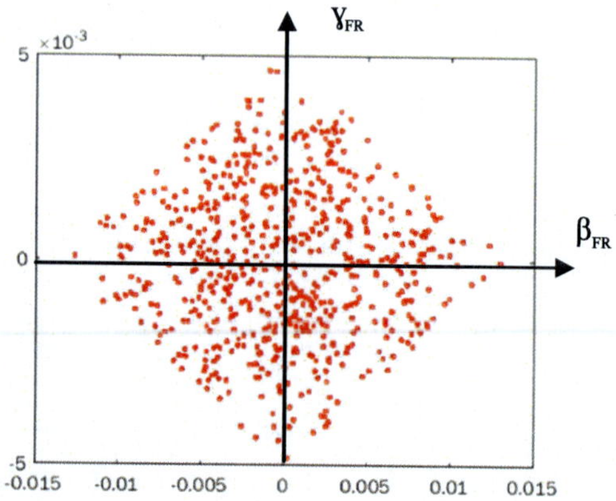

Fig. 6. Statistical tolerancing by taking flatness defect into account

3.3 Calculation of the non-conformity rate %R

Figure 7 demonstrates the tolerancing with and without flatness defects, with assemblies outlined in blue conforming exclusively to parallelism tolerances, and assemblies marked in red conforming by adding flatness defects. All assemblies outside the blue zone are considered defective when tolerancing is used without taking flatness defects into account. After running the simulation 10 times to show the influence of flatness defect on the research and measuring the rate of nonconformity, it was decided that the average rate of defective assemblies is set at: 5%.

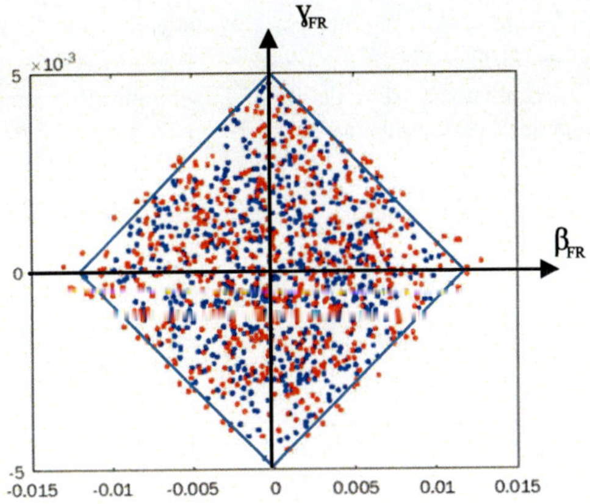

Fig. 7. Tolerancing with and without flatness defects

4 Conclusion

For the purpose of this study, we made use of a statistical method to compare the results of an analysis of orientation tolerances conducted with and without the impact of flatness defect. We produced a non-conformity rate that indicates the amount of defective components by including flatness defect in order to give a comparison between the statistical methodology with and without taking flatness defect into account. This allowed us to see which method was more effective. This will make it possible for us to demonstrate how the two approaches are distinct from one another. In comparison to previous models, the modified Jacobian torsor model, which takes flatness defect into account, is more conducive to tolerance, design, and manufacturing.

References

Shen, Z., Ameta, G., Shah, J., Davidson, J.: Comparative study of tolerance analysis method. J. Comput. Inf. Sci. Eng. (2015)

Chen, H., Jin, S., Zhimin, L., Xinmin, L.: A modified method of the unified Jacobian-Torsor model for tolerance analysis and allocation (2015)

Chahbouni, M.: Contribution to the design of mechanisms: Tolerance analysis with influence of form deviations (2016)

Ghie, W., Laperrière, L., Desrochers, A.: Statistical tolerance analysis using the unified Jacobian-Torsor (1999)

Zeng, W., Rao, Y., Wang, P., Yi, W.: A solution of worst-case tolerance analysis for partial parallel chains based on the Unified Jacobian-Torsor model (2017)

Frédéric, G.: Tolérancement statistique tridimensionnel, intégration en CFAO (2007)

El Feki, M.: Analyse et synthèse de tolérance pour la conception et le dimensionnement des systèmes mécatroniques (2011)
Lafond, P., Laperrière, L.: Jacobian-based modeling of dispersions affecting predefined functional requirements of mechanical assemblies (1999)

Industrial Engineering and Digital Factory

Environmental Benefits of Resources Pooling Applied on Hydrocarbon Supply Chain

Youness El Bouazzaoui[1](✉), Mourad Abouelala[1], S. Abdoudrahamane Kebe[2], and Fayçal Mimouni[2]

[1] Processes and Mechanical and Thermal Controls (PMTC) ENSAM, Mohammed V University, Rabat, Morocco
elbouazaoui.youness@gmail.com,
mourad.abou-el-ala@ensam.um5.ac.ma

[2] SMARTiLAB Moroccan School of Engineering Sciences (EMSI), Rabat, Morocco

Abstract. The notion of pooling logistics resources has been the topic of multiple studies in recent years in a variety of activity areas (transport, storage, manufacturing, etc.) in order to improve the performance of logistics chains in a sustainable development framework. Several articles deal with the subject of pooling of resources. However, due to data security and sector sensitivity, the pooling of resources (warehouses and transportation modes) for hydrocarbon supply chain reasons is a seldom covered topic in literature review. The principles of "resources pooling" will be explained in this article, as well as a literature evaluation of new organization tactics for sustainable logistics. Furthermore, we start our proposed topic by describing the existing petroleum products supply chain model in Morocco (diesel gasoil and gasoline) as well as a DES (Discrete event simulation) using ARENA simulation software, to quantify CO_2 emissions fluctuation rates and estimate the environmental effect for assessing the environmental efficiency of the pooling resources proposed approach, by focusing on a tactical level. The simulation results of this study have a positive impact on CO_2 emissions in the sustainable supply chain.

Keywords: Sustainable development · Hydrocarbon · Resources pooling · ARENA simulation · CO_2 emissions · Environmental efficiency

1 Introduction and Context Study

In line with the environmental objectives launched by the various Conference of the Parties (COP), the national strategy "Morocco logistics 2030", launched in September 2016 by the Moroccan Government, supports a dynamic of logistics in Morocco based on a societal responsibility approach companies and integrated sustainable requirements by all stakeholders into their supply chain. Sustainable objectives, in theory, include all three dimensions economic, environmental, and social that originate from the needs of the relevant partners (Mardani et al. 2020; Kannan 2018). Within a supply chain, there are two types of possible cooperation, horizontal cooperation and vertical cooperation.

L. Azrar et al. (Eds.): CIP 2022, LNME, pp. 49–58, 2023.
https://doi.org/10.1007/978-3-031-23615-0_6

Horizontal cooperation concerns manufacturers, located at the same level of the supply chain. They share data, resources and ultimately added value for the logistics service in question. The literature on horizontal cooperation is still scarce enough to be in an emerging phase (Leitner et al. 2011). The reasons given by the authors (Cruijssen et al. 2007) for adopting an horizontal cooperation are reducing logistics costs and improving customer service. The vertical cooperation, for its part, is a prerequisite both for obtaining information from distributors, necessary for the calculations of production, storage, transport of goods each manufacturer, but also to obtain common delivery niches and thus optimize the last phase of logistics pooling (Pomponi et al. 2015).

Regardless of government subsidies for renewable energy, fossil fuels still account for a large portion of the global energy mix (81% of world consumption in 2015). In fact, Morocco's sustainable development policy, which relies on renewable energy sources such as wind, hydropower, and solar, is in the same boat. Fossil fuels, particularly petroleum products, account for the majority of the country's energy output (88% in 2017).[1] Because these items manufacture, distribution, and use are substantial sources of CO_2, steps may be done to mitigate their environmental effect. Our research focuses on identifying Moroccan hydrocarbon supply networks in order to pool transportation and distribution resources. The goal is to estimate the CO_2 emission effects before and after transportation and storage pooling using simulations approach as an indicator which contributes to the achievement of sustainability development objectives.

After the shutdown of the only Moroccan refinery in August 2015, the petroleum products consumption characterized by the entire dependence on international hydrocarbon markets. In 2017, Moroccan imports of fossil fuels accounted for 93.3% of all petroleum products with an annual demand rise of around 5%, and the energy cost came to 54 billion Dirhams (The Economist 2017). In the same year, Moroccan consumption of hydrocarbons reached 6.4 million tons, as shown in Fig. 1 (The Parliamentary Committee on the Hydrocarbons Sector 2018).

In fact, the Moroccan government has defined an integrated national logistics sector development strategy for the year 2030. One of the strategy's goals is to decrease CO_2 emissions from road freight transportation by 35% (AMDL 2016). Pooling transportation resources is thus an intriguing concept to investigate, particularly in the case of moving petroleum products. We shall offer a synopsis of major studies dealing with resources pooling approach before moving on to the comprehensive Moroccan hydrocarbons logistics chain diagnostic.

2 Literature Review

In contrast to the connection of cooperation between upstream and downstream, downstream is a topic that has been discussed in recent literature, the supply chain may be thought of as a corporate network of companies that includes customers, suppliers, and other players in a company's manufacturing process (Mazzola et al. 2015). Furthermore, according to some academics, sustainable supply chain networks allow supply chain participants to connect with one another in order to discover competent partners with

[1] World Energy Balances preliminary edition (2019) The International Energy Agency IEA. www.iea.org/statistics/.

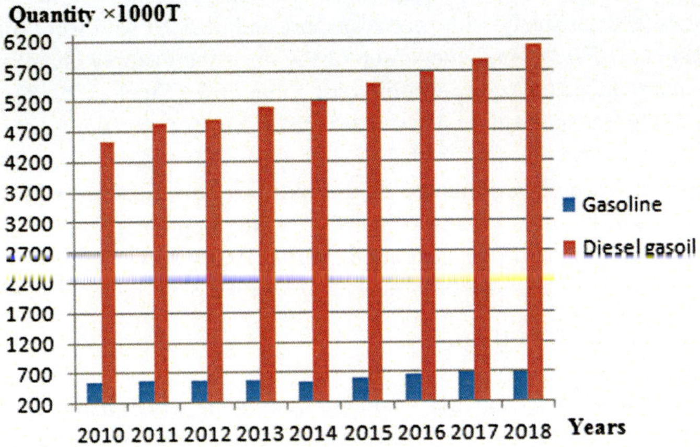

Fig. 1. Consumption evolution of petroleum products (The Parliamentary Committee on the Hydrocarbons Sector 2018)

whom to form long-term partnerships (Darbari et al. 2019). Logistics pooling, according to authors (Pan et al. 2013) relates to partners co-designing logistics networks (suppliers, customers, carriers, etc.) with a shared goal of sharing logistic networks (warehouses, platforms, modes of transportation, etc.) and providing management with essential data. Pooling is ideal for logistic chains working in the same sectors while pursuing a shared aim and being nearby to facilities. In addition, the bulk of studies imply that pooling resources is a solution to numerous logistical difficulties that can improve the performance of logistics chains. The authors emphasize the predicted benefits of pooling in terms of economic, environmental, and political performance in their paper. Summary of work based on sector activity and pooling goal. In the context of the supply chain, resource pooling has been used in several research. A study proposed multiple terminals and shared resources of the Hunter Valley coal export system in Newcastle, Australia was optimized by sharing a rail network for inbound and outgoing operations and increasing throughput at a single coal terminal while sharing resources at the same time; the results showed 17% better on average than those achieved using the existing approach (Rocha et al. 2019). Furthermore, the authors (Li et al. 2018) analyzed in their recent paper, the benefit of storage sharing by quantifying the benefit of multi-depot vehicle routing problem with fuel consumption and they expected that the benefit depends on instance characteristics (depot–customer geographic distribution, maximum route distance, and number of depots). The authors (Silbermayr et al. 2017) discussed the inventory pooling with environmental constraints using copulas method by modeling of the trade-off quantitatively between economic performance and environmental sustainability in the transport and warehouse sectors. The result showed a decreasing emissions with increasing demand dependence also the copulas' dependency structure among the demands can have a considerable influence on the best strategy. In the same way, adopting pooling supply chains may be reduced greenhouse gas emissions, especially CO_2 emissions from freight transportation; two significant results confirmed which are: reducing the size of the problem and obtaining solutions with a gap less than 3% and decreasing of CO_2

emissions of 14% exclusively with road transport and of 52% with joint road and rail transport (Pan et al. 2013). Implementation of the horizontal supply chain by applying pooling scheme reduced the transportation cost about 26% (Xu et al. 2012).

In general, the transportation of hydrocarbons in Morocco accounts for about 14% of the total logistics expenditures, with a fleet of over 1200 tanker trucks traversing the country (The Parliamentary Committee on the Hydrocarbons Sector 2018). Thus, what is the benefits expected by adopting resources pooling approach, especially transport and storage to decrease the CO_2 gas emitted by companies fleet?

3 Problematic Description

The Moroccan hydrocarbon logistics chain has undergone a big change after the shutdown of the only oil refinery located in Mohammedia city, the consumption of hydrocarbons has become totally linked to external petroleum suppliers, who are mainly concentrated in Europe and the Gulf countries. Before this stop, all the actors in the hydrocarbons market had an internal supply to optimize the charges due to maritime transport and the port customs fees, a situation which motivated us to deeply study this new logistics chain, more precisely the transport and storage resources by proposing the "pooling" of these resources between two major importers and distributors of hydrocarbons present in the market, also by focusing on a dynamic destination linking the petroleum port of Mohammedia and the region of Tangier contains the most important cities of the Moroccan north, which are: Tangier, Tetouan and Larach.

The data of our model were obtained directly from a real data bases for a period of 78 months from the two studied companies A and T. Otherwise, the trucks make an empty journey from the various towns in the Tangier region. Arriving at Mohammedia petroleum port, loading process started and the average time to load a truck is estimated at 35 min. Furthermore, during the resupply process, the fleet used to transport the full quantities of hydrocarbons emits greenhouse gas (GHG) emissions due to the combustion of truck engine fuel. We can then distinguish two emission channels: Emissions produced when trucks are loaded and during an empty journey.

4 Methodology for Calculating CO_2 Emissions

There are many references in the literature that use different approaches and units to calculate gas emissions, making it difficult to compare all of the results. Moreover, greenhouse gases (GHG) or CO_2 alone are commonly measured with three units: Carbon equivalent (kg eq C); CO_2 equivalent (kg CO_2eq) and emission of CO_2 alone (kg CO_2). In this paper, we'll focus just on CO_2 emissions. It is the quantity of CO_2 that is directly released. Because the period for loading and unloading activities is very brief, we may overlook the quantity of polluting gases released, particularly CO_2, during the start-up of the truck engine. We emphasize that the worldwide fleet investigated in this article consists of a HDV (Heavy Duty Vehicle) tanker truck for delivering petroleum products with a capacity of 25 tons. We used the MEET methodology (Methodologies for Estimating Air Pollutant Emissions from Transportation) to calculate the CO_2 emissions, which defines a vehicle's global pollutant emissions as the sum of three emissions:

Emissions when the vehicle is warm after the start (E_{hot}); emissions when starting the vehicle ($E_{start\ up}$) and emissions due to fuel evaporation ($E_{evaporation}$). (Hickman et al. 1999).

$$E_{total} = E_{hot} + E_{evaporation} \tag{1}$$

In our research, transportation is always over a long distance, allowing us to ignore the vehicle's starting emissions, which occur only while the engine is cold.

$$E_{total} = E_{hot} \tag{2}$$

Furthermore, we compute CO_2 emissions by taking into account that the amount emitted is entirely dependent on the vehicle loading rate, distance traveled, and truck average speed. The following simplified formula allows us to determine CO_2 emissions in function just of loading rate while neglecting the gradient component (Pan 2010).

$$\varepsilon = E_{hot} \cdot \Phi(\gamma, v) \tag{3}$$

$$\Phi(\gamma, v) = k + n \cdot \gamma + p \cdot \gamma^2 + q \cdot \gamma^3 + r \cdot \gamma + s \cdot v^2 + t \cdot v^3 + u \div v \tag{4}$$

$$\varepsilon = \left(1576 - 17.6v + 0.00117v^3 + \frac{36067}{v^2}\right) \cdot [1 + \alpha(0.43 - 0.916/v)] \tag{5}$$

$$\varepsilon = 0.418\alpha + 0.753 \tag{6}$$

where:

ε is the mission factor when loaded (g/km); E_{hot} is the emission factor when unloaded (g/km); $\Phi(\gamma, v)$ is the load correction factor function; γ is the gradient (%); v is the mean velocity of the vehicle (km/h); k is a constant; n and u are coefficients which depend of the vehicle type; α is the truck loading rate.

With $v = 70$km/h and $\alpha \in [0, 1]$ ($\alpha = 1$ for a full load regime and $\alpha = 0$ for an empty load regime). Tables A27, A40, and A46 of the project report MEET contain all of the coefficients and constants needed to calculate the loaded vehicle factor (Hickman et al. 1999).

The purpose of using the simulation is to allow us to study several scenarios of the transport of petroleum products in Morocco. It will allow us to compare:

- The distribution system can be studied more easily;
- With a smaller data collection, it's easier to generate results;
- Effortless implementation.

5 Simulation

In order to improve the proposed approach "resources pooling" performance and to explore the interaction between various parameters integrating in our supply chain; we adopted a Discrete Event Simulation (DES), using "Arena simulation" offers the user many advantages. It has an integrated framework for creating simulation models, also requires many functions (modelling, animation, model verification, analysis of inputs and outputs data and results of analysis). Moreover, it may be used to simulate any production or service-related system (logistics operations; such as storage and transport, vehicle planning and scheduling, supply chain and business process).

5.1 Case Study

The data model obtained from real data sources from the two firms under consideration A and T. There are many petroleum product providers for importers A and T, who have a temporary storage facility at Mohammedia petroleum port. Two hundred and seventy trucks are assigned to the two companies A and T, whose truck deliveries from the port are contingent on demand expressed in Tangier storage center. The objective of our simulation will be to obtain the distance traveled by the carriers and their load factors in order to measure the CO_2 emission for different fuel distribution companies.

In this part, we propose to diagnose the ecological hydrocarbon transport impact. For this reason, we will extract from the simulation the total CO_2 emission before and after testing resources pooling approach. To measure this emission, we will have as input data for our model: The demand of different stations, the capacity of each carrier, the distance to be traveled between the different places of interest and the means of transport used (Table 1).

5.1.1 CO_2 Emissions Before Resources Pooling

We consider two main importers (A and T) and their deliveries from Tangier storage center and Mohammedia petroleum port as resupplying sources for three studied destinations (Tangier, Tetouan and Larach). Moreover, we calculate the CO_2 emissions using Eq. (6). Thus, our simulation made on two cases:

- Go route emission: The trucks always had a full load ($\alpha > 0$).
- Return route emission: The trucks always had an empty load regime ($\alpha = 0$).

For computing the global quantity emitted by all fleet, we are just going to sum up the quantities emitted whole the way (go and return routes) and multiply it by the total number of deliveries.

$$\varepsilon = (0.418\alpha + (2 \times 0.753)) \times L_{total} \quad (7)$$

With L_{total} is the number of total deliveries from Mohammedia port to Tangier city.

5.1.2 CO_2 Emissions After Resources Pooling

In this part, the simulation is done by regrouping the transport and storage capacity resources, for global trucks and tanks, respectively 267 and 105,000 tons as a common resource between the two companies. The results of the simulation are given in the Tables 2 and 3.

Table 1. Flows related to the different processes of the supply chain of the hydrocarbon importers A and T.

Process	Flows	Values
Storage	Storage capacity-Tangier city (ton × 10^3)	Importers: A T 71 34
Transportation (trucks)	Fleet used between Mohammedia port - Tangier storage center	Importers: A T 200 67
	Truck capacity (ton)/Average speed (km/h)	25/70
Distances (km)	From Tangier storage center to Tangier/Tetouan/Larach	5/53/128

6 Results and Discussions

The results of the simulation are given in the next table, we have allocated storage capacity and transport in proportion to each importer's vehicle registration share. Additionally, to calculate the emissions; we used the following formula: $E_{kg/km} = 0.418\alpha + 0.753$.

The findings of this research also reveal that in the case of local resupplying from Tangier storage center to Tangier, Tetouan and Larach destinations, the total CO_2 emission decreased by 79% after using resources pooling. Otherwise, the CO_2 emission from Mohammedia petroleum port to Tangier storage center increased by 8% also an increasing about 32% on the average stock. This result may caused by imbalance between pooled resources and market requests to achieving a sustainable supply chain. Thus, adopting transport and storage pooling have a positive impact on CO_2 emissions in the sustainable supply chain.

7 Conclusion

In the context of supply chain sustainability, this study aims to explore the importance of pooling resources approach. The research model is tested by utilizing the data bases from two largest importer companies in Morocco by testing resources pooling technique, using DES (Discrete event simulation). Furthermore, we create a simulation model for each of the two scenarios investigated: in the first, we simulate our supply chain model before to resource pooling, and in the second, we simulate our supply chain model after resource pooling. In the case of local replenishment after using resources pooling from Tangier storage center to the various studied destinations, the total CO_2 emission decreased by 79%. In addition, the CO_2 emission from Mohammedia petroleum port to Tangier storage center increased by 8% also an increasing about 32% on the average stock. Moreover, this simulation enabled us to estimate the principal variation factors needed to calculate CO_2 emissions, such as total deliveries and filling rate. The simulation results indicate that adopting resources pooling technique have a significant impact on CO_2 emissions in the sustainable supply chain. However, the model presented remains limited to understanding all the present constraints as road gradient and vehicle age.

Table 2. Simulation results for the importers A and T before and after resources pooling

Output	Before resources pooling						After resources pooling				
	Importer A			Importer T			Importers (A + T)			GAP	GAP (%)
	Tangier	Tetouan	Larach	Tangier	Tetouan	Larach	Tangier	Tetouan	Larach		
Total deliveries	20461	16480	8288	32950	20259	16475	11312	8254	4513		
CO_2 emission (kg)	193,899	1,655,431	2,010,655	292,958	1,909,301	3,749,868	103,663	801,780	1,058,743		
Total CO_2 emission (kg) (from Tangier storage center to Tangier, Tetouan and Larach)	3,859,985			5,952,127			1,964,186			– 7,847,926	– 79
CO_2 emission (kg) (from Mohammedia to Tangier storage center)	62,592,142			23,348,405			92,677,290			+ 6,736,743	+ 7.8
Total CO_2 emission (kg)	66,452,127			29,300,532			94,641,476			– 1,111,183	– 1.2

Table 3. Simulation results for the importers A and T before and after resources pooling

Output	Before resources pooling		After resources pooling		
	Importer A	Importer T	Importers (A + T)	GAP	GAP (%)
Filling rate(α)	0.9314	0.6513	0.7819	– 0.0094	– 1.2
Average stock (ton) of tank storage center	64,119	11,111	99,623	+ 24	+ 32

In perspective, our next work will suggest a mathematical model of petroleum supply chain, composed by objective function; decision variables and constraints, as an operational research problem to minimize the CO_2 emission generated by the global fleet for various destinations, also to validate our proposed approach for applying it in large and various industrial actors in the future.

References

Cruijssen, F.C., Cool, M., Dullaert, W.: Horizontal cooperation in logistics: Opportunities and impediment. Transp. Res. Part E: Logistics Transp. Rev. **43**(2), 129–142 (2007)

Darbari, J.D., Kannan, D., Agarwal, V., Jha, P.C.: Fuzzy criteria programming approach for optimising the TBL performance of closed loop supply chain network design problem. Ann. Oper. Res. **273**(1–2), 693–738 (2019). https://doi.org/10.1007/s10479-017-2701-2

Hickman, J., Hassel, D., Joumard, R., Samaras, Z., Sorenson, S.: Methodology for calculating transport emissions and energy consumption. Report for the Project MEET, Transport Research Laboratory (1999)

Kannan, D.: Role of multiple stakeholders and the critical success factor theory for the sustainable supplier selection process. Int. J. Prod. Econ. **195**, 391–418 (2018)

Leitner, R., Meizer, F., Prochazka, M., Sihn, W.: Structural concepts for horizontal cooperation to increase efficiency in logitsics. CIRP J. Manuf. Sci. Technol. **4**(3), 332–337 (2011)

Li, J., Wang, R., Li, T., Lua, Z., Pardalos, P.M.: Benefit analysis of shared depot resources for multi-depot vehicle routing problem with fuel consumption. Transp. Res. Part D **59**, 417–432 (2018)

Mardani, A., Kannan, D., Hooker, R.E., Ozkul, S., Alrasheedi, M., Tirkolaee, E.B.: Evaluation of green and sustainable supply chain management using structural equation modelling: a systematic review of the state of the art literature and recommendations for future research. J. Clean. Prod. **249**, 119383 (2020)

Mazzola, E., Bruccoleri, M., Perrone, G.: Supply chain of innovation and new product development. J. Purch. Supply Manag. **21**(4), 273–284 (2015)

Moroccan Agency for Development of Logistics AMDL.: Moroccan green logistics: Sustainable development at the heart of the logistics dynamics of Morocco. Report, Marrakech COP22, Morocco (2016)

Moroccan Company of Refining Industry SAMIR.: Logistics department. Archive, Mohammedia, Morocco (2018)

Pan, S.: Contribution à la définition et à l'évaluation de la mutualisation de chaînes logistiques pour réduire les émissions de CO_2 du transport : Application au cas de la grande distribution. Gestion et management, École Nationale Supérieure des Mines de Paris, pp. 93–101 (2010)

Pan, S., Ballot, E., Fontane, F.: The reduction of greenhouse gas emissions from freight transport by pooling supply chain. Int. J. Prod. Econ. **143**(1), 86–94 (2013)

Pomponi, F., Fratocchi, L., Rossi Tafuri, S.: Trust development and horizontal collaboration in logistics: a theory based evolutionary framework. Supply Chain Manag.: Int. J. **20**(1), 83–97 (2015)

Rocha, D.P., Boland, M.N., Ernst, A.T., Mendes, A., Savelsbergh, M.: Throughput optimization in a coal export system with multiple terminals and shared resources. Comput. Ind. Eng. **134**, 37–51 (2019)

Silbermayr, L., Jammernegg, W., Kischka, P.: Inventory pooling with environmental constraints using copulas. Eur. J. Oper. Res. **263**, 479–492 (2017)

The Economist.: Hydrocarbons, salt bill, N 5082, 8 Aug 2017, pp. 4 (2017)

The Parliamentary Committee on the Hydrocarbons Sector.: Arabic version report, pp. 24–53 (2018)

Xu, X., Pan, S., Ballot, E.: Allocation of transportation cost and CO2 emission in pooled supply chains using cooperative game theory. In: Proceedings of the 14th IFAC Symposium on Information Control Problems in Manufacturing, Bucharest, Romania, 23–25 May (2012)

Lean, Green, Resilient Supply Chain and Sustainable Performance: Practices and Measruesements Review

Ikram Ait Hammou[1](✉), Salah Oulfarsi[1], Ali Hebaz[1], Samah Mahmah[1], and Anass Cherrafi[2]

[1] ENCG El Jadida, Chouaib Doukkali University, El Jadida, Morocco
aithammou.ikram@gmail.com
[2] ENSAM-Meknes, Moulay Ismail University, Meknes, Morocco
a.cherrafi@ensam.umi.ac.ma

Abstract. The complexity of the environment driven by global climate change, social responsibility, global pandemic i.e., COVID-19 and demand uncertainty has caused the supply chain to be inefficient and vulnerable. In this regard, the integration of lean, green and resilient supply management practices, jointly or progressively, have been reported by researchers and practitioners as strategies to enhance competitiveness while facing environmental uncertainty and decreasing social and environmental harm. However, only few studies have investigated their integration within the supply chain. Consequently, this research conducts an in-depth literature review of 37 research articles in order to present lean, green and resilient practices that have been used in the context of supply chain management and also to classify them into different levels. The results of this study show that the dominant paradigm is lean supply chain with 27 practices identified in the literature, followed by resilient supply chain with 20 practices, and finally green supply chain with 16 practices. In addition,60% are all related to the internal supply chain level. Although the literature offers a significant number of social performance measurements, the results indicates that the social performance appears to be the least studied compared to the economic and environmental performances.

Keywords: Lean · Green · Resilient · Supply Chain Management · Sustainability

1 Introduction

The rapid growth of industualization along with the excessive unrational use of resources over the past decades has generated severe environmental degradation that we had already witnessed its impacts all over the globe, as for exemple, the increased levels of carbon dixiod (CO2), pollution, greenhouse gas emissions (GHG) and global warming which present a serious threat for the mankind and its existance. As a consequence, an increasing concern of governments, individuals and various stakeholders for the environment and sustainability have led industries to mitigate their environmental impacts as being considered as a primary actor in making environmental and ecological issues resulting

L. Azrar et al. (Eds.): CIP 2022, LNME, pp. 59–76, 2023.
https://doi.org/10.1007/978-3-031-23615-0_7

not only from their manufacturing operations but also by their supply chains activties (Hebaz and Oulfarsi 2021). Moreover, today's COVID-19 outbreak appears to affect not only the global economy but also global supply chains as being considered as an unexpected event that disturbe industries normal activities (El Baz and Ruel 2021; Ivanov and Dolgui 2020). All these concerns cause the supply chain and its components at different levels to be dysfunctional and vulnerable (Sharma et al. 2021). Whithin this context, industies have started to develop and implement alternative strategies allowing them to sync their environmental goals with operational ones by integrating emerging approaches combining lean, green and resilient practices within their supply chains so as to maintain competitivness along with the effeciency and simultaneously becoming more resilient to distruption meanwhile responding to their stakeholders and achieving sustainability in their activities (Agarwal 2021; Hammou et al. 2022). A large number of researchers have studied lean, green and resilient paradigms jointly through the incorporation of other paradigms such as agile or sustainble, forming LARG or LARGS models (Sharma et al. 2021). Nevertheless, only a handful of studies have investigated their integration in the context of the supply chain (Agarwal 2021; Govindan et al. 2013). Despite these efforts, the body of research on the integration of lean, green and resilient practices in the supply chain context is still poor. Therefore, the present research aims to enhance the current body of knowledge through the identification, ranking, and classification of the three paradigms, lean, green and resilient upon the different levels of the supply chain as such the upstream, internal, downstream and common practices. Also, to identify sustainable performance measurments in the context of the supply chain. The paper is organized as follows: introduction followed by a review of the extant literature, then the methodology is highlited in the Sect. 3 followed by findings and discussion in the Sect. 4 and finally the conclusion in the Sect. 5.

2 Literature Review

2.1 Lean, Green and Resilience Paradigms

According to previous studies, lean management stands for a set of tools and activities implemented by organizations to elimitate several waste sources that do not contribute to the creation of an added value for the final customer. In the context of supply chain management (SCM), the scope of lean approach becomes wider, since the concern to minimize waste will be shared with the organization's supply networks. Therefore, LSCM requires a high level of collaboration between the different supply chain actors in order to obtain mutual benefits in terms of productivity, quality, efficiency, customer satisfaction, profitability and competitiveness (Alqudah et al. 2020; Campos and Vazquez-Brust 2016; Carvalho et al. 2010; Garcia-Buendia et al. 2021).

In recent years, companies are facing new challenges due to changes in the environment and the appearing of new expectations and requirements from their various stakeholders, especially in terms of the protection of the environment. Therefore, the pressure has increased from community, consumers and government regarding the respect of stringent environmental regulations, driving industries to adopt environmental practices into their management and production processes (Carvalho et al. 2017; Sharma et al. 2019; Zhu and Sarkis, 2004), hence the emergence of the green paradigm which was initially introduced by Michigan State University in 1996 (Singh et al. 2020). As stated by Wu et al. (2015), green supply chain management (GSCM) has the same target as the green production adopted only at the level of the company's own operations, while sharing these goals with external partners through joint strategic decisions. Thus, GSCM is defined as an organizational philosophy aiming to incorporate environmental thinking within all operations and exchanges that exist between the supply chain partners. This objective can only be achieved if the partners work together, taking into consideration the environmental concerns, in order to reduce the harmful effects of their operations as well as increasing efficiency (Azevedo et al. 2011; Cabral et al. 2012; Carvalho et al. 2010; Engin et al. 2019; Martínez-jurado and Moyano-fuentes, 2013; Singh et al. 2020; Zhu and Sarkis 2004).

Moreover, today's marketplace is characterized by disruptive elements, described as unexpected events that disturb the company's normal activity. These disturbances might come from many sources, such as environmental disasters, political instability, physical events, information crises, personnel events and terrorist actions. In order to minimize the consequences of these events, resilience is one of the important characteristics that supply chains need to adopt to cope with disruptive events efficiently (Christopher and Peck 2004; Rice and Caniato 2003). Therefore, resilient supply chain management (RSCM) appears to offer a way for companies to avoid the limits of old approaches related to protection strategies and risk prevention, and to deal with the complexities of global supply chains (Sheffi 2015). Moreover, resilience is defined as not only the ability to maintain control at disruptions, but also as being adaptive and capable of sustained response to sudden shifts in the environment in the figure of uncertain demands. It represents the capacity to proactively plan and design the supply chain network for anticipating unexpected events and to respond adaptively to disruptions (Sharma et al. 2021).

2.2 Lean, Green, Resilience and Sustainable Performance

Over the past decade, sustainability has become a popular and a key issue for regulators, companies and even for researchers and academics. It represents a goal for companies, however, it is difficult to measure the degree of the company's sustainability. In the mid-1990s, John Elkington strove to measure sustainability by proposing a new framework called Triple Bottom Line (TBL). It provides a framework for measuring the performance of the company and the success of the organization using three lines: economic, social, and environmental (Goel, 2010). According to (Zhu and Sarkis 2004), environmental performance is the ability to reduce air emissions, solid waste, wastewater, consumption of harmful materials, frequency of environmental accidents and so on. While social performance represents a configuration of social responsibility principles, social response processes, policies, programs, and observable outcomes. Therefore, the firm must respond to workers' expectations and satisfy their needs, providing reasonable incomes and creating a favorable work climate for them. Concerning the economic performance, it focuses primarily on the company's profitability and growth (Alqudah et al. 2020). Indeed, the literature has presented a set of studies that demonstrate the impact of lean, green and resilience on performance dimensions. (Vais et al. 2006) have used lean tools, such as Kaizen, 5S and autonomous maintenance to reduce natural resources consumption and production outputs. The positive impact of lean practices on facilitating the achievement of environmental goals has also been demonstrated by (Vinodh et al. 2011). Thus, in most of the studies, lean practices have been found to be inherently capable of facilitating the achievement of environmental goals and improvements. However, the findings are still not conclusive as long as negative relationships have been found to exist (Rothenberg et al. 2001). Moreover, Yang et al. (2011) have found a positive impact of lean implementation on financial performance through enhanced productivity and inventory leanness. For green practices, Zhu and Sarkis (2004) investigated the relationships between internal environmental management, external GSCM, investment recovery, and eco-design, and their impact on environmental performance. Their results show a positive relationship between green and environmental performance. In addition, when green practices are successfully implemented, the company's negative environmental ratings will be reduced (Azevedo et al. 2012). For the economic performance, Zhu and Sarkis (2004) have found that green practices are unable to generate better economic performance. In contrast, recent studies mainly report a positive effect of GSCM on economic performance (Geng et al. 2017). In addition to this, the implementation of environmental management practices impacts environmental and social performance through reducing resource consumption and improving stakeholder relations. De Giovanni (2012) and (Geng et al. 2017) have found that internal environmental management has a positive effect on social performance. Regarding the impact of RSCM on sustainable performance, the literature has not been abundant on this topic. In 2015, Govindan has conducted a study on the influence of resilient supply chain on company's performance, and he found that flexible transportation leads to a reduction in CO2 emissions which improves environmental performance. Ruiz-Benítez et al. (2018) have confirmed that resilient practices help achieving improvements in economic performance. Wang et al. (2021) conducted a study on the performance of the resilient supply chain in COVID 19, they confirmed that the use of robotics in the current pandemic situation can ensure the safety of humans by performing work processes without problems, which allows to fight against the spread of infection using a minimum of human workforces which implies an improvement in social performance.

3 Research Methodology

An in-depth review of the extant literature on the lean, green and resilient practices integration from a SCM perspective was conducted using Google Scholar, allowing the use of different databases such as Elsevier ScienceDirect, Emerald, Springer, Wiley, MDPI, ResearchGate, Clarivate Web of Science and EBSCO. The keywords searched i.e., "green" "lean" "resilient" "supply chain" "sustainable performance". In the process of searching the articles, a combination of paired keywords was used along with "sustainable performance". Other articles related to "green supply chain management" e.g., (Qorri et al. 2018) were gathered separately. The search resulted in 20.200 articles using Google Scholar. Articles selected are based on the following inclusion criteria: only top ranked journals, conference proceedings were included to ensure quality. Along with analysis of their abstract, over a 147 article were gathered using the keywords aforementioned, only 79 were used to be included in the study, these articles are published between 1991 and 2021. The rest of irrelevant articles were excluded. While reviewing the literature, the articles were categoriezed by topics, methodology, and findings forming a database through the use of excel. Around 37 articles were used to extract data concerning lean, green and resilient practices, meanwhile only 17 articles were used to reveal sustainable performance indicators.

4 Findings and Discussion

Table 1 presented in the appendix outlines the main lean, green and resilient practices found in the literature of supply chain management, classified into four categories: upstream, internal, downstream and common practices. Upstream practices stand for those practices used within a matter of managing and improving the relationship between the company and its suppliers, while internal practices include lean, green and resilient practices used within the organization in order to improve its operational and managerial processes. The third category includes downstream practices, which refer to the set of practices used by the company in order to get a better relationship with its customers. Finally, while classifying the different practices, we have found that there are some common practices that can be part of the three previous levels. According to this classification, it has been found that the internal practices have taken the lead. Thus, 60% of the lean, green and resilient practices identified are all part of the internal supply chain level, i.e. 44 practices, which shows that previous studies have mainly focused on the analysis of practices used within the company. For the other levels, we have have identified 13 upstream and 12 downstream practices with almost the same importance. As for the practices shared between the 3 levels mentioned above, we have identified four practices that we have classified under the category of common practices. Moreover, the practices presented in Table 1 have been also grouped according to their nature. Therefore, while there are some practices which belong to one of the paradigms: Lean (L), Green (G) or Resilient (R), there are other practices that are shared between many of the paradigms: Lean and Green (L&G), Lean and Resilient (L&R), as well as Lean and Green and Resilient (L&G&R). Thus, the dominant paradigm is lean supply chain with 27 practices identified from the literature, followed by the resilient supply chain paradigm with 20 practices, and finally the green supply chain with 16 practices. Furthermore, the set of practices beloging to each of the supply chain levels have been ranked following their frequency of citations. Consequently, JIT/Pull Philosophy is the most significant internal practice, nevertheless it is undergoing a great debate because it gives rise to a conflict between operational and environmental concerns. The majority of studies stated that this practice is always beneficial for the SCM (Azevedo et al. 2011; Campos and Vazquez-Brust 2016; Carvalho et al. 2010; Cherrafi, Garza-reyes, et al. 2018; Dües et al. 2013; Engin et al. 2019; Govindan et al. 2015; Hajmohammad et al. 2013; Sawhney et al. 2007; Singh et al. 2020; Wu et al. 2015). However, many authors have demonstrated that this practice could also have a negative effect on the SC, especially when it comes to the environmental aspects (Azevedo et al. 2012; Carvalho et al. 2017; Zhu and Sarkis 2004). In addition, the most important upstream practice identified from the literature is Supplier development, which is considered as having a great impact on improving SCM. However, only Sawhney et al. (2007) stated that this practice could have a mixed impact while being implemented in the SC. In addition, Cooperation with suppliers has also the same weight as Supplier development, which proves that the cooperative mindset is a must in order to manage the SC efficiently. Regarding the downstream practices, the most important ones are Customer relationship and Cooperation with customers, confirming the same issue of the great need for cooperation in SCM. Finally, Information spreading through the network is the most important practice as being common between all the SC levels. After the identification of lean, green and resilient practices in different levels of the supply chain, the next step was the assessment of the different measurements used to evaluate their impact on the three dimensions of sustainable performance, namely: environmental, economic and social performances. These measurements were gathered from the previous studies and summarized in Table 2 in the appendix. Thus, 17 articles have studied environmental performance and have presented nine measurements to evaluate this performance dimension, 10 articles have studied the economic performance and have offered 14 measurements, while only five articles have studied the social performance, but despite this, they have presented a significant number of measurements (13 measurements). Hence, the environmental and economic dimensions are the most widely studied, then the social dimension was found to be the least studied in terms of the number of articles that focused on this aspect of performance in the supply chain.

5 Conclusion

The major contribution of our research is the identification and classification of the main practices of green, lean and resilient supply chain paradigms into three main categories i.e., upstream, internal and downstream, we have also considered other practices to be part of all three levels at the same time, and we have classified them under a fourth category named common practices. As a result, it has been found that most of the practices are included into the lean paradigm, and also the supply chain level that covers the largest number of practices is the internal one. The study also discusses and synthesizes the measures presented in the literature concerning sustainable performance. It has been found that he economic and environmental dimensions are the most studied in the literature. Thus, our work invites researchers to engage in the social side as we confirm that this dimension is not studied enough. Previous studies have also focused on internal practices, which requires further investigation into practices to improve upstream relationships with suppliers, as well as downstream relationships with customers. Moreover, this article provides a reference for researchers in this field, offering them a large database of the articles that have dealt with the integration of the three paradigms lean, green and resilient supply chain, and also the measurements that previous studies have used in order to evaluate these three aspects of sustainable performance. This could be very useful, especially with the COVID 19 pandemic, whose consequences are still present especially when it comes to the impact on the global supply chain. Therefore, companies should seriously think about integrating strategies to strengthen their supply chains against such events that could threaten several sustainability aspects (Crane and Matten 2020; Luckstead et al. 2021; Majumdar et al. 2020). This research has several limitations. The search has revealed only on the extant published literature, at a particular time frame, using available articles, while the body of knowledge continue to grow.

Moreover, this research has not considered the use of a systematic literature review or an empirical investigation focusing on a particular research question. Also, synergies and trade-offs between lean, green and resilient practices in the supply chain are outside of the scope of this research. Thus, calling future research to be conducted in this area.

Appendix

See Tables 1 and 2.

Table 1. Lean, green and resilient practices in the supply chain

		Practices	1	2	3	4	5	6	7	8	9	10	11	12	13	14	15	16	17	18	19	20	21	22	23	24	25	26	27	28	29	30	31	32	33	34	35	36	37
Upstream	L	Supplier development	✓				✓		✓			✓					✓					✓	✓		✓			✓			✓		✓			✓	✓		
		Supplier LT relationship					✓	✓		✓		✓								✓		✓	✓		✓					✓									
		JIT delivery by suppliers			✓		✓			✓		✓				✓	✓					✓							✓						✓				
		Geographic concentration from suppliers					✓					✓				✓						✓		✓					✓										
		Suppliers reduction										✓				✓						✓		✓															
	G	EMS mandatory for suppliers										✓					✓	✓	✓	✓		✓		✓				✓					✓						
		Green/less packages from suppliers										✓				✓								✓					✓										
		Environmental risk sharing with suppliers										✓											✓																
	R	Sourcing strategies			✓					✓							✓			✓															✓	✓	✓		
		Visibility of upstream inventories								✓																												✓	
		Backward linkage																		✓																			
	L&R	Suppliers lead time reduction								✓										✓																			
	L&G&R	Cooperation with suppliers			✓		✓			✓		✓					✓	✓	✓			✓	✓	✓									✓			✓		✓	

(continued)

Table 1. *(continued)*

		Practices	1	2	3	4	5	6	7	8	9	10	11	12	13	14	15	16	17	18	19	20	21	22	23	24	25	26	27	28	29	30	31	32	33	34	35	36	37
Internal	L	JIT/Pull	✓	✓		✓	✓	✓	✓	✓		✓		✓			✓		✓			✓		✓				✓	✓		✓	✓	✓						
		QMS	✓				✓	✓				✓			✓				✓	✓		✓						✓				✓			✓				
		TPM	✓				✓					✓			✓							✓								✓	✓	✓							
		VSM		✓	✓				✓			✓						✓		✓										✓									
		Kaizen										✓			✓			✓												✓		✓		✓					
		Set-up time reduction				✓	✓			✓		✓										✓						✓											
		Cellular manufacturing			✓	✓	✓															✓									✓	✓							
		Total time reduction					✓					✓								✓		✓						✓				✓							
		5S system			✓		✓					✓						✓												✓									
		Kanban			✓		✓					✓																				✓	✓						
		Lot sizing					✓			✓												✓									✓	✓							
		Quick changeover					✓															✓									✓	✓							
		Work standardization													✓			✓				✓								✓									
		Poka-yoke			✓		✓																								✓								
		Six sigma					✓					✓																											
		Order consolidation					✓															✓																	
		Innovation management	✓																																				
		Replenishment frequency reduction		✓																																			
	G	LCA tools		✓	✓	✓	✓	✓				✓			✓				✓	✓	✓	✓			✓				✓				✓						
		EMS					✓	✓				✓			✓			✓	✓	✓		✓						✓	✓				✓		✓				

(continued)

Table 1. *(continued)*

		Practices	1	2	3	4	5	6	7	8	9	10	11	12	13	14	15	16	17	18	19	20	21	22	23	24	25	26	27	28	29	30	31	32	33	34	35	36	37
		Cleaner production					✓					✓							✓	✓		✓						✓		✓		✓	✓		✓				
		3R					✓	✓				✓						✓	✓			✓		✓				✓	✓			✓							
		Green purchasing		✓			✓								✓					✓		✓		✓									✓						
		Green technology					✓					✓			✓							✓																	
		Sustainable VSM		✓	✓		✓					✓																											
		Efficiency of resource consumption		✓								✓										✓																	
		Energy efficiency					✓											✓				✓																	
		Environmental risk sharing		✓			✓															✓																	
	R	Strategic stock			✓			✓						✓																								✓	
		Balancing between redundancy and efficiency											✓							✓			✓														✓		
		Contingency planning															✓			✓																			
		Postponement											✓										✓																
		SC engineering			✓																															✓			
		Hedging and security																		✓																✓			
		Use of information control systems															✓																						

(continued)

Table 1. *(continued)*

		Practices	1	2	3	4	5	6	7	8	9	10	11	12	13	14	15	16	17	18	19	20	21	22	23	24	25	26	27	28	29	30	31	32	33	34	35	36	37
		Forecasting																		✓																			
		Readiness training																		✓																			
		Impact reduction																		✓																			
		Flexible manufacturing processes																																			✓		
	L&G	Waste minimization		✓		✓	✓	✓				✓			✓			✓		✓		✓		✓	✓			✓		✓		✓	✓	✓	✓				
		Inventory minimization		✓	✓	✓	✓	✓				✓			✓						✓	✓			✓									✓					
		Transport modes optimization					✓																																
	L&R	HRM			✓		✓					✓			✓		✓	✓				✓			✓	✓				✓	✓	✓		✓		✓	✓		
		Lead time reduction		✓		✓	✓	✓				✓	✓									✓	✓	✓														✓	
Downstream	L	Customer relationship	✓				✓					✓						✓		✓		✓	✓							✓		✓	✓						
		Delivery time reduction																		✓																			
	G	Reverse logistics		✓		✓	✓					✓								✓		✓	✓										✓		✓				
		Environmental risk sharing with customers										✓											✓																
		Green or less packages (to costumers)										✓																											

(continued)

Table 1. *(continued)*

		Practices	1	2	3	4	5	6	7	8	9	10	11	12	13	14	15	16	17	18	19	20	21	22	23	24	25	26	27	28	29	30	31	32	33	34	35	36	37
	R	Alternative transportation routing						✓									✓			✓															✓		✓		✓
		Quick response			✓					✓										✓																			
		Alternative market																		✓																			
		Visibility on demand conditions																																				✓	
	L&G	Distribution network configuration					✓													✓																			
	L&R	Transportation lead times reduction		✓		✓		✓				✓																											
	L&G&R	Cooperation with customers			✓		✓					✓						✓	✓			✓	✓	✓									✓			✓		✓	
Common	L	Information management		✓		✓	✓				✓	✓				✓		✓			✓	✓	✓	✓	✓	✓												✓	
		Electronic-enabled SCs															✓																						
	R	SC risk management culture			✓																		✓												✓				
		Creating total SC visibility											✓																									✓	

References

Singh et al. (2020); Dües et al. (2013); Sharma et al. (2021); Cherrafi et al. (2018); Engin et al. (2019); Govindan et al. (2013); Martínez-Jurado and Moyano-Fuentes (2013); Cabral et al. (2012); Scholten and Schilder (2015); Campos and Vazquez-Brust (2016); Carvalho et al. (2012); Iakovou et al. (2007); Hussain et al. (2019); Sharma et al. (2019); López and Ruiz-Benítez (2020); Duarte and Cruz-Machado (2019); Zhu and Sarkis (2004); Alqudah et al. (2020); Kainuma and Tawara (2006); Carvalho et al. (2010); Espadinha-Cruz et al. (2011); Azevedo et al. (2012); Duarte and Cruz-Machado (2015); Pettit et al.(2013); Ponomarov and Holcomb (2009); Hajmohammad et al. (2013); Carvalho et al. (2017); Verrier et al. (2016); Sawhney et al. (2007); Wu et al. (2015); Azevedo et al. (2011); Zhan et al. (2018); Govindan et al. (2014); Ali et al.(2017); Rice and Caniato (2003); Christopher and Peck (2004); Zsidisin and Wagner (2010)

Table 2. Sustainable performance measurements in the supply chain

	Performance measurements	4	6	7	8	10	14	15	18	20	22	26	27	29	30	31	32	38
Environmental performance	LCA-Quantity of energy, water and materials usage and environmental releases	✓		✓		✓	✓	✓	✓	✓	✓	✓	✓	✓	✓	✓	✓	✓
	Environmental costs									✓						✓		
	Business wastage		✓		✓			✓		✓	✓		✓			✓		
	Green image		✓				✓		✓	✓	✓		✓		✓			✓
	Environmental revenues									✓								
	Environmental accidents; Environmental situation							✓	✓								✓	
	Environmental impacts; Carbon footprint																	✓
Economic performance	Cost		✓		✓			✓	✓	✓	✓		✓		✓	✓		
	Economic Value Added (EVA)									✓							✓	
	Net operating profit								✓	✓								
	Return on Assets (ROA)									✓					✓			
	Cash-to-cash cycle				✓					✓								
	Efficiency								✓	✓			✓			✓	✓	
	Return on Sales (ROS)														✓		✓	
	Return on Investment (ROI); Declining debt																✓	
	Income distribution; Market competitiveness; Order fill lead time; Defect rate; Level of productivity								✓									

(continued)

Table 2. *(continued)*

	Performance measurements	4	6	7	8	10	14	15	18	20	22	26	27	29	30	31	32	38
Social performance	Corruption risk										✓							
	Supplier screening										✓							
	Local suppliers										✓							
	Employee's safety and health							✓	✓					✓				
	Social reputation; Social awards														✓			
	Worker's motivation and participation; Working environment stress							✓										
	Worker's skills and training							✓	✓									
	Worker's retribution							✓	✓									
	Society impact; Vendor assessment includes social factors; Career growth opportunities								✓									
References	Cherrafi et al. (2018); Govindan et al. (2013); Martínez-Jurado and Moyano-Fuentes (2013); Cabral et al. (2012); Campos and Vazquez-Brust (2016); Sharma et al. (2019); López and Ruiz-Benítez (2020); 18-Alqudah et al. (2020); Carvalho et al. (2010); Azevedo et al. (2012); Hajmohammad et al. (2013); 27-Carvalho et al. (2017); Sawhney et al. (2007); Wu et al. (2015); Azevedo et al. (2011); Zhan et al. (2018); Farias et al. (2019)																	

References

Agarwal, V.: Analysis of lean, green, and resilient practices for Indian automotive supply chain performance using best–worst method. In: Kapur, P.K., Singh, G., Panwar, S. (eds.) Advances in Interdisciplinary Research in Engineering and Business Management. AA, pp. 349–358. Springer, Singapore (2021). https://doi.org/10.1007/978-981-16-0037-1_27

Hammou, I.A., Oulfarsi, S., Hebaz, A.: The impact of lean and green supply chain practices on sustainability: Literature review and conceptual framework. Logforum **18**(1), 1–13 (2022). https://doi.org/10.17270/j.log.2022.684

Ali, A., Mahfouz, A., Arisha, A.: Analysing supply chain resilience: Integrating the constructs in a concept. Supply Chain Manag.: Int. J. **22**(1), 1–49 (2017)

Alqudah, S., Shrouf, H., Suifan, T., Alhyari, S.: A moderated mediation model of lean, agile, resilient, and green paradigms in the supply chain. Int. J. Supply Chain Manag. **9**(4), 158–172 (2020)

Azevedo, S.G., Carvalho, H., Cruz Machado, V.: The influence of green practices on supply chain performance: A case study approach. Transp. Res. Part E: Logistics Transp. Rev. **47**(6), 850–871 (2011). https://doi.org/10.1016/j.tre.2011.05.017

Azevedo, S.G., Carvalho, H., Duarte, S., Cruz-Machado, V.: Influence of green and lean upstream supply chain management practices on business sustainability. IEEE Trans. Eng. Manag. **59**(4), 753–765 (2012). https://doi.org/10.1007/11548669_27

Cabral, I., Grilo, A., Cruz-Machado, V: A decision-making model for lean, agile, resilient and green supply chain management. Int. J. Prod. Res. **50**(17), 4830–4845 (2012). http://run.unl.pt/handle/10362/6620

Campos, L.M., Vazquez-Brust, D.: Lean and green synergies in supply chain management. Supply Chain Manag.: Int. J. **21**(5) (2016)

Carvalho, H., Azevedo, S.G., Cruz-Machado, V.: Agile and resilient approaches to supply chain management: Influence on performance and competitiveness. Logistics Res. **4**(1–2), 49–62 (2012). https://doi.org/10.1007/s12159-012-0064-2

Carvalho, H., Govindan, K., Azevedo, S.G., Cruz-Machado, V.: Modelling green and lean supply chains: An eco-efficiency perspective. Resour. Conserv. Recycl. **120**, 75–87 (2017). https://doi.org/10.1016/j.resconrec.2016.09.025

Carvalho, H., Azevedo, S.G., Cruz-Machado, V.: Supply chain performance management : lean and green paradigms. Int. J. Bus. Perform. Supply Chain Model. **2**(3/4), 304–333 (2010)

Cherrafi, A., Garza-reyes, J.A., Kumar, V., Mishra, N., Garza-reyes, J.A.: Lean, green practices and process innovation: A model for green supply chain performance. Int. J. Prod. Econ. (2018). https://doi.org/10.1016/j.ijpe.2018.09.031

Christopher, M., Peck, H.: Building the resilient supply chain. Int. J. Logistics Manag. **15**(2), 1–14 (2004). https://doi.org/10.1108/09574090410700275

Crane, A., Matten, D.: COVID-19 and the future of CSR research. J. Manag. Stud. **58**(1), 278–282 (2020). https://doi.org/10.1111/joms.12642

De Giovanni, P.: Do internal and external environmental management contribute to the triple bottom line? Int. J. Oper. Prod. Manag. **32**(3), 265–290 (2012). https://doi.org/10.1108/01443571211212574

Duarte, S., Cruz-Machado, V.: Investigating lean and green supply chain linkages through a balanced scorecard framework. Int. J. Manag. Sci. Eng. Manag. **10**(1), 20–29 (2015). https://doi.org/10.1080/17509653.2014.962111

Duarte, S., Cruz-Machado, V.: Green and lean supply-chain transformation: A roadmap. Prod. Plann. Control **30**(14), 1170–1183 (2019)

Dües, C.M., Tan, K.H., Lim, M.: Green as the new Lean: How to use Lean practices as a catalyst to greening your supply chain. J. Clean. Prod. **40**, 93–100 (2013). https://doi.org/10.1016/j.jclepro.2011.12.023

El Baz, J., Ruel, S.: Can supply chain risk management practices mitigate the disruption impacts on supply chains' resilience and robustness? Evidence from an empirical survey in a COVID-19 outbreak era. Int. J. Prod. Econ. **233**(November), 107972 (2021). https://doi.org/10.1016/j.ijpe.2020.107972

Engin, B.E., Martens, M., Paksoy, T.: Lean and green supply chain management: A comprehensive review. In: Paksoy, T., Weber, G.-W., Huber, S. (eds.) Lean and Green Supply Chain Management. ISORMS, vol. 273, pp. 1–38. Springer, Cham (2019). https://doi.org/10.1007/978-3-319-97511-5_1

Espadinha-Cruz, P., Grilo, A., Puga-Leal, R., Cruz-Machado, V.: A model for evaluating lean, agile, resilient and green practices interoperability in supply chains. Int. J. Embed. Syst. **6**(2–3), 167–175 (2011). https://doi.org/10.1504/IJES.2014.063814

Farias, L.M.S., Santos, L.C., Gohr, C.F., de Oliveira, L.C., da Silva Amorim, M.H.: Criteria and practices for lean and green performance assessment: Systematic review and conceptual framework. J. Clean. Prod. **218**, 746–762 (2019)

Garcia-Buendia, N., Moyano-Fuentes, J., Maqueira-Marín, J.M.: Lean supply chain management and performance relationships : What has been done and what is left to do. CIRP J. Manuf. Sci. Technol. **32**, 405–423 (2021). https://doi.org/10.1016/j.cirpj.2021.01.016

Geng, R., Mansouri, S.A., Aktas, E.: The relationship between green supply chain management and performance: A meta-analysis of empirical evidences in Asian emerging economies. Int. J. Prod. Econ. **183**, 245–258 (2017). https://doi.org/10.1016/j.ijpe.2016.10.008

Govindan, K., Azevedo, S.G., Carvalho, H., Cruz-Machado, V.: Lean, green and resilient practices influence on supply chain performance: Interpretive structural modeling approach. Int. J. Environ. Sci. Technol. **12**(1), 15–34 (2013). https://doi.org/10.1007/s13762-013-0409-7

Govindan, K., Azevedo, S.G., Carvalho, H., Cruz-Machado, V.: Impact of supply chain management practices on sustainability. J. Clean. Prod. **85**, 212–225 (2014)

Hajmohammad, S., Vachon, S., Klassen, R.D., Gavronski, I.: Reprint of Lean management and supply management: Their role in green practices and performance. J. Clean. Prod. **56**, 86–93 (2013). https://doi.org/10.1016/j.jclepro.2013.06.038

Hebaz, A., Oulfarsi, S.: The drivers and barriers of green supply chain management implementation: A review. Acta Logistica **8**(2), 123–132 (2021). https://doi.org/10.22306/al.v8i2.211

Hussain, M., Al-Aomar, R., Melhem, H.: Assessment of lean-green practices on the sustainable performance of hotel supply chains. Int. J. Contemp. Hosp. Manag. **31**(6), 2448–2467 (2019). https://doi.org/10.1108/IJCHM-05-2018-0380

Iakovou, E., Vlachos, D., Xanthopoulos, A.: An analytical methodological framework for the optimal design of resilient supply chains. Int. J. Logistics Econ. Globalisation **1**(1), 1–20 (2007)

Ivanov, D., Dolgui, A.: Viability of intertwined supply networks: extending the supply chain resilience angles towards survivability. A position paper motivated by COVID-19 outbreak. Int. J. Prod. Res. **58**(10), 2904–2915 (2020). https://doi.org/10.1080/00207543.2020.1750727

Kainuma, Y., Tawara, N.: A multiple attribute utility theory approach to lean and green supply chain management. Int. J. Prod. Econ. **101**(1), 99–108 (2006)

López, C., Ruiz-Benítez, R.: Multilayer analysis of supply chain strategies' impact on sustainability. J. Purch. Supply Manag. **26**(2), 100535 (2020)

Majumdar, A., Shaw, M., Sinha, S.K.: COVID-19 debunks the myth of socially sustainable supply chain: A case of the clothing industry in South Asian countries. Sustain. Prod. Consumption **24**, 150–155 (2020). https://doi.org/10.1016/j.spc.2020.07.001

Martínez-Jurado, P.J., Moyano-Fuentes, J.: Lean management , supply chain management and sustainability : A literature review. J. Cleaner Prod. 1e17 Contents (2013). https://doi.org/10.1016/j.jclepro.2013.09.042

Pettit, T.J., Croxton, K.L., Fiksel, J.: Ensuring supply chain resilience: Development and implementation of an assessment tool. J. Bus. Logistics **34**(1), 46–76 (2013)

Ponomarov, S.Y., Holcomb, M.C.: Understanding the concept of supply chain resilience. Int. J. Logistics Manag. **20**(1) (2009). https://doi.org/10.1108/09574090910954873

Qorri, A., Mujkić, Z., Gashi, S., Kraslawski, A.: Green supply chain management practices and company performance: A meta-analysis approach. Procedia Manuf. **17**, 317–325 (2018). https://doi.org/10.1016/j.promfg.2018.10.052

Rice, J.B., Caniato, F.: Building a secure and resilience supply chain. Supply Chain Manag. Rev. **5**(September/October), 22–30 (2003)

Rothenberg, S., Pil, F.K., Maxwell, J.: Lean, green, and the quest for superior environmental performance. Prod. Oper. Manag. **10**(3), 228–243 (2001). https://doi.org/10.1111/j.1937-5956.2001.tb00372.x

Ruiz-Benítez, R., López, C., Real, J.C.: The lean and resilient management of the supply chain and its impact on performance (2018). https://doi.org/10.1016/j.ijpe.2018.06.009

Sawhney, R., Teparakul, P., Bagchi, A., Li, X.: En-Lean: A framework to align lean and green manufacturing in the metal cutting supply chain. Int. J. Enterp. Netw. Manag. **1**(3), 238–260 (2007). https://doi.org/10.1504/IJENM.2007.012757

Scholten, K., Schilder, S.:. The role of collaboration in supply chain resilience. Supply Chain Manag.: Int. J. (2015)

Sharma, G., Singhi, R., Mittal, A.: Lean and green supply chains—key practices, inter linkages and effects on sustainability—a case study with reference to automobile industry. Int. J. Mech. Eng. Technol. **10**(03), 317–330 (2019)

Sharma, V., Raut, R.D., Mangla, S.K., Narkhede, B.E., Luthra, S., Gokhale, R.: A systematic literature review to integrate lean, agile, resilient, green and sustainable paradigms in the supply chain management. Bus. Strat. Environ. **30**(2), 1191–1212 (2021). https://doi.org/10.1002/bse.2679

Sheffi, Y.: Preparing for disruptions through early detection. MIT Sloan Manag. Rev. **57**(1), 31–42 (2015)

Singh, J., Singh, H., Kumar, A.: Impact of lean practices on organizational sustainability through green supply chain management—an empirical investigation. Int. J. Lean Six Sigma **11**(6), 1035–1068 (2020). https://doi.org/10.1108/IJLSS-06-2017-0068

Trautrims, A., Schleper, M.C., Cakir, M.S., Gold, S.: Survival at the expense of the weakest? Managing modern slavery risks in supply chains during COVID-19. J. Risk Res. **23**(7–8), 1067–1072 (2020). https://doi.org/10.1080/13669877.2020.1772347

Vais, A., Miron, V., Pedersen, M., Folke, J.: "Lean and Green" at a Romanian secondary tissue paper and board mill—putting theory into practice. Resour. Conserv. Recycl. **46**(1), 44–74 (2006). https://doi.org/10.1016/j.resconrec.2005.06.005

Verrier, B., Rose, B., Caillaud, E.: Lean and green strategy: The lean and green house and maturity deployment model. J. Clean. Prod. **116**, 150–156 (2016)

Vinodh, S., Arvind, K.R., Somanaathan, M.: Tools and techniques for enabling sustainability through lean initiatives. Clean Technol. Environ. Policy **13**(3), 469–479 (2011). https://doi.org/10.1007/s10098-010-0329-x

Wang, Y., Iqbal, U., Gong, Y.: The performance of resilient supply chain sustainability in COVID-19 by sourcing technological integration. Sustainability **13**(11) (2021). https://doi.org/10.3390/su13116151

Wu, L., Subramanian, N., Abdulrahman, M.D., Liu, C., Lai, K.H., Pawar, K.S.: The impact of integrated practices of lean, green, and social management systems on firm sustainability performance-evidence from Chinese fashion auto-parts suppliers. Sustainability (Switzerland) **7**(4), 3838–3858 (2015). https://doi.org/10.3390/su7043838

Yang, M.G., Hong, P., Modi, S.B.: Impact of lean manufacturing and environmental management on business performance: An empirical study of manufacturing firms. Int. J. Prod. Econ. **129**(2), 251–261 (2011). https://doi.org/10.1016/j.ijpe.2010.10.017

Zhan, Y., Tan, K.H., Ji, G., Chung, L., Chiu, A.S.F.: Green and lean sustainable development path in China: Guanxi, practices and performance. Resour. Conserv. Recycl. **128**, 240–249 (2018). https://doi.org/10.1016/j.resconrec.2016.02.006

Zhu, Q., Sarkis, J.: Relationships between operational practices and performance among early adopters of green supply chain management practices in Chinese manufacturing enterprises. J. Oper. Manag. **22**(3), 265–289 (2004). https://doi.org/10.1016/j.jom.2004.01.005

Zsidisin, G.A., Wagner, S.M.: Do perceptions become reality? The moderating role of supply chain resiliency on disruption occurrence. J. Bus. Logistics **31**(2), 1–20 (2010)

Recent Development Techniques on Digital Twins for Manufacturing: State of the Art

Ghayth Gandouzi[1](✉), Imen Belhadj[1], Moncef Hammadi[2], Nizar Aifaoui[1], and Jean-Yves Choley[2]

[1] Mechanical Engineering Laboratory, National Engineering School of Monastir, University of Monastir, 05 Av. Ibn Eljazzar, 5019 Monastir, Tunisia
Gandouzi.Ghaith@gmail.com

[2] Quartz Laboratory, ISAE-Supmeca, 3 rue Fernand Hainaut, 93400 Saint-Ouen, France
moncef.hammadi@isae-supmeca.fr

Abstract. The current industrial trend is toward the fourth industrial revolution. One of the main important areas that are affected by this revolution is the manufacturing sectors. As a result, a new concept for the digitalisation of manufacturing which is the digital twin concept are currently used. It is a virtual representation of a physical model using Industry 4.0 technologies in order to check the flexibility of the manufacturing layout and the efficiency of the production in real time. The digital twin system allows us to test virtual scenarios, optimize and improve working conditions, reduce prototyping and waste, it predict also what will happen in the real world. Due to the importance of digital twins in terms of performance and significant gains, researchers contributions have pointed to the implementation of new methods to exploit the benefits of digital twinning in a large collaborative community. In order to contribute digital twins for small and medium-sized enterprises, it is necessary to detail the main technologies used in the state of the art. In this paper, the main technologies used in the three layers of a digital twin system, such as the physical layer, the network layer, and the service layer are reviewed in detail. Their requirements, problems, and proposed solutions are also discussed. At the same time, considering the ability of small and medium-sized enterprises to evolve to new working environments, as well as their revenues to apply digitalisation, a digital twin simulation based on open-source solutions is proposed.

Keywords: Digital twin · Manufacturing · Simulation · CAD model · IoT · Sensors · Small and medium-sized enterprises

L. Azrar et al. (Eds.): CIP 2022, LNME, pp. 77–86, 2023.
https://doi.org/10.1007/978-3-031-23615-0_8

1 Introduction

Today, we are in the fourth industrial revolution (or I4.0) of emerging technologies such as the Internet of Things (IoT), Artificial Intelligence (AI), Machine Learning (ML), Big Data (BD), Cloud Computing (CC), Blockchain, Virtual and Augmented Realities (VR/AR), etc. It is transforming the way we learn, work, and interact. One of the major benefits of I4.0 is the high-performance computing level which allow us to move from digital simulation to physical replicas (Glaessgen and Stargel 2012). The virtual model of a physical object is what we call a digital twin (DT) (Grieves and Vickers 2017). The DT aims to create a high-fidelity virtual model for each physical entity (Semeraro et al. 2021). It is used in manufacturing, aerospace, urban management, healthcare, shipping, and navigation (Semeraro et al. 2021). In the industrial fields, the DT system came-up to solve the real time monitoring problems in order to analyze the interacting behaviors and to detect errors and defects among productions planning throughout the manufacturing phase. Moreover, the Human-Physical (i.e. the human perception of the physical space) interaction is always not sufficient to monitor systems and make decisions. For this purpose, researchers are interested in moving towards more efficient and accurate perception methods while relying on new I4.0 technologies that are around the perception performed by Cyber-Physical DT systems. Small and medium-sized enterprises (SMEs) are faced with the application of DT for accelerated product development, real-time decision making, machine condition prediction, optimization, and customer needs identification (Barricelli et al. 2019). At the same time, the ability to move from the current state to the I4.0 concept, particularly DT monitoring systems, is surrounded by many issues and challenges. One of the main key considerations is to minimize the digitalisation costs. Another key issue is the potential of human resources to be integrated into an intelligent environment driven by DT technologies. In this context, it is necessary to ask the following question: how to choose the most reliable, efficient and cost-effective means of DT techniques for SMEs?

Many studies have been devoted to the analysis of DT techniques for manufacturing. Therefore, each application of DT varies in terms of requirements and perspectives. In this context, this paper has mainly detailed the DT techniques, their architectures, requirements, components, problems and proposed solutions in the literature. These areas have been examined in detail to highlight and explore the key features, research and technical challenges of designing and implementing DT in manufacturing. As a result, a formal and conceptual analysis was performed to define the phases of the DT life cycle and its features, followed by the architecture of the DT and its various components and functionality. The outcomes of this paper are a multi-perspective picture of the DT in manufacturing, which is an emerging paradigm for SMEs.

The content of this paper is structured as follow. After the introduction, Sect. 2 presents a state of the art of recent development techniques on DT of manufacturing. Section 3 presents the outcomes of the literature review. Section 4 proposes a DT simulation platform for SMEs. The last section of this paper presents a conclusion and some perspectives.

2 State of the Art: Digital Twin in Manufacturing

Given the significance of DT concept, there are over 16 definitions that have already been introduced by scholars (Rabah et al. 2018) was definition of DT was introduced in 2003 by Barricelli et al. as a 3D model that contains information about the physical product, the digital product, and the interactions between them during its life cycle (Barricelli et al. 2019). In addition, the general definition of a DT has been developed by NASA. It is an integrated multi-physics, multi-scale probabilistic simulation of an as-built product, system, or process that can mirror the life of its corresponding twin using available physical models, historical data, real-time data, etc. (Glaessgen and Stargel 2012). The DT is executed as a new paradigm in simulation (Rosen et al. 2015). In DT cases, simulation is founded for design decision-making, validation, and tests, not only for a generic device but also for monitoring complete systems. DT requires implementing three main layers: the physical layer, the network layer, and the computing layer (Semeraro et al. 2021). In the following section, the requirements and components of DT, including the list of DT development problems and proposed solutions, at the three main layers, are discussed in detail (Fig. 1).

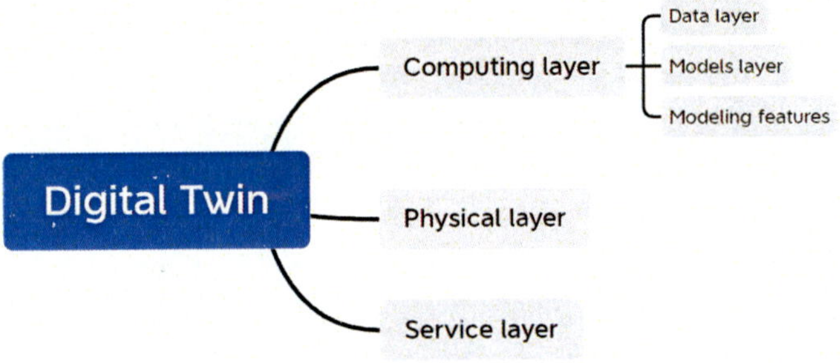

Fig. 1. The main layers of a DT system

2.1 Physical Layer

In manufacturing context, the literature concurs that the data type, and therefore the data sources, are contingent on the physical layers chosen (Qi and Tao 2018). In DT, the physical layer (or level) includes various sensory subsystems and features that capture data and operational parameters. The main problem and challenge is the selection of integrated sensors and communication systems to respond to real-time states such as temperature, vibration, force, torque, and speed in physical system equipment (Cai et al. 2017; Ruppert et al. 2018). The main proposed technologies for the physical layer in DT are Radio Frequency Identification (RFID), RFID sensor systems and Wireless Sensor Systems (WSNs) (Semeraro et al. 2021). Cai et al. have proposed a DT implementation of a 3-axis milling machine coupled with a Hall effect current sensor and an accelerometer (Cai et al. 2017). Tao et al. focused on a DT scheme that integrates IoT sensors, RFID tags, and embedded systems (Tao et al. 2019). Fera et al. divided the data collectors into two main groups as the optical systems since they are installed with cameras in the test environment that capture the sensors on the body and allow the movements of the reproducer (Fera et al. 2020). Then the non-optical systems are proposed; it is an electromagnetic system, mainly composed of portable suites with wires and joints, magnetic sensors, and the inertial motion unit (IMU) to capture the orientation of each segment of the human body (Fera et al. 2020). Ding et al. concentrated on a RFID tag for real-time control in a DT-based cyber-physical production system (DT-CPPS) (Ding et al. 2018). Kousi et al. proposed a DT environment that maps and locates resources and enables collision-free navigation by installing distance sensors and a 2D laser scanner in the building (Kousi et al. 2019). Zhuang et al. have proposed to build an industrial LAN, wireless and mobile network, Bluetooth and RFID sensor networks in the shop-floor to realize real-time perception and data acquisition (Zhuang et al. 2018). Also, Liu et al. focused on real-time data perception and acquisition by integrating RFID tag and bar coder for machining process planning based on DT (Liu et al. 2019). Wang and Wang aimed to develop a DT system of manufacturing operation controlled with smart monitors, wireless sensor systems and other IoT sensors (Wang and Wang 2019). The components that belong to the physical layer perform real-time synchronization of the virtual model with their corresponding physical model in order to detect errors, predict, regulate and optimize capabilities.

2.2 Computing Layer

The computer layer consists of virtual models that simulate the corresponding physical entities. It is thus taken as a set of connected layers that contain the data, models and modeling features. (Semeraro et al. 2021).

- **Data layer**

The data layer consists of all the various types of data defined in the physical layer (Semeraro et al. 2021). This data is transferred from physical spaces to virtual spaces and contains sensing and measurement results. The main requirement is that the data layers have heterogeneity and data resources; therefore, they cannot be used directly in decision-making. Under these conditions, data analysis and processing are the main aspects of the literature discussion. Cai et al. concentrated on transferring data to a local PC using python scripts and saved in a text file that shows the current and 3-axis vibration data uploaded after each predefined interval (Cai et al. 2017). Fera et al. focused on developing a plugin in the virtual framework that loads sensor data stored during the test (Fera et al. 2020). Graessler et al. aimed to collect the data of the DT framework with the computing devices; thus, based on the recorded properties, a database is used for DT decision making (Graessler and Pöhler 2017). Likewise, Kousi et al. have focused on the ability of robots to conceive their environment in terms of process requirements and human interaction, based on an infrastructure that involves all the data on the shop floor and combines it into a unified common environment (Kousi et al. 2019). Bouhlin et al. focused on the data flows from the physical layers needed for discrete and continuous control actions. They are used to make many decisions fully automatic (Bohlin et al. 2017). Similarity, Liu et al. focused on acquiring, organizing, and managing heterogeneous and multi-source physical machine condition data with DT-based machining process planning evaluation (DT-MPPE) technologies, which combine machining condition and process information through the evaluation of DT data for decision making (Liu et al. 2019).

The main data-driven methods used in DT are ML, artificial neural network (ANN) and deep learning (DL). ML techniques applied in DTs are supervised and unsupervised learning (Jain et al. 2017). The main application of ANN and DL in DT are health assessment, error detection (Xu et al. 2019), performance prediction (Jain and Bhatnagar 2019). A reinforcement learning algorithm, have been integrated to control the robot motion planing and to replace the traditional kinematics robot definition method in a DT robotic peg-in-hole application (Li et al. 2022).

- **Models and modeling features**

DT involves building and applying numerical models representing resources and process knowledge elements (Semeraro et al. 2021).

The main requirement of this sub-layer is to move from the traditional CAD model to the digital CAD model to define behaviors in a DT manufacturing simulation environment. DT is faster and more compelling for adding new information technologies. For this reason, many companies have been adding new information technologies to their DT applications, such as Siemens, Dassault System, General Electric, PTC, and Tesla (Schleich et al. 2017).

In this context: Cai et al. aimed to develop a CAD model with true simplification that has already been built with Solidworks software and transferred in STL format (digital CAD format) (Cai et al. 2017). Fera et al. have defined separatory and behaviors when assembling parts with the open GL library, coupled with the graphically realistic virtual milling machine using Tecnomatix Process simulate V15.0.1 by Siemens (Fera et al. 2020). Kousi et al. focused on a DT environment based on CAD files of Machine Shop for the layout representation, ROS (Robot Operating System) virtual controllers of Universal Robots (UR) arms, and mobile platform is used (Kousi et al. 2019). In addition, methods for generating digital CAD models for manufacturing can be supported by extensive research. In the first place, Lu et al. interested in a new modeling method of a digital CAD model using vector representation, transformation matrix, and spatial meshing theory for CNC machining simulation (Lu et al. 2014). Gian et al. studied the transformation of a CAD model into a Unified Robot Description Format (URDF) model using Computer-Aided Design (CAD) tools (Qian et al. 2014). Lee et al. have presented a method that creates engineering design data from existing parts using 3D laser scanners to STL (generate Stereo Lithography) files imported as meshes into URDF (Lee et al. 2001). Collada (collaborative design activity) data format is based on XML file schema; media that can be imported as STL files into the URDF (Miyahara and Okada 2009). Many communities focus on the simulation of machining. Robot Operating System (ROS) provides Gazebo and RViz environments to simulate robots in a three-dimensional world. Meng et al. aimed to combine Unity 3D and ROS to create a 3D simulation system for miniature unmanned aerial vehicles (UAVs) (Meng et al. 2015).

2.3 Network Layer

The network layer involves the connections and the interactions between the physical and the virtual areas. The key enabling technologies discussed in the literature review are Middleware, Application Programming Interfaces, Communication Protocol/Interface Conversion, Communication Protocol Analysis, and Wireless Communication (Semeraro et al. 2021). The protocol and communication interface transformation converts various elements into a uniform form (Semeraro et al. 2021). The most popular middleware architecture used in DT is the Service Oriented Architecture (SOA) approach. The SOA approach allows the disaggregation of complex and monolithic systems into applications composed of an ecosystem of simpler and well-defined elements (Gubbi et al. 2013). Wireless communication allows wireless networking of DT entities, which increases the flexibility of data transmission, such as IoT protocols. Open Platform Communications Unified Architecture (OPC UA) and MT-Connect are the most used protocols in DT applications to access and transfer data in real time (Redelinghuys et al. 2018). The network layer involves many technologies of connection. Each technology can be adapted to a specific application of DT in manufacturing.

3 Outcomes of the Literature Review

In the three key layers, the DT techniques depend on many factors such as interoperability, communication, flexible format, and protocols. In this context, the techniques used in the literature are treated in detail according to the three main layers.

In the physical layer, the main challenges are capturing, cluster, sharing data and related information from the physical and virtual worlds. In this context, data collection technologies based on RFID sensor networks, RFID tags, barcodes, IoT sensors and 2D lasers are presented.

The computing layer faces many issues related to the implementation of data-driven models and the selection of analytical models. Some related works involving analytical model techniques, combined with how much information is presented. The leading data-driven models' technologies used in DT are introduced, such as ML, deep learning, and neural network. Then, some related work is presented to model the virtual copy of DT.

A transition from the CAD model to the digital CAD models methods is presented. Then, some simulation environments to make behaviors, kinematics, and 3D visualization of the physical system in real-time are defined. Finally, the network layer involves many technologies of interconnection, interactions, and communication protocols.

4 Proposed DT Simulation Platform for Small and Medium-Sized Manufacturing Enterprises

Based on the previous literature review, a new DT simulation method is presented in this paper. The proposed approach takes into account the needs of SMEs in terms of technologies, resources and stepwise integration of DTs. In the toward of DT for SMEs, our approach started with the description of the CAD components of the equipment and machines, which is the first level of integration. The second level of integration is the definition of behavioral data. Simulation level is the third level of integration, in which we define process and transition state between equipment. The proposed approach aims to create an open and collaborative DT environment for manufacturing. For this, the use of open source solutions is proposed. For the first level of integration, FreeCAD software includes many features and macro development capabilities for the generation of the URDF format. For the definition of behavioral layer, ROS as an open-source software contains an open source simulators such as Gazebo and a visualization tool such as RViz is proposed. Figure 2 illustrates the proposed DT platform for SMEs. After defining the real features of the physical world, which have been defined at the first three levels of integration, SMEs can be able to capture, analyze and manage data about their equipment. This data, will be used in the next levels of integration in order to conduct information and knowledge in the towards of DT development.

5 Conclusion and Future Works

This paper presents the main technologies of DT in manufacturing, which have been divided into three main layers: physical layer, computing layer, and network layer. However, to be more informed about the technologies used in DT, some technologies that are used in many applications are detailed. These technologies may inspire other ideas in future works to develop DT tools for SMEs. Based on the literature results, a proposed simulation environment for DT is suggested. The proposed approach is based on the coupling between FreeCAD and ROS tools, which are two open-source software, the combination of FreeCAD and ROS requires an adaptive format, like XML, which is a standard exchange format.

In future work, the proposed approach will be detailed and implemented. A case study of drill machine will be developed to highlight the DT interoperability process between physical and virtual models.

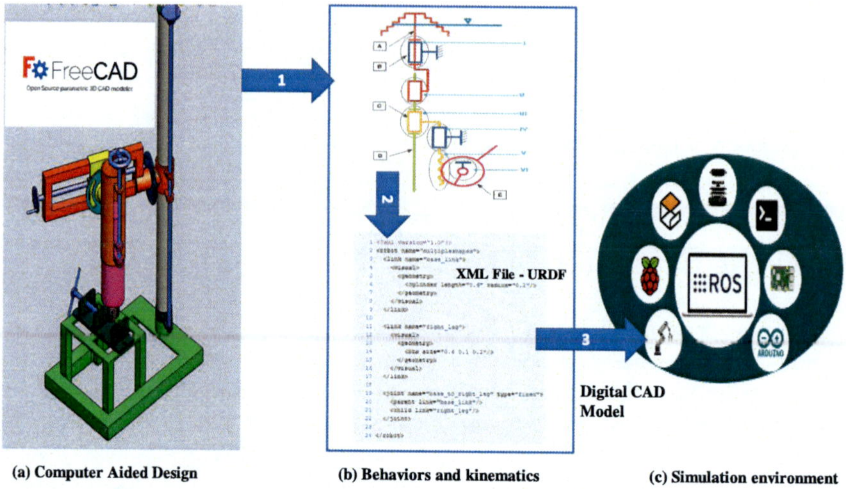

Fig. 2. Proposed development platform of DT for SMEs

Acknowledgements. This work is carried out within the Hubert Curien "Utique" partnership of the French Ministry of Europe and Foreign Affairs, and the Tunisian Ministry of Higher Education and Scientific Research (CMCU_21G1112). The authors gratefully acknowledge their financial support.

References

Bao, J., Guo, D., Li, J., Zhang, J.: The modelling and operations for the digital twin in the context of manufacturing. Enterpr. Inf. Syst. **13**(4), 534–556 (2019)

Barricelli, B.R., Casiraghi, E., Fogli, D.: A survey on digital twin: definitions, characteristics, applications, and design implications. IEEE Access **7**, 167653–167671 (2019)

Bohlin, R., Hagmar, J., Bengtsson, K., Lindkvist, L., Carlson, J.S., Söderberg, R.: Data flow and communication framework supporting digital twin for geometry assurance. In: ASME International Mechanical Engineering Congress and Exposition, vol. 58356, p. V002T02A110. American Society of Mechanical Engineers (2017)

Brettel, M., Friederichsen, N., Keller, M., Rosenberg, M.: How virtualization, decentralization and network building change the manufacturing landscape: an industry 4.0 perspective. FormaMente 12 (2017)

Cai, Y., Starly, B., Cohen, P., Lee, Y.-S.: Sensor data and information fusion to construct digital-twins virtual machine tools for cyber-physical manufacturing. Procedia Manuf. **10**, 1031–1042 (2017)

Clarke, B., Fokoue, E., Zhang, H.H.: Principles and Theory for Data Mining and Machine Learning. Springer Science & Business Media (2009)

Ding, K., Jiang, P., Su, S.: RFID-enabled social manufacturing system for inter-enterprise monitoring and dispatching of integrated production and transportation tasks. Robot. Comput.-Integr. Manuf. **49**, 120–133 (2018)

Drath, R., Luder, A., Peschke, J., Hundt, L.: AutomationML—the glue for seamless automation engineering. In: 2008 IEEE International Conference on Emerging Technologies and Factory Automation, pp. 616–623. IEEE (2008)

Fera, M., Greco, A., Caterino, M., Gerbino, S., Caputo, F., Macchiaroli, R., D'Amato, E.: Towards digital twin implementation for assessing production line performance and balancing. Sensors **20**(1), 97 (2020)

Glaessgen, E., Stargel, D.: The digital twin paradigm for future NASA and US air force vehicles. In: 53rd AIAA/ASME/ASCE/AHS/ASC Structures, Structural Dynamics and Materials Conference 20th AIAA/ASME/AHS Adaptive Structures Conference 14th AIAA, p. 1818 (2012)

Graessler, I., Pöhler, A.: Integration of a digital twin as human representation in a scheduling procedure of a cyber-physical production system. In: 2017 IEEE International Conference on Industrial Engineering and Engineering Management (IEEM), pp. 289–293. IEEE (2017)

Grieves, M., Vickers, J.: Digital twin: mitigating unpredictable, undesirable emergent behavior in complex systems. In: Transdisciplinary Perspectives on Complex Systems, pp. 85–113. Springer (2017)

Gubbi, J., Buyya, R., Marusic, S., Palaniswami, M.: Internet of things (IoT): a vision, architectural elements, and future directions. Future Gener. Comput. Syst. **29**(7), 1645–1660 (2013)

Jain, R., Bhatnagar, R.: Applications of machine learning in cyber security—a review and a conceptual framework for a university setup. In: International Conference on Advanced Machine Learning Technologies and Applications, pp. 599–608. Springer (2019)

Jain, S., Shao, G., Shin, S.-J.: Manufacturing data analytics using a virtual factory representation. Int. J. Prod. Res. **55**(18), 5450–5464 (2017)

Kousi, N., Gkournelos, C., Aivaliotis, S., Giannoulis, C., Michalos, G., Makris, S.: Digital twin for adaptation of robots' behavior in flexible robotic assembly lines. Procedia Manuf. **28**, 121–126 (2019)

Lee, K., Woo, H., Suk, T.: Data reduction methods for reverse engineering. Int. J. Adv. Manuf. Technol. **17**(10), 735–743 (2001)

Liu, J., Zhou, H., Liu, X., Tian, G., Wu, M., Cao, L., Wang, W.: Dynamic evaluation method of machining process planning based on digital twin. IEEE Access **7**, 19312–19323 (2019)

Lu, H., Liu, Z., Wang, S.: Digitization modeling and CNC machining for enveloping surface parts. Int. J. Adv. Manuf. Technol. **73**(1–4), 209–227 (2014)

Meng, W., Hu, Y., Lin, J., Lin, F., Teo, R.: ROS+ unity: an efficient high-fidelity 3D multi-UAV navigation and control simulator in GPS-denied environments. In: IECON 2015-41st Annual Conference of the IEEE Industrial Electronics Society, pp. 002562–002567. IEEE (2015)

Miyahara, K., Okada, Y.: Collada-based file format supporting various attributes of realistic objects for VR applications. In: 2009 International Conference on Complex, Intelligent and Software Intensive Systems, pp. 971–976. IEEE (2009)

Qi, Q., Tao, F.: Digital twin and big data towards smart manufacturing and industry 4.0: 360 degree comparison. IEEE Access **6**, 3585–3593 (2018)

Qian, W., Xia, Z., Xiong, J., Gan, Y., Guo, Y., Weng, S., Deng, H., Hu, Y., Zhang, J.: Manipulation task simulation using ROS and Gazebo. In: 2014 IEEE International Conference on Robotics and Biomimetics (ROBIO 2014), pp. 2594–2598. IEEE (2014)

Rabah, S., Assila, A., Khouri, E., Maier, F., Ababsa, F., Maier, P., Mérienne, F., et al.: Towards improving the future of manufacturing through digital twin and augmented reality technologies. Procedia Manuf. **17**, 460–467 (2018)

Redelinghuys, A., Basson, A., Kruger, K.: A six-layer digital twin architecture for a manufacturing cell. In: International Workshop on Service Orientation in Holonic and Multi-Agent Manufacturing, pp. 412–423. Springer (2018)

Ríos, J., Hernandez, J.C., Oliva, M., Mas, F.: Product avatar as digital counterpart of a physical individual product: literature review and implications in an aircraft. In: ISPE CE, pp. 657–666 (2015)

Rosen, R., Von Wichert, G., Lo, G., Bettenhausen, K.D.: About the importance of autonomy and digital twins for the future of manufacturing. IFAC-PapersOnLine **48**(3), 567–572 (2015)

Ruppert, T., Jaskó, S., Holczinger, T., Abonyi, J.: Enabling technologies for operator 4.0: a survey. Appl. Sci. **8**(9), 1650 (2018)

Schleich, B., Anwer, N., Mathieu, L., Wartzack, S.: Shaping the digital twin for design and production engineering. CIRP Ann. **66**(1), 141–144 (2017)

Semeraro, C., Lezoche, M., Panetto, H., Dassisti, M.: Digital twin paradigm: a systematic literature review. Comput. Ind. **130**, 103469 (2021)

Semeraro, C., Panetto, H., Lezoche, M., Dassisti, M., Cafagna, S.: A monitoring strategy for industry 4.0: Master Italy s.r.l case study. In: INSIGHT-International Council on Systems Engineering (INCOSE), vol. 22, no. 4, pp. 20–22 (2019)

Tao, F., Qi, Q., Wang, L., Nee, A.: Digital twins and cyber-physical systems toward smart manufacturing and industry 4.0: correlation and comparison. Engineering **5**(4), 653–661 (2019)

Wang, X.V., Wang, L.: Digital twin-based WEEE recycling, recovery and remanufacturing in the background of industry 4.0. Int. J. Prod. Res. **57**(12), 3892–3902 (2019)

Xu, Y., Sun, Y., Liu, X., Zheng, Y.: A digital-twin-assisted fault diagnosis using deep transfer learning. IEEE Access **7**, 19990–19999 (2019)

Zhuang, C., Liu, J., Xiong, H.: Digital twin-based smart production management and control framework for the complex product assembly shop-floor. Int. J. Adv. Manuf. Technol. 1149–1163 (2018). https://doi.org/10.1007/s00170-018-1617-6

Progress and Trends in Industry 4.0 and Lean Six Sigma Integration

Dounia Skalli[1](✉), Abdelkabir Charkaoui[1], and Anass Cherrafi[2]

[1] Laboratory of Engineering, Industrial Management and Innovation, Faculty of Sciences and Technology, Hassan 1St University, PO Box 577, Settat, Morocco
d.skalli@uhp.ac.ma

[2] EST- Safi, Cadi Ayyad University, Marrakech, Morocco
a.cherrafi@uca.ac.ma

Abstract. The main concern of companies is to increase productivity and efficiency with less resources usage and wastes. In this context, Lean Six Sigma 'LSS' is considered as the widely best used methodology to improve performance for a high efficient processes and has been adopted in manufacturing for several decades. Industry 4.0 'I4.0' refers to various technology-driven changes in an organization's manufacturing systems. The I4.0 related technologies are driving companies to a new level of operational excellence. The growing awareness of technological advances and the competitiveness of industries has resulted to an important transition in the way of thinking about optimization, costs reduction production and continuous improvement. The effect of Industry 4.0 on LSS is not well explored in literature. The present paper aims to address the related works and discussion on the interactions between the two approaches I4.0 and LSS based on literature review. The study is relevant to both scholars and practitioners by highlighting the ongoing challenges of implementing LSS in the Industry 4.0 era.

Keywords: Industry 4.0 · Lean · Six sigma · Lean six sigma · Digitalization · Literature review

1 Introduction

Industry 4.0 is a new strategic vision to integrate advanced technologies to enable communication between people, products and processes. I4.0 refers to the digital transformation of an organization's manufacturing systems based on the advanced technologies (Lasi et al. 2014). The landscape of industrial manufacturing will be modified with the integration of I4.0 technologies, which will affect the concepts of operations and quality management (Skalli et al. 2022). Lean aim to improve process through eliminating waste and Six sigma focus on reducing process variation and eliminating root causes of defects.

Advanced technologies of I4.0 support improved production and reduced machine downtime, rejects and rework, improved quality and more productive maintenance operations. (Mayr et al. 2018). They connect people, devices and products and enable the exchange of information through the Internet of Things (Wagner et al. 2017).

L. Azrar et al. (Eds.): CIP 2022, LNME, pp. 87–95, 2023.
https://doi.org/10.1007/978-3-031-23615-0_9

Currently, research is focusing more on the application of Industry 4.0, which is of great interest to several researchers (Ghobakhloo, 2018, p. 0; Lasi et al. 2014; Raj et al. 2020; Rosin et al. 2022; Skalli et al. 2023). There are few studies discussing the integration of Lean Six Sigma and Industry 4.0 (Anass et al. 2021; Anvari et al. 2021; Belhadi et al. 2020; Bhat et al. 2020; Tissir et al. 2022), and only certain aspects of Industry 4.0 are discussed. Furthermore, the link between Industry 4.0 and lean principles is widely discussed in the literature(Mayr et al. 2018; Rosin et al. 2020; Sony, 2018; Tortorella et al. 2019, 2021; Wagner et al. 2017). (Anass et al. 2021) in his survey conducted in Moroccan context found that Lean Six Sigma and Industry 4.0 are synergetic and compatible. In the same way (Anvari et al. 2021) states that LSS and I4.0 mutually support each other.

An in-depth study in which the potential benefits, enablers, inhibitors, and critical success factors (CSFs) of a proposed integrated model is missing,

Motivated by this research gap, we are interested in studying the relationship between LSS and I4.0. Given this objective, the following research question arises: Is there a relationship between lean six sigma and Industry 4.0 tools that could help manufacturing companies implement an integrated framework?

To address this question, the paper is organized as follows. We first explore the relationship between LSS and I4. In Sect. 3, we describe the research methodology, Sect. 4 presents a descriptive analysis, and Sect. 5 illustrates the research flows. Next, Sect. 6 highlights the future direction of the research. Finally, the conclusion and limitations are presented in Sect. 7.

2 LSS and I4.0 Relationship

Lean six sigma as an integrated method includes the velocity of lean and the resilience of six sigma through a systematic approach to problem solving (Antony et al. 2018). Lean focus on eliminating waste and non-value-added activities. According to ASQ "Six Sigma is a method that provides organizations tools to improve the capability of their business processes. This increase in performance and decrease in process variation helps lead to defect reduction and improvement in profits, employee morale, and quality of products or services" (ASQ Six Sigma, 2019). The DMAIC methodology is used as a proxy for Lean Six Sigma to investigate its relationship to Industry 4.0.

There are few studies on the relationship between LSS and Industry 4.0. (Anass et al. 2021; Anvari et al. 2021; Arcidiacono and Pieroni, 2018; Belhadi et al. 2020; Dogan and Gurcan, 2018; Gupta et al. 2020; Sodhi 2020). However, among these studies, we found that LSS is studied with specific Industry 4.0 technology, including big data (Belhadi et al. 2020; Gupta et al. 2020; Koppel and Chang, 2021; Rifqi et al. 2021), IoT (Ito et al. 2020), simulation(Bhat et al. 2020) and Cyber physical system 'CPS' (Anvari et al. 2021; Sony 2020). The most significant papers are Gupta et al. 2020) that highlighted the applications of big data analytics in various phases of LSS for effective decision making, also (Sodhi 2020) discussed the integration of Internet of thing 'IoT' and big data with LSS tools for faster and better decision making while (Sony, 2020) has examined in detail the 8 C architecture of the cyber physical system and its possible integration with the LSS methodology.

Many partial frameworks for Lean Six Sigma and Industry 4.0 have been proposed by researchers, covering various concepts and techniques(Ali et al. 2020; Burggräf et al. 2020; Cherrafi et al. 2016; Ghobakhloo 2018; Schumacher et al. 2016; Sony 2018), yet we have noted the absence of a structured integrated model for Lean Six Sigma and Industry 4.0. Literature reflects that the lean-six sigma approach is closely related to the current Industry 4.0 concept and share a common objectives (Sodhi 2020).

3 Methodology of Research

We conducted a literature review following (Tranfield et al. 2003) as described in Fig. 1 to better understand current research trends, synthesize the best knowledge, and build on existing research literature to foster new perspectives and directions for future research. Literature review is a systematic process for exploring and assessing an emerging research topic.

We first define the research questions, then the main keywords and the inclusion and exclusion criteria (Table 1) for papers collection.

Papers were searched on Scopus and Web of Science, the two leading online databases used by researchers, as they provide multidisciplinary scientific publications with broad coverage of journals, conference proceedings, and books for abstracts and citations. The research period is from 2011 to December 2021.

Table 1. Research criteria.

Inclusion criteria	Peer-reviewed Journal articles and conference papers published in English from 2011 to December 2021
Exclusion criteria	Full text not available, Papers not relevant to studied topic

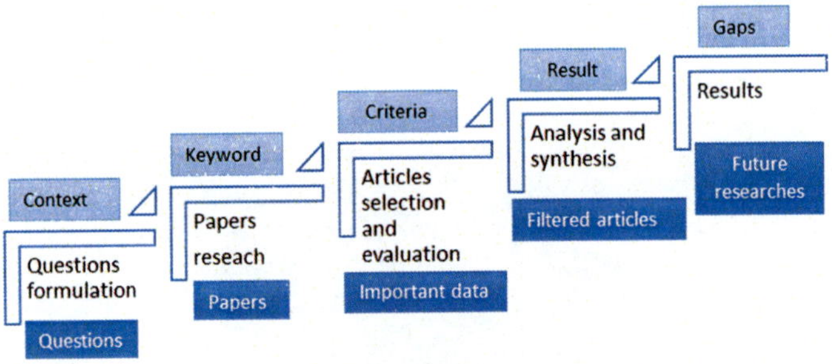

Fig. 1. Research process.

4 Analysis and Results

4.1 Distribution by Year

Figure 2 presents an analysis of the annual distribution to illustrate the progression of research. Although the Industry 4.0 paradigm emerged in 2011, the first publication in this area was in 2015. We conclude that the LSS and I4.0 topic has gained special interest and popularity within the research community since 2019 as 82% of the publications are released from 2019 to 2021 which indicate the novelty of the research topic.

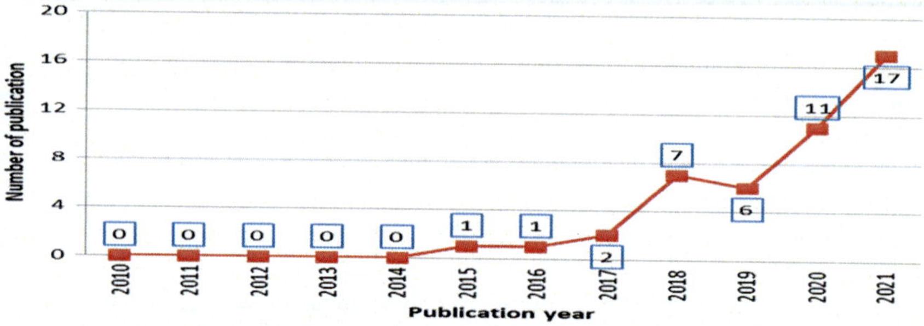

Fig. 2. Distribution of papers over the research time period.

4.2 Distribution Across the Geography

According to the distribution of papers over continents shown in Fig. 4, Europe is by far the continent predominant in the number of publications (28/46). In the second range came Asia (11/46) represented by India. Figure 3 illustrates the four European countries involved in the research field by 28 publications which are respectively published by Italy (10), UK (8), Germany and Norway (5 papers).

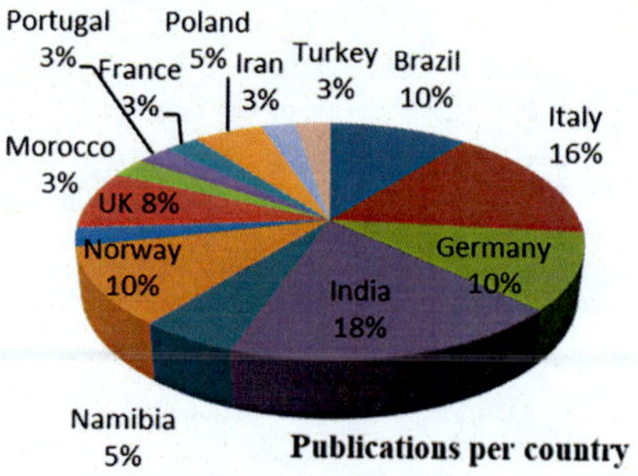

Fig. 3. Distribution of papers per country

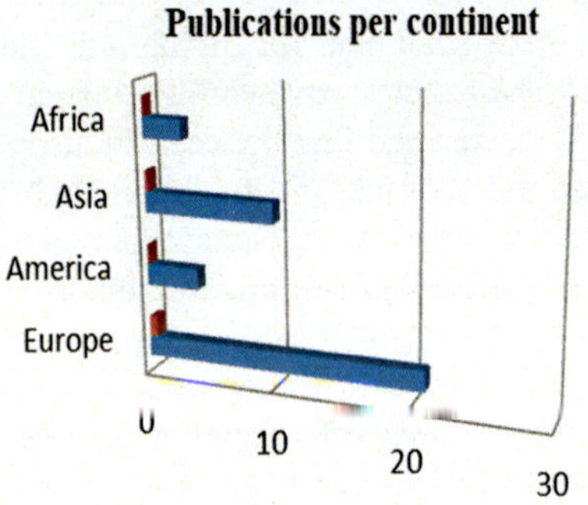

Fig. 4. Geographical application area.

4.3 Distribution Across the Mains Resources:

Out of 39 papers evaluated, journals papers have a predominant aspect when looking at the types of the publications. IJPR played an important role in this area as it published 18% of the papers reviewed. We can consider that research on the integration of Lean Six Sigma and Industry 4.0 is destined to be published in a highly specialized journal (e.g., IJPR International journal of Production research, JMTM for meaningful manufacturing research associate to Technology Management, and Procedia CIRP) and in high-index international conferences (e.g., IFAC online, Procedia manufacturing, OEOM) (Fig. 5).

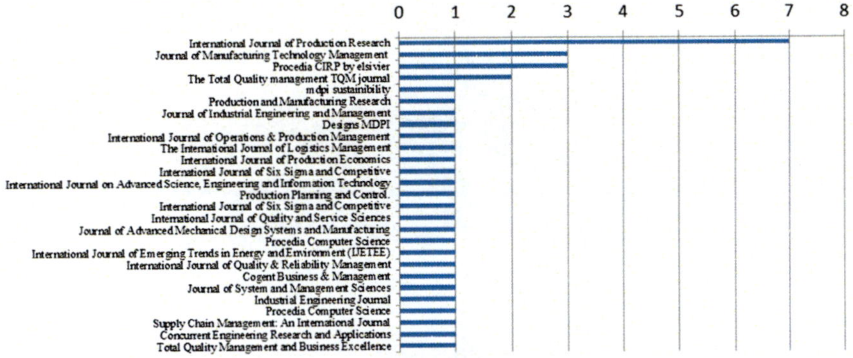

Fig. 5. Distribution of reviewed papers by journal

5 Research Streams

The statistical report shows that 72% of papers studied are related to the link between lean and industry 4.0 and few papers (20%) discuss the integration of LSS with I4.0. This suggests that the integration of Industry 4.0 and lean s a developing topic that has gained and continues to gain attention from academics and practitioners, while research in the topic lean six sigma and industry 4.0 still limited. Jayaram (2016) is the first article to explore the link between Lean Six Sigma and Industry 4.0 for improving supply chain performance. The article provides little insights into the research topic. Kolberg and Zühlke (2015) discuss the link between Industry 4.0 techniques and lean manufacturing and indicate the benefits of integration such as smart operator, smart product, smart machine, and smart planner. The authors highlight the need to develop a framework for integrating the two paradigms. In their study (Rosin et al. 2020) the authors studied the impact of Industry 4.0 technologies on Lean principles and explore how they support the application of Lean principles. (Vinodh et al. 2020) conducted a review and defined a conceptual framework of continuous improvement and I4.0. While (Sony 2018) developed an integration framework model that outlines the steps needed for an organization to integrate both I4.0 and Lean concepts. The model of integration uses the Lean Principles as a guide, focusing on the application of five Lean principles to define the best possible combination of sequences.

6 Discussion

We found a lack of a structured model for Lean Six Sigma and industry 4.0. We highlight the need for a structured integrated model based on the theoretical elements emerged from the literature review. The 'define- measure- analyse- improve- control' DMAIC methodology is used as a proxy for Lean Six Sigma to investigate its relationship to Industry 4.0. Figure 6 illustrates the I4.0 and LSS integration model based on CPS industry 4.0 solutions inspired from (Sony 2020) findings.

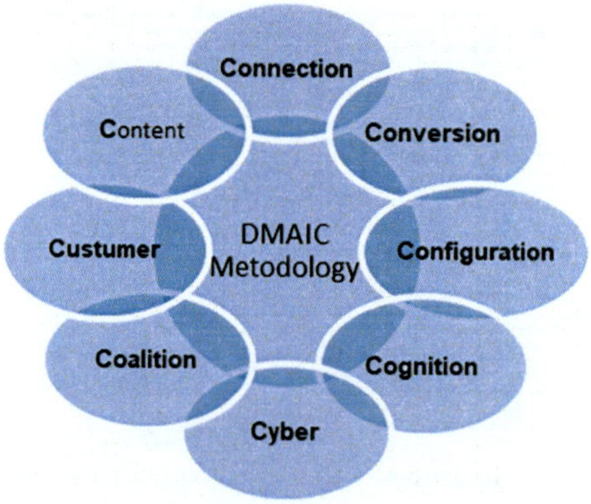

Fig. 6. LSS and I4.0 Integration model based 8C-CPS "(Sony 2020)"

Companies must consider the integrated LSS4.0 model as a roadmap for operational excellence which will lead to improve economic performance, efficient and efficiency by means of LSS tools and techniques. Managers need to know the drivers, challenges and barriers which can impede the integration. They could be economical, cultural as well as technological. There is a need for further research to identify and assess the drivers, barriers, challenges and benefits, synergies and conflicts are linked the integrated model which can be measured through indicators.

7 Conclusion

This review has enabled us to assess the current state of the art regarding the integration of LSS and I4.0. There are limitations to our study, represented by the small number of articles reviewed, as well as the use of Scopus and Web of Sciences only, which may affect the omission of relevant articles. We suggested that future research expand their search sources by including other bibliographic databases. Our results remain to be discussed and strengthened in future studies. The present study can provide information for both scholars and practitioners. In terms of future research, the LSS4.0 framework needs to be developed. Additional research is required to assess and evaluate the relative importance of the steps in the framework, to be implemented and tested.

References

Ali, S.M., et al.: Barriers to lean six sigma implementation in the supply chain: An ISM model. Comput. Ind. Eng. **149**, 106843 (2020). https://doi.org/10.1016/j.cie.2020.106843

Anass, C., Amine, B., Ibtissam, E.H., Bouhaddou, I., Elfezazi, S.: Industry 4.0 and lean six sigma: Results from a pilot study. In: Saka, A., et al. (eds.) CPI 2019. LNME, pp. 613–619. Springer, Cham (2021). https://doi.org/10.1007/978-3-030-62199-5_54

Antony, J., Gupta, S., Sunder M.V., Gijo, E.V.: Ten Commandments of lean six sigma: A practitioners' perspective. Int. J. Prod. Perform. Manag. **67**, 1033–1044 (2018). https://doi.org/10.1108/IJPPM-07-2017-0170

Anvari, F., Edwards, R., Yuniarto, H.A.: Lean Six Sigma in Smart Factories based on Industry 4.0. Int. J. Emerg. Trends Energy Environ **1**, 1–26 (2021)

Arcidiacono, G., Pieroni, A.: The revolution lean six sigma 4.0. Int. J. Adv. Sci. Eng. Inf. Technol. **8**, 141–149 (2018). https://doi.org/10.18517/ijaseit.8.1.4593

Belhadi, A., Kamble, S.S., Zkik, K., Cherrafi, A., Touriki, F.E.: The integrated effect of big data analytics, lean six sigma and green manufacturing on the environmental performance of manufacturing companies: The case of North Africa. J. Clean. Prod. **252**, 119903 (2020). https://doi.org/10.1016/j.jclepro.2019.119903

Bhat, V.S., Bhat, S., Gijo, E.V.: Simulation-based lean six sigma for Industry 4.0: An action research in the process industry. Int. J. Qual. Reliab. Manag. (2020). https://doi.org/10.1108/IJQRM-05-2020-0167

Burggräf, P., Lorber, C., Pyka, A., Wagner, J., Weißer, T.: Kaizen 4.0 Towards an Integrated framework for the lean-industry 4.0 transformation. In: Arai, K., Bhatia, R., Kapoor, S. (eds.) FTC 2019. AISC, vol. 1070, pp. 692–709. Springer, Cham (2020). https://doi.org/10.1007/978-3-030-32523-7_52

Cherrafi, A., Elfezazi, S., Chiarini, A., Mokhlis, A., Benhida, K.: The integration of lean manufacturing, Six Sigma and sustainability: A literature review and future research directions for developing a specific model. J. Clean. Prod. **139**, 828–846 (2016). https://doi.org/10.1016/j.jclepro.2016.08.101

Dogan, O., Gurcan, O.F.: Data Perspective of lean six sigma in industry 4.0 era: A guide to improve quality, vol. 12 (2018)

Ghobakhloo, M.: The future of manufacturing industry: a strategic roadmap toward industry 4.0. J. Manuf. Technol. Manag. **29**, 910–936 (2018). https://doi.org/10.1108/JMTM-02-2018-0057

Gupta, S., Modgil, S., Gunasekaran, A.: Big data in lean six sigma: A review and further research directions. Int. J. Prod. Res. **58**, 947–969 (2020). https://doi.org/10.1080/00207543.2019.1598599

Ito, T., Abd Rahman, M.S., Mohamad, E., Abd Rahman, A.A., Salleh, M.R.: Internet of things and simulation approach for decision support system in lean manufacturing. J. Adv. Mech. Des. Syst. Manuf. **14** (2020). https://doi.org/10.1299/jamdsm.2020jamdsm0027

Jayaram, A.: Lean six sigma approach for global supply chain management using industry 4.0 and IIoT. Presented at the proceedings of the 2016 2nd international conference on contemporary computing and informatics, IC3I 2016, pp. 89–94 (2016). https://doi.org/10.1109/IC3I.2016.7917940

Kolberg, D., Zühlke, D.: Lean automation enabled by industry 4.0 technologies. Presented at the IFAC-PapersOnLine, pp. 1870–1875 (2015). https://doi.org/10.1016/j.ifacol.2015.06.359

Koppel, S., Chang, S.: MDAIC—a six sigma implementation strategy in big data environments. Int. J. Lean Six Sigma **12**, 432–449 (2021). https://doi.org/10.1108/IJLSS-12-2019-0123

Lasi, H., Fettke, P., Kemper, H.-G., Feld, T., Hoffmann, M.: Industry 4.0. Bus. Inf. Syst. Eng. **6**(4), 239–242 (2014). https://doi.org/10.1007/s12599-014-0334-4

Mayr, A., Weigelt, M., Kühl, A., Grimm, S., Erll, A., Potzel, M., Franke, J.: Lean 4.0—a conceptual conjunction of lean management and Industry 4.0. Presented at the Procedia CIRP, pp. 622–628 (2018). https://doi.org/10.1016/j.procir.2018.03.292

Raj, A., Dwivedi, G., Sharma, A., Lopes de Sousa Jabbour, A.B., Rajak, S.: Barriers to the adoption of industry 4.0 technologies in the manufacturing sector: An inter-country comparative perspective. Int. J. Prod. Econ. **224**, 107546 (2020). https://doi.org/10.1016/j.ijpe.2019.107546

Rifqi, H., Zamma, A., Ben Souda, S.: Lean 4.0, six sigma-big data toward future industrial opportunities and challenges: A literature review. In: Saeed, F., Al-Hadhrami, T., Mohammed, F., Mohammed, E. (eds.) Advances on Smart and Soft Computing. AISC, vol. 1188, pp. 201–210. Springer, Singapore (2021). https://doi.org/10.1007/978-981-15-6048-4_18

Rosin, F., Forget, P., Lamouri, S., Pellerin, R.: Enhancing the decision-making process through industry 4.0 Technologies. Sustainability **14**(1), 461 (2022)

Rosin, F., Forget, P., Lamouri, S., Pellerin, R.: Impacts of industry 4.0 technologies on lean principles. Int. J. Prod. Res. **58**, 1644–1661 (2020). https://doi.org/10.1080/00207543.2019.1672902

Schumacher, A., Erol, S., Sihn, W.: A maturity model for assessing industry 4.0 readiness and maturity of manufacturing enterprises. Procedia CIRP **52**, 161–166 (2016). https://doi.org/10.1016/j.procir.2016.07.040

Skalli, D., Charkaoui, A., Cherrafi, A., Garza-Reyes, J.A., Antony, J. and Shokri, A.: Industry 4.0 and lean six sigma integration in manufacturing: A literature review, an integrated framework and proposed research perspectives. Qual. Manage. J. **30**(1), 16–40 (2023). https://doi.org/10.1080/10686967.2022.2144784

Skalli, D., Charkaoui, A., Cherrafi, A.: The integration of industry 4.0 in operations management: Toward smart lean six sigma. In S., Motahhir, B., Bossoufi, (eds.) Digital Technologies and Applications. ICDTA 2022. Lecture Notes in Networks and Systems, vol. 454. Springer, Cham. (2022). https://doi.org/10.1007/978-3-031-01942-5

Sodhi, H.: When industry 4.0 meets lean six sigma: A review. Ind. Eng. J. **13** (2020). https://doi.org/10.26488/IEJ.13.1.1214

Sony, M.: Design of cyber physical system architecture for industry 4.0 through lean six sigma: Conceptual foundations and research issues. Prod. Manuf. Res. **8**, 158–181 (2020). https://doi.org/10.1080/21693277.2020.1774814

Sony, M.: Industry 4.0 and lean management: A proposed integration model and research propositions. Prod. Manuf. Res. **6**, 416–432 (2018). https://doi.org/10.1080/21693277.2018.1540949

Tissir, S., Cherrafi, A., Chiarini, A., Elfezazi, S., Bag, S.: Lean six sigma and industry 4.0 combination: scoping review and perspectives. Total Qual. Manag. Bus. Excell. 1–30 (2022). https://doi.org/10.1080/14783363.2022.2043740

Tortorella, G., Miorando, R., Mac Cawley, A.F.: The moderating effect of industry 4.0 on the relationship between lean supply chain management and performance improvement. Supply Chain Manag. **24**, 301–314 (2019). https://doi.org/10.1108/SCM-01-2018-0041

Tortorella, G.L., Fogliatto, F.S., Cauchick-Miguel, P.A., Kurnia, S., Jurburg, D.: Integration of industry 4.0 technologies into total productive maintenance practices. Int. J. Prod. Econ. **240**, 108224 (2021). https://doi.org/10.1016/j.ijpe.2021.108224

Vinodh, S., Antony, J., Agrawal, R., Douglas, J.A.: Integration of continuous improvement strategies with Industry 4.0: A systematic review and agenda for further research. TQM J. **33**, 441–472 (2020). https://doi.org/10.1108/TQM-07-2020-0157

Wagner, T., Herrmann, C., Thiede, S.: Industry 4.0 impacts on lean production systems. In Tseng, M.M., Tsai, H.Y., Wang, Y. (eds.) Manufacturing Systems 4.0, pp. 125–131. Elsevier Science Bv, Amsterdam (2017). https://doi.org/10.1016/j.procir.2017.02.041

Framework for the Integration of Cookware into Life Cycle Assessment: Case Study

Zineb El Haouat[1(✉)], Fatima Bennouna[2], and Driss Amegouz[1]

[1] Higher School of Technology, Sidi Mohammed Ben Abdellah University, 30050 Fez, Morocco
{zineb.elhaouat,Driss.amegouz}@usmba.ac.ma

[2] Ecole Nationale Des Sciences, Appliquées Sidi Mohammed Ben Abdellah University, 30050 Fez, Morocco
Fatima.bennouna@usmba.ac.ma

Abstract. The development aluminum industry, as well as the aluminum cookware industry's major contribution to global consumption and CO_2 (Green-house Gas GHG) emissions, highlights the need to employ approaches that reduce environmental consequences while minimizing waste. This article provides an overview and critical analysis of the recent life cycle assessment (LCA) literature, moreover a list of the primary sources of aluminum cookware's environmental impacts. The goal of this research is to gain a comprehensive understanding of the stages of the product life cycle and the appropriate methodology. However, only a few studies of this type have been conducted to far to develop an LCA that meets the requirements of the ISO standards and too quantify the different outgoing and incoming raw material flows and also polluting emissions from an aluminum container. Our results indicate that the aluminum pan's end-of-life phase accounts for the majority of CO_2 emissions, accounting for 36% of the total impact on the end-of-life, approximately 2.48 kg CO_2 equivalent. To contribute to components and to follow the primary environmental tendencies to reduce the main consequences on the environment, a practical framework for the integration of aluminum cookware LCA/end of life study has been proposed to us.

Keywords: LCA · LCI · Impact category · LCI analysis · Aluminum production · Eco-conception · Environmental impact

1 Introduction

Aluminum is a good conductor of heat which makes it a great choice for cook-ware with its thermal conductivity (Alabi and Adeoluwa 2020). Since its origin in the late 1960s, the life cycle assessment (LCA) method has attracted growing interest from industry, government, and the public as a holistic environmental system analysis method, and tremendous methodological progress has been achieved (Guo et al. 2019).

LCA of aluminum cooking utensils, hasn't been enough treated in the literature (Liu et al. 2017; Farjana et al. 2019; Guo et al. 2019) and beyond my intervention for more studies to clearly clarify the stages of cooking utensil's life cycle as well as to quantify the consumption of materials and energy, discharges and emissions into the atmosphere

L. Azrar et al. (Eds.): CIP 2022, LNME, pp. 96–104, 2023.
https://doi.org/10.1007/978-3-031-23615-0_10

and waste production. The framework of this article will first focus on a literature review, followed by a detailed description and means based on LCA, including the normative methodology and the relationship between Aluminum cooking utensils and LCA. The second part will discuss the methodology and application of analytical epistemology in a case study on aluminum cooking utensils. The consumption of materials and energy, as well as discharges and emissions into the atmosphere and waste creation, are all quantified through the life cycle analysis of cookware. In this last part, which revolves around the approach followed, The article follow the implementation of the proposed method; the four main stages will allow us first of all to define what the objectives of the LCA are, then the life cycle inventory which will allow us to draw up the flows of materials and incoming, outgoing energies associated with the stages of the life cycle related to the selected functional unit. The LCI is carried out using the ADEM product assessment software, followed by the evaluation of potential impacts. This is the most important step of the LCA since it aims to convert a flow inventory into a series of potential impacts. As a result, the final phase, called interpretation of the results, is used to combine the inventory results with the assessment of the life cycle's impact, as well as the interpretation. A conclusion and recommendations are included at the end of the document.

2 Literature Review

The Life Cycle Assessment (LCA) method has attracted growing interest from industry, government, and the public as a holistic environmental system analysis method, and tremendous methodological progress has been achieved (Guinee et al. 2011). It is a reliable tool for measuring and evaluating the potential energy and environmental performance of a product (Bouyarmane et al. 2019). This multi-criteria approach, initiated in the 1970s during the oil shocks, has developed significantly in terms of its methodology and its application since the beginning of the 2000s (Guinee et al. 2011). It includes a complete analysis covering the extraction of the material, the manufacturing, the transport, the use and the end of life as well as the operations which depend on the supply of energy, transport and secondary material. It is a methodological approach that aims to assess the socio-positive and negative impacts of a product or service throughout its life cycle (Cadena et al. 2019). It is an analysis that gives us a bottom-up perspective of not only direct emissions from the manufacturing process but also indirect emissions from upstream raw materials and energy, which has great advantages in identifying pollution sources in the production chain and make targeted improvements to reduce emissions (Guo et al. 2019).

The first phase of the LCA is decisive for all the other phases of the LCA. A clear and initial definition of the objective is therefore essential for a correct later interpretation of the result. In this part, it is necessary to define the function of the studied system and the functional unit (FU). The environmental effects of the system will subsequently be documented on the basis of UF. This phase is also used to identify the product system bounds, such as the amount of detail and procedures that are taken into account in the study. Inventory analysis (phase 2) plays a crucial role in all LCA studies (Pradhan and Mbohwa 2017). It provides impact category indicators for environmental and social emissions, but neglects energy and water consumption due to the lack of quality data at present and the difficulty of quantifying impacts (Blaauw et al. 2020). Phase three assesses the quantities of materials and energy extracted as well as the impact of polluting emissions in the air, water and soil. In this phase, characterization methods, in other words, impact calculation methods are used, such as: Eco-Indicator99, CML 2001 (Lv et al. 2019), Recipe (Goedkoop et al. 2009), etc. These approaches combine pollutant emissions that have similar environmental effects, then group them into the same impact category. Finally, the results must be interpreted (phase 4).

3 Methodology and Approach: Case Study (Aluminum Pan)

The ISO standard differentiates the methodological framework of LCA from its different applications, which are multiple such as product development, eco-labelling, carbon footprint and other footprints (Hauschild 2017). The standard organization's LCA applications are covered in its own publications.

Guo et al. (2019) went to discuss the current use of the LCA method, the aluminum industry's potential for reduction, and the relationship between carbon efficiency and total emissions in the first section. The authors introduced the offered materials and methods in the second section, including system boundaries, data sources, and measurements, and the LCA results were presented in the third section. In the fourth part, several aluminum-producing provinces in China are chosen, and several scenarios incorporating technical advancement and energy structure optimization are created to investigate the sources, distribution, and possibilities for reducing carbon emissions in China. A sanitary ware manufacturing was selected as a case study by (Lv et al. 2019). They started with an analysis of material flows, which was completed using data provided by the people in charge of the operations. This analysis includes the first phase of the LCA, which is the determination of the objectives and the research field, as well as inventory. Following that, the CML2001 method was used to evaluate environmental impacts. The results of the LCA are then subjected to a sensitivity analysis to investigate the variations in the results produced by changes in the inputs and waste recycling. Finally, there are regulations and suggestions for reducing environmental impacts.

According to the various approaches used, we applied the normative framework to aluminum cooking utensils, which would be used in the case study section that follows (Fig. 1).

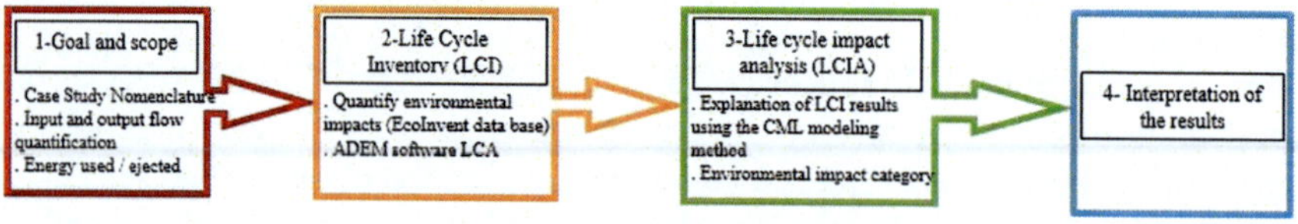

Fig. 1. Methodology proposed

3.1 Goal and Scope Phase

The goal of this research is to create an LCA that complies with ISO 14040 and 14044 standards and to determine an aluminum pan's environmental profile. This article covers a description of the product system and its boundaries, as well as data collection and inventory of emissions and extractions of an aluminum pan used in a Moroccan company. The study looks at an aluminum saucepan made by an aluminum kitchen utensil manufacturer in Morocco, with a diameter of 200mm, a height of 93mm, a weight of 0.58kg, and a maximum capacity of 2900 L. The aluminum pan's nomenclature is shown in Table 1.

Table 1. Pan nomenclature.

Component	Coef	Materiel	Mass	Unit
Disk	1	Aluminum	503	g
Handle	1	Bakelite	74	g
Support	1	Stainless steel	5	g
Rivet	2	Aluminum	0.8	g
Screw	1	Stainless steel	3	g
Pan cover	1	Aluminum	3	g

The system considers the energy required in the manufacturing of the pan as well as the raw materials, the energy used in transportation, the product's use, and the type of end-of-life. Figure 3 was created with the ADEM software and depicts the system's border, incoming flows, and outgoing flows.

Some flows of services, materials, and energy are not directly related to the analyzed product during its life cycle since they do not manufacture or convey it. Non-attributable processes are a good description. Transformation, transportation, and storage procedures are all involved in the creation of the saucepan. Environmental flows (extracted natural resources, contaminants, etc.) will be the source of incoming and outgoing flows, or environmental loads. As a result, waste streams heading out the door are diverted to treatment facilities. Then, to quantify all of the system's components, including aluminum and Stainless steel, an inventory study of the aluminum pan's life cycle is required The second phase of the LCA study is believed to be the most time and effort-intensive.

3.2 Life Cycle Inventory Phase

To calculate the inventory results in units of CO_2 equivalent, a GWP factor must be applied to the GSE emissions and removals data. The total CO_2 inventory results per unit of analysis should be quantified, which includes all emissions and removals from biogenic sources, non-biogenic sources, and land use change impacts. To model the aluminum saucepan in this study, we used the ADEM product balance sheet. This technology allows us to quantify environmental impacts across the course of a product's life cycle (from manufacturing to end of life). The data for this project came from the Eco invent version 2.0 database, which was just introduced to the software.

Product life cycle GHG accounting is a subset of life cycle analysis (LCA), which aims to quantify and address environmental aspects and potential impacts throughout the life cycle of a product. Product, from raw material extraction to end-of-life waste treatment. This indicator indicates the global warming potential induced by greenhouse gas (GHG) emissions and represents classification I. Manufacturing accounts for 29% of climate change, distribution and packaging7.8%, use 28% and the end of life constitutes 36% of this potential impact (MidPoint) expressed in kg CO_2 equivalent. Based on the LCA, the environmental impacts of aluminium production, particularly carbon emissions, were assessed at the regional or industry-average level, and the reduction potential, driving forces and strategy proposals were further proposed (Farjana et al. 2019).

The company has included a process map in the Life Cycle Inventory Report (Fig. 2). A process map describes the services, materials, and energy required to move the product under investigation through its life cycle Specific facts, on the other hand, are considered sensitive and only for internal use because they form the basis for data collection, which is why a reduced version of the report was prepared. At a minimum, the process map identified established lifecycle stages, generalized attribution methods at each level, and the movement of the investigated product through its lifecycle. These are the negative impacts of pollutants with a photochemical origin. The release of volatile organic compounds such as carbon monoxide or nitrogen oxides is the first step in the creation of photochemical ozone (MVOCs). Class I particle pollution (illness incidence), where the pollutant can be a single or several substances. This MidPoint category indicator is measured by the number of cases of various diseases, the number of years of life lost, or the number of DALYs (Disability adjusted life years) that travel through the system.

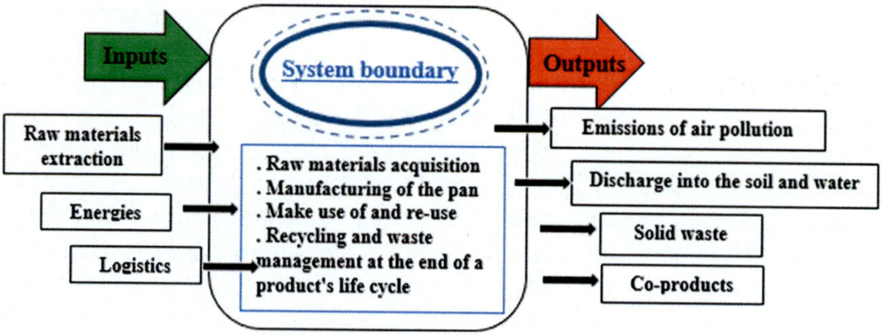

Fig. 2. Quantification flows

3.3 Life Cycle Impact Analysis Phase

During the simulation, the CML technique in ADEM's product assessment software was used to characterize the impact. This modeling method corresponds to first-order effects: quantifiable and relatively direct, so it is the most widely used method with few additional uncertainties. It is often considered the most complete method because it includes new categories of environmental impacts such as acidification, climate change, eutrophication, ozone layer depletion, and so on.

When it comes to environmental dangers, many publications criticize the insufficiency of both preventive and curative measures, as well as fines against those responsibilities for pollution and nuisances. These weaknesses in environmental protection can be traced back to legislation, which is frequently content to encourage risky activity by just imposing means responsibilities. Then, at each stage of the life cycle, the consumption of materials and energy, discharges and emissions into the air, water, and soil, and waste creation are quantified and expressed in terms of indicators of potential environmental impacts. Climate change, primary and secondary particulate matter (PM) emissions, and ozone layer depletion are the three types of consequences that are constantly or regularly examined in association to production, distribution, use, and end-of-life flows. Depending on the effect categories to be examined, the CML technique is the most usually applied by practitioners. Their goal is to summarize and explain the inventory results (ICV).

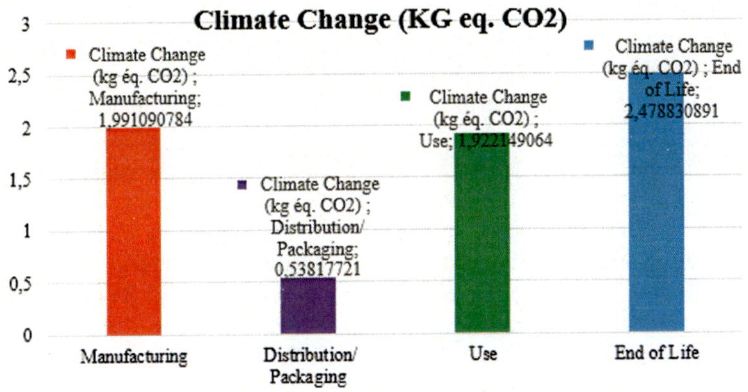

Fig. 3. Diagram of climate change's potential impact

Furthermore, this study highlights a number of strict aims; for example, the IPCC 2007 approach is nearly the only one employed for the greenhouse effect. The CML technique picks it up. Only the CML approach is nearly employed for eutrophication and resource consumption.

3.4 Interpretation Phase

The assessment of the aluminum saucepan allowed us to become connected with ADEM's product evaluation program and gain an understanding of the saucepan's environmental consequences. The first step in interpreting the results is to choose and examine the aluminum pan's life cycle. The effect of global warming or equivalent kilogram CO_2 emitted is one of the common and standard factors in the topic of energy consumption with reference to GHG emissions. In the second step of the LCA, the results were expressed in terms of environmental impact category using a similar LCI with nine different families of impact categories. Aluminum manufacturing is typically connected to lessening the environmental effect while increasing capacity and quality, due to the worldwide production of aluminum, which amounts to 64.336 million tons per year (in 2018). The production stage begins when product components are delivered to the manufacturing facility and concludes when the finished product is delivered to the customer. The phrases "site" and "gate" are used metaphorically because a product may move through a number of procedures and intermediary facilities before leaving the production stage as a final product. Processes related with co-products and/or the treatment of waste created during production may also be included in this phase. The impact categories' key (MidPoint) pilot phases are the production and use of the aluminum pan. The manufacturing of these raw materials refers to environmental repercussions and impacts. As a result, LCA is being used to investigate several types of impacts, such as carbon emissions and CO_2 sources, resulting from the production of an aluminum household item in a Moroccan company.

4 Conclusion and Future Directions

LCA is a tool that throws new insight on how organizations perform and how products and services are designed. In the context of LCA, data quality is a critical issue. Many of the aluminum cooking utensils in this article are used all over the world, with many of them being created in Morocco, as is the case with our aluminum pan analysis. Previous studies have not adequately addressed the analysis of household products, and it is from this that our study and evaluation to estimate the environmental implications of an aluminum pan in the first place occur. The fixed purpose and field of study for a later correct interpretation of the data are represented in the first steps of the life cycle analysis of the pan. The first expectation is that this research will provide a complete LCA of aluminum kitchen utensils, which will include all stages of the methodology, including the compilation, quantification, and qualification of the input and output data related to our study; this is the most time and effort-intensive step of any LCA. This step, referred to as ICV, is the most delicate of the entire study. It was completed with the help of the ADEM product balance sheet. After doing in-depth research of the various materials and components of the pan, the wear of one of the components brings us back to the end-of-life study phase. The end-of-life stage starts when a consumer deletes a used product and concludes when it is returned to nature or assigned to the life cycle of another (recycled) product. Because the method used to treat the product (landfill, incineration, etc.) is the key attributable end-of-life process, LCA users must know or assume the product's fate in order to map this stage. Requirements and guidance for end-of-life recycling are discussed below. A new development in kitchen utensil manufacturing procedures is to produce and manufacture kitchen utensils from PTFE, also known as Teflon. This, combined with the increase in green application to these areas related to lean manufacturing, will create more value with fewer resources, all while replacing these resources with others that are more eco sustainable. However, LCA appears to be a technique capable of contributing to technical components of reflection as well as following significant environmental trends in order to reduce major environmental and health impacts (end of life and use of the pan already cited). The wear of a component or section of the pan takes us back to the end of its life cycle.

- Outlooks:
  - How can we incorporate integrated management information systems into the production of a "complete" LCA for aluminum cookware in Moroccan sectors?
  - How can we create an integration module for information systems that will lead to the evaluation of environmental elements associated with the production of aluminum cookware?

References

Journal article

Cadena, E., Rocca, F., Gutierrez, J.A., Carvalho, A.: Social life cycle assessment methodology for evaluating production process design: Biorefinery case study. J. Clean. Prod. **238**, 1–14 (2019). https://doi.org/10.1016/j.jclepro.2019.117718

Guo, Y., Zhu, W., Yang, Y., Cheng, H.: Carbon reduction potential based on life cycle assessment of China's aluminum industry-a perspective at the province level. J. Clean. Prod. **239**, 1–10 (2019). https://doi.org/10.1016/j.jclepro.2019.118004

Hauschild, M.Z.: Introduction to LCA methodology. In: Hauschild, M.Z., Rosenbaum, R.K., Olsen, S.I. (eds.) Life Cycle Assessment, pp. 59–66. Springer, Cham (2017). https://doi.org/10.1007/978-3-319-56475-3_6

Lv, J., Gu, F., Zhang, W., Guo, J.: Life cycle assessment and life cycle costing of sanitary ware manufacturing: A case study in China. J. Clean. Prod. **238**, 1–16 (2019). https://doi.org/10.1016/j.jclepro.2019.117938

Pradhan, A., Mbohwa, C.: Development of life cycle inventory (LCI) for sugarcane ethanol production in South Africa. In: 2017 International Renewable and Sustainable Energy Conference (IRSEC), pp. 1–4 (2017). https://doi.org/10.1109/IRSEC.2017.8477418

Guinee, J.B., et al.: Life cycle assessment: past, present, and future. Environ. Sci. Technol. **45**(1), 90–96 (2011). https://doi.org/10.1021/es101316v

Bouyarmane, H., El Amine, M., Sallaou, M.: Environmental assessment in the early stages of product design. In: 2019 5th International Conference on Optimization and Applications (ICOA), pp. 1–5 (2019). https://doi.org/10.1109/ICOA.2019.8727611

Blaauw, S.A., Maina, J.W., Grobler, L.J.: Life cycle inventory of bitumen in South Africa. Transp. Eng. **2**, 1–7 (2020). https://doi.org/10.1016/j.treng.2020.100019

Farjana, S.H., Huda, N., Mahmud, M.A.P.: Impacts of aluminum production: A cradle to gate investigation using life-cycle assessment. Sci. Total Environ. **663**, 958e970 (2019). https://doi.org/10.1016/j.scitotenv.2019.01.400

Liu, S., Wang, S., Wang, K., Yue, H., Liu, L., Yang, S., Zhang, P., Zhang, R.: Energy consumption and GHG emission for regional aluminum industry: A case study of Henan province, China. Energy Procedia **105**, 3391–3396 (2017). https://doi.org/10.1016/j.egypro.2017.03.777

Journal article only by DOI

Alabi, O.A., Adeoluwa, Y.M.: Production, usage and potential public health effects of aluminum cookware: A review. Ann. Sci. Technol. **5**(1), 20–30 (2020). https://doi.org/10.2478/ast-2020-0003

Book

Goedkoop, M., Heijungs, R., Huijbregts, M., Schryver, A., Struijs, J., Van Zelm, R.: A life cycle impact assessment method which comprises harmonised category indicators at the midpoint and the endpoint level. ReCiPe 2008 (2009)

Robust Optimization of Pipe Extrusion Process by Using Gum and Monte Carlo Methods with Factorial Designs

Abdelaziz Ait Elkassia[1](✉), Abdelouahhab Salih[1], Abdelillah Jalid[1], and Abderrahim Zegoumou[2]

[1] ENSAM of Rabat, Mohamed V University, Rabat, Morocco
aziz_iaa@hotmail.com, {a.salih,a.jalid}@um5r.ac.ma
[2] ENAM Meknes, Meknes, Morocco

Abstract. In this work, robust process optimization method is proposed, with application on plastic pipe extrusion process, following Guide of Uncertainty in Measurement GUM based on law of propagation of uncertainties and then Monte Carlo simulations. The parameter to optimize is the Standard Diameter Ratio SDR of a plastic tube and this after obtaining a model adjusting to the experimental data (good model), here by the methodology of design of experiments, factorial fractional and full designs. As a source of uncertainty are considered the variation's domains of the input factors and then the experimental variance. The results obtained made it possible to determine the optimal configuration of the input factors for robust optimization, this configuration does not change or slightly considering the experimental variance. Comparing result of both fractional factorial and full factorial design, in this example, optimum has minimum variability or more robust with full design rather than fractional design.

Keywords: DOE · GUM and MCM · Robust optimization · SDR

Abbreviations

DOE Design of experiments
GUM Guide of Uncertainty in Measurement
MCM Monte Carlo Method
SDR Standard Diameter Ratio

1 Introduction

Robustness is not always achieved after optimization, for example optimizations got by the methodology of design of experiments DOE (J. Goupy 1997 and 2006) where the satisfaction optimization can be obtained from several points of the response surface. (J. García and A. Peña 2018). In effect, output variation depends on the input variations

L. Azrar et al. (Eds.): CIP 2022, LNME, pp. 105–116, 2023.
https://doi.org/10.1007/978-3-031-23615-0_11

due to controllable and uncontrollable factors (Fig. 1) (Wu and Hamada 2009, ReliaSoft Corporation 2015).

Admittedly, optimality criteria minimize, in part, the variability of the predicted response, through the minimization of variance due to the model (lack of fit) but does not depend on experimental variability (pure error).

In the case studied, the source of variation considered are variability of input factors on their variation's domain and then experimental variability due to uncontrollable factors that are difficult or impossible to modify (R. A. McLean and V. L. Anderson 1984).

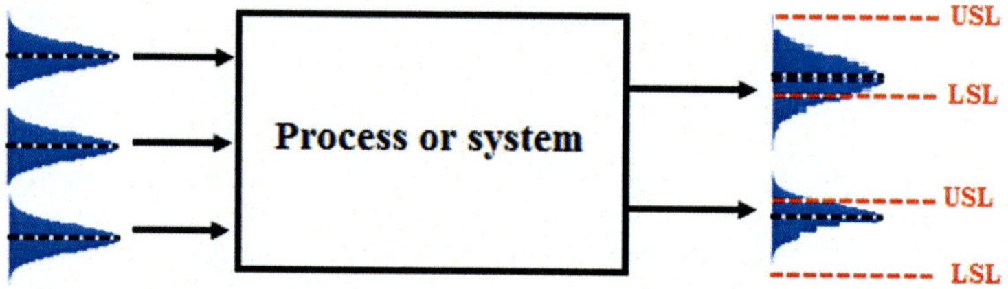

Fig. 1. Output variation depends on the input variations

Additive model is considered with factorial design:

$$\mathbf{y} = \mathbf{f(x)} + \mathbf{e} = a_0 + \sum a_i x_i + \sum\sum a_{ij} x_i x_j + \cdots + e \quad (1)$$

The response is the Standard Diameter Ratio SDR and the objective is to search optimum and minimize its variability (T. Kiatcharoenpol and T. Vichiraprasert, 2018)

$$\mathbf{SDR} = \mathbf{D/e}$$

D: external diameter, e: thickness. D, e and SDR depend on input factors and experimental variation.

For extrusion process capability, we seek to ensure that the SDR is within the range [10, 20; 10, 63] (Fig. 2). Our optimum will be the target value: SDR $= \frac{(10.20+10.63)}{2} =$ 10.42. As a result, we consider any piece obtained with dimensions outside this interval is out of specification (O.S.). (NF EN 12201-1 and -2: 2011),

The experimenter indicates that there are four factors, which are the most preponderant (Table 1) (P. Alexis 2015).

2 Methodology

The methodology described in (A. Ait Elkassia et al. 2020) can be followed specially to evaluate uncertainty using GUM or MCM methods. However, for main aim of this study, this general scheme is applied (Fig. 3).

To obtain a relationship (model) between the output and inputs, we apply DOE.

Validation of the model depend on data collected from the precedent phase. So, using an estimation of experimental variance on one or each trial will affect validation model.

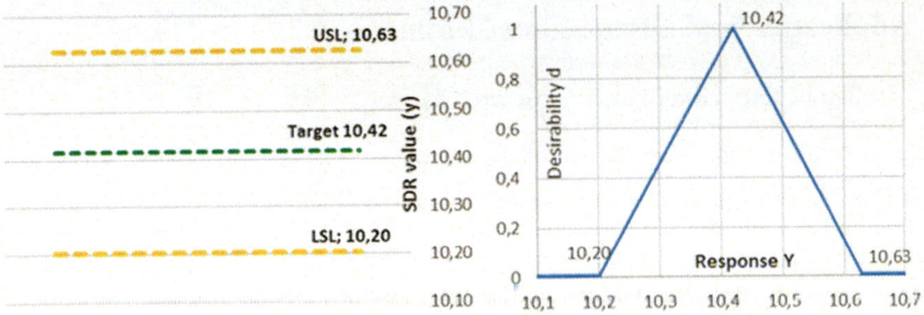

Fig. 2. Domain of response variation desired by experimenter

Table 1. Input variables and their variations

Study Factors	Domain of factor		Mean	Expanded uncertainty U	Standard deviation SD	SD of coded variable u(x)
	Real variable	Coded variable				
Cooling temperature X_1	(45–55) °C	(−1;1)	50	5	2	0.4
Pulling speed X_2	(10–15) cm/s	(−1;1)	12.5	2.5	1	0.4
Die temperature X_3	(180–200) °C	(−1;1)	190	10	4	0.4
Caliber X_4	(91–92) mm	(−1;1)	91.5	0.5	0.2	0.4

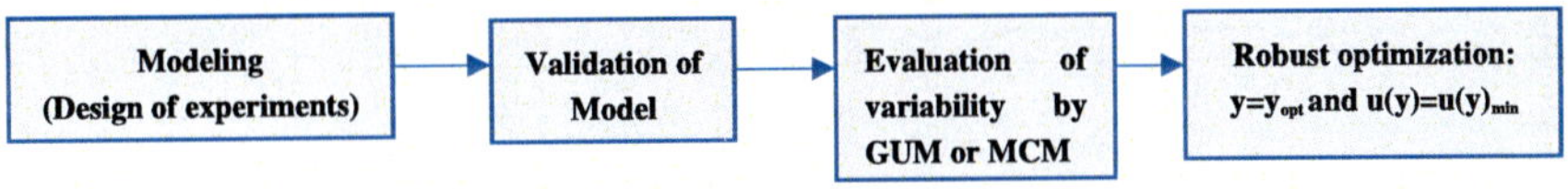

Fig. 3. General scheme for application of GUM or MCM for robust optimization

After obtaining a good model, we use it to evaluate variability (uncertainty) u(y) by low of propagation of uncertainty (JCGM 100:2008) GUF. Interaction term is exploited to find robustness. An optimization tool or application will be used.

MCM simulation (JCGM 101:2008) will be used instead of GUF or just to confirm/validate the result obtained by GUF methodology.

3 Application and Discussion of Results

3.1 Design of Experiment and Associated Model

This methodology will be applicated on the example, we have treated on (A. Ait elkassia et al. 2020).

For optimization study, a fractional design was established first, with $2^{4-1} = 8$ trials in gray, then full design will be developed for comparison, completing trials $2^4 = 16$. Both are used for the same problem, to have robust optimization (Table 2).

Table 2. Fractional (trials in gray) and full factorial design and results

Trial	X_1	X_2	X_3	X_4	Y
1	− 1	− 1	− 1	− 1	**10.40**
2	1	− 1	− 1	− 1	**11.20**
3	− 1	1	− 1	− 1	**10.22**
4	1	1	− 1	− 1	**9.87**
5	− 1	− 1	1	− 1	**10.76**
6	1	− 1	1	− 1	**10.93**
7	− 1	1	1	− 1	**9.96**
8	1	1	1	− 1	**10.14**
9	− 1	− 1	− 1	1	**10.13**
10	1	− 1	− 1	1	**9.60**
11	− 1	1	− 1	1	**10.76**
12	1	1	− 1	1	**9.68**
13	− 1	− 1	1	1	**10.58**
14	1	− 1	1	1	**9.78**
15	− 1	1	1	1	**10.93**
16	1	1	1	1	**9.69**

Fractional design is obtained by aliasing/confounding factor X_3 with the third order interaction: $X_1X_2X_4$ (initial alias: $X_3 = X_1X_2X_4$ and generating ratio: $I = X_1X_2X_3X_4$) (J. Goupy and L. Creighton 2006).

With the experimental results obtained, effect of the factor X_3: Die temperature, will be neglect. But we preserve it in the fractional associated model, but not its interactions.

Considering the hypotheses of fractional design interpretation:

$$\begin{aligned} y = & h_0(1 \text{ ou } x_1x_2x_3x_4) + h_1(x_1 \text{ou } x_2x_3x_4) + h_2(x_2\text{ou } x_1x_3x_4) \\ & + h_3(x_3\text{ou } x_1x_2x_4) + h_4(x_4\text{ou } x_1x_2x_3) \\ & + h_{12}(x_1x_2\text{ou } x_3x_4) + h_{24}(x_1x_3\text{ou } x_2x_4) + h_{14}(x_1x_4\text{ou } x_2x_3) \end{aligned} \quad (2)$$

Thus, the postulated model (1) of the fractional design changes on Table 3.

Table 3. Studied factorial DOE and their associate models

Experimental design	Associate polynomial model	
Fractional design Fr	M_f (n = p = 8)	$y = h_0 + h_1x_1 + h_2x_2 + h_3x_3 + h_4\,x_4 + h_{12}\,x_1x_2 + h_{24}\,x_2x_4 + h_{14}\,x_1x_4$
Full design Fu	M_c (n = p = 16)	$y = a_0 + a_1x_1 + a_2x_2 + a_3x_3 + a_4x_4 + a_{12}\,x_1\,x_2 + a_{13}\,x_1\,x_3 + a_{14}\,x_1\,x_4 + a_{23}\,x_2\,x_3 + a_{24}\,x_2\,x_4 + a_{34}\,x_3\,x_4 + a_{123}x_1\,x_2\,x_3 + a_{124}x_1\,x_2\,x_4 + a_{134}x_1\,x_3\,x_4 + a_{234}x_2\,x_3\,x_4 + a_{1234}\,x_1\,x_2\,x_3\,x_4$

3.2 Model Validation

To have estimation of experimental variance, six (6) replications on the CenterPoint were performed: mean and standard uncertainty are: $y_c = 10.280$ and $\sigma_{exp} = 0.100$.

Table 4 gives the results obtained from fractional factorial design.

Table 4. Results of estimation of contrasts h_k and coefficients a_k

Contrast	Effect	Value	u(h)	t	p-value df = 5 (%)	p-value df = 30 (%)
h_0	$a_0 + a_{1234}$	10.251	0.035	289.94	0.00	0.00
h_1	$a_1 + a_{234}$	– 0.171	0.035	– 4.83	0.48	0.00
h_2	$a_2 + a_{134}$	– 0.127	0.035	– 3.58	1.59	0.12
h_3	$a_3 + a_{124}$	0.037	0.035	1.05	**34.03**	**30.05**
h_4	$a_4 + a_{123}$	– 0.096	0.035	– 2.72	4.20	1.09
h_{12}	$a_{12} + a_{34}$	– 0.060	0.035	– 1.69	**15.18**	**10.14**
h_{13}	$a_{13} + a_{24}$	0.193	0.035	5.46	0.28	0.00
h_{14}	$a_{14} + a_{23}$	– 0.340	0.035	– 9.62	0.02	0.00

The model validation study, in the case of fractional ad full design (n = p), is done by means comparison test. The t-student test can be used after validation of the model in order to assess the significance of the model coefficients.

Despite the insignificance of certain coefficients, according to the experimental variance, we keep the standard models associated with each DOE, they are often used by the experimenter (Table 5).

Table 5. Model validation test

	Fractional design	Full design
Model	**MFr** (n = p = 8)	**MFu** (n = p = 16)
Value at centerpoint ($y_c = 10.28$; $\sigma_{exp} = 0.100$)	10.25	10.30
Standard deviation	0.035	0.025
Deviation Model	− 0.03	0.02
t-value	0.273	0.223
p-value (%)	79.55	83.23
Statistical validation: mean comparison	Yes	Yes

3.3 Prediction of the Optimum

Basing on the guide of uncertainty framework GUF, the aim is first to know how many changes each point of the domain variation factors makes on response variability Y.

With GUF, a general form of uncertainty/variability on y is (GUF1):

$$u(y)^2 = \sum\left(\frac{\partial y}{\partial x_i}\right)^2 u^2(x_i) \tag{3}$$

Without correlation between X. u(x; x) = 0.

If the linearity of y = f(x) is not proved, the immediately term to be added is (GUF2) (Fig. 4):

$$\sum\sum\left[\frac{1}{2}\left(\frac{\partial^2 y}{\partial x_i \partial x_j}\right)^2 + \frac{\partial y}{\partial x_i}\frac{\partial^3 y}{\partial x_i \partial x_j^2}\right]u^2(x_i)u^2(x_j) \tag{4}$$

The initial point (center):

	X_1	X_2	X_3	X_4	Y	u(y)	O.S.*
Coded variables	0	0	0	0	*Fractional design*		
					10.25	**0.094**	**29.5%**
Real variables	50.0 °C	12.50 m/s	190.0 °C	91.50 mm	*Full design*		
					10.30	**0.104**	**16.3%**

***O.S.**: Out of Specification (Rate)

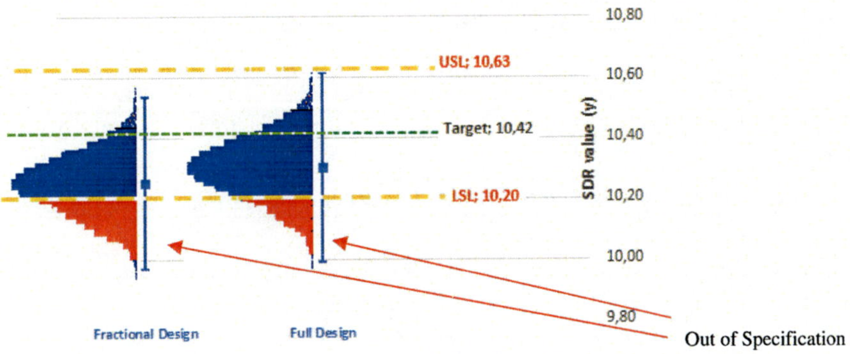

Fig. 4. Critical process produce out of specification in center point (k = 3)

3.4 Maximizing Desirability

To maximize desirability, we have to search combined point from experimental domain that meet a target. Some examples, with coded variables X:

	X_1	X_2	X_3	X_4	Y	u(y)	O.S. (%)
Optimum 1							
Fractional design	– 0.925	– 1	– 0.925	– 1	10.42	0.141	12.8
Full design					10.44	0.186	25.3
Optimum 2							
Fractional design	– 0.798	0.446	1	– 0.144	10.35	0.129	13.4
Full design					10.42	0.145	13.9

3.5 Robust Optimization

The system equation is:

$$\begin{cases} y_{opt} = a_0 + \sum a_i x_i + \sum\sum a_{ij} x_i x_j + \cdots \\ u(y)^2_{min} = \sum\left(\frac{\partial y}{\partial x_i}\right)^2 u^2(x_i) \quad \text{with } u(x;\ x) = 0 \end{cases} \tag{5}$$

	X_1	X_2	X_3	X_4	Y	u(y)	O.S
Robust optimum	− 0.693	− 0.662	− 0.089	− 0.195	10.42	0.058	0.02%
Fractional design	**46.5 °C**	**10.85 cm/s**	**189.1 °C**	**91.40 mm**			
	X_1	X_2	X_3	X_4	Y	u(y)	O.S
Robust optimum	− 0.880	− 0.303	− 0.244	− 0.280	10.42	0.047	0.00%
Full design	**45.6 °C**	**11.74 cm/s**	**187.6 °C**	**91.36 mm**			

In this case, we have to search X_1, X_2, X_3 and X_4 that produce $u(y)^2 = u(y)^2_{min}$ were the $y_{opt} = 10.42$; $x \in [-1; 1]$ and $u(x) =$ constant.

We find the following factor settings, giving the robust optimum for each associated design (Fig. 5):

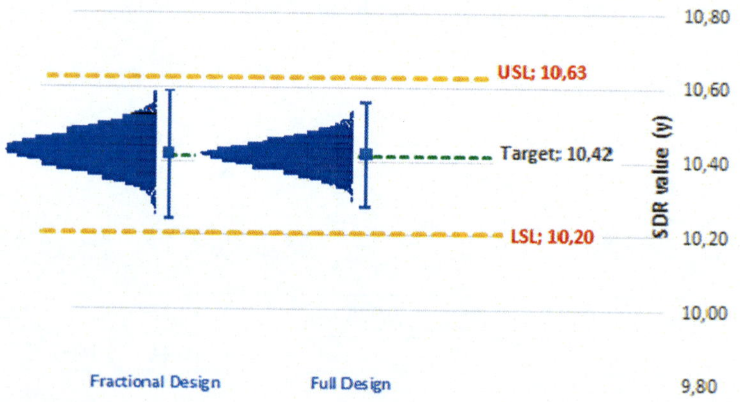

Fig. 5. Illustration of robust optimization with factor variations/uncertainties $u(x) = 0.4$ ($k = 3$) in coded variables

The factor settings in real variables X, giving the robust optimums by both fractional and full factorial design are close.

Instead of propagating standard uncertainties (GUM method), Monte Carlo simulation will propagate distributions assigned to factors X, in order to obtain probability density function (PDF) of output (G. Wübbeler, M. Krystek, 2008).

d_L and d_H are compared in the coverage probability of 99.73%. So for GUF, coverage factor is $k = 3$ and for MCM y_{low} and y_{high} taken for coverage intervals 99.73%.

We used here Monte Carlo simulations to confirm or not the results obtained by GUF method (Table 6):

Table 6. Comparison between GUF1, GUF2 and MCM when comparing results of fractional and full design

Robust optimum	Mean Y	Fractional design				Robust optimum	Mean Y	Full design			
		u(y)	O.S. (%)	d_L	d_H			u(y)	O.S. (%)	d_L	d_H
GUF1	10.42	0.058	0.02	−0.15	0.18	GUF1	10.42	0.047	0.00	−0.11	0.19
MCM	10.42	0.085	2.2	–	–	MCM	10.42	0.083	2.1	–	–
GUF2	10.42	0.087	1.36	−0.06	0.09	GUF2	10.42	0.083	0.94	−0.02	0.09

The sum of O.S. rate obtained by GUF2 is close to normal distribution adjusted to result output by MCM.

MCM result validate/confirm that obtained by GUF2 and not GUF1. Even if the output doesn't adjust to a normal PDF, the O.S. rates stay acceptable. We note that the output obtained by both fractional and full design model are closely adjustable to Loglogistic PDF (Fig. 6) (Palisade 2020).

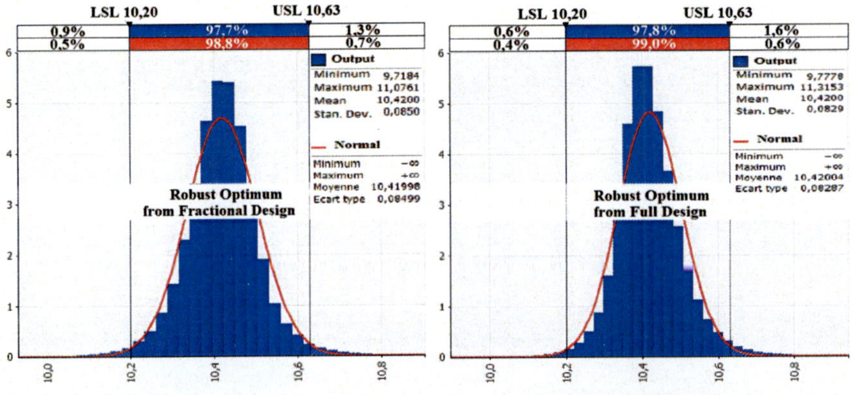

Fig. 6. Adjustment comparison between outputs and normal distribution

3.6 Robust Optimization with Non-zero Experimental Variability

We trying to incorporate the noise due to uncontrollable factors and random errors, so we add experimental variability as sources of variation.

The system precedent equation become:

$$\begin{cases} y = a_0 + \sum a_i x_i + \sum\sum a_{ij} x_i x_j + \cdots \\ u(y)^2 = \sum\left(\frac{\partial y}{\partial x_i}\right)^2 u^2(x_i) + \sum\left(\frac{\partial y}{\partial a_k}\right)^2 u^2(a_k) \quad \text{with} u(x;\ x) = 0 \text{ and } u(a,\ a) = 0 \end{cases} \tag{6}$$

	X_1	X_2	X_3	X_4	Y	u(y)	O.S
Robust optimum	– 0.670	– 0.668	– 0.072	– 0.161	10.42	0.078	0.58%
Fractional design	46.7 °C	10.83 cm/s	189.3 °C	91.42 mm			
	X_1	X_2	X_3	X_4	Y	u(y)	O.S
Robust optimum	– 0.830	– 0.271	– 0.208	– 0.264	10.42	0.059	0.03%
Full design	45.9 °C	11.82 cm/s	187.9 °C	91.37 mm			

Looking to real variable, the factor settings obtained with or without considering experimental errors are very close.

In order to have more control over the extrusion process and produce fewer O.S. rate, experimental variance must be reduced by reducing random errors.

4 Conclusions

This study proposes a methodology to success robust optimization applied to process extrusion, where the prediction variability is estimated from GUM method and/or Monte Carlo simulations. The models used are built by fractional and full factorial DOE. These DOE minimize just a part of variability on prediction, basing on D-optimality criteria. This methodology allows to explore the variation on the input factors and their effects on the output/prediction with exploitation of factor's interactions on the postulated and validated model, it allows to take into consideration experimental error variation too, and then incorporate noise due to unknown factors as source of variation in the output.

Estimation of variability obtained by GUF2, was validated by Monte Carlo simulations according to Monte Carlo procedure. Result optimum and its variability are now very close to specification and avoids producing defects.

References

Goupy, J.: Plans d'expériences. Techniques de l'ingénieur (1997)

Goupy, J., Creighton, L.: Introduction Aux Plans D'expériences. Dunod, Paris (2006)

García, J., Peña, A.: Robust optimization: Concepts and applications. In: Ser, J.D., Osaba, E. (eds.) Nature-Inspired Methods for Stochastic, Robust and Dynamic Optimization. InTech (2018)

Tiplica, T.: Contributions à la maitrise statistique des processus industriels multivariés. Université d'Angers (2002)

McLean, R.A., Anderson, V.L.: Applied Factorial and Fractional Designs. M. Dekker, New York

Cherfi, Z.: Ingénierie robuste. In: La Qualité : Démarche, Méthodes Et Outils, pp. 115–154. Lavoisier, Paris (2002)

MacKay, R.J., Steiner, S.H.: Strategies for variability reduction. Qual. Eng. **10**(1), 125–136 (1997). https://doi.org/10.1080/08982119708919115

Mori, T., Tsai, S.-C.: Taguchi Methods: Benefits, Impacts, Mathematics, Statistics and Applications. ASME Press (2011)

JCGM 100: Évaluation des données de mesure—guide pour l'expression de l'incertitude de mesure. www.bipm.org (2008)

JCGM 101: Evaluation of measurement data—supplement 1 to the "guide to the expression of uncertainty in measurement—propagation of distributions using a Monte-Carlo method. www.bipm.org (2008)

JCGM 104: Évaluation des données de mesure—une introduction au «guide pour l'expression de l'incertitude de mesure» et aux documents qui le concernent Evaluation. www.bipm.org (2009)

JCGM 200: Vocabulaire international de métrologie—concepts fondamentaux et généraux et termes associés (VIM). www.bipm.org (2012)

Wübbeler, G., Krystek, M., Elster, C.: Evaluation of measurement uncertainty and its numerical calculation by a Monte Carlo method. Meas. Sci. Technol. **19**(8), 084009 (2008). https://doi.org/10.1088/0957-0233/19/8/084009

Cox, M.G., Forbes, A.B., Harris, P.M., Smith, I.M.: The classification and solution of regression problems for calibration. NPL Report CMSC 24/03 (2004)

Ait Elkassia, A., Salih, A., Fagroud, M., Zegoumou, A.: Evaluation and comparison of uncertainties on fractional and full factorial design: case of a plastic pipe extrusion process. IJARET **11**, 249–267 (2020)

NF EN 12201-1 and -2: Systèmes de canalisations en plastique pour l'alimentation en eau et pour les branchements et les collecteurs d'assainissement avec pression-Polyéthylène (PE) (2011)

Palisade Corporation: @Risk. https://www.palisade.com/risk/fr/ (2020)

Ait Elkassia, A., Salih, A., Jalid, A.: The law of propagation of uncertainties GUM and Monte Carlo applied to experimental design; case of a Taguchi design. IJMET **9**, 832–843 (2019)

Kiatcharoenpol, T., Vichiraprasert, T.: Optimizing and modeling for plastic injection molding process using Taguchi method. J. Phys.: Conf. Ser. **1026**, 012018 (2018). https://doi.org/10.1088/1742-6596/1026/1/012018

Alexis, P.: Cours de plan d'expérience (2015)

Product Lifecycle Management Sustainable Approach to Automotive Logistics Services

Narjiss Tilioua(✉), Fatima Bennouna, and Zakaria Chalh

LISA Laboratory, Sidi Mohamed Ben Abdellah University, Fez, Morocco
{narjiss.tilioua,zakaria.chalh}@usmba.ac.ma

Abstract. The automotive industry is characterized by a very strong competitive context, and it represents today a strong sector, and logistics holds in it a preponderant place at all levels. From the design stage to distribution. In this context, the logistics function has a big importance at all levels: logistics plays an important role in the automotive sector, especially since it involves the most widespread products in the world. Therefore, we need to improve the collaboration between stakeholders along the entire value chain of automotive products, which is becoming a necessity today. PLM systems have been implemented to give more information about products, and between this information are logistics tasks from raw material transportation to finished product transportation, logistics integration, and supplier part delivery issues in PLM. Our goal in this paper, is to have most product information, and eliminate waiting times and all unnecessary movements in all product lifecycle stages, as well as the analysis of delays in the delivery of parts in the development phase and therefore to propose a strategy for managing the logistics of suppliers of automotive parts of rank 1 and 2 products, with the solution of digitalization of tasks by using Cameras in production lines and IOT techniques, that help the employee to decide himself to improve and therefore increase the effectiveness and efficiency of collaboration.

Keywords: PLM · Logistic · Green product Lifecycle · Sustainability · Collaboration · Digitalization · Automotive · Fournisseur · Production

1 Introduction

Product Lifecycle Management (PLM) is between the most recent solutions, which continue to evolve over the time, thanks to the technology, and PLM's objective is to share information in the best moment with precise quality and regulating products and product information flows processes [1]. Before being a solution, PLM is a strategic approach that aims to optimize creation processes, manufacture and maintenance of industrial products. Indeed, the PLM will allow the management of a product and its evolutions, from the design until its recycling stage.

PLM has become also an extremely useful concept that helps understand the relationships between a specific physical product (at the item level) and its fundamental information throughout its life cycle [2].

L. Azrar et al. (Eds.): CIP 2022, LNME, pp. 117–124, 2023.
https://doi.org/10.1007/978-3-031-23615-0_12

The huge need for an innovative solution to the problems of parts availability and logistics, from supplier to customer has become very important for "OEM" Original equipment manufacturers (of automotive sector), especially in the early life "BOL" phase of the development of automotive parts, to push organizations to adapt their strategies to the context and new challenges such as the evolution of technology, globalization of digitization, large flow of information and data, intensive logistics resources and difficult to manage and control, etc.

This work describes the constraints that need to be considered to solve the availability issues and the on-time delivery of tier 1 and tier 2 parts according to the customers' needs. In addition, the paper outlines a proposal for the digitization of order processes between suppliers and other collaborators, as well as logistics support activities. A systematic and complete analytical study through all the BOL phases of the product development life cycle, in order to reach the availability and maintainability objectives.

Therefore, the need to digitalize the actions in an automotive company has become very important. So, we can say that Sect. 2 is a literature review in order to identify the concepts used and the context, and Sect. 3 is where we mention our paper problem and the methodology we used, for arriving finally to discussion stage in Sect. 4 that offers other authors results, for giving finally our proper propositions.

2 Literature Review

This section presents the state of the art of the methods and practices used in the automotive sector, in the context of improvement and efficiency of demand processing, first of all to know the forecast of the demands especially the irregular demands, and therefore it is necessary to find an adaptable forecasting method to face the unpredictability and the uncertainty of the demand of parts Moreover, the application of Lean manufacturing in the automotive sector by eliminating waiting times at the MOL phase "Middle of life" and in the BOL phase "Beginning of life" also to freeze the design of the product before its industrialization.

Product Lifecycle Management is composed of several processes, methods, technologies and software (PLM tool) to manage the entire lifecycle of a product, which in turn is made up of 3 major stages: The beginning of life "BOL": design and development, the middle of life "MOL": production and manufacturing and services (use(s)) and the end of life "EOL": which is the end of life of the product. The objective is ma inly to simplify the management of automotive products, which has become increasingly complex due to competition and the high use of 4.0 technologies that has become paramount in this sector.

In the following subsections, we will define the basic concepts and discuss their correlations as main components for the implementation of an effective digitalized product lifecycle management (PLM) system. This paper is based on several scientific publications, a literature review of several sectors that use PLM, and our goal is to compare the solutions already conveyed by other authors and suggest the best solution that will be based on digital technology to finally improve the efficiency of life cycle management in the automotive sector.

Logistical tasks have changed profoundly due to information and communication technologies revolution. This revolution has a continuous improvement, which transforms the traditional supply chain into a digital one [3]. The main technologies currently in development for digital supply chains among all the industry 4.0 tools are: big data analytics, Internet of Things, cloud technologies, cognitive computing, and most recently, block chains [4].

On the other hand, C. Vila described a green PLM approach, which looks at product lifecycle management from an environmental point of view, and he proposed a frame work for sustainable product development that takes into account the entire product lifecycle and will allow companies to be more resource efficient. For him, the factory of the future is a set of companies that can compete and must be available to collaborate when needed by sharing knowledge and resources. With customers, stakeholders and suppliers and thus enable sustainability through green products and processes (certified green factory).

In addition, to achieve our goal of efficiency in life cycle management, it would be an asset to have digitalized solutions in this sector. Therefore, the role of PLM software is to manage the life cycle of its products by providing the company with a platform of information related to these products [5].

Simulation is a technique frequently used automotive manufacturer, as modeling as well as dynamics are required nowadays, to make modifications and verifications with a simulation analysis before approving them for final manufacturing design. A simulation model is a presentation of the work process or system.

In the context of improving responsiveness, simulation tools that we propose here are not expensive and very efficient, we use The best and cheapest solution, as using camera s in front of the production lines, also we can temporarily hire 2 expert analysts to eliminate unnecessary logistics movements, and to propose improvements in the system, if not propose hours of sharing ideas "brainstorming" between the different operators and employees in front of these videos of their work during the day, and thus motivate them to improve together with the least cost.

This following picture is used in Mohamed-Amine Abidi's paper, and this technology can offer many advantages, like allowing users to interact with simulation model in real-time and in a 3D format, which complete our proposed idea [6] (Fig. 1).

It is thanks to the rapid evolution of information technology that the automotive and other sectors are competing now in high levels. Internet technology has contributed also to virtually connecting staffs and processes even with distant actors also to reduce product's industrialization times, to satisfy finally the customer.

3 Problem Statement and Methodology

This work was developed based on a comparison between author's research papers especially in industrial sectors, and our own observation on the development of manufacturing and technology to 4.0 stage, and the details will be explained in another paper.

The problem that we aim to resolve is to arrive to the effective integration of suppliers and a best communication between logistics and production engineers on projects and problems encountered, so we propose to integrate an additional step in the process of

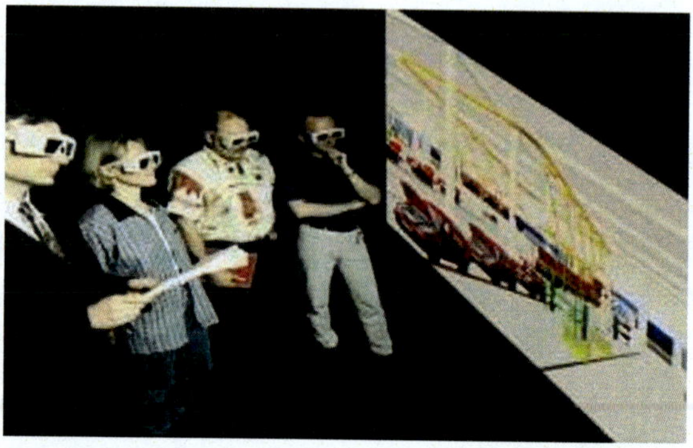

Fig. 1. Visualization in VR of manufacturing system's simulation

the PLM tool which is the effective communication between each department with its counterpart with large screens, and these steps are described as follows (Fig. 2):

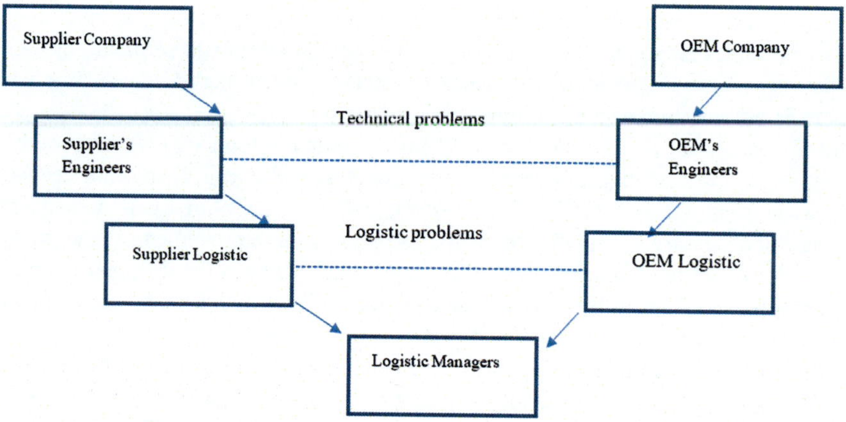

Fig. 2. Production and logistic problems correlation proposition

4 Results

After implementing these solutions between all co-workers in the same status, in OEM company and the other collaborating company, they can validate or not the part modification, innovation or logistics, in the real moment and showing that, to all other members of all companies in PLM software system. So that can minimize waiting times.

Digitalizing production and logistical tools, to improve and have excellent quality, by using cameras in lines and all places for detecting any inappropriate actions, wasted time and excessive movement can be avoided.

5 Discussion

Organizations work on many programs to improve their working manner and to digitalize all their tools, to have an excellent production and for improving the quality of their production.

Using of MFIB (Model Factory box) as mini model of the company's production lines where employees can implement their improvement propositions and where they can detect wastes and trying to eliminate them from the process, and this solution was mentioned in another paper of other author [7].

And if we move to the other paper of Mitja VARL "Application of lean methods into the customized product Development Process of large Power Transformer" [8], we can say that Lean thinking in organizations is important, because it demonstrates how lean manufacturing can aid to push the process development to efficiency basing on Process and tools, People and Knowledge. Lean thinking contains also supply chain activity composed by: Customers, suppliers, manufacturers, transporters, warehousemen. And has as purpose arriving to a total profitability and an efficient management during all supply chain stage, and this was descripted in Magdalena Ramirez-Pena in the paper Sustainability in the aerospace, Navale, and Automotive Supply chain 4.0: Descriptive Review [9].

The digitization of the processes between OEMs (Original equipment manufacturers) and their suppliers allows all the actors of the product value chain, and specifically the logisticians and production engineers, to be involved in the different problems related to the production and transportation of parts between suppliers and their customers. Our document is therefore in the direction of optimizing the procurement time of manufactured products that are going to be tested in the assembly plants.

At this level, we are led to give digital solutions that can be implemented in the form of effective and optima l tools. Communication of these problems of product technology, or transportation of parts. For example by putting large screens in the factories of OEMs and suppliers with the possibility of visual and remote collaboration between the different engineers and collaborators to face any production problem at the supplier and at the OEMs as well, same thing for the transportation of the parts we can visualize with the techniques of digitalization RFID + IOT + large screens in the different factories, the blockages and logistic problems in real time and thus optimize the different unnecessary exchanges between the different interlocutors. Also, we propose the same collaboration with the different suppliers and this part must be studied separately because there will be problems of confidentiality between two competitors.

And so we are in front of a solution that we propose in order to improve the exchanges during the management of the life cycle of the automotive products, while minimizing the externa l interventions that require a lot of displacements and therefore we minimize the impact on the environment, same thing for the screens that we must use, They must be powered by sola r panels for example in Morocco, so we are in front of a sustainable development in our factories while keeping the availability in time and quantity on all products and all problems, globally alert in real time on all problems in the value chain.

The collaboration is very important to have a management of production problems and logistics that is very effective. The literature on the effectiveness of the integration of suppliers and OEMs in all internal problems and communication between suppliers and stakeholders and OEMs, among the objectives of this paper is to facilitate the understanding and flow of topics between the two parties, using the screens of large format plants. See Fig. 3.

Fig. 3. Factory information screen

Because of this proposition there will be an efficiency in dealing with sup- plier's production problems, and in every logistic point between suppliers and clients. Digitalized exchange of productions and logistic information will certainly improve the collaboration.

The hyper-connected world of products, human actors, and operating environments, enabled by new technologies [10], supply chains have traditionally contributed to value cha ins through efficient cost control and fast and efficient delivery of a desirable assortment of goods. Several case studies suggest that digital supply cha ins are a way to extend this function by creating consumer circumstances that surprise and intrigue and suggest a need for quality service [11].

Using internet of things "IOT" in suppliers' factories by using sensors and actuators' is in order to limit delays in OEM manufacturers [12].

Pushing employees to react quickly in front of all manufacturing problems, by establishing the following graph which pushes to a great efficiency of work and collaboration between subcontractors and keeps the traceability of each activity by emails, as you see on Fig. 4.

Managing the life cycle of automotive products, and having them faster, with a higher quality in all markets, is our objective in this paper by digitalizing all processes from the Beginning of Life "BOL" to the End of life 'EOL" of all parts in the automotive sector, It is by no means an exaggeration to say that PLM is a digital process that is crucial to the proper functioning of manufacturers; without it, there would be no information infrastructure linking each department and each stage of a product, making the whole thing function in complete isolation [13].

And as result of our comparison between all cited solutions, we found that using camera s is cheapest and will give also the opportunity to employees to visualize their

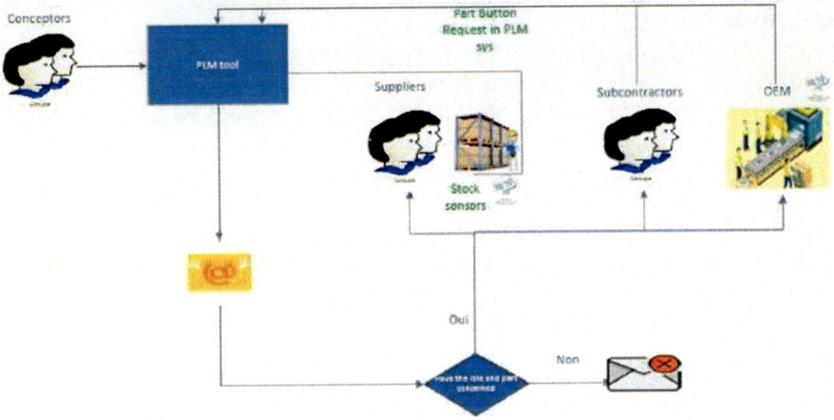

Fig. 4. Development process digitalization graph [12].

movements and to better guide the parts logistics into factories, discuss and give solutions by themselves. So we observed after a small discussions with some companies, that they reacted positively with this solution that we proposed, first for the cost and for rising employees innovation by brainstorming.

6 Conclusion

This work provides the possibility of better transportation and logistical side of the parts optimization, especially among collaborators of PLM first phase. The optimization and integration of all actors in the value chain of an automotive company as well as the application of the principle of sustainable development, become necessary conditions for efficient product management.

The article highlights how the concepts of collaboration and integration (by optimizing and positively influencing the environment especially in logistics side) can help the work efficiency among suppliers and customers.

Contributing to solve problems through collaboration with other suppliers. The principles presented for supporting decision making in PLM context to provide a basis for effective PLM decision support with significant potential in the future, so using cameras and rising employee's innovation and collaboration is the best.

As a perspective, we are finalizing a survey on the digitalization of automotive parts problems detection, So we can focus also on the "MOL" of PLM cycle which is not treated by most researchers. Who focus on "BOL" beginning of life, and "EOL" end of life phases?

Acknowledgements. This work is partially supported by Mme. Fatima Bennouna and Mr Zakaria Chalh, I also gratefully acknowledge the helpful comments and suggestions of the reviewers, which have improved the presentation.

References

1. Saaksvuori, A. et al.: Twenty Years of PLM–the Good (2008)
2. Grieves: Industry 4.0 and Industrial IoT in Manufacturing: A Sneak Peek (2011)
3. Ageron, et al.: Defining Product Lifecycle Management (2020)
4. Attaran: Defining Product Lifecycle Management (2020)
5. Xin, Y. et al.: Dealing with knowledge management practices in different product lifecycle phases within product-service systems. Procedia CIRP **83**, 111–117 (2019)
6. Sodhro, A.H.: Convergence of IoT and product lifecycle management in medical healthcare (2018)
7. Koshechkin, K.: Implementation of Digital Technologies in Pharmaceutical Products Lifecycle (2020)
8. Leal, A.G., Santiago, A.: Integrated environment for testing IoT and RFID technologies applied on intelligent transportation system in Brazilian scenarios (2014)
9. Abidi, M.-A.: Contribution of Virtual Reality for Lines Production's Simulation in a Lean Manufacturing Environment (2014)
10. https://www.planzone.fr/blog/definition-plm-mise-en-place Consulté le 30 Nov 2021
11. Burroughs, B.: Digital logistics: Enchantment in distribution channels (2020)
12. Tilioua, N.: Using internet of things to increase efficient collaboration in PLM. In: ICDTA'21 Conference ENSA Fes, Morocco (2021)
13. Lenoble, N.: Optimisation de la préparation de commandes dans les entrepôts de distribution (2021)

Analysis and Comparison Between Artificial Neural Network Models in Image Recognition

Sara Belattar[1](✉), Otman Abdoun[2], and El khatir Haimoudi[1]

[1] Computer Science Department, Advanced Science and Technologies Laboratory, Abdelmalek Essaâdi University, Polydisciplinary Faculty, Larache, Morocco
Sarah.belattar12@gmail.com

[2] Computer Science Department, Faculty of Science, Abdelmalek Essaâdi University, Tetouan, Morocco

Abstract. Nowadays, face (or image) recognition represents a challenging problem and an important area in various applications due to issues that can be encountered in different domains such as monitoring operations, security, criminal identification, identifying a person, and so on. For that reason, many methods are proposed to develop the intelligent face recognition system, and each method provides a solution to a specific situation. Firstly, the Kohonen Neural Network (KNN) is proposed to create an intelligent system for face recognition to solve problems related to the identification of human faces. This type is a well-known neural network model and is based on an unsupervised learning mode; the goal of using it is to classify and cluster data, either numerical or image. Secondly, the convolutional neural network (CNN) is suggested in this study thanks to its capabilities in image classification and recognition tasks. The main objective of choosing these artificial neural network models is to analyze and compare their performances in learning, image classification, and recognition tasks. The experimental tests demonstrate that the KNN and CNN have some limitations and advantages in specific problems, and each of them gave good results in different cases. Therefore, this study provides an appropriate model for a particular issue.

Keywords: Intelligent face recognition system (IFRS) · Kohonen Neural Network (KNN) · Convolutional Neural Network (CNN) · Image classification · Computer vision

1 Introduction

In Computer Vision (CV), face or image recognition is still one of the challenging tasks and plays an important part in many different areas. The principle of face recognition systems (FRS) is to identify and detect a human face based on available face images in the database and give a decision about the tested face images (i.e., known or unknown). Sometimes we can encounter some images that are not available on the database. Therein lies the strength of the face-recognition systems. The main objective of the FRS is to extract the facial features of people to facilitate the verification and identification of any human face. Many previous studies suggested different methods

L. Azrar et al. (Eds.): CIP 2022, LNME, pp. 125–133, 2023.
https://doi.org/10.1007/978-3-031-23615-0_13

and techniques for developing and enhancing face recognition systems, including SRC and 2D multi-color fusion, were used to propose face recognition versus occlusion. In this study, the authors suggested fused color information to increase the accuracy of the face recognition rate without occlusions detection. The experimental tests were based on four different databases to show the accuracy of the face recognition rate through their proposed method, and their results showed an interesting performance (Alrjebi et al., 2017). Three-dimensional face recognition using local shape descriptors and geometric was employed to counteract distortions produced by facial expressions. The authors used four steps to solve this problem (model 3D face, feature extraction, geometrics information on the 3D surface, and finding feature vectors on any scale). For comparing the results, the authors used two datasets (i.e., GavabDB and Bosphorus). The rate of identification was achieved at 98.9% on the GavabDB dataset (Abbad et al., 2018). The self-organizing map (SOM) or Kohonen Network was applied to develop an intelligent face recognition system. The purpose of using it is to train the database and make a simulation of face recognition systems (RS). For feature extraction, they used the techniques of discrete wavelet transform (DWT), Sobel edge detection (SED), and discrete cosine transform (DCT). These techniques are the most popular in face recognition. In the tests, they were based on 30 face image datasets in order to show the performance of their proposed approach. The validation results showed that the rate of recognition was achieved at 100%, and for feature extraction, the DWT technique gave better performance compared to DCT and SED (Soni et al., 2013). The convolution neural network was suggested to create an embedded face recognition system. In this study, the authors used a combination of two algorithms; the first is called an "optimized Multitask Cascaded Convolutional Network (OMTCNN)", and the other is a "lightweight face recognition based on CNN (LCNN)". This combination was proposed for crop preprocessing and simulation transformation of the face image, and their second objective is to minimize the complexity of the computational. Their results showed that the OMTCNN gave good accuracy in training (95.78%) and a better performance in identifying faces than classical MTCNN. Furthermore, the LCNN reduces the complexity and speeds up the computational time (Lv et al., 2021). From the literature review cited above, we found many techniques used in face recognition systems. In our study, we limit ourselves just to ANN models such as Kohonen Neural Network (KNN) and Convolutional Neural Network (CNN) due to their abilities in classification problems because the face recognition system is based on classification performance. The purpose of choosing these models is to make a comparison between them at the level of their learning, classification, and recognition rates. Our paper contributes to giving the important criteria that can help us choose the suitable model of ANN (i.e., KNN, or CNN) according to an issue that we have to resolve.

The remainder of this research paper is organized as follows: the proposed models of ANN (Kohonen Neural Network and Convolutional Neural Network) are described in the Materials and methods section. The experimental tests and results are illustrated in the Results section. Finally, we conclude our work with a conclusion and an outlook on the future works.

2 Materials and Methods

2.1 Image Dataset

An example of the image dataset was used in our work to train and test the performance of the KNN and CNN in face recognition systems, particularly in image classification capability. This image dataset example is available on the web, and it contains various human face images with various facial expressions, as shown in Fig. 1. This dataset contains 72 images that are divided into two sets (i.e., training (75%) and testing (25%) phases).

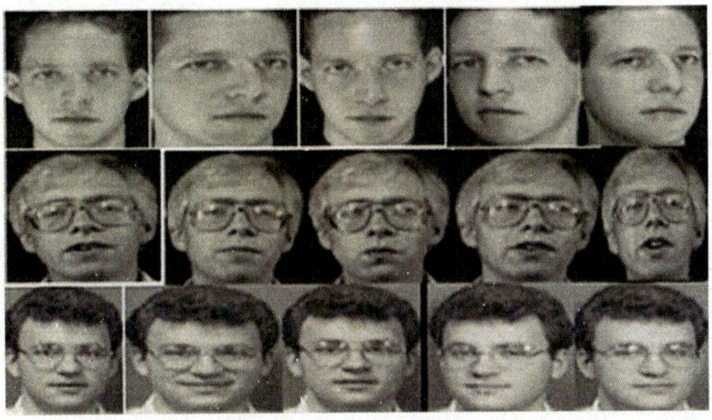

Fig. 1. The example of dataset used.

2.2 Kohonen Neural Network (KNN)

The Kohonen Neural Network (KNN) (Kohonen 1990) is among the best-known and most widely used neural network models. This model is also called the Self-organizing map or Kohonen feature map (Kohonen 1991), and it was developed in 1980 by Teuvo Kohonen. The architecture of the Kohonen map neural network is a two-dimensional lattice of connected nodes (or neurons) (Vracko 2005), and it is based on the unsupervised learning mode. The robustness of this model is characterized by its ability to solve the issues of complex systems (i.e., data exploration, classification, and clustering) (Kohonen 2013), which leads to developing intelligent systems in various application domains with high classification accuracy and lower cost of computation. Generally, the KNN consists of two layers: input and output layers. The purpose of the first one is to distribute the input data (X) to the learning procedure of the network, and the second one represents the outcomes of the network (i.e., winning nodes). The neurons on the map are represented in the m-dimensional lattice, and each input layer is linked with neurons by synaptic weight in order to produce the outputs. The synaptic weight is initialized randomly between 0 and 1.

The learning algorithm of the Kohonen Neural Network is summarized in the following steps:

Step 1: The KNN begins with the normalization step for training the network; the purpose of this technique is to normalize the input data (*IP* (or *X*)). This technique has a good effect on the acceleration of the learning time, and it is given by the formula below (Belattar et al., 2020):

$$IP_i = IP_i \cdot / \sqrt{\sum_{j=0}^{n-1} IP_j^2} \tag{1}$$

where: *IP* is the input signal.

Step 2: The second step is to determine the winning nodes (or neurons) that represent the outputs of the Kohonen Neural Network. The winning node *wn(x)* is the smallest distance between the input vector (*X*) and synaptic weight (*w*). In this study, the Euclidean distance (ED) criterion is used to calculate the distance (see Eq. (2)) (Mohan et al., 2018, Belattar et al., 2021). The outputs of the winning node are 1, and 0 for all other nodes.

$$wn\,(x) = \arg\min \left\| x(n) - w_j \right\|, j = 1,\ 2 \ldots n \tag{2}$$

Step 3: In the last step, the synaptic weight vectors of the winning node and all its neighbors are updated and changed (see Eq. 3) (Belattar et al., 2021).

$$w_{ij}(t+1) = w_{ij}(t) + \delta(t) \cdot h(t)\left(x_i - w_{ij}(t)\right) \tag{3}$$

where $w_{ij}(\mathrm{t}+1)$ and $w_{ij}(t)$ are the synaptic weight vectors at t and t + 1 iteration, $\delta(\mathrm{t})$ is the learning rate, and $h(\mathrm{t})$ is the neighbourhood function.

2.3 Convolutional Neural Network (CNN)

The Convolutional Neural Network (CNN) is among the Artificial Neural Network models applied in computer vision, and it exceeds the performance of humans in classification on the dataset of ImageNet in 2015 (He et al., 2015). Additionally, CNN has a powerful ability in image recognition tasks due to its capabilities in feature extraction, segmentation, and detection, which allows employing it in several applications with success, in particular for the computer vision field. Overall, the Convolutional Neural Network comprises various combinations of layers (e.g., the convolution layer, pooling layer, and fully connected layer) (Ortac and Ozcan, 2021). The convolution layer represents the important building block used in CNN. The pooling layer is used for reducing the dimension, and this operation is often included after the convolution layer. The CNN ends with certain fully connected layers. We discovered numerous architectures in the Convolutional Neural Network (e.g., MobileNet, VGG, ResNet, DenseNet, GoogLeNet, and so on). In our study, we used the DenseNet 169 architecture due to its competency in image classification, visual object recognition, and reducing the parameter numbers.

3 Results and Discussion

3.1 Performance of the Kohonen Neural Network

In this section, we will describe our facial image recognition system using the Kohonen Neural Network. This system is developed using the Java language based on Netbeans

IDE 8.2. The first step is to train the network from the image dataset described in Fig. 1 for our system to be capable of having the ability to identify a human face based on this dataset. Before beginning the training phase, we must insert the values of the training parameters such as map dimension, iteration number, learning error, and learning rate (see Table 1). The input image was resized to (170 × 170 × 3).

Table 1. The values of the training parameters used

Training parameters	Value
Map dimension	8
Max iteration	50
Learning rate	0.5
Learning error	0.01

Therefore, our facial image recognition system was trained with success. In the following step, we will test some new face images to show the ability of the proposed network (KNN) in face recognition (see Fig. 2).

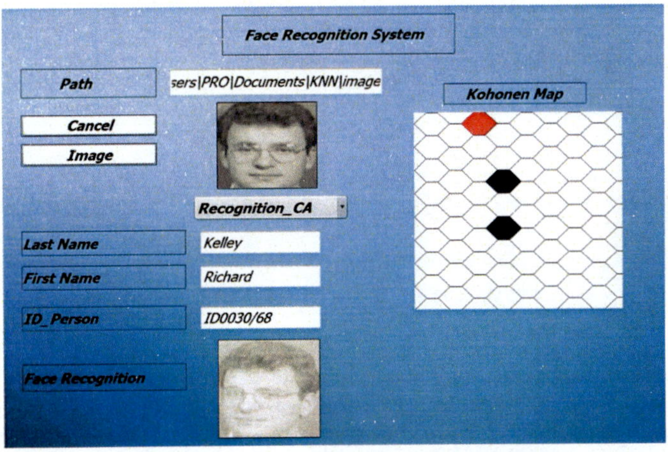

Fig. 2. Face recognition system using the KNN

From Fig. 2, it can be seen that the Kohonen Neural Network was able to recognize the human face inserted with high accuracy (100%) due to giving a similar image of the same human face based on the training database. Firstly, to test the performance of the KNN in face recognition, we must insert the path of the image to be recognized. After that, we click on the button Recognition_CA in order to show the answer from our system. In Fig. 2, the red winning node on the Kohonen map represents the image output of the system, and when we click it, we obtain the answer about face recognition. Furthermore, our system gave information about the human face detected (e.g., first name, last name, and number ID of the person). If the inserted image is not recognized, our system attempts to provide a close image of a human face from the training database, utilizing KNN's generalization capability. In addition, the performance of the KNN in the test set reached nearly 100%.

3.2 Performance of the Convolutional Neural Network

To test the performance of the CNN in the face recognition system, we use the same image dataset (as KNN). In the learning procedure of CNN, we resized the input image to (224 × 224 × 3). Then we used some parameters like batch size, epoch numbers, optimizer method, learning rate, and dropout (see Table 2). The experimental test shows that the CNN (i.e., DenseNet 169) is well-performing in this example of the dataset, despite this dataset containing few images. The CNN performance during 10 epochs is described in Fig. 4. In our case, we use the Softmax function because we have a multiclassification problem (i.e., 9 classes). Thus, the architecture of DenseNet 169 is presented in Fig. 3.

```
Layer (type)                 Output Shape              Param #
=================================================================
densenet169 (Functional)     (None, 7, 7, 1664)        12642880

flatten_2 (Flatten)          (None, 81536)             0

dropout_2 (Dropout)          (None, 81536)             0

dense_2 (Dense)              (None, 9)                 733833

=================================================================
Total params: 13,376,713
Trainable params: 733,833
Non-trainable params: 12,642,880
```

Fig. 3. The architecture of DenseNet 169.

The DenseNet 169 architecture achieved a rate of 94.44% in the test set. Hence, in the prediction phase, the DenseNet 169 model correctly predicted the face image with an accuracy of 100% (see Fig. 5) due to recognizing it in the correct class (Figs. 4 and 5).

From the experimental tests of CNN and KNN in face recognition, we can conclude some advantages and disadvantages of using them in specific situations as follows:

- CNN is a well-known architecture in computer vision due to its power to train large image datasets. Although CNN gives good results in our case (i.e., few images), it can produce poor performance in a small number of images in other problems as well. For that reason, the KNN can work in a few more images than CNN.
- For a large number of images, CNN has been considered the appropriate tool because it is pre-trained on ImageNet, which can manipulate large image datasets in training and testing phases. The authors of this research (Bhupendra et al., 2022) trained a

Table 2. The parameters used for training the CNN (DenseNet 169)

Parameters	Value
Batch size	18
Epochs	10
Optimizer	Adam
Learning rate	0.1
Dropout	0.2
Classifier	Softmax
Classes	9

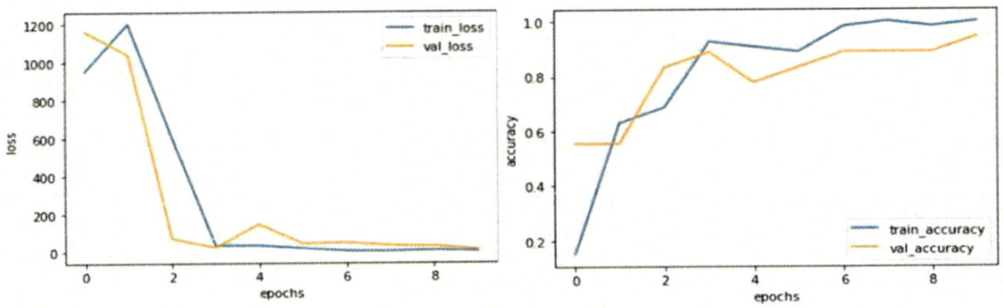

Fig. 4. The performance of the CNN (DenseNet 169) during 10 epochs

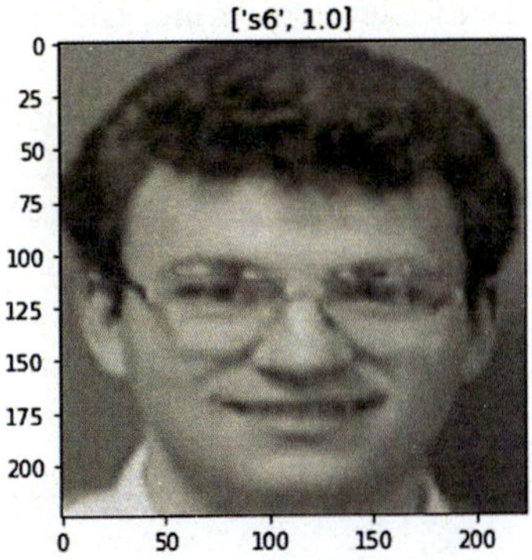

Fig. 5. The recognition face image by CNN (DenseNet 169).

variety of deep convolutional neural networks on a dataset including 8048 images, and the results showed an impressive level of accuracy.

- In the training phase, the KNN is incapable of training a large number of images because it can only train the image dataset one by one due to its two-dimensional lattice. Whereas CNN has the robust ability to train a large number of images in input owing to its three-dimensional lattice (or multidimensional) (Ortac and Ozcan,

2021). Therefore, the classification task using a high number of images becomes more complicated in the two-dimensional lattice.
- If we have a blurry image, the KNN has difficulties identifying it due to its poor performance in feature extraction compared to CNN, which is strong in feature extraction due to its deep learning.

The face recognition system depends on the classification performance. For that reason, we suggested the convolutional neural network and the Kohonen neural network due to their capabilities in resolving classification problems. As a result, we can conclude from the comparison performance between CNN and KNN that each of them has the power to solve specific problems.

The Kohonen neural network is ideal for data segmentation, visualization, clustering, and more. Hence, the KNN is most often used for solving problems related to numerical data (Belattar et al., 2021), (Belattar et al., 2022).

In computer vision, it is better to use the convolutional neural network than the KNN because the former is ideal for working with large images and has a good capacity for feature extraction from images.

4 Conclusion

In this paper, we propose two artificial neural network models (i.e., Kohonen Neural Network (KNN) and Convolutional Neural Network (CNN)) to build an intelligent face recognition system. The main objective of choosing these models is their abilities in learning and classification issues. The KNN model is the most popular in solving problems regarding data (i.e., classification, clustering, and more), and CNN is the most well-known Neural Network model in image classification. Each one of these models has advantages in certain situations. Although both the KNN and CNN performed well in a few images (our case study), it is preferable to use CNN for image classification because it is more dedicated to computer vision than the KNN. In addition, CNN has a higher ability to extract features from images compared to KNN. The KNN is less powerful in computer vision, particularly in large image datasets, because it cannot train on large datasets and has difficulty with feature extraction and image analysis. Therefore, the KNN is ideal for data segmentation, clustering, regrouping, and numerical data analysis. Meanwhile, CNN is perfect for image analysis and classification.

References

Alrjebi, M., Pathirage, N., Liu, W., Li, L.: Face recognition against occlusions via colour fusion using 2D-MCF model and SRC. Patt. Recog. Lett. **95**, 1339–1351 (2017). doi.org/https://doi.org/10.1016/j.patrec.2017.05.013

Abbad, A., Abbad, K., Tairi, H.: 3D face recognition: multi-scale strategy based on geometric and local descriptors. Comput. Electr. Eng. (2018). doi.org/https://doi.org/10.11445/3177148.3180087

Soni, N., Kumar, M., Mathur, G.: Face recognition using SOM neural network with different facial feature extraction techniques. Int. J. Comput. Appl. **76**(3), 7–11 (2013). https://doi.org/10.5120/13225-0647

Lv, X., Su, M., Wang, Z.: Application of face recognition method under deep learning algorithm in embedded systems. Microprocess. Microsyst. 104034 (2021). https://doi.org/10.1016/j.micpro.2021.104034

Kohonen, T.: The self-organizing map. Proc. IEEE **78**(9), 1464–1480 (1990). https://doi.org/10.1109/5.58325

Kohonen, T.: Self-organizing maps: optimization approaches. Artif. Neural Netw. 981–990 (1991)

Vracko, M.: Kohonen artificial neural network and counter propagation neural network in molecular structure-Toxicity studies. Curr. Comput. Aid.-Drug Design **1**, 73–78 (2005). https://doi.org/10.2174/1573409052952224

Kohonen, T.: Essentials of the self-organizing map. Neural Netw. **37**, 52–65 (2013). https://doi.org/10.1016/j.neunet.2012.09.018

Belattar, S., Abdoun, O., Haimoudi, E.K.: Intelligent Management of Using Natural Resources in Agriculture. In: Ezziyyani, M. (ed.) AI2SD 2019. AISC, vol. 1103, pp. 224–238. Springer, Cham (2020). https://doi.org/10.1007/978-3-030-36664-3_26

Mohan, P., Kumari Patil, K.: Weather and crop prediction using modified self organizing map for Mysore region, REVA University, Computer Engineering, India, December 23–2017. Int. J. Eng. Intell. Syst. **11**(2) (2018)

Belattar, S., Abdoun, O., El khatir, H.: Toward an Intelligent Hybrid System Based on Data Analysis and Preprocessing Method. In: Ben Ahmed, M., Mellouli, S., Braganca, L., Anouar Abdelhakim, B., Bernadetta, K.A. (eds.) Emerging Trends in ICT for Sustainable Development. ASTI, pp. 45–52. Springer, Cham (2021). https://doi.org/10.1007/978-3-030-53440-0_6

He, K., Zhang, X., Ren, S., Sun, J.: Delving deep into rectifiers: surpassing human-level performance on imageNet classification (2015). https://doi.org/10.1109/ICCV.2015.123

Ortac, G., Ozcan, G.: Comparative study of hyperspectral image classification by multidimensional convolutional Neural Network approaches to improve accuracy. Expert Syst. Appl. **182**, 115280 (2021)

Bhupendra, Moses, K., Miglani, A., & Kumar Kankar, P (2022) Deep CNN-based damage classification of milled rice grains using a high-magnification image dataset. Comput. Electr. Agricult. **195**(3), 106811. doi.org/https://doi.org/10.1016/j.compag.2022.106811

Belattar, S., Abdoun, O., Haimoudi, E.K.: New learning approach for unsupervised neural networks model with application to agriculture field. Int. J. Adv. Comput. Sci. Appl. **11**(5), 360–369 (2020). https://doi.org/10.14569/IJACSA.2020.0110548

Belattar, S., Abdoun, O., Haimoudi, E.K.: A novel strategy for improving the counter propagation artificial neural networks in classification tasks. J. Commun. Softw. Syst. **18**(1), 17–27 (2022). https://doi.org/10.24138/jcomss-2021-0121

Joint Design of an Eco-product and Its Supply Chain: A Literature Review

Mohamed Barane(✉), Latifa Ouzizi, and Mohammed Douimi

Department of Industrial Engineering, Ecole Nationale Supérieure d'Arts Et Métiers(ENSAM), Moulay Ismail University of Meknès, Meknes, Morocco
m.barane@edu.umi.ac.ma, l.ouzizi@ensam-umi.ac.ma, m.douimi@umi.ac.ma

Abstract. The problem of the simultaneous design of the product and its supply chain is becoming one of the famous subject trends discussed in the literature. We study the simultaneous design within an optimization approach to find the optimal design with its corresponding supply chain configuration under certain number of constraints. That's why, the objective of this paper is to present a state-of-art review of the problem of the simultaneous design of the product and its supply chain under cost and environmental impact constraints. This paper handles the problem within available literature, focus on the recent researches, highlights and proposes the potential future researches. It gives also, an insight about the topic and how can be solved with different methods existed in the literature. At the end of this paper, different approaches within the literature are illustrated for the resolution of the bi-objective problem with directions to adopt for our future contribution.

Keywords: Product redesign · Supply chain design · Supply chain optimization · Green Product · Mixed integer linear programming

1 Introduction

Currently Companies are evolving within a competitive environment which allow them to have a small margin of benefits.

In order to respond literally and efficiently to customer need the company has to optimize its processes to respond quickly and overcome products returns issues. Besides, the new trend of reducing the impact of greenhouse gases and its weight on the international market in one hand and within the country policies on the other hand, researches and engineers work hard to find effective cure to this issue.

Back to the history, The United Nation Framework conventional on Climate Change (UNFCCC) is an agreement that was established in Rio De Janeiro in 1992. This agreement was activated on Mar. 21, 1994 and approved by 197 countries (Kuh 2018). UNFCCC comes to bring a solution for the growing concern to the climate change and with aims to stabilize greenhouse gas concentration in the atmosphere a level to prevent dangerous anthropogenic interference with the climate system (MichaelStephenson 2018).

L. Azrar et al. (Eds.): CIP 2022, LNME, pp. 134–141, 2023.
https://doi.org/10.1007/978-3-031-23615-0_14

That's why this work has come to bring to light the major interest of the environmental criteria within the company's policy to introduce a competitive product in term of cost and responding to the new trend of preserving the environment. It provides also which are methods used to find between proposed product alternative the optimal one responding perfectly to the trade-off between the cost and the environment impact constraint.

This paper is organized as follow: the next section presents a literature review of the eco-product. The third part presents an insight about the simultaneous design of an eco-product and its supply chain. The fourth part illustrates and presents the bi-objective problem under mathematical model approach. The fifth part is dedicated for a problem description with its mathematical model. A numerical example is also presented. And to conclude, a conclusion and a list of references are mentioned at the end.

2 Eco-product Design

The topic of the product design becomes an interesting problem. It is in this phase that all of the client requirements and criteria must be defined and taken into account before the development phase.

One of criteria which is becoming more conclusive, is the environmental impact criterion. As a matter of fact, the integration of the environmental criteria from the beginning (design phase) illustrate interesting results in papers dealing with eco-design principles. (Schöggl et al., 2016) affirm that the environmental impact of a product in its future life cycle phases is determined from its design phase. The same study confirms that approximatively the half of damaging environmental impact can be eliminated from the same phase.

Besides, another powerful method, uses an approach devised on tree major steps (Rought-cut LCA, AHP (Analytical Hierarchy Process), ER (Evidential Reasonning)) to evaluate design's alternative environmental impact. The key point of this method, that it deals with the uncertain information throughout the product life cycle (NG C.Y. 2016).

Although, the fact that considering the design phase process separately to the configuration of the supply chain is a horrible mistake to make. New product design implies, certainly, the configuration of its corresponding supply chain. We mention that 85% of logistics costs are determined directly by the design concepts (Laurentie et al. 2006). Consequently, the integration of the product architecture decision with the corresponding supply chain configuration must be done from the earlier stages of the product development (Nepal et al., 2012).

3 Simultaneous Design of an Eco-product and Its Supply Chain

The concept of the simultaneous design is based on the integration of the product constraints when designing the supply chain network and the integration of the supply chain attributes when designing the product. At first, the simultaneous design of the product and its supply chain often propose to optimize the cost of the supply chain of each product selected within the design phase. But recently, some works have suggested to integrate environmental constraints in the design of the product and the supply chain. (Kremer 2015), shows the importance of the environmental constraints integration to obtain the product sustainability development. (Lavigne 2012) highlights the influence of product and supply chain design decisions towards the environment, especially at the level of production and the distribution. (Labbi et al., 2015) Bring to light the idea of the recycling product and the reverse logistic.

Moreover, considers that dealing with the simultaneous design of the product and its supply chain is a tough task, and to deal with this difficult, a new optimization approach by levels, Genetic Algorithm and Tagushi experimental design method has been proposed to obtain, therefore, feasible solutions in a short computational time.

Indeed, there exist several approaches for the treatment of the simultaneous design of the product and its supply chain under two constraints (Cost, Environmental impact), and the majority of literature uses the famous:

Linear programming models: It concerns the minimization of the supply chain's cost and the environmental impact of the product. The resolution of this type of problem is usually based on the search of a set of Pareto efficiency.

Ballouki et al. (2021), present a new model that integrate the environmental constraints in product selection based on multi-agent system. First, the selection of friendly product is to be done by the TOPSIS Method in order to determine the environmental performance of each product alternative, then the product having the minimum cost of the supply chain is selected.

The next section will be dedicated to the bi-objective problem, its properties and some of its applications relating to the problem of the simultaneous design of the product and the supply chain.

4 The Simultaneous Design and the Bi-objective Problem: A Mathematical Models

Usually, the problem of the simultaneous design of product and the supply chain is presented in the form of mathematical model. This modeling leads to the resolution of an optimization problem. This problem is defined by a couple $(P(t), f)$ such that $P(t)$ is the set of candidate design at the "t" moment, f is the objective function that optimize a criteria for the design of the supply chain relevant to a product alternative. The function defined by:

$$f : P(t) \to IR^{+} \tag{1}$$

It's usually a function to minimize (e.g. Cost) or maximize (e.g. Benefit). Thus the problem is mathematically defined by:

$$\min / \max_{\substack{p \in P(t) \\ t \in [0, T]}} f(p, t) \tag{2}$$

Although, dealing with a problem with two objectives is a bi-objective problem and to solve it we are invited to find the set of Pareto-optimal solution.

One of methods to solve a problem of this type is an $\varepsilon - constraint$ method for a bi-objective optimization problems with integer bi-objective values. The principle of this method is to switch from a bi-objective problem to a mono-objective one by maintaining one objective function and transforming the others into constraints (Jean-François et al. 2007).

For the bi-objective case, the $P_i(\varepsilon_j)$ problems are:

$$\begin{cases} \min f_i(x) \\ subject\ to : x \in X \\ f_i(x) \leq \varepsilon_j\ ,\ i,j = 1,2\ \ i \neq j \end{cases} \tag{3}$$

His method is used in order to transform the bi-objective problem to a mono-objective one (A priori method).

Another a priori method consist on the transformation of the bi-objective functions (f_1, f_2) to a linear sum using the weighted sum method (a function with a unique objective) (Collette et al., 2003). Generally, experts determine by experience the values of weights to be affected foe each objective function. A method like AHP and ANP can be used in order to determine those weights. For the case of bi-objective function, two weights to be determined such that:

$$\begin{cases} w1\ w2 \geq 0 \\ w1 + w2 = 1 \end{cases} \tag{4}$$

Giving two objective functions, the final objective to be solved is giving as:

$$\text{Min}F = w1 * f1 + w2 * f2 \tag{5}$$

The posteriori method (Exploratory method), It's a specific method, of which the designer interfere at the end of the optimization process to choose between a huge number of solutions the suitable one according to a list of preferences. We can cite some famous algorithms such that: Genetic Algorithm (GA) and Ant Colony Algorithm (ACA).

(Labbi et al. 2020) uses the NSGA-II Algorithm to determine between several suppliers the suitable one to find the best trade-off between the cost of the supply chain and the quality compliance of the products.

(Jack et al., 2012) present an optimization approach in two level. At first level, possible combinations of assembly structures and sequences will be searched according to the evolutionary GA algorithm. At the second level, for each combination of assembly structure and sequence under evaluation in GA, the corresponding supply chain structure is optimized by dynamic programming. At the end, to determine the environmental impact of each assembly design, a life cycle assessment (LCA) is used.

The next chapter is dedicated to our proposed mathematic model with some literature treating the resolution of bi-objective with same interest (bi-objective problem resolution).

5 Problem Description

In this paper we will present a first approach of our problem dealing with the simultaneous design of a product and its supply chain taken in consideration the optimization of the cost and the environmental impact.

For the first objective function-function of Cost- is taken from (Ballouki et al. 2021)

$$f_{\cos t}\text{: min} \sum_{f\in F}\sum_{c\in Ca}\left(\sum_{r\in Ra} Density_r * N_{r,c,a} * V_{c,a} * CA_r + CP_{c,f} + CT * D_f)\right) * Q_{c,f}$$
$$+CF_f * Z_{c,f}\sum_{i=1}^{n}(X_i - \bar{X})^2 + \sum_{a} CA_a * Q_a + \sum_{a}\sum_{s} Q_{a,s} * \left(CS_{a,s} + CT_s * D_s\right)$$
$$+Z_{a,s} * CF_{a,s} \tag{6}$$

The second Function-Environmental impact function- is presented below, such that:

$$f_{CO_2}\text{:min} \sum_{f\in F}\sum_{c\in Ca} P_c * D_f * Q_{c,f} * \alpha + \sum_{f\in F}\sum_{c\in Ca} EP_{c,f} * Q_{c,f}$$
$$+\sum_{f\in F}\sum_{c\in Ca}\sum_{r\in Ra} Ext_{r,f} * N_{r,c,a} * Q_{c,f} + \sum_{a\in A} EA_a * Q_a + \sum_{a\in A}\sum_{s\in S} EA_{a,s} * Q_{a,s} \tag{7}$$

For those functions, a set of: Assumptions, datas and constraints are provided:

a. Problem assumptions:

- We prioritize the optimization of the procurement, production, subcontracting and the delivery,
- The demand for components is known,
- Components transportation unit cost is in the range of minimum and maximum capacity of each supplier,
- All suppliers selected for a component must deliver it with the same quality index to have homogeneous quantities,
- Only one mode of transportation is considered.

b. Problem sets and data

We propose the following definition:

- AF: The Assembly center,
- F: The set of component's supplier,
- Ca:The set of alternative components indexed by 'a',
- Ra: The set of raw material of the alternative 'a'.

The model include the following data:

- $\boldsymbol{Density_r}$: Density of Raw material « r» $\left(\mathrm{Kg/m^3}\right)$,
- $\boldsymbol{P_c}$: Weight of the component (Kg),
- $\boldsymbol{V_{c,a}}$: Volume of the component « c» of the alternative « a»,

- $\boldsymbol{N_{r,c,a}}$: Binary variable for the selected raw material r, such that $N_{r,c,a} = 0$ if the raw material doesn't exist in the component "c", $N_{r,c,a} = 1$ otherwise,
- $\boldsymbol{Ext_{rf}}$: Emission of the extraction of a raw materiel r,
- $\boldsymbol{EP_{cf}}$: Emission of the production of a component "c" derive from the supplier "f".
- $\boldsymbol{EA_a}$: Emission of the alternative assembly "a",
- $\boldsymbol{EA_a}$: Emission of the alternative assembly "a" from the supplier "S",
- $\boldsymbol{CA_r}$: Purchasing cost of the Raw material « r»,
- $\boldsymbol{CP_{cf}}$: Production cost of component « c» by the supplier « f»,
- $\boldsymbol{CT}$: Transport cost,
- $\boldsymbol{D_f}$: Distance between supplier « f» and the assembly center « AF»,
- $\boldsymbol{CA_a}$: Assembly cost of the alternative « a»,
- $\boldsymbol{CS_{a,s}}$: Sub-constractor cost of the alternative « a» from the sub-constractor **« s»,**
- $\boldsymbol{CT_s}$: Transport cost from the sub -constractor « s»,
- $\boldsymbol{D_s}$: Distance from « s»,
- $\boldsymbol{Dem_c}$: Demand of the component « c»,
- $\boldsymbol{capmax_{cf}}$: Maximum capacity of « f» for the component « c»,
- $\boldsymbol{D_a}$: Demand of alternative « a»,
- $\boldsymbol{capmax_{a,s}}$: Maximum capacity of « s» for the alternative « a»,
- $\boldsymbol{capmax_a}$: Maximum capacity of the AF for « a».
- $\boldsymbol{\alpha}$: is a fixed value such that $\alpha = 1{,}5438$ $KgCO_2$/Ton-Km; for a truck of at less than 3, 5t (*Source: DoItProTM; A Decision Support System to Estimate the Carbon Emission and Cost of Product Designs*)

c. Decision variables:

- $\boldsymbol{Q_{cf}}$: Quantity of « c» delivered by « f»,
- $\boldsymbol{Q_{a,s}}$: Quantity of « a» delivered by « s»,
- $\boldsymbol{Q_a}$: Quantity of assembled alternative « a»,
- $\boldsymbol{Z_{cf}}$: Binary variable for the selected supplier « f» to insure the assembly alternative « a»,
- $\boldsymbol{Z_{a,s}}$: Binary variable for the selected sub-constractor « s» to insure the assembly alternative « a».

d. Constraints:

- Subassembly demand satisfaction:

$$\sum_s Q_{a,s} + Q_a \geq D_a \tag{10}$$

- Subcontractor's capacity:

$$Q_{a,s} \leq capmax_{a,s} * Z_{a,s} \tag{11}$$

- Sub-assembly capacity:

$$Q_a \leq capmax_a \tag{12}$$

- Non-negativity:

$$Q_{a,s} \geq 0 \tag{13}$$

$$Q_a \geq 0 \tag{14}$$

$$Q_{cf} \geq 0 \tag{15}$$

e. Numerical examples:

The resolution of the two objective models will be the subject of the next publication. As mentioned above, this paper will be the opportunity to bring to light literature dealt with the resolution of the bi-objective problems under the cost and the environmental impact objective.

The first example (Jack et al., 2012) is related to a real world computer chair case study where the approach used is described in the previous section.

Results founded in the paper give two optimal BOM alternative designs (BOM1; BOM2), but the BOM1 is the one which optimize simultaneously the cost and the carbon emission of the supply chain.

The Second example is taken from (Labbi et al. 2020). The mainly objective as mentioned above is to find the product that respond efficiently to the cost and environmental impact objective. The study used a NSGA-II algorithm with fixed parameter:

- Population size: Variable,
- Selection: Tournament,
- Crossover probability: 0.8,
- Mutation probability: 0.01,
- Number of generation: [100; 500].

From the graphical presentation of the GA with population size 200 for the 3 alternatives (Readers are invited to consult the paper to see the graphical presentation), we conclude that the Product1 (Alternative product) is the optimal one because it represents the best fitness function.

6 Conclusion

Due to customer's rising awareness of environmental sustainability and government's strict environmental policies, environmental dimension becomes one of the major topic and design criterion. This paper captures the problem of the simultaneous design of the product and its supply chain and tackles literature dealing with the same subject. It studies the dilemma that exists between the cost of the supply chain and the environmental impact of the product throughout its life-cycle and how this factor may affect the company strategy.

This paper tackled also manners and methods for the resolution of the bi-objective problems, and how those methods are powerful in term of high computational time and result accuracy.

Our next work will focus on the improvement of the two mathematical models of each objective, and we will try to validate our model basing on validated models (Case study already confirmed).

References

Ballouki, I., Douimi, M., Ouzizi, L.: A distributed and collaborative model for product design selection considering its supply chain costs and environmental footprint. Int. J. Syst. Sci. (2021). http://mc.manuscriptcentral.com/tsys E-mail: TSYS-peerreview@journals.tandf.co.uk

Bérubé, J. –F., Gendreau, M., Potvin, J. –Y., CIRRELT.: An exact method fo Bi-objectif combinatorial Optimization Problems-Application to the traveling Salesman Problem with Profits. Bibliothèque nationale du Québec. (2007)

Collette, Y., Siarry, P.: Multiobjective optimization, principles and case studies. 1st ed. Corr 2nd printing, X, 293 p. 153 illus., Hardcover, (2003). ISBN: 978–3–540–40182–7

Jack C. P. Su, Chu, C. H., Wang, Y. T.: A decision support system to estimate the carbon emission and cost of product designs. Int. J. Prec. Eng. Manuf. **13**(7), 1037-1045 (2012)

Kremer, G. E., Haapala, K., Murat, A., Chinnam, R. B., Kim, K. -Y.: Directions for instilling economic and environmental sustainability across product supply chains. J. Clean. Prod. **112**(3), 2066–2078 (2015)

Kuh, K. F.:The Law of Climate Change Mitigation: An Overview. Hofstra University, Hempstead (2018)

Labbi, O., Ouzizi, L., Douimi, M.: Simultaneous design of a product and its supply chain integrating reverse logistic operations: an optimization model. Internationale: Conception et. (2015) https://hal.archives-ouvertes.fr/hal-01260795/

Labbi, O., Ahmadi, A., Ouzizi, L., Douimi, M.: A non-dominant sorting genetic algorithm for optimization of a product design and selection of its suppliers. J. Adv. Manuf. Syst. **19**(1), 167–188 (2020). https://doi.org/10.1142/S0219686720500092

Laurentie, J., Berthelemy, F., Grégoire, L., Terrier, C.: Logistic processes and methods, Supply chain Management, AFNOR (2006)

Lavigne, B., Agard, B., Penz, B.: Mutual impacts of product standardization and supply chain design. Int. J. Prod. Econ., Adv. Optim. Design Supply Chains **135**, 50–60 (2012)

MichaelStephenson.: Chapter 4 - The Coming Industrial Revolution? Fossil Fuels and Developing Countries. Energy and climate change (2018)

Nepal, B., Monplaisir, L., Famuyiw, O.: Matching product architecture with supply chain design. Eur. J. Oper. Res. **216**(2), 312–325 (2012)

NG, C. Y.: An evidential reasoning-based AHP approach for the selection of environmentally-friendly designs. Environ. Impact Assess. Rev. **61**, 1–7 (2016)

Schöggl, J. -P., Baumgartner, R. J., Hofer, D.: Improving sustainability performance in early phases of product design: A checklist for sustainable product development tested in the automotive industry. J. Clean. Prod. **140**(3), 1602–1617 (2016)

Computing and Data-Driven Digital Industry

Performance Evaluation of Diagnostic and Classification Systems Using Deep Learning on Apache Spark

Chaymae Taib[1](✉), Otman Abdoun[2], and Elkhatir Haimoudi[1]

[1] Computer Science Department, Polydisciplinary Faculty, Abdelmalek Essaadi University, Larache, Morocco
taib.chaymae@etu.uae.ac.ma

[2] Computer Science Department, Faculty of Science, Abdelmalek Essaadi University, Tetouan, Morocco

Abstract. Respiratory disorders endanger people's lives. A novel Sars species was discovered for the first time in 2019 in Wuhan when it was isolated from three people suffering from respiratory sickness, indicating that early discovery can help save lives. The purpose of this study is to implement and evaluate six hybrid architectures for multiclass chest X-ray images classified into four classes (Covid19, Normal, Lung Opacity, and Viral Pneumonia) using an open dataset. For feature extraction, we employed three CNN approaches (VGG19, ResNet50, and InceptionV3), along with two machine learning algorithms as classifiers (Logistic Regression and Random Forest). According to the outcomes of this study, the design using Logistic Regression as a classifier and VGG19 as a feature extractor was the best-performing architecture, with an accuracy value of 94.12%. Aside from the great value-added by deploying Apache Spark, which allows for speedier training and testing time.

Keywords: Covid-19 · Classification · Machine learning · Hybrid · Features Extraction · CNN techniques · Apache Spark

1 Introduction

In 2019, a newly severe acute respiratory syndrome coronavirus was discovered in Wuhan, China. Then spread to the rest of the world, on 11 March 2020 this virus was Declared as a pandemic by the WHO (Baloch et al., 2020), it caused a global shutdown leads to a big impact on economic statue, normal social live, education. Every day, the number of infected people of coronavirus rise up. Therefore, this provides to us more data to work with and to train our model to make the best prediction. (*Coronavirus Disease(COVID-19)* (2022)) As of 5:29 pm CEST on July 8, 2021, the World Health Organization reported 185,291,530 confirmed cases and 4,010,834 deaths worldwide. The dashboard shows that America has first place in confirming cases then Europe, South-East-Asia, Eastern Mediterranean, (Karlinsky and Kobak, 2021) Africa. This new species of SARS has three kinds of symptoms: the first has a symptoms like the flue's

L. Azrar et al. (Eds.): CIP 2022, LNME, pp. 145–154, 2023.
https://doi.org/10.1007/978-3-031-23615-0_15

such as fever, body ache, dry cough, loss of smell, fatigue, chills, headache, sore throat, and for the second kind has a more aggressive symptoms that indicate often as pneumonia and the last kind is that the person infected doesn't shows any symptoms (Chaudhary et al., 2021; Gelfand et al., 2021). The world has changed in this period and depends on the evolution of technology to continue the usual life use the platforms like zoom and Microsoft teams and google meet in education fields and companies for their meeting (Hiselius and Arnfalk, 2021). The evolution of the technology helped the world to adjust with the virus (Chaymae et al., 2022). Deep learning is one of the techniques that is helped to detect the virus in accurate and less time than the blood test that is required 5H to 24H (Song et al., 2020). Chest X-ray plays a big role in the detection of covid 19, as shown in several studies that they use chest x-ray images or CT images by applying CNN techniques for detecting the new SARS (Pinter et al., 2020; Song et al., 2020; Wang et al., 2020). Big data has a significance role in controlling and tracking the spread of the virus covid19 and provides a lot of information that can help the doctors to predict the virus moreover, it can help to limit the range of infected people (Haleem et al., 2020).

This study develops and evaluates six HB architectures with Apache spark using three deep learning techniques as feature extractors (VGG19, ResNet50 and InceptionV3) and two machine learning algorithms as classifiers (Logistic regression and Random Forest) for multiclass Covid19 classification. The six architectures were trained on 5845 X-ray images divided on four classes (Covid19, Normal, Lung opacity and Viral Pneumonia) and their performance was assessed using accuracy and confusion matrices.

The following research questions (RQs) are addressed in this study:

- (RQ1): In Covid19 multiclassification, what is the total performance of hybrid architectures built on Apache Spark?
- (RQ2): Exists a DL feature extraction method that, when used in a hybrid architecture, clearly outperforms the rest?
- (RQ3): Is there any improvement using Apache spark?

The following are the primary contributions of this empirical study:

1. Creating six hybrid architectures with logistic regression and random forest classifiers. Three CNN feature extraction algorithms were used in covid 19 multiclassification classification: VGG19, Inception V3, and ResNet50.
2. Using accuracy performance metrics, evaluate the six hybrid designs.
3. A comparison of six hybrid designs' performance.

The following is how this paper is organized: The Sect. 2 presents an overview of the three CNN approaches employed in this research, as well as the Apache Spark environment. The Sect. 3 discusses related work. Section 4 presents the suggested architecture. Section 5 presents the results and discussion. Conclusions and future works are presented in Sect. 6.

2 Background

2.1 Apache Spark

Apache Spark is an open-source framework that combines a core engine for parallel programming and innovative programming for in-memory computation. It has emerged as the de facto standard for big data analytics after Hadoop's MapReduce, and it is faster and easier to use. It also appears to come with rich APIs in many languages for performing complex distributed operational processes on distributed data.

2.2 ML and DL

- Random Forest

Random forest is a robust machine learning algorithm capable of fitting complex that can be used in classification and regression, it combined the predictions of a vast number of little decision trees referred to as estimators. This algorithm is powerful and accurate also is very useful when trying to determine feature or variable importance.

- Logistic Regression

Logistic Regression is the simplest and most commonly used machine learning algorithm used for classification problems for the two types Binary and Multilinear classes, The main purpose of this process is to discover a relationship between characteristics and outcome likelihood. This function, known as the sigmoid function.

- VGG

VGG was founded by a Visual Geometry Group at Oxford, and hence the moniker VGG is a successor to The AlexNet. That was improved on the traditional Convolutional neural networks in 2012.

- InceptionV3

The idea of Inception v3 was proposed by (Szegedy et al., 2015) in 2015 and it's basically focused on burning less computational power to be more efficient in terms of parameters generated number by the network and the economic cost of the resources and memory.

- ResNet50

ResNet is an acronym for Residual Networks, a conventional CNN architecture. This design won the ImageNet Large Scale Visual Recognition Challenge (ILSVRC) in 2015. (He et al., 2015). ResNet 50 is a deep neural network with 50 layers that can categorize photos from the ImageNet dataset into 1000 different classes. The default input size for this network is 224 × 224, and it is influenced by the VGG designs (VGG16 and VGG19) with filter sizes of 3 × 3. ResNet used a method known as residual mapping to solve this problem.

3 Related Works

In this section, we discuss prior research that used convolution neural networks and machine learning methods to identify covid 19:

Joshi et al., (2021) employed chest X-ray images to diagnose COVID 19 by developing a robust deep learning-based system for an accurate diagnosis to classify Three classes (COVID, normal, and pneumonia) that obtained 97.1% AUC and 99.81% for binary class. Wang et al., (2020) used CNN on 13800 CXR images to detect covid-19 cases; the goal of their studies was to improve the evaluation efficacy of the proposed covid-NET, and they performed both analyses to achieve a better understanding; the accuracy was 92.6%, and the PPV was 96.4%, indicating that the model was performing well despite the limitation of a small data set. Wang et al., (2021) construct a three-process architecture. 1st, input preprocessing 2-feature extraction of ROI images and training 3-classification with fully connected network and prediction of various classifiers So they basically extract ROI images using CV model and extract features using a modified Inception network Their model achieved an accuracy of 89.5%, and AUC of 93.5%, and sensitivity of 88.5%. Shi et al. (2020) had three contributions: first, they offered the infection size-aware approach, secondly, they proposed a location-specific feature extraction procedure that was assessed on a large-scale dataset; accuracy is 87.9%, sensitivity is 90.7%, and specificity is 83.3%. Cheng J et al. created an AI system for quick identification of covid-19 from x-ray pictures, and their model has a high-performance accuracy of 97.91% and a specificity of 95.47% (Jin et al., 2020). Ghoshal and Tucker, (2020) pre-trained the ResNet50V2 model and collected data to limit model fault range, hence the association between model uncertainty and prediction accuracy was strong in this experiment. Among the other models inceptionV3 and ResNetV2, pre-trained ResNet50 has the greatest accuracy with 98% (Narin et al., 2021). Ibrahim et al. demonstrate in their study a DL-based approach for detecting Using a mix of data from various sources, three classifications were identified: COVID-19, pneumonia, and lung cancer. Images from CT scans and X-rays. VGG19-CNN, Gated Recurrent Unit (GRU) + ResNet152V2, and ResNet152V2 + Bidirectional GRU are the four architectures employed in this study (Bi-GRU) The maximum accuracy was reached by the VGG19 + CNN model, which was 98.05% (Ibrahim et al., 2021). Based on the previous reading, there have been several studies on image x-ray and CT-images to make accurate and faster detection to covid-19, however, the data was a big challenge for these studies, especially because traditional tools couldn't handle a large amount of data. To address this issue, we proposed an architecture based on DL Databricks pipeline on Apache spark cluster.

4 Materials and Methods

4.1 Dataset Description

This study's empirical evaluation was done by using a publicly available dataset titled "COVID-19 RADIOGRAPHY DATABASE." A database was built by a research team in consultation with medical specialists of chest X-ray images for COVID-19 positive cases, as well as Nor-mal, Lung Opacity, and Viral Pneumonia im-ages. This dataset contains images of 3616 COVID-19 positive patients, 6012 Lung Opacity patients, 10,192 Normal patients, and 1345 Viral Pneumonia (Chowdhury et al., 2020; Rahman et al., 2021). Figure 1 shows various examples from our dataset. Due to environmental and resource constraints, such as RAM and processing units, we gathered samples from each class: Covid19, Normal and Lung Opacity has 1500 images, and Viral Pneumonia has 1345 images.

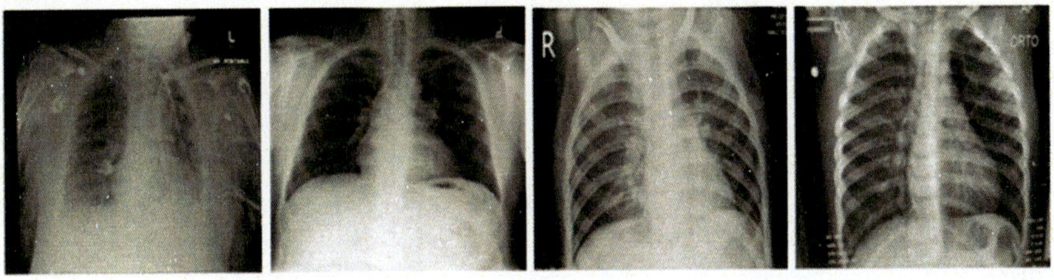

Fig. 1. Samples of the three respiratory disease.

4.2 Proposed Approach

The vast amount of data was defying the usual DL platform; therefore, we devised an architecture that allows us to employ a big amount of data. The steps of this approach are represented in Fig. 2.

- Storing Stage

The first stage of our approach is the storing stage which is based on Deep Learning Pipelines, which help us to load the data to RDD. We can load millions of images into a Spark Data Frame and decode them automatically in a distributed fashion, allowing manipulation of scale with the help of utility functions included in Deep Learning Pipelines. After loading our dataset containing 5845 CTR images divided into four classes (pneumonia, lung opacity, normal, covid19) we split the data into 80%, 20% ratios for the training set and test set randomly.

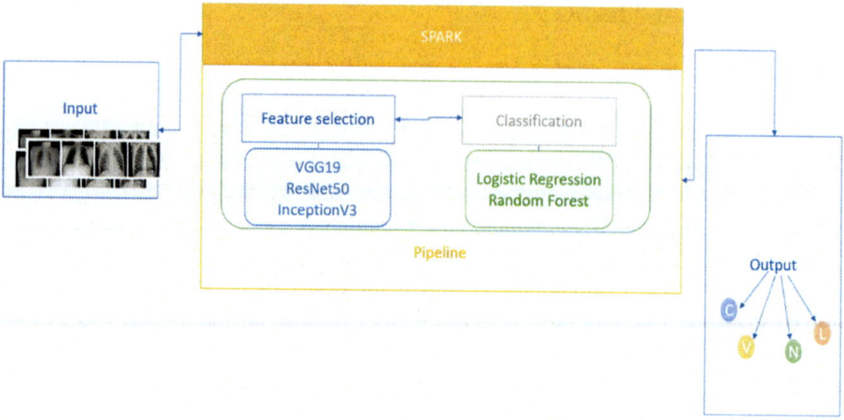

Fig. 2. Architecture of DL pipeline on apache spark

- Processing Stage

Spark-deep-learning library built by Databricks to provide two capabilities for spark, notably efficiency of using APIs to enable deep learning in very few lines of code and the provision of a robust distributed engine to scale up a big quantity of data and deep learning.

This second stage consists of two steps: feature extraction and classification.

5 Results and Discussion

In this section shows and discusses the findings of our exploratory study over CRD, in which the performance of various designs was tested using accuracy, F1-score, precision, and recall performance standards First and foremost, the performances were assessed in terms of the accuracy of each classifier's hybrid models (RQ1): In Covid19 multiclassification, what is the total performance of hybrid architectures built on Apache Spark (RQ2): Exists a DL feature extraction method that, when used in a hybrid architecture, clearly outperforms the rest (RQ3): Is there any improvement using Apache spark.

The empirical evaluations of the hybrid architectures were carried out in Colab Notebook using Python 3.7.4 with the two DL frameworks Keras and Tensorflow as the backends, as well as SparkMlib, and were running on a GBU processing unit with 8 cores, 25 GB of RAM, and a Linux-based OS provided by Google.

- (RQ1): In Covid19 multiclassification, what is the total performance of hybrid architectures built on Apache Spark?

This section provides and assesses the performance of each classifier's hybrid architecture on the CXR dataset.

$$Accuracy = (TP + TN)/(TN + TP + FP + FN) \tag{1}$$

$$Precision = TP/(TP + FP) \tag{2}$$

$$Recall = TP/(TP + FN) \tag{3}$$

$$F1 = 2 \times \frac{Recall \times precision}{Recall + precision} \tag{4}$$

where:

TP: the number of cases that were predicted positively and turned out to be true positive.

TN: the number of cases that were projected negative but turned out to be true negative.

FP: the number of cases that were expected to be positive but turned out to be false.

FN: the number of cases that were projected as negative but turned out to be correct.

Table 1. The four metrics of the six hybrid architectures.

CNN techniques	ML	Accuracy (%)	F1-score (%)	Precision (%)	Recall (%)
VGG19	RF	85.70	85.84	86.07	85.70
	LR	94.12	94.12	94.13	94.12
InceptionV3	RF	84.79	84.90	85.16	84.79
	LR	84.60	84.69	85.66	84.60
ResNet50	RF	85.20	85.33	85.71	85.20
	LR	93	93.01	93.02	93

- When using logistic regression as the classifier in hybrid architectures:
  - The highest accuracy was 94.12% when VGG19 was used as the feature extractor.
  - The less accuracy value obtained when using InceptionV3 as feature extraction: 84.60%.
- When using Random Forest as the classifier in hybrid architectures:
  - The highest level of accuracy was attained when VGG19 was used for feature extraction: 85%.
  - The less accuracy value obtained when using InceptionV3 as feature extraction: 84.79%.
- (RQ2): Exists a DL feature extraction method that, when used in a hybrid architecture, clearly outperforms the rest?

In order to identify FE technique that enhance classification performance, this section examines the effects of the three DL algorithms on the functionality of the two classifiers.

The findings demonstrate that the DL method VGG19 performed well during the feature extraction phase, which has an impact on how well hybrid architectures work. The best accuracy value, as indicated in Table 1, was obtained while adopting VGG19 at a ratio of 94.12% to 84.60%. Irrespective of the classifier.

- (RQ3): Is there any improvement using Apache spark?

Apache Spark is a distributed processing engine that plays a major role in this research by shortening the training and testing stages which are especially important when dealing with enormous amounts of data. As shown in Table 2 the hybrid architecture using res-net50 as classifier combining with the logistic regression takes only 2380.875/s for training 4676 images and takes 963.59/s over 1.169 images in testing stages.

Table 2. The time of training and testing per seconds.

CNN techniques	ML	Time of training/s	Time of testing/s
VGG19	RF	10780.092	1020.80
	LR	7831.886	1827.466
InceptionV3	RF	11228.212	1501.5
	LR	2617.011	1005.422
ResNet50	RF	9202.284	1300.04
	LR	2380.875	963.59

6 Conclusion and Future Work

To summarize, we present six hybrid architectures that combine three CNN-techniques (InceptionV3, Resnet50, VGG19) with two classifiers, LR and RF applied on an Apache Spark cluster, allowing us to use 5845 X-ray images previously unavailable to traditional systems. The hybrid architecture that performed the best was LRV19 (logistic regression with VGG19) achieving accuracy of 94.12%, and VGG19 performed well in feature extraction independent of the machine learning technique. Finally, the key value-added by using Apache Spark to provide rapid training and enhance prediction. In the future, we will test this architecture in other datasets with a considerable amount of data, as well as employ different machine learning methods and compare it to the current study.

Acknowledgments. For their assistance, the authors would like to thank "the Moroccan Ministry of Higher Education and Scientific Research", CNRST, Advanced Science and Technology Lab, and FPL. Abdelmalek Essâadi University in Morocco financed this research.

References

Baloch, S., Baloch, M.A., Zheng, T., Pei, X.: The coronavirus disease 2019 (COVID-19) pandemic. Tohoku J. Exp. Med. **250**(4), 271–278 (2020). https://doi.org/10.1620/TJEM.250.271

Chaudhary, R., et al.: Thromboinflammatory biomarkers in COVID-19: systematic review and meta-analysis of 17,052 patients. Mayo Clinic Proc. Inno. Qual. Outcomes **5**(2), 388–402 (2021). https://doi.org/10.1016/J.MAYOCPIQO.2021.01.009

Chaymae, T., Elkhatir, H., Otman, A. A.: Comparative study of recent advances in machine learning and deep learning in vehicular ad-hoc networks. In Bendaoud, M., Wolfgang, B., Chikh, K. (eds) The International Conference on Electrical Systems and Automation Proceedings. ICESA 2021. Singapore: Springer (2022). https://doi.org/10.1007/978-981-19-0039-6_1

Chowdhury, M.E.H., et al.: Can AI help in screening viral and COVID-19 pneumonia? IEEE Access **8**, 132665–132676 (2020). https://doi.org/10.1109/ACCESS.2020.3010287

Coronavirus disease (COVID-19). (n.d.). Retrieved 24 March 2022, from https://www.who.int/emergencies/diseases/novel-coronavirus-2019

Gelfand, M.J., et al.: The relationship between cultural tightness–looseness and COVID-19 cases and deaths: a global analysis. Lancet Planetary Health **5**(3), e135–e144 (2021). https://doi.org/10.1016/S2542-5196(20)30301-6

Ghoshal, B., Tucker, A. Estimating Uncertainty and Interpretability in Deep Learning for Coronavirus (COVID-19) Detection. (2020). https://doi.org/10.48550/arxiv.2003.10769

Haleem, A., Javaid, M., Khan, I.H., Vaishya, R.: Significant applications of big data in COVID-19 Pandemic. Indian J. Orthopaedics **54**(4), 526–528 (2020). https://doi.org/10.1007/s43465-020-00129-z

He, K., Zhang, X., Ren, S., Sun, J.: Deep Residual Learning for Image Recognition. In: Proceedings of the IEEE Computer Society Conference on Computer Vision and Pattern Recognition, 2016-December, pp. 770–778 (2015). https://doi.org/10.48550/arxiv.1512.03385

Hiselius, L.W., Arnfalk, P.: When the impossible becomes possible: COVID-19's impact on work and travel patterns in Swedish public agencies. Eur. Transp. Res. Rev. **13**(1), 1 (2021). https://doi.org/10.1186/S12544-021-00471-9/TABLES/11

Ibrahim, D.M., Elshennawy, N.M., Sarhan, A.M.: Deep-chest: Multi-classification deep learning model for diagnosing COVID-19, pneumonia, and lung cancer chest diseases. Comput. Biol. Med. **132**, 104348 (2021). https://doi.org/10.1016/J.COMPBIOMED.2021.104348

Jin, C., Chen, W., Cao, Y., Xu, Z., Tan, Z., Zhang, X., Deng, L., Zheng, C., Zhou, J., Shi, H., Feng, J. (2020). Development and evaluation of an AI system for COVID-19 diagnosis. MedRxiv. https://doi.org/10.1101/2020.03.20.20039834

Joshi, R.C., et al.: A deep learning-based COVID-19 automatic diagnostic framework using chest X-ray images. Biocybernetics Biomed. Eng. **41**(1), 239–254 (2021). https://doi.org/10.1016/J.BBE.2021.01.002

Karlinsky, A., Kobak, D.: Tracking excess mortality across countries during the covid-19 pandemic with the world mortality dataset. ELife **10**(2021). https://doi.org/10.7554/ELIFE.69336

Narin, A., Kaya, C., Pamuk, Z.: Automatic detection of coronavirus disease (COVID-19) using X-ray images and deep convolutional neural networks. Patt. Anal. Appl. **24**(3), 1207–1220 (2021). https://doi.org/10.1007/s10044-021-00984-y

Pinter, G., Felde, I., Mosavi, A., Ghamisi, P., Gloaguen, R.: COVID-19 pandemic prediction for hungary; a hybrid machine learning approach. Mathematics **8**(6), 890 (2020). https://doi.org/10.3390/MATH8060890

Rahman, T., et al.: Exploring the effect of image enhancement techniques on COVID-19 detection using chest X-ray images. Comput. Biol. Med. **132**, 104319 (2021). https://doi.org/10.1016/J.COMPBIOMED.2021.104319

Shi, F., Xia, L., Shan, F., Wu, D., Wei, Y., Yuan, H., Jiang, H., Gao, Y., Sui, H., & Shen, D.: Large-scale screening of COVID-19 from community acquired pneumonia using infection size-aware classification. Phys. Med. Biol. **66**(6) (2020). https://doi.org/10.1088/1361-6560/abe838

Song, J., et al.: End-to-end automatic differentiation of the coronavirus disease 2019 (COVID-19) from viral pneumonia based on chest CT. Eur. J. Nucl. Med. Mol. Imaging **47**(11), 2516–2524 (2020). https://doi.org/10.1007/s00259-020-04929-1

Szegedy, C., Vanhoucke, V., Ioffe, S., Shlens, J., Wojna, Z.: Rethinking the inception architecture for computer vision. In: Proceedings of the IEEE Computer Society Conference on Computer Vision and Pattern Recognition, 2016, pp. 2818–2826 (2015). https://doi.org/10.48550/arxiv.1512.00567

Wang, L., Lin, Z.Q., Wong, A.: COVID-Net: a tailored deep convolutional neural network design for detection of COVID-19 cases from chest X-ray images. Sci. Reports **10**(1), 1–12 (2020). https://doi.org/10.1038/s41598-020-76550-z

Wang, S., et al.: A deep learning algorithm using CT images to screen for Corona virus disease (COVID-19). Eur. Radiol. **31**(8), 6096–6104 (2021). https://doi.org/10.1007/s00330-021-07715-1

Sensitivity and Uncertainty Analysis of SLM Process Using Artificial Neural Network

Shubham Chaudhry(✉) and Azzeddine Soulaimani

Department of Mechanical Engineering, École de Technologie Supérieure, Montreal, Canada
shubham.chaudhry.1@ens.etsmtl.ca, azzeddine.soulaimani@etsmtl.ca

Abstract. Selective Laser Melting (SLM) is a one-of-the-metal-based Additive Manufacturing (AM) process where the successive powder layers are fused with a high-intensity laser beam. The continuous melting and cooling of powder layers lead to a rise in localized tension and compression. This causes a significant increase in residual stress in the building part and becomes a primary factor for cracks, alteration in shape, etc. Many factors contribute to the SLM output, including machine and material parameters. In such a scenario, the analysis of the causal effects becomes critical. Examining such effects through numerical simulation and experimental methods can be time-consuming and expensive. This research presents an architecture to study the uncertainty and sensitivity in SLM process. The proposed work combines numerical simulations and machine learning methods to study the SLM process more quickly and efficiently. A total of five input parameters: Young modulus, Poisson ratio, hatch spacing, laser speed, and layer thickness were considered, and we predicted the normal strains in the building part using a workbench additive software. We also constructed a surrogate model using a feed-forward neural network with this dataset and generated more outputs on large numbers of input samples. The data generated with this Surrogate model helped in analyzing the uncertainty and sensitivity in the SLM.

Keywords: Additive manufacturing · Selective Laser Melting · Machine learning

L. Azrar et al. (Eds.): CIP 2022, LNME, pp. 155–164, 2023.
https://doi.org/10.1007/978-3-031-23615-0_16

1 Introduction

Selective Laser Melting (SLM) is one of the Additive Manufacturing (AM) technique where a material powder is converted into a 3D object by successively fusing the powder layers together. The process requires a high-intensity laser power to the powder, making SLM available to various materials, including metals. This latest approach is capable of producing lighter, more complex, and robust geometries with less material wastage. Because of such qualities, the SLM has grown and created its way to the bigger manufacturing industries like medical implants, automobiles, aerospace etc. [1]. However, many challenges with the SLM end product, like premature process termination, part distortion, dimensional accuracy, and mechanical properties, require further analysis and research [1, 2].

To improve the SLM quality, we must find the source of uncertainty at each SLM process level and how it affects the product quality. To do that, there are two ways, 1) Sensitivity Analysis (SA), and 2) Uncertainty Analysis (UA), that we have considered to solve these issues and to build the trust in our simulation model [3]. Uncertainty analysis will help us investigate the model output uncertainty because of the uncertainties in the model input. The sensitivity analysis will give us direction for the relation of uncertainties in the model outputs and inputs. Many of the reported SA and UA frameworks in additive process are experiment-driven. Which leads to wastage of material and increases the process costly [4]. An alternative way to perform the SA and UA is using numerical models. But a numerical model can be computationally expensive and require hours to days to complete [5]. Besides, these models are deterministic, and some methods don't even address the uncertainties in the input parameters [6]. In this paper, we propose an alternative approach using statistical and numerical methods to understand and quantify the uncertainty in the SLM process.

This work is divided in four parts. The first part briefly introduces the SLM process and UA and SA models used in additive manufacturing. Section 2 gives mathematical models for SA and surrogate modeling. Section 3 discusses the results. Finally, the concluded remarks are given in Sect. 4.

2 Model and Formalism

This paper aims to provide a framework for using numerical and surrogate models to study UA and SA in the SLM process. The SLM is modeled using workbench additive software 2019R3, a residual stress model that uses a thermo-mechanical couple method. The thermal analysis gives the transient temperature field in this model, which then transfers to the mechanical analysis model as input [4]. The workbench additive model was validated using a publicly available benchmark test case (Fig. 1) published by NIIST in 2018 [7]. The benchmark case comprises a total of four bridge structures on a plate. All bridges were constructed with the IN625 material and using a powder bed fusion technique. During the manufacturing process, the structures had enough space between them, so the manufacturing process didn't affect each other's properties. As one bridges' building process doesn't affect another's properties, we simulated only one bridge on the base plate for the current research work. This consideration has significantly lowered the simulation time. The final dimensions for the substrate and the bridge were 81 * 12.7 * 11 and 75 * 5 * 12.5, respectively. The research's first objective was to analyze the strain and stress for the built part in the validation case and then perform the UA and SA study using a surrogate model.

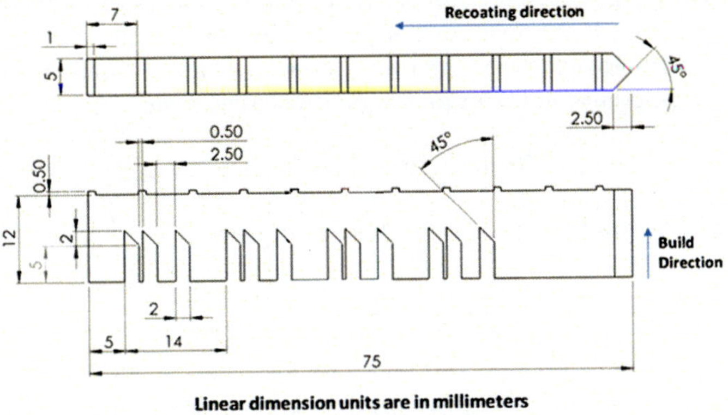

Fig. 1. The representation of the AMB2018-01 bridge structure. Geometry from plan (top) and elevation (bottom) [8].

2.1 Mathematical modeling

An artificial neural network (ANN) is a surrogate method that is used to construct a surface between input and their corresponding outputs. The ANN is built similar to a brain with multiple neurons connected like a web [9]. ANN has hundreds or thousands of artificial neurons, known as processing units. These processing units are interconnected by nodes, and each unit is formed by input and output units [10]. In this case, we have used a feed-forward neural network [11]. Our model considered a six hidden layers network with 800 neurons in the first and the end layer and 500 each in all the four middle layers. This ANN model is then used to conduct sensitivity and uncertainty analysis in the SLM system, as explained below.

A sensitivity analysis considers the input parameters' uncertainty and their influence on the output to find the significance of every input variable. A variance-based sensitivity model is used in this research work, which decomposes the output variance among the model inputs [3]. For more explanation, let us consider the following equation:

$$Y = f(X_1, \ldots, X_p) \tag{1}$$

Here, Y is the process output and $X_1, \ldots, X_p$, represents the independent input parameters, and we can represent it by a known probability distribution. For comparison, the above equation in our case, the strain values are represented by Y, whereas laser speed, layer thickness, etc. are the input parameters that are represented by $X_1, \ldots, X_p$. The effect of each input variables X_n on the variance of outputs Y is calculated with the assumption that the actual value of X_n is x_n^*. So, the change in the variance of Y is given by the following conditional variance equation.

$$O_{X_{-n}}(Y|X_n = x_n^*) \tag{2}$$

In Eq. (2), $O_{X_{-n}}$ gives the conditional variance for the input variable space (p−1), considering every input variable other then x_n^*. Also, the actual value for X_n is not known, so, we will take an average on every possible values of X_n that is presented by $E_{X_n}(O_{X_{-n}}(Y|X_n))$. Using the law of total variance, we can write:

$$O(Y) = O_{X_n}\left(E_{X_{-n}}(Y|X_n)\right) + E_{X_n}\left(O_{X_{-n}}(Y|X_n)\right) \tag{3}$$

After the normalization, the Eq. (3) can be represented as:

$$1 = \frac{O_{X_n}\left(E_{X_{-n}}(Y|X_n)\right)}{O(Y)} + \frac{E_{X_n}\left(O_{X_{-n}}(Y|X_n)\right)}{O(Y)} \tag{4}$$

The term $S_n = \frac{O_{X_n}\left(E_{X_{-n}}(Y|X_n)\right)}{O(Y)}$ gives the first-order sensitivity index for the variable X_n, and the rest of the terms in Eq. (4) will help calculate the total order index. Furthur Eq. (1) is written in the dimension with increasing order as:

$$f(X_1, .., X_p) = f_0 + \sum_{i=1}^{p} f_i(X_i) + \sum_{1 \le i < j \le p} f_{ij}\left(X_i X_j\right) + \ldots + f_{1,\ldots,p}\left(X_1, ., X_p\right) \tag{5}$$

For the above equation, if all the input parameters are mutually independent, then there exists a decomposition where every term is mutually orthogonal. With this assumption, the output variance ($O(Y)$) is given as follow:

$$O(Y) = \sum_{i=1}^{p} O_i + \sum_{1 \le i < j \le p} O_{ij} + \ldots + O_{1,\ldots,p} \tag{6}$$

where $O_i, O_{i,j}, \ldots, O_{1,\ldots,p}$ represent the variance of $f_i, f_{ij}, \ldots, f_{1,\ldots,p}$, respectively. The first-order sensitivity index is the first p term of the above decomposition, $S_i = \frac{O_i}{O(Y)} = \frac{O(E(Y|X_i))}{O(Y)}$. We can find the other indexes like second-order indexes or the any other higher order indexes for the sensitivity indices with a similar approach.

3 Results

3.1 Machine Parameters

The bridge and substrate were built with the Nickel alloy IN625, where AM part was constructed in the middle of the substrate. The benchmark experimental study considered the horizontal and vertical scan strategies to melt the odd and even layers, respectively. But because of the software restriction, we considered only one approach for all the layers (Table 1).

Table 1. Machine variables.

Machine variables	Value
Base plate temperature	80 °C
Chamber temperature	40 °C
Layer thickness	0.02 mm
Laser power	195 W
Lase speed	800 mm/s
Hatch spacing	0.02 mm

3.2 Mesh Convergence Study

The mesh of 0.3 mm size was chosen for the simulation studies, which can be explained using the mesh convergence graph. We considered three types of mesh sizes and compared the numerical residual strain with the experimental strains at z = 9.536 mm. Table 2 shows the comparison of error norm for different values of the mesh size. From the Table 2 we can present that error between the numerical and experimental outputs is decreasing with the decrease in mesh dimension. We have used 32 CPUs and an Intel E5-2683 v4 processor for each simulation on the Graham cluster, Compute Canada. Table 2, proves that with the reduction in mesh size leads to a decrease in error, but it also leads to an increase in computation cost. So, a mesh size of 0.3 mm was selected for all the simulations because it is not as computationally expensive as 0.2 mm and gives results better than the 0.4 mm mesh size.

Table 2. Results for different mesh sizes.

Mesh size (mm)	Relative L2 error norm	Computation cost (Min)
0.2	0.8588	155
0.3	0.9642	60
0.4	0.9924	45

3.3 Validation Study of Ansys Workbench Additive Simulation

A numerical model of the additive manufacturing process was constructed using Ansys workbench additive software. The additive model was validated using a benchmark case study published in 2018 [7]. To validate the numerical studies, the normal residual elastic strains from the simulation and the experiment were compared at two locations (z = 2.216 and 9.536) in the bridge. Figure 2a, b and c compare directional strains at z = 9.536 for x, y and z directions, respectively, whereas Fig. 3a, b and c represent results for z = 2.216. The comparison study at both places gives a very good equivalence with the benchmark results. The relative l2 norm at z = 9.536 are 0.4041, 0.6202, and 0.7820 for x, y, z direction strains respectively. It is important to mention that the benchmark strains were available only till 60 mm of bridge length. The minor difference in the comparison may occur because of the different types of scan strategies applied in the simulation and experimental studies.

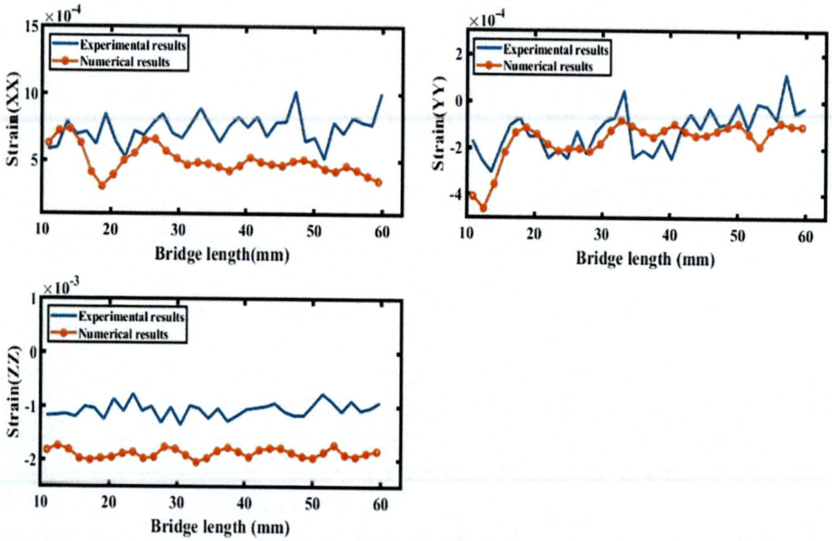

Fig. 2. Fig. 2(a), (b) and (c) show the comparison of directional strain for z = 9.536 in the x, y, and z, respectively, for the experimental and simulation cases.

3.4 Artificial Neural Network

In the research, we used five variables: three belong to the machine parameters, and two are material parameters. The inputs are hatch spacing, Young's modulus, laser speed, layer thickness, and Poisson ratio. A sample space of 360 samples was generated with the Sobol sampling method. Initially, the sample space was solved using a workbench additive, and the directional strains were extracted at z = 9.536. After that, a feed-forward neural network was trained with results from the 360 samples space. The 70% of the data was utilized to train the ANN. In addition, the other 15% and 15% of the data were used to validate and test the model. Figure 4 presents the decrease in error during different stages of ANN model. The model was constructed to predict the normal directional strains at z = 9.536. Once the model is ready, it is used with the different number of samples set to find the convergence plot of standard deviation (Std). A Monte Carlo method was used to achieve the convergence, where different numbers of samples set like: 12 48, 84, 120, 192, 264, 336, 516, 732, 1020, 1524, 2028, 2532, 3000, 40008, 5016. 8004 and 9588 were used. The analysis shows that the results rarely change after 4000 samples. As a result, 4000 samples were selected to conduct the uncertainty and

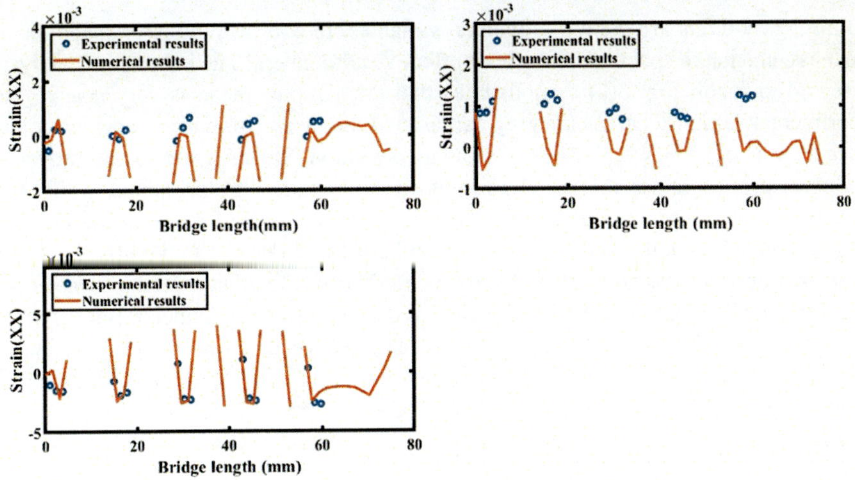

Fig. 3. Fig. 3(a), (b) and (c) show the comparison directional strain for z = 2.216 in the x, y, and z, respectively, for the experimental and simulation cases.

sensitivity studies with the stains as output data generated from the ANN model at the corresponding points (Fig. 5).

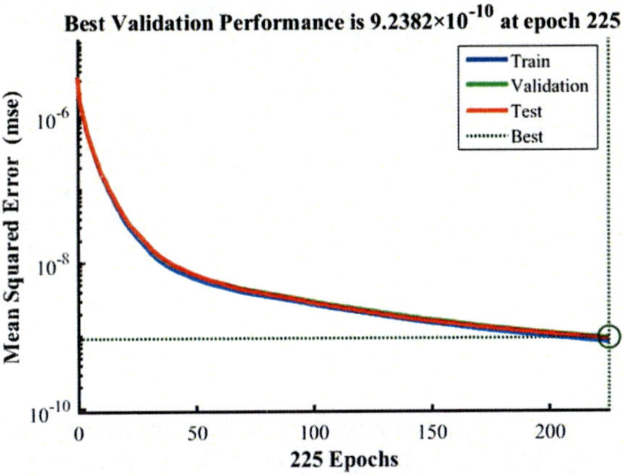

Fig. 4. Error graph ANN model building process.

3.5 Sensitivity and Uncertainty Analysis

A total of 4000 samples and their corresponding x-directional strains were used to perform the sensitivity analysis. The normal strains for the 4000 input variables were calculated using the trained neural network. Figure 6(a) shows the variation of Sobol indices across the length of the bridge. The Sobol indices graph represents that the Poisson ratio and Young modulus are the top influential variables. In contrast, the laser speed, layer thickness, and hatch spacing are the least effective parameters. The uncertainty analysis presented in graph Fig. 6(b) demonstrates that the samples mean is very near to the experimental values, whereas the grey part around the mean curve represents the confidence interval of 95%, which is ±2 of standard deviation (Std). Another same study was conducted using y and z strains. The results for both studies are presented in Fig. 6(c to f). Figure 6(c) represents that the y directions strains depend more on the Poisson ratio, whereas the young modulus is the second critical parameter in the process. If we consider strains in z direction, then the Young modulus is the most crucial parameter for the z directional strains, and the Poisson ratio is the next critical parameter. But the other three parameters are the least important for each directional strain. Figure 5 presents a Pie chart for each input variable's sensitivity indices. From Fig. 5, we can conclude that the effect of hatch spacing, laser speed, and layer thickness is still significant, so we can't ignore them during the SLM process.

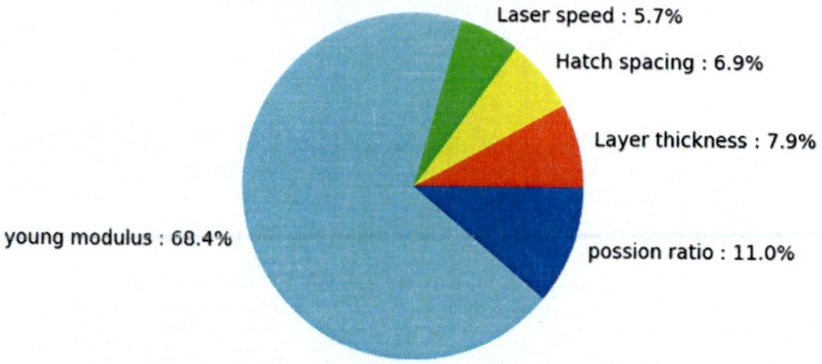

Fig. 5. The pie chart for sensitivity analysis in an SLM process for z strains.

4 Conclusion

This paper presented a framework that can be used to study the SLM process as an alternative to the traditional numerical and experimental methods. We used an artificial neural network and SLM simulation model to analyze the uncertainty and sensitivity in the AM technology. In the beginning, we trained our neural network on the data generated by using a numerical model. After this we used Monte Carlo method with ANN model to find the correct number of samples required to conduct the uncertainty and sensitivity studies for SLM process. The study found that the 4000 samples are good enough to analyze as the results did not change. The sensitivity analysis helped us find that the Poisson ratio and Young modulus are the critical variables during the process. In contrast, laser speed, layer thickness, and hatch spacing are the least important. But as these variables still hold a significant value (5% to 10%), so we cannot ignore them during the process. We used the same framework to optimize the SLM process, and more details can be found in [12].

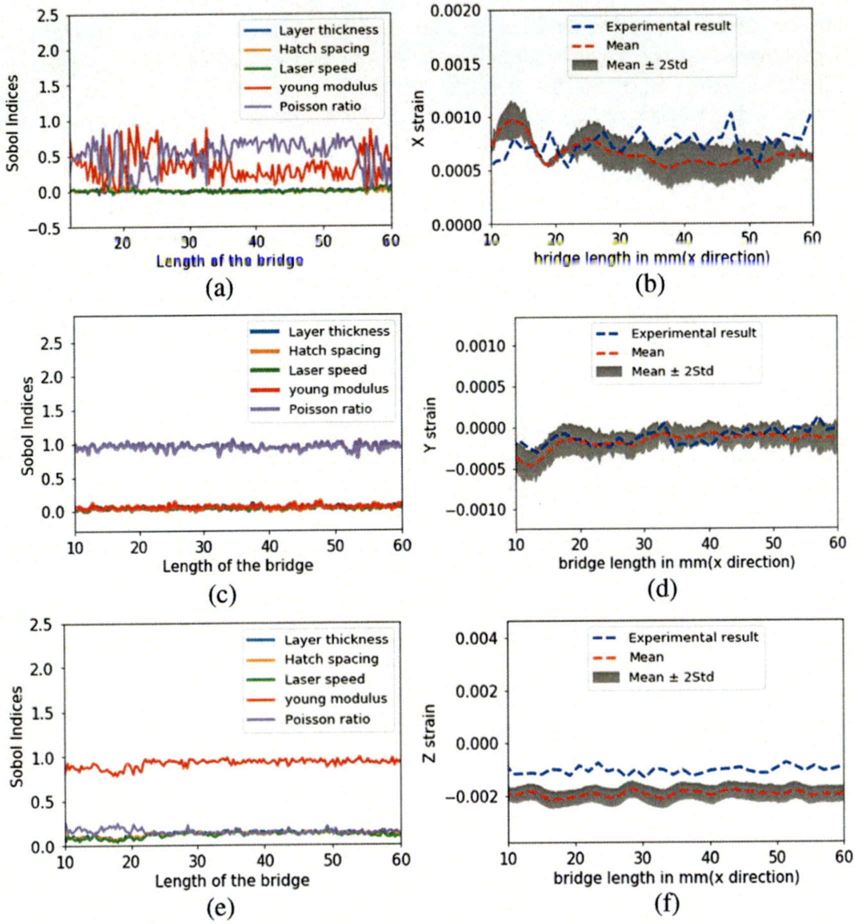

Fig. 6. Sensitivity indices over the bridge length on z = 9.536 are given in Fig. 6(a), (c), and (e) for the directional strains. Whereas Fig. 6(b), (d), and (f) presents the uncertainty analysis for directional strains.

References

1. Ngo, T.D., Kashani, A., Imbalzano, G., Nguyen, K.T.Q., Hui, D.: Additive manufacturing (3D printing): A review of materials, methods, applications and challenges. Compos. B Eng. **143**, 172–196 (2018). https://doi.org/10.1016/j.compositesb.2018.02.012
2. Abdulhameed, O., Al-Ahmari, A., Ameen, W., Mian, S.H.: Additive manufacturing: challenges, trends, and applications. Adv. Mech. Eng. **11**(2), 168781401882288 (2019). https://doi.org/10.1177/1687814018822880
3. Saltelli, A., Annoni, P., Azzini, I., Campolongo, F., Ratto, M., Tarantola, S.: Variance based sensitivity analysis of model output. Design and estimator for the total sensitivity index. Comput. Phys. Commun. **181**(2), 259–270 (2010). https://doi.org/10.1016/j.cpc.2009.09.018

4. Hu, Z., Mahadevan, S.: Uncertainty quantification and management in additive manufacturing: current status, needs, and opportunities. Int. J. Adv. Manuf. Technol. **93**(5–8), 2855–2874 (2017). https://doi.org/10.1007/s00170-017-0703-5
5. Borchani, H., Varando, G., Bielza, C., Larrañaga, P.: A survey on multi-output regression: Multi-output regression survey. WIREs Data Mining Knowl. Discov. **5**(5), 216–233 (2015). https://doi.org/10.1002/widm.1157
6. Asserin, O., Loredo, A., Petelet, M., Iooss, B.: Global sensitivity analysis in welding simulations—What are the material data you really need? Finite Elem. Anal. Des. **47**(9), 1004–1016 (2011). https://doi.org/10.1016/j.finel.2011.03.016
7. Phan, T.Q., et al.: Elastic residual strain and stress measurements and corresponding part deflections of 3D additive manufacturing builds of IN625 AM-bench artifacts using neutron diffraction, synchrotron X-Ray diffraction, and contour method. Integrat. Mater. Manuf. Innov. **8**(3), 318–334 (2019). https://doi.org/10.1007/s40192-019-00149-0
8. torey.liepa@nist.gov, "Benchmark Test Data," NIST, May 17, 2018. https://www.nist.gov/ambench/benchmark-test-data (2018). Accessed 22 Dec 2021
9. Wang, C., Tan, X.P., Tor, S.B., Lim, C.S.: Machine learning in additive manufacturing: state-of-the-art and perspectives. Addit. Manuf. **36**, 101538 (2020). https://doi.org/10.1016/j.addma.2020.101538
10. Francis, J., Bian, L.: Deep learning for distortion prediction in laser-based additive manufacturing using big data. Manuf. Lett. **20**, 10–14 (2019). https://doi.org/10.1016/j.mfglet.2019.02.001
11. Meng, L., et al.: Machine learning in additive manufacturing: a review. JOM **72**(6), 2363–2377 (2020). https://doi.org/10.1007/s11837-020-04155-y
12. Chaudhry, S., Soulaïmani, A.: A comparative study of machine learning methods for computational modeling of the selective laser melting additive manufacturing process. Appl. Sci. **12**(5), 2324 (2022)

Artificial Intelligence for Colorectal Polyps Classification Using 3D CNN

Khadija Hicham[1(✉)], Sara Laghmati[2], and Amal Tmiri[1]

[1] Mohammed V University (UM5), Rabat, Morocco
hicham.khadija101@gmail.com
[2] Chouaib Doukkali University (UCD), El Jadida, Morocco

Abstract. Convolutional Neural Network (CNN) has made remarkable progress in the medical field. The use of CNN is widely necessary to extract highly representative characteristics in the case of acute medical pathology. Composed of fully connected layers, the CNN allows the classification of the data. The classification process is done among the network layers by filtering, selecting, and applying these features at the last layers. CNN offers a better prognosis, especially in the case of colorectal cancer (CRC) prevention. CRC develops from cells that line the inner lining of the colon. Mostly, it comes from a benign tumor, called a polyp, which slowly grows with time to develop into malignant cells. However, classification of 3D scan images of the abdomen based on the presence or absence of polyps is necessary to increase the chance of early detection of the disease and thus guide it to the appropriate treatment. In this work, we present and study a 3D CNN model for the processing and classification of polyps. The results show promising performances for a 12 layers 3D CNN model.

Keywords: Artificial intelligence AI · Deep learning · 3D-convolutional neural networks 3D-CNN · Polyps · Colorectal cancer CRC

1 Introduction

Over the decade, Computer-Aided Diagnosis (CAD) has become a crucial diagnostic tool for various diseases affecting humankind. Worldwide, colorectal cancer (CRC) is the fourth leading cause of cancer death [1,2]. The intestinal/colonic polyp is the precursor to Most CRCs. Early diagnosis and elimination of the polyp is the most effective in preventing CRC [3].

Most of the material on the abdomen aims to locate and segment organs, mainly the liver, kidneys, bladder, and pancreas. The prime modality is imagery magnetic resonance (MRI) for prostate analysis and computed tomography (CT) for all other organs. The colon is the only abdominal surface where several applications were addressed, in a simple manner: CNN extract features Which are used afterward for classification [4].

L. Azrar et al. (Eds.): CIP 2022, LNME, pp. 165–174, 2023.
https://doi.org/10.1007/978-3-031-23615-0_17

Artificial intelligence and deep learning techniques have made remarkable progress in the medical field. These modern techniques can provide a better prognosis of the polyp using a convolutional neural network [3]. 3D data is an important asset of computer vision because it supplies valuable information on the complete geometry of the detected objects and scenes [5]. Recently, with the increase in the availability and computing power of large sets of data, deep learning can be considered today to learn from specific tasks such as recognition, segmentation, and correspondence on 3D data.

The incorporation of 3D data can be necessary to obtain good performance in classification tasks of medical imaging objects. Since computer procession of images tends to be natural 2D images, networks developed in these scenarios do not directly take advantage of 3D information [4]. Thus, localization in medical imaging often necessitates the analysis of 3D volumes. For 3D data analysis using deep learning algorithms, various approaches have been proposed, which treat 3D space as a composition of 2D orthogonal planes (slices) [6].

In this study, we propose a 12-layer CNN 3D model. It allows the processing and classification of three-dimensional CT images of polyps acquired by the colo-scanner. This study aims to develop an artificial intelligence model based on 3D CNN. The model is dedicated to the classification of polyps while integrating 3D scan images of the abdominal area. The present article is organized as follows. Section 2 presents some relevant related work about diverse 3D-CNN and their use in the medical field. Section 3 details the data, the materials, the methods, and the predictive system utilized to conduct this study. Whereas Sect. 4 provides details of the training, the results, and a discussion. We conclude the paper in Sect. 5.

2 Related Work

Many studies are interested in the classification of diseases using artificial intelligence and precisely the 3D-CNN models. However, different approaches are in use to effectively integrate 3D images into personalized architectures [7]. Their ultimate goal is early detection with the highest performance possible.

Zhou et al. developed a 3D deep learning model for the classification of breast cancer and the location of malignancy from MRI. It processed high-dimensional data and 3D volumetric images by taking exploiting 3D DenseNet. The performance of the model has shown an accuracy of 83.7%; Which gives promising results compared to the radiologist's measures [8].

Kawahara et al. applied a CNN architecture to a brain connectivity graph derived from Diffusion Tensor Imaging (DTI). They formed a network that aims to predict brain development by elaborating several new layers. This app outperforms existing methods for assessing cognitive and motor scores [9].

Nie et al. adopted a 3D-CNN to predict the survival time OS time of patients with a brain tumor. Results show that the methods can achieve an accuracy of 89.9%. They also concluded that the features learned from DTI and fMRI have a fundamental role in effectively predicting the OS time [10].

Payan and Montana advanced a pattern classification model that associated between convolutional neural networks and sparse auto encoders. The authors tested the system on Alzheimer's disease dataset. They also provided a benchmarking between 2D-CNN and 3D-CNN where HF + fc6-PCA takes the upper hand with an accuracy of 89.8% [11].

3 Materials and Methods

3.1 Dataset

To conduct this study, we used the CT colonography database provided by ACRIN [12–14]. It contains 825 cases which 243 cases are polyp-free. Polyps are bumps like cell growths on the inner colon tissues as shown in Fig. 1, mostly benign but since they arise abnormally, they can eventually turn into cancerous cells. The CT-Colonography dataset includes three-dimensional (3D) images made of numerous two-dimensional 2D slices in DICOM format of a dimension of 512 × 512 pixels each, with a varying depth size.

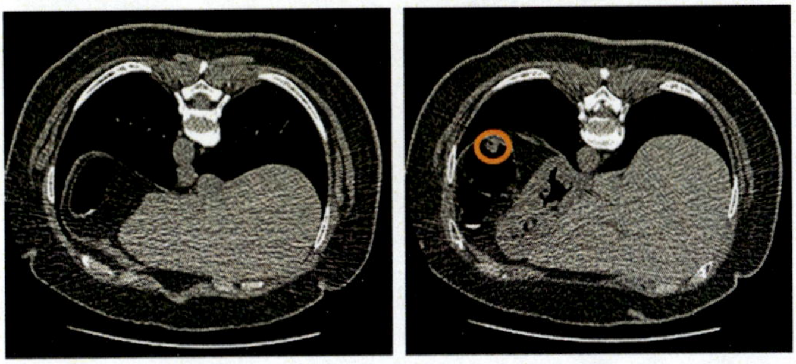

Fig. 1. Examples of normal image (a) and abnormal image (b). The yellow line enclose the polyp in abdominal 2D CT scan (b).

3.2 3D-CNN

Convolutional Networks have proved their value in image classification problems in several domains and in the medical field specifically [8–11]. We built a 3D-CNN model, which we implemented with polyps scans. After testing several layouts, we settled on the most optimal architecture; an artificial neural network which contains 12 connected layers with a kernel size of (3, 3, 3), featuring four convolutional 3D layers, with two 3D-CNN 64, 3D-CNN 128, and 3D-CNN 256. Each convolutional layer is followed by a 3D Max Pooling layer, with a stride of 2 and ReLU activation.

The feature extraction block consists of four CONV-MAXPOOL modules. A global pooling layer was used to reduce the dimensionality of the final output of the feature extraction block and passed through the following a fully connected layer with 512 neurons. The function of the activation sigmoid was applied to the last layer of binary classification of a single neuron. Figure 2 shows the network architecture.

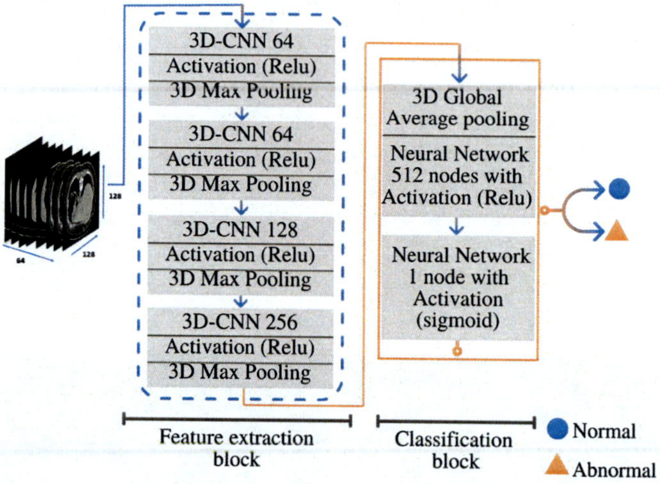

Fig. 2. 12 layers 3D convolutional neural network (3D-CNN) architecture.

3.3 Performance Evaluation

To assess the quality of classification of the 3D-CNN model we used the confusion matrix. In statistical classification, the confusion matrix is a table of rows (Predicted class) and columns (reference class) and each case reports the number of True Positive (TP), True negative (TN), False positive (FP), and False Negative (FN). Where TP designates the number of cases that actually have polyps, TN denotes the number of correctly classified cases who are healthy, FP Indicates the number of cases classified with the disease but who actually are polyp-free, and FN represents the number of scans classified as normal but who actually have a polyp [15].

For a further evaluation of the performance of the 3D-CNN model, we used the metrics derived from the confusion matrix, which are accuracy, specificity, recall, and F1 score. Accuracy represents the percentage of correctly classified cases among the total number. Precision represents the percentage of correctly classified cases among the total predicted cases as abnormal scans. Specificity states the rate of true negatives among healthy Cases, it indicates the ability to detect all situations where there are no polyps. The sensitivity indicates the number of correctly classified cases among the total cases that are actually malignant, the recall or sensitivity represents the capacity to correctly detect each positive case. F1 score represents the equilibrium between the precision and the recall [16].

$$Accuracy = \frac{TP + TN}{TP + TN + FP + FN} \tag{1}$$

$$Precision = \frac{TP}{TP + FP} \tag{2}$$

$$Recall = \frac{TP}{TP + FN} \tag{3}$$

$$Specificity = \frac{TN}{FP + TN} \tag{4}$$

$$F - measure = 2 \times \frac{Precision \times Specificity}{Precision + Specificity} = \frac{2TP}{2TP + FP + FN} \tag{5}$$

3.4 Predictive System

The system is built-in TensorFlow 1.5.2 with the support of the Nvidia Tesla P100 graphics processing unit (GPU) and a 12 GB memory. The first step consists of data preprocessing; after resizing images to the size $128 \times 128 \times 64$, we proceeded to standardize the data which goal is to transform data so that all data points can be on the same scale. A typical normalization process is to reduce digital data to a scale of zero to one. In this regard, to make all the images compatible and the same scale, we imported our image data into arrays, converting them from uint64 to float32 and then dividing them by 255 while maintaining lossless image detail. The output was a number between 0 indicating normal scan (absence of polyps) for a value less than 0.5 and 1 indicating abnormal scan (presence of polyps) for a value equal or superior to 0.5.

After data preprocessing, the next step consists of training the 3D-CNN algorithm. For the training phase, we used the Adam optimizer with a learning rate of 0.0001 with a batch size of 5 for 35 epochs. Then the fitting model uses the validation data set for the tunning. Once the 3D-CNN is set with the best parameters, it undergoes the test providing measurements based on the confusion matrix. Figure 3 presents the phases of our predictive system.

4 Results and Discussion

In this section, we evaluate the performance of the 3D-CNN. To avoid any disruption during the training of the model, we ought to make sure the data are the same size, height, width, and depth. We specify that we used 201 images of 825 patients. Among the selected scans, we dedicated 185 for training (115 cases are polyp-free and 70 are positive cases). The validation set represents 30% of the training cases. We reserved 16 images for the test dataset (9 are positive scans and 7 are negative scans).

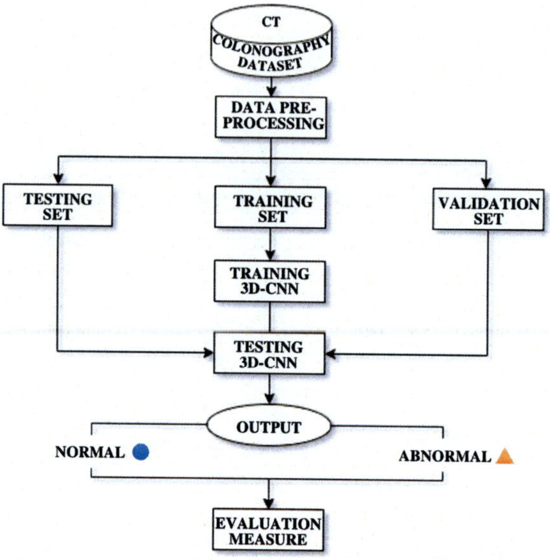

Fig. 3. The proposed predictive system framework.

4.1 Training and Validation of the 3D-CNN

4.1.1 Loss Function The loss function is a valuable method for an effective CNN model training evaluation. In Fig. 4, the loss function for the training keeps on decreasing, nearly reaching zero at the 28 epochs. However, the model was not as successful in the validation step. Indeed, the curve shows a minimal loss function validation value of around the 18th epoch, which increases remarkably from the 27 epochs.

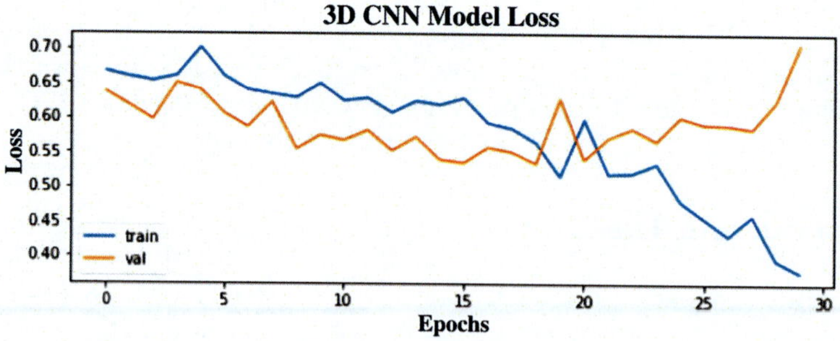

Fig. 4. 3D CNN loss function of training and validation data by epochs.

4.1.2 Accuracy Apart from the loss function, accuracy is mainly of importance. The ideal would be for the accuracy to reach 1, which is generally impossible without overfitting. What prompts us to look for the best compromise between both the values of training and validation. Figure 5 displays the accuracy curve for training and validation according to the epochs. The training accuracy keeps increasing with each epoch, and the validation accuracy reaches its peak of 0.75 at the 14th epoch. The two curves intersect at the 20th epoch with a value of 0.65. However, at the 30th epoch, the training and validation accuracy reaches 0.83 and 0.65 respectively.

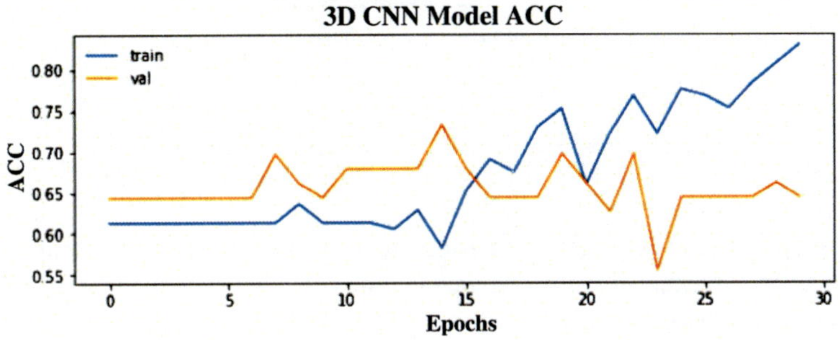

Fig. 5. 3D CNN accuracy of training and validation data by epochs.

4.2 Testing Results

4.2.1 Confusion Matrix The confusion matrix in Fig. 6 shows that, in the testing phase, the 3D CNN model returned 7 TP (patients who have polyps), 3 TN (healthy patients), 4 FN (misclassified cases with polyps), and 2 FN (misclassified cases who are polyp-free). Overall, the model correctly classified 10 out of 16 cases.

4.2.2 Evaluation Metrics Table 1 gathers the performance measures. The 3D CNN model achieved an accuracy of 62.5% for the test dataset, so the model detected 77.8% of the positive cases where polyps are present, expressly the rate of true positives, and 63% of positive cases in the entered positive cases. For a trade-off between the evaluation of the recall and precision, we calculated the harmonic mean, which gave 70%.

4.3 Discussion

After a profound analysis of the results, we established two main points. Beforehand this study uses a 3D CNN model built from scratch for the classification of 3D abdominal CT scans based on the presence or the absence of polyps. Secondly, the 3D CNN model provides high precision for the training set. Results

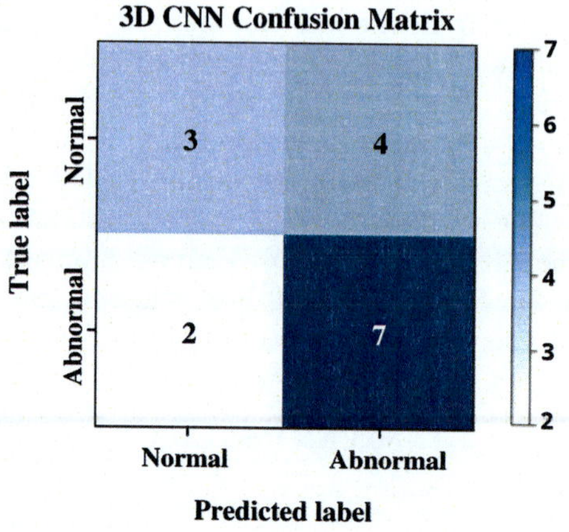

Fig. 6. Confusion matrix of the 3D-CNN model.

are overall promising, and there is room for improvement. Aforementioned in Sect. 2, multiple studies introduced various 3D CNN models to predict different diseases. The models obtained outstanding results, especially auteur's [10] system reached an accuracy of 89.9% using cerebral brain MRI to predict the OS time of patients with a brain tumor. Many factors influence the performance of the system, such as the dimension, the number of data, and the nature of the disease.

Table 1. 3D CNN performance evaluation measures

Model	Accuracy	Recall	Specificity	Precision	F-measure
3D CNN	62.5%	77.8%	42.8%	63%	70%

5 Conclusion

Considering CRC prediction, Deep learning plays a primordial role in polyp prognosis. Polyp early detection can change the outcome since they are one of the principal sources of CRC. In summary, this paper developed a 12 layers 3D-CNN model composed of a features extraction block that automatically learns features followed by a classification block. The findings of this study provide an accuracy of 62.5%, a recall of 77.8%, and an f-measure of 70%. Ideally, the system should be replicated in an advanced environment taking into consideration a larger amount of data for the training. Despite the limitations, the model shows encouraging results. Exploring other possibilities such as adding cross-validation can take this study further.

Acknowledgment. The authors would like to express their special thanks of gratitude to the late Professor Nabil ELMARZOUQI, who contributed considerably to this work.

References

1. Ferlay, J., Soerjomataram, I., Dikshit, R., et al.: Cancer incidence and mortality worldwide: sources, methods and major patterns in GLOBOCAN 2012: Globocan 2012. Int. J. Cancer **136**(5), 359–386 (2015). https://doi.org/10.1002/ijc.29210
2. Zhang, R., Zheng, Y., Poon, C., et al.: Polyp detection during colonoscopy using a regression-based convolutional neural network with a tracker. Pattern Recognit. **83**, 209–219 (2018). https://doi.org/10.1016/j.patcog.2018.05.026
3. Ribeiro, E., Uhl, A., Wimmer, G., et al.: Exploring deep learning and transfer learning for colonic polyp classification. Comput. Math. Methods Med. **2016**, 1–16 (2016). https://doi.org/10.1155/2016/6584725
4. Litjens, G., Kooi, T., Bejnordi, B.E., Setio, A.A.A., Ciompi, F., Ghafoorian, M., van der Laak, J.A.W.M., van Ginneken, B., Sánchez, C.I.: A survey on deep learning in medical image analysis. Med. Image Anal. **42**, 60–88 (2017). https://doi.org/10.1016/j.media.2017.07.005
5. Chang, A.X., Funkhouser, T., Guibas, L., et al.: ShapeNet: an information-rich 3D model repository. arXiv:1512.03012 [cs] (2015)
6. Yang, D., Zhang, S., Yan, Z., et al.: Automated anatomical landmark detection on distal femur surface using convolutional neural network. In: 2015 IEEE 12th International Symposium on Biomedical Imaging (ISBI), pp. 17–21. IEEE, Brooklyn, NY, USA (2015). https://doi.org/10.1109/ISBI.2015.7163806
7. Singh, S.P., Wang, L., Gupta, S., et al.: 3D deep learning on medical images: a review. Sensors **20**(18), 5097 (2020). https://doi.org/10.3390/s20185097
8. Zhou, J., Luo, L.-Y., Dou, Q., et al.: Weakly supervised 3D deep learning for breast cancer classification and localization of the lesions in MR images. J. Magn. Reson. Imaging **50**(4), 1144–1151 (2019). https://doi.org/10.1002/jmri.26721
9. Kawahara, J., Brown, C.J., Miller, S.P., et al.: BrainNetCNN: convolutional neural networks for brain networks; towards predicting neurodevelopment. NeuroImage **146**, 1038–1049 (2017)
10. Nie, D., Zhang, H., Adeli, E., et al.: 3D deep learning for multi-modal imaging-guided survival time prediction of brain tumor patients. In: Medical Image Computing and Computer-Assisted Intervention: MICCAI. International Conference on Medical Image Computing and Computer-Assisted Intervention, vol. 9901, pp. 212–220 (2016)
11. Payan, A., Montana, G.: Predicting Alzheimer's disease: a neuroimaging study with 3D convolutional neural networks. arXiv:1502.02506 (2015)
12. Smith, K., Clark, K., Bennett, W., et al.: Data from CT_COLONOGRAPHY. The cancer imaging archive. https://doi.org/10.7937/K9/TCIA.2015.NWTESAY1 (2015)
13. Johnson, C.D., Dachman, A., Cheema, J.I., et al.: Accuracy of CT colonography for detection of large adenomas and cancers. N. Engl. J. Med. 11 (2008)
14. Clark, K., Vendt, B., Smith, K., Freymann, J., Kirby, J., Koppel, P., Moore, S., Phillips, S., Maffitt, D., Pringle, M., Tarbox, L., Prior, F.: The cancer imaging archive (TCIA): maintaining and operating a public information repository. J. Digit. Imaging **26**(6), 1045–1057 (2013). https://doi.org/10.1007/s10278-013-9622-7

15. Laghmati, S., Cherradi, B., Tmiri, A., et al.: Classification of patients with breast cancer using neighbourhood component analysis and supervised machine learning techniques. In: 2020 3rd International Conference on Advanced Communication Technologies and Networking (CommNet), pp. 1–6. IEEE, Marrakech, Morocco (2020). https://doi.org/10.1109/CommNet49926.2020.9199633
16. Laghmati, S., Tmiri, A., Cherradi, B.: Machine learning based system for prediction of breast cancer severity. In: 2019 International Conference on Wireless Networks and Mobile Communications (WINCOM), pp. 1–5. IEEE, Fez, Morocco (2019). https://doi.org/10.1109/WINCOM47513.2019.8942575

Toward Natural Language Processing Approaches for Text

Rkia Bani[1(✉)], Samir Amri[2], Lahbib Zenkouar[1], and Zouhair Guennoun[1]

[1] Electrical Department, Mohammadia School of Engineering, Rabat, Morocco
rkya.bani@gmail.com

[2] Department of Mathematics and Informatics, ENSAM School of Engineering, Meknes, Morocco

Abstract. Natural language processing is the field of study that includes multiples domains such as linguistic, informatics and artificial intelligence. In this paper, we explore the development of this field and introduce the different approaches used to deal with several task of natural language processing of the text. Using the tools offered by the field cited above and choosing the appropriate approach, the computers analysis texts, so this can be exploited in numerous use-cases. This field has become very important and popular in research in all its five levels: morphological, syntactic, semantic, discourse sentence and pragmatic. These researches had given us fundamental task such as part of speech tagging, named entity recognition. This leads us to make a comparison of these different approaches on different tasks. Also we evaluate their accuracy on different languages, and we show that the lack of data-sets influences negatively the results whatever the language is or the chosen approach especially the machine learning ones.

Keywords: Language processing · Linguistic · Machine learning · Artificial intelligence

1 Introduction

With the expansion of the world of information and the billions of data circulating in the net, natural language processing (NLP) becomes a very interesting field for researchers to use machine learning (ML) and artificial intelligence (AI) for processing this huge data for a specified goal. In this introduction, we will present the levels of NLP, the basic tools, and some of its applications. Then, in the next section, we introduce the approaches used in the NLP research and make a comparison of some interesting works done on each one. Finally, in the last section, we will put in light the different AI and ML methods used in this field and compare the result from interesting research done on different NLP tasks before concluding the paper.

Samir Amri, Lahbib Zenkouar and Zouhair Guennoun are contributed equally to this work.

L. Azrar et al. (Eds.): CIP 2022, LNME, pp. 175–183, 2023.
https://doi.org/10.1007/978-3-031-23615-0_18

NLP uses five hierarchical processing levels for analysis (Fig. 1): morphological, syntactic, discourse sentence and pragmatic:

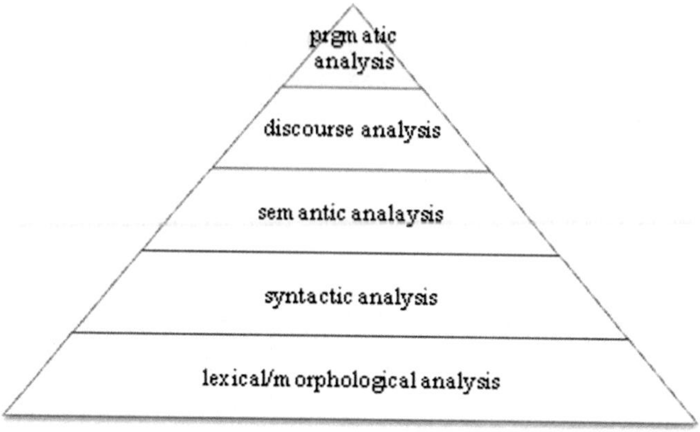

Fig. 1. Natural language processing levels

2 NLP Basic Tools

In these last years, NLP has been a very interesting field for research that gives birth to a lot of tools that make NLP what is today.

2.1 Tokeniser

A tokenizer divides text into a sequence of tokens, which approximately correspond to "words". Tokenization is considered as a standard pre-processing step for multiples NLP tasks.

2.2 Part of Speech Tagger

Part of speech tagger (POS tagger) is a process that assigns each word of a text its convenient grammatical information.

2.3 Named Entity Recognition

Named entity recognition (NER) is the task of finding named entities like: name of person, biological protein, name of organization and location, drug's name, clinical procedure, time, etc. in text.

3 NLP Applications

Natural language processing can be applied into multiples areas like sentiment analysis, spam filtering and information extraction.

3.1 Sentiment Analysis

Sentiment analysis or opinion mining is the computational study of people's opinions, sentiments, emotions, appraisals, and attitudes towards entities such as products, services, organizations, individuals, issues, events, topics, and their attributes (Liu 2012).

3.2 Spam Filtering

Spam filtering protects the user from the huge spam circulating in the net by using different techniques.

3.3 Information Extraction

Information extraction (IE) is the task of converting unstructured text into a maximum structured data base involving selected information of the text.

4 NLP Text Approaches

There are three different NLP approaches: rules based approach, probabilistic approach and hybrid approach:

4.1 Rules Based Approach

This approach uses the proprieties and features of the language to establish rules to be implemented in algorithm, these rules help the model to process the text in the desired goal.

4.2 Probabilistic-statistic Approach

Starting from an annotated corpus, the chosen model estimates its parameters based on probabilities and statistics.

4.3 Hybrid Approach

Hybrid approach is a combination of both rules based and statistical approaches.

Table 1. List of diffents researchs for each approach

Approach	Authors	Task	Method	Accuracy	Year
Rules based	Plisson et al. (2004)	Word Lemmatization	If then rules, ripple down rules	77	2004
	Shaalan (2010)	Machine translation	Transfer based machine translation system	Not indicated	2010
	Shaalan and Raza (2007)	Name entity recognition	Gazetteer+Grammar	85.5	2007
Statistical	Ratnaparkhi (1996)	Part of speech tagging	Maximum entropy model	96.6	1996
	Fersini et al. (2014)	Sentiment analysis	Bayesian model	88	2014
Hybrid	Hajic et al. (2007)	Part of speech tagging (POS tagging)	Hidden Markov Model, Maximum-Entropy, Averaged Perceptron, hand writing rules	95.68	2007
	Amri et al. (2019)	Named entity recognition	Conditional random fields and rule based	95.5	2019

4.4 Comparison of Approaches

Table 1 provides a list of different methods used by approach for different NLP task: NER, POS tagging, IR and for different language: English, Arabic, Czech and Amazigh as a low resource language. As we can see the rule based approach don't show great result comparing with the hybrid approach. However, statistical approach showed good result since 1996 in POS tagging.

Each approach gives different results depending on the chosen task and the language to process. Rules based approach gives a great result in Arabic (Shaalan 2010) and the hybrid one shows greatest results in part of speech tagging of the Czech language (Plisson et al. 2004), also the availibilty of language resources makes a difference in the results whatever the chosen approach.

5 Artificial Intelligence and Machine Learning Approaches

There are four approaches of AI and ML used in NLP, which we detail below.

5.1 Supervised Learning

Supervised learning uses labeled DATASETS to train algorithms to classify data or predict outcome accurately, these training datasets composed of input and its correct corresponding outputs so that the model can learn over time. Various algorithms are used in the field of natural language processing:

5.1.1 Support Vector Machine Support vector machine is a popular model for learning used for both data classification and regression but commonly very used in classification problems.

5.1.2 Naïve Bayes Naïve bayes is a classification approach that uses the principle of class conditional independence from the Bayes theorem.

5.1.3 Linear Regression Linear regression is used to determine the relationship between a dependent variable with one or more independent variables to make predictions about future outcome.

5.1.4 K-nearest Neighbor K-nearest neighbor is a non-parametric algorithm that distributes data points into class depending on their proximity.

5.1.5 Neural Network Neural networks process training data by using the principle of the inter-connectivity of human brain through layers of nodes.

5.1.6 Random Forest Random forest is a supervised machine learning algorithm used for both classification and regression purposes.

Table 2 presents a list of different supervised algorithms for various NLP tasks.

Table 2. List of supervised learning algorithm for various NLP tasks

Algorithm	Authors	Task	Year
SVM	Ahmad et al. (2017)	Sentiment analysis	2017
	Hernault et al. (2010)	Discourse parser	2010
	Bunescu et al. (2006)	Named entity disambiguation	2006
Naïve Bayes	Granik and Mesyura (2007)	Fake news detection	2017
	Sang-Bum et al. (2007)	Text classification	2007
Bayesian linear ridge regression	Phandi et al. (2015)	Domain adaptation	2015
K-nearest neighbor	Zhang and Zhou (2005)	Multi-label classification	2005
Neural network: Bidirectional long short-term memory	Plank et al. (2016)	Part of speech tagging	2016
Convontional and recurrent neural network	Wang et al. (2016)	Sentiment analysis	2016
Random forest	Xu et al. (2012)	Text categorisation	2012

5.2 Unsupervised Learning

Unsupervised machine learning algorithm is used to analyze, cluster unlabeled data-sets through discovering hidden patterns or data grouping without the intervention of human. Unsupervised learning models are utilized in three main tasks: clustering, association and dimensionality reduction.

Table 3 provides unsupervised learning algorithms for different NLP tasks:

Table 3. List of different unsupervised learning algorithms for NLP task

Authors	Algorithm	Task	Year
Pon et al. (2009)	Log linear model	Morphological segmentation	2009
Goldwater and Griffiths (2007)	Bayesian model	Part of speech tagging	2007
Ko and Seo (2000)	Naïve bayes	Text categorization	2000
Paltoglou and Thelwall (2012)	Algorithm lexicon based	Sentiment analysis	2012

5.3 Semi-supervised Learning

Semi-supervised learning is a learning paradigm associated with constructing models that use labeled and unlabeled data. The goal of semi-supervised learning is to understand how combining labeled and unlabeled data could change the learning behavior, and to take advantages of both this combination.

Table 4 presents different semi-supervised algorithms for both basique and complicated NLP tasks.

Table 4. List of semi-supervised learning algorithm for NLP tasks

Authors	Algorithm	Task	Year
Toutanova and Johnson (2007)	Bayesian model	Part of speech tagging	2007
Wang et al. (2013)	Gibbs sampling inference algorithm (Geman and Geman 1984)	Named entity recognition	2013
Carlson et al. (2010)	Coupled Pattern Learner, Coupled SEAL, Meta-bootstrap learner	Information extraction	2010
Zhou et al. (2013)	Active deep network	Sentiment analysis	2013

5.4 Reinforcement Learning

Reinforcement learning aims to train machine learning models to make sequence of decisions in a given complex environment. The learning is done by establishing a system of rewards and penalties and the goal of the artificial intelligence is to get the maximum of rewards.

Table 5 provides reinforcement learning methods used in different tasks.

Table 5. List of different methods used in reinforcement learning by task

Authors	Method	Task	Year
Zhang et al. (2018)	Convontional neural network, long schort term memory	Text classification	2018
Paulus et al.(2017)	Recurrent neural network	summarization	2017
Cuayahuitl (2017)	Deep Q-learning, (Mnih et al. 2013)	Dialogue generation	2016

5.5 Comparison and Discussion

From the work cited above in Tables 2, 3 and 4, performance depends on the learning approach used: (Goldwater and Griffiths 2007) unsupervised POS tagger with 86.8% of accuracy, (Toutanova and Johnson 2007) proposed a semi-supervised POS tagger with an accuracy of 93.4%, compared with (Plank et al. 2016) supervised POS tagger with an accuracy of 97.22%, Plank's tagger seems to be better.(Ko and Seo 2000) in his proposed unsupervised method of text classification acknowledeges that it is less accurate with 3.8% than the supervised learning systems. In sentiment analysis, the performance of the semi-supervised learning algorithm proposed by (Zhou et al. 2013) achieves a 78.2% of accuracy on KIT (kitchen appliance) corpus. (Paltoglou and Thelwall 2012) reaches 86, 5% on Twitter's corpus with his unsupervised learning algorithm lexicon based. The supervised learning algorithm proposed by (Wang et al. 2016) achieves 89, 95% on SST2(Stanford sentiment Treebank with neutral reviews removed and binary labels) corpus. Which means the results remain comparable between the three learning methods, and the quality of corpus impress the accuracy of the chosen learning approach.

6 Conclusion

With the development of machine learning and artificial intelligence, the choice of the NLP approach and the learning method to train the algorithm is the key to a success project. This task faces more challenges in case of working with low resources language where the corpus is very small. And that's our case in our future work aiming to choose the appropriate approach for Amazigh language. The supervised learning approach in the natural language processing of text offers satisfying and trustful results. However it requires a good amount of well labeled corpus which is not always available, moreover it's time and resources consuming. In Amazigh language, there is no labeled data, some researcher(S. Amri) was able to annotate 60000 tokens in texts which make the NLP of Amazigh language an interesting challenge.

References

Liu, B.: Sentiment analysis and opinion mining (Introduction and survey). Morgan & Claypool (2012)

Shaalan, K.: Rule-based approach in Arabic natural language processing. IJICT **3**(3), 11–19 (2010). Jun

Plisson, J., Lavrac, N., Mladenic, D.: A rule based approach to word lemmatizatio. In: Proceedings of IS, pp. 83–86 (2004)

Shaalan, K., Raza, H.: NERA: named entity recognition for Arabic. JASIST **60**(8), 1652–1663 (2009). Aug

Ratnaparkhi, A.: A maximum entropy model for part-of-speech tagging. In: Proceedings of Conference on Empirical Methods in NLP (1996)

Fersini, E., Messina, E., Pozzi, F.A.: Sentiment analysis: bayesian ensemble learning. Decis.Support Syst. **68**, 26–38 (2014). Dec

Hajic, J., Votrubec, J., Krbec, P., Květoň, P.: The best of two worlds: Cooperation of statistical and rule-based taggers for Czech. In: Proceedings of the Workshop on Balto-Slavonic Natural Language Processing, pp. 67–74 (2007)

Amri, S., Zenkouar, L., Benkhouya, R.: A hybrid statistical approach for named entity recognition for amazighe language. In: Proceedings of the 4th International Conference on Big Data and Internet of Things, pp. 1–6 (2019)

Ahmad, M., Aftab, S., Ali, I.: Sentiment analysis of tweets using SVM. Int. J. Comput. Appl. **177**(5), 25–29 (2017)

Hernault, H., Prendinger, H., du Verle, D.A.: Mitsuru Ishizuka, HILDA: a discourse parser using support vector machine classification. Dialogue Discourse **1**(3), 1–33 (2010)

Bunescu, R., Pasca, M.: Using encyclopedic knowledge for named entity disambiguation. In: Proceedings of 11th Conference European Chapter Association Computational Linguistics, pp. 9–6 (2006)

Granik, M., Mesyura, V.: Fake news detection using naive bayes classifier. In: Proceedings of IEEE First Ukraine Conference Electrical Computer Engineering (UKRCON),2017, pp.900–903

Kim, S.-B., Han, K.-S., Rim, H.-C., Myaeng, S.H.: Some effective techniques for naive bayes text classification. IEEE Trans. Knowl. Data Eng. **18**(11), 1457–1466 (2006)

Phandi, P., Chai, K.M.A., Ng, H.T.: Flexible domain adaptation for automated essay scoring using correlated linear regression. In: Proceedings of Conference on Empirical Methods in NLP, pp. 431–439 (2015)

Zhang, M.-L., Zhou, Z.-H.: A k-nearest neighbor based algorithm for multi-label classification. In: Proceedings of IEEE International Conference on Granular Computer, vol. 2 , pp. 718–721 (2005)

Plank, B., Søgaard, A., Goldberg, Y.: Multilingual part-of-speech tagging with bidirectional long short-term memory models and auxiliary loss. In: ACL (2016)
Wang, X., Jiang, W., Luo, Z.: Combination of convolutional and recurrent neural network for sentiment analysis of short texts. In: Proceedings of COLING2016, 26th International Conference Computational Linguistics: Technical Papers, pp. 2428–2437 (2016)
Baoxun, X., Guo, X., Ye, Y., Cheng, J.: An improved random forest classifier for text categorization. J. Comput. **7**(12), 2913–2920 (2012)
Poon, H., Cherry, C., Toutanova, K.: Unsupervised morphological segmentation with log-linear models. In: Proceedings of Human Language Technical: Annual Conference North American Chapter Association for Computational Linguistics, pp. 209–217 (2009)
Goldwater, S., Griffiths, T.: A fully bayesian approach to unsupervised part-of-speech tagging. In: Proceedings of 45th Annual Meeting Associat. Computational Linguistics, pp.744–751 (2007)
Ko, Y., Seo, J.: Automatic text categorization by unsupervised learning. In: Proceedings of COLING 2000 Volume 1: 18th International Conferene on Computational Linguistics (2000)
Paltoglou, G., Thelwall, M.: Twitter, MySpace, Digg: Unsupervised sentiment analysis in social media. ACM Trans. Intell. Syst. Tech. (TIST) **3**(4), 1–19 (2012)
Toutanova, K., Johnson, M.: A Bayesian LDA-based model for semi-supervised part-of-speech tagging. Adv. Neural Inf. Proces. Syst. **20**, 1521–1528 (2007)
Wang, M., Che, W., Manning, C.: Effective bilingual constraints for semi-supervised learning of named entity recognizers. In: Proceedings of AAAI Conference Artificial Intelligence **27**(1), 919–925 (2013)
Carlson, A., Betteridge, J., Wang, R.C., Hruschka, Jr, E.R., Mitchell, T.M.: Coupled semi-supervised learning for information extraction. In: Proceedings of The third ACM International Conference Web Search Data Mining, pp. 101–110 (2010)
Zhou, S., Chen, Q., Wang, X.: Active deep learning method for semi-supervised sentiment classification. Neurocomputing **120**, 536–546 (2013)
Zhang, T., Huang, M, Zhao, L.: Learning structured representation for text classification via reinforcement learning. In: Proceedings of Thirty-Second AAAI Conference Artificial Intelligence, pp. 6053–6060 (2018)
Paulus, R., Xiong, C, Socher, R.: A deep reinforced model for abstractive summarizatio. arxiv2017
Cuayáhuitl H.: SimpleDS: a simple deep reinforcement learning dialogue system. In: Jokinen, K., Wilcock, G. (eds.) Dialogues with Social Robots. Lecture Notes in Electrical Engineering, vol. 427. Springer, Singapore (2017)

Solving the Traveling Salesman with the Rat Swarm Optimization Algorithm (RSO)

Toufik Mzili[1(✉)], Mohammed Essaid Riffi[1], and Ilyass Mzili[2]

[1] Department of Computer Science, Faculty of Science, Chouaib Doukkali University, El Jadida, Morocco
toufikmzili95@gmail.com

[2] Department of Management, Faculty of Economics, Hassan First University, Settat, Morocco

Abstract. The new Rat Swarm Optimization (RSO) metaheuristic is inspired by the behavior of rats in nature. This algorithm was recently designed to solve continuous optimization problems. This algorithm was inspired by the hunting behavior of rats and the aggressive behavior of arguing. The results given by this optimizer in the continuous domain are better than the eight other optimizers. In this paper, we attempt for the first time to adapt this bio-inspired method to solve discrete optimization problems as well. To justify its ability to solve such problems, we chose the famous traveling salesman problem. An NP-hard problem that researchers have not yet found an efficient method to solve. While taking into account the natural behavior of rats. We used some instances of TSPLIB to test the efficiency of our discrete version. This adaptation was able to give good results in comparison with other metaheuristics. The experimental results show that the proposed RSO method is quite efficient in solving optimization problems compared to other well-known optimization algorithms.

Keywords: Rat Swarm Optimization · Swarm Intelligence · Travelling Salesman Problem · Nature-inspired · Combinatorial optimization · Metaheuristic

1 Introduction

In practice, the frequent use of computers leads to a large number of optimization problems in various fields, which are very complex and difficult to solve. The metaheuristic defines an algorithmic framework that can approximately solve such problems by combining construction methods with local and global search strategies and local optimal avoidance strategies. Metaheuristics represent a compromise between exact methods, which guarantee an optimal solution, but usually at the cost of large computational resources, while greedy methods are very fast, but usually give poor or unsatisfactory solutions.

Classical metaheuristics such as iterative local search hill climbing, genetic algorithms, simulated annealing, tabu search, and ant colony optimization have been shown to solve complex scheduling, space allocation, and clustering problems. The number of meta-heuristics is increasing, and there has been growing interest in it in recent years by

L. Azrar et al. (Eds.): CIP 2022, LNME, pp. 184–191, 2023.
https://doi.org/10.1007/978-3-031-23615-0_19

the scientific community, which employs techniques from artificial intelligence, operations research, soft computing, and more. New metaheuristics provide more powerful and efficient ways to solve traditional optimization problems, while also discovering new ways to solve unsolved problems.

Despite the variety of classical approaches to solving optimization problems, they do not always solve practical and applied optimization problems. It is generally accepted that these problems are qualified as difficult and very complex. The TSP (Reinelt 2003) is a classic and an NP-Hard (Hochba et al. 1997) example of a very difficult combinatorial optimization problem. A TSP can be characterized as an undirected and weighted graph, where the cities are the vertices of the graph, the paths are the edges of the graph, and the path distances are the edge lengths. It is a minimized problem beginning at a given node and terminating after visiting another node exactly once. Metaheuristics can solve multidimensional problems quickly. Heuristics can be classified into two categories: specific heuristics and meta-heuristics. Customize and develop specific heuristics to solve specific problems and/or instances.

TSP has many applications, including scheduling, logistics, microchip manufacturing, and DNA sequencing.

To solve this problem in an efficient way and in an acceptable time, researchers have proposed several approximation algorithms such as Tabu search (Cai et al. 2022), local search (Mendes et al. 2023), and simulated annealing (SA) (Venkateswaran et al. 2022), as well as metaheuristics such as the ant colony (Yang et al. 2008), Genetic Algorithm (GA) (Han 2022), Cuckoo Search (CS) optimization (Abed-alguni et al. 2022) and hybrid methods Particle Swarm Optimization (PSO) (Cui & Qiao, 2022), Firefly Algorithm (HA) (Xu & Lai, 2022), Bat Algorithm (BA)(Saji & Riffi, 2016).

The objective of this study was to apply the rat swarm optimization (RSO) method to the traveling salesman problem (TSP), a classical combinatorial optimization problem known for its high complexity and its use in many fields.

In order to prove the ability of this algorithm to solve discrete problems, we will compare the results given by this adaptation with those of a very well-known algorithm in the discrete which is the ACO (ant colony optimization).

This work will be presented according to the following structure: in the second section we will present the traveling salesman problem, in the third section we will present the RSO algorithm, and in the fourth section, we will present the approach to adopting RSO to solve TSP, In the fifth section, we will see the results using the TSPLIB instances and compare them with those of the ACO, and we finish by a discussion and a conclusion.

2 The Travelling Salesman Problem

The Traveling Salesman Problem (TSP) is a discrete optimization problem that consists in finding the shortest path between a set of places or cities to visit while visiting each city only once and returning to the starting city.

The objective of this problem is to minimize the distance traveled by a commercial traveler while also minimizing travel costs, fuel consumption, delivery time, and fatigue.

The TSP is a problem that can be used in various fields such as business, engineering, and mathematics. Its formulation has its roots in optimization and game theory. The problem consists of finding a traveling salesman who visits each stop consecutively so that in the end, the salesman has visited all the stops (Fig. 1).

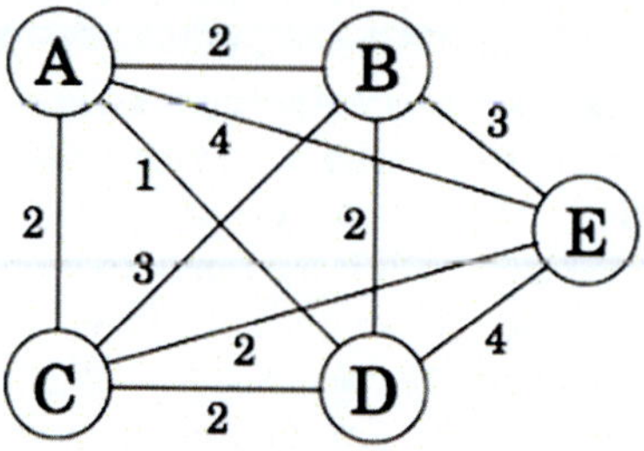

Fig. 1. Example of a TSP with five cities.

3 RSO Algorithm

The RSO algorithm was developed in 2020 to address problems involving continuous optimization (Dhiman et al., 2020).

The aggression and tracking of rats were used as a model for this program. Rats are inherently socially intelligent animals.

They participate in a variety of activities in pairs of two males and two females. They enjoy running, jumping, tumbling, and boxing. However, they are incredibly violent and frequently kill animals. This algorithm was inspired by this predator's aggressive propensity to chase and fight its prey. To create and improve RSO algorithms, mathematical models of rat chasing and fighting behavior must be used.

3.1 Pursuit of the Prey

Rats search for their prey in packs because they are social creatures. The location of the prey is considered to be known by the best search agent.

The other search agents can change their location. The proposed equations to simulate this scenario are as follows.

$$Y = A \times Y_i + C \times (Y_{Best} - Y_i) \tag{1}$$

where Y_{Best} is the best ideal solution and Y_i represents the positions of the rats. C and are random values ranging from [0, 2] to [1, 5], respectively, they are responsible for good exploration and exploitation of the search space during the iterations.

$$A = R - x\left(\frac{R}{\text{Max}_{Iteration}}\right), 1 \leq R \leq 5 \tag{2}$$

The x in [0, 1, 2…Max-iteration].

3.2 Fight with the Prey

The fight between the rats and their prey is mathematically described by the formula below:

$$Y_{i+1} = |Y_{best} - Y_i| \tag{3}$$

where Y_{i+1} represents the most recently updated location of the rat. The ideal solution is saved, and the locations of the other search agents are changed relative to it.

3.3 The Rat Swarm Optimizer Algorithm

Begin:

- **Step 1**: Produce the initial population of rats Yi, where I = 1, 2, and so on.
- **Step 2**: Initialize the RSO parameters.
- **Step 3**: Determine the suitability of each search agent.
- **Step 4**: Choose the best search agent.
- **Step 5**: Use Eq. (1) to update the positions of the search agents.
- **Step 6**: Check and update a search agent if it has exceeded the search space limit.
- **Step 7**: If there is a better solution than the already discovered optimal solution, calculate the updated fitness value of the search agent and update the Ybest.
- **Step 8**: Terminate the algorithm when the stopping conditions are satisfied. Otherwise, return to step five.
- **Step 9**: Return the best-found solution.

4 Modification of the RSO Algorithm for the Discrete TSP

It is impossible to solve discrete combinatorial optimization problems using the RSO approach because it is designed to solve continuous optimization problems, whereas discrete combinatorial optimization problems are represented by orders that can be modeled as vectors or a sequence of numbers. In order to adapt RSO to discrete combinatorial optimization problems, it is crucial to do so while respecting the real behavior of the Rat.

- The position of a Rat symbolizes a solution and is expressed mathematically as a collection of integers, where each integer is a city P = [1, 2, 3, 4, 5].
- The list of permutations to apply to the solution Y (vector) to produce the solution X is given by the difference between each of the two positions (X - Y).
- The product of the multiplication (*), which is an operation between a real k [0; 1] and a list of permutations Q, is a subset of Q'.
- Between the solution X and the set of permutations Q, the addition operation is a legal operation, and the result is a new solution X'. In order to create a new X, this process consists in applying the permutation of Q' to X.

5 Experimental Results and Evaluation

5.1 Test Results

The demonstration was performed on a computer with a CORE i7-3210 M CPU running at 2.5 GHz and 8 GB of RAM, and the OSR was developed in C++ (Table 1). Table 2 shows the results of running the algorithm on several instances of the TSPLIB benchmark. The parameters are defined as follows: 100 rats will be used, with between 3000 and 4000 iterations depending on the instance. The following data are presented in the Table 1:

Table. 1. Comparison criteria.

Information	Description
Inst	The name of the TSPLIB library's benchmark instance
Nb.node	The number of nodes
Optimum	The instance's most popular solution
Best	The algorithm's best result after numerous iterations is referred to as "best"
Worst	The incorrect solution that the algorithm discovered after numerous iterations
Mean	The mean of ten distinct algorithm executions
Time	Shows the average execution time of the algorithm over 10 different executions, expressed in seconds
Error (%)	Represents the relative error percentage $\frac{\text{average}-\text{opt}}{\text{opt}} \times 100$
PDbest (%)	Measures the best solution length's percentage departure from the ideal $\frac{\text{Solution length}-\text{best known solution length}}{\text{best known solution length}} \times 100$

5.2 Evaluation and Discussion

In this section, we will attempt to compare the results of the RSO application to those of ant colony optimization, an even more well-known metaheuristic. The information about ant colony optimization (ACO) is taken from (Bao, 2015).

Table 2 represents the experimental results of the proposed algorithm for TSPLIB instances, and it shows that RSO is able to achieve optimum for several TSPLIB instances in extremely respectable execution times.

Table 3 compares the average solution of the RSO algorithm to other algorithms that use the TSPLIB library to demonstrate the efficiency and performance of the algorithm.

According to the results shown in the Tables 2 and 3, Fig. 2, it is clear that the RSO algorithm produces better results than the other approaches. This is demonstrated by the RSO curve, which is smaller than that of other methods.

Table. 2. Result obtained by applying RSO to solve TSP.

Instance	Optimum	Best	Mean	Worst	Error (%)	PDbest	Time (s)
Eil51.Tsp	426	**426**	432.57	442	1.54	0	6.76
Berlin52.Tsp	7542	**7542**	7788.7	8111	3.27	0	3.84
St70.Tsp	675	**675**	685.46	699	1.55	0	9.39
Oliver30.Tsp	420	**420**	421.62	426	0.38	0	0.39
Eil76.Tsp	538	549	569.5	591	5.85	2.00	13.35
Kroa100.Tsp	21282	21353	21748.4	21986	2.19	0.33	18.76
Krob100.Tsp	22141	22337	23165.5	23929	2.62	0.87	9.50
Eil101.Tsp	629	653	672.62	696	6.93	3.67	0.39
Ch130.Tsp	6.110	6275	6507	6816	6.49	2.62	13.53
Rat99.Tsp	1211	1229	1274.44	1308	5.23	1.46	0.39
D198.Tsp	15780	16045	16517.8	16994	4.67	1.65	19.89

Table. 3. Comparison of experimental results of RSO with ACO.

Instance	Method	Best	Mean	Worst	Time (s)
Oliver30.Tsp	**RSO**	**420**	**421.62**	**426**	**0.39**
	ACO	423.74	424.68	429.36	35.20
Eil51.Tsp	**RSO**	**426**	**432.57**	**442**	**6.76**
	ACO	450.59	457.86	463.55	112.11
Berlin52.Tsp	**RSO**	**7542**	**7788.79**	**8111**	**3.84**
	ACO	7548.99	7659.31	7681.75	116.67
St70.Tsp	**RSO**	**675**	**685.46**	**699**	**9.39**
	ACO	696.05	709.16	725.26	226.06
Eil76.Tsp	**RSO**	**549**	**569.5**	**591**	**13.35**
	ACO	554.46	561.98	568.62	271.98
Kroa100.Tsp	**RSO**	**21353**	**21748.4**	**21986**	**18.76**
	ACO	22455.89	22880.12	23365.46	615.06
Eil101.Tsp	**RSO**	**653**	**672.62**	**696**	**0.39**
	ACO	678.04	693.42	705.65	527.42

This is due to the simplicity of RSO and its specifications, which allow for faster convergence and good exploitation and exploration of the search species compared to other techniques.

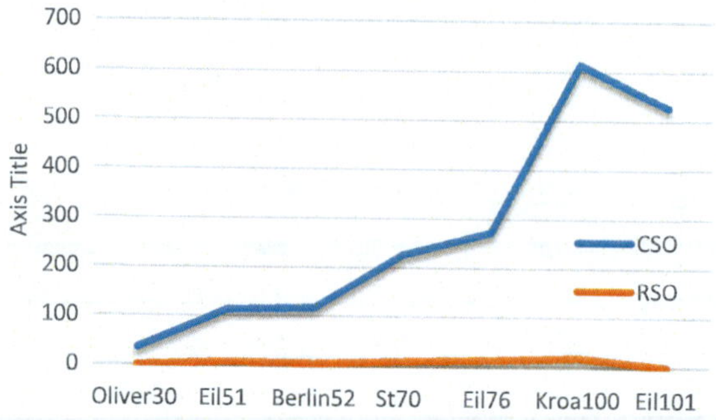

Fig. 2. Comparison of execution time between RSO and ACO.

6 Conclusion and Perspectives

To address the challenges of continuous optimization, the Rat Swarm optimization algorithm, a metaheuristic based on group intelligence and nature-inspired artificial intelligence, was recently introduced. This metaheuristic is inspired by the way rats attack and hunt their prey. When it was first introduced, it was intended to solve the continuous problems mentioned in the article (Dhiman 2020). The results of the continuous are far superior to those of the majority of robust metaheuristics. This prompted us to test this optimization in the discrete domain and show that it can solve discrete problems in addition to continuous problems. In this study, we used our optimizer to solve the TSP, a well-known NP-hard combinatorial optimization problem. The results of this optimizer surpass those of ACO, a very popular metaheuristic used in the discrete sector (ant colony optimization).

Future research on this algorithm will aim at improving its efficiency by avoiding local minima, expanding its use cases, and solving all discrete optimization problems such as topology control, scheduling, transit problems, vehicle routing, and electronic industrial facilities.

Acknowledgements. The authors would like to thank the editors and anonymous referees for their insightful and helpful comments and suggestions for improving this publication, as well as for emphasizing the importance of research guidance.

References

Cai, L., Wang, X., Luo, Z., Liang, Y.: A hybrid adaptive large neighborhood search and tabu search algorithm for the electric vehicle relocation problem. Comput. Ind. Eng. **167**, 108005 (2022)

Yang, J., Shi, X., Marchese, M., Liang, Y.: An ant colony optimization method for generalized TSP problem. Prog. Nat. Sci. **18**(11), 1417–1422 (2008)

Hochba, D.S. (Ed.).: Approximation algorithms for NP-hard problems. ACM Sigact News **28**(2), 40–52 (1997)

Sequeiros, J.A., Silva, R., Santos, A.S., Bastos, J., Varela, M.L.R., Madureira, A.M.: A Novel Discrete Particle Swarm Optimization Algorithm for the Travelling Salesman Problems. In: Machado, J., Soares, F., Trojanowska, J., Ivanov, V. (eds.) icieng 2021. LNME, pp. 48–55. Springer, Cham (2022). https://doi.org/10.1007/978-3-030-78170-5_5

McCall, J.: Genetic algorithms for modelling and optimisation. J. Comput. Appl. Math. **184**(1), 205–222 (2005)

Reinelt, G.: The traveling salesman: computational solutions for TSP applications, vol. 840. Springer (2003)

Mzili, I., Riffi, M.: Discrete Penguins search optimization algorithm to solve the traveling salesman problem. J. Theor. Appl. Inf. Technol. **72**, 331–336 (2015)

Abed-alguni, B. H., Paul, D.: Island-based Cuckoo search with elite opposition-based learning and multiple mutation methods for solving optimization problems. Soft Comput. **26**(7), 3293–3312 (2022)

Han, S., Xiao, L.: An improved adaptive genetic algorithm. In: SHS Web of Conferences, vol 140, p. 01044. EDP Sciences (2022)

Mendes, A.B., e Alvelos, F.P.: Iterated local search for the placement of wildland fire suppression resources. Eur. J. Oper. Res. **304**(3), 887–900 (2023)

Merz, P., Freisleben, B.: Genetic local search for the TSP: new results. In: Proceedings of 1997 IEEE International Conference on Evolutionary Computation (Icec'97), pp. 159–164. IEEE (1997, April)

Johnson, D., Aragon, C., McGeoch, L., Schevon, C.: Optimization by simulated annealing: an experimental evaluation. Part I. Graph Partitioning Oper. Res. **37**, 865–892 (1989). https://doi.org/10.1287/opre.37.6.865

Hossam, A., Bouzidi, A., Riffi, M.E.: Elephants Herding Optimization for Solving the Travelling Salesman Problem. In: Ezziyyani, M. (ed.) AI2SD 2018. AISC, vol. 912, pp. 122–130. Springer, Cham (2019). https://doi.org/10.1007/978-3-030-12065-8_12

Gündüz, M., Kıran, M., Özceylan, E.: A hierarchic approach based swarm intelligence to traveling salesman problem. Turk. J. Electr. Eng. Comput. Sci. **23** (2013). https://doi.org/10.3906/elk-1210-147

Gündüz, M., Kiran, M.S., Özceylan, E.: A hierarchic approach based on swarm intelligence to solve the traveling salesman problem. Turk. J. Electr. Eng. Co. **2015**(23), 103–117 (2015)

Nowicki, E., Smutnicki, C.: A fast taboo search algorithm for the job shop problem. Manage. Sci. **42**(6), 797–813 (1996)

Dhiman, G., Kumar, V.: Spotted hyena optimizer: a novel bio-inspired based metaheuristic technique for engineering applications. Adv. Eng. Softw. **114**, 48–70 (2017)

Discrete novel hybrid particle swarm optimization to solve travelling salesman problem. 17–20. https://doi.org/10.1109/WCCCS.2014.7107912

Dhiman, G., Garg, M., Nagar, A., Kumar, V., Dehghani, M.: A novel algorithm for global optimization: rat swarm optimizer. J. Ambient. Intell. Humaniz. Comput. **12**(8), 8457–8482 (2020). https://doi.org/10.1007/s12652-020-02580-0

Ye, H., Fu, Q., Ye, J., Zhong, C.: An improved chicken swarm optimization for TSP. In: International Conference on Applications and Techniques in Cyber Security and Intelligence, pp. 211–220. Springer, Cham (2019, June)

BAŞ, E., ÜLKER, E.: Dıscrete socıal spıder algorıthm for the travelıng salesman problem. Artif. Intell. Rev. **54**(2), 1063–1085 (2020). https://doi.org/10.1007/s10462-020-09869-8

Bao, H.: A two-phase hybrid optimization algorithm for solving complex optimization problems. Int. J. Smart Home **9**, 27–36 (2015). https://doi.org/10.14257/ijsh.2015.9.10.04

Venkateswaran, C., Ramachandran, M., Ramu, K., Prasanth, V., Mathivanan, G.: Application of simulated annealing in various field. Mater. Character. **1**(1), 01–08 (2022)

Cui, Y., Meng, X., Qiao, J.: A multi-objective particle swarm optimization algorithm based on two-archive mechanism. Appl. Soft Comput. **119**, 108532 (2022)

Xu, G.H., Zhang, T.W., Lai, Q.: A new firefly algorithm with mean condition partial attraction. Appl. Intell. **52**(4), 4418–4431 (2022)

Saji, Y., Riffi, M.E.: A novel discrete bat algorithm for solving the travelling salesman problem. Neural Comput. Appl. **27**(7), 1853–1866 (2015). https://doi.org/10.1007/s00521-015-1978-9

Mechanical Engineering and Mechatronics

Numerical Computation of Plasticity in Large Deformations Using the Asymptotic Numerical Method

El Kihal Chafik[1](✉), Askour Omar[1], Belaasilia Youssef[1], Hamdaoui Abdellah[1], Braikat Bouazza[1], Damil Noureddine[1], and Potier-Ferry Michel[2]

[1] Laboratoire d'Ingénierie Et Matériaux LIMAT, Faculté Des Sciences Ben M'Sik, Hassan II University of Casablanca, B.P 7955 Sidi Othman, Casablanca, Morocco
Chafik.elkihal@gmail.com

[2] Université de Lorraine, CNRS, Arts Et Métiers ParisTech, LEM3, F-57000 Metz, France

Abstract. The primary goal of this study is to simulate finite transformation elasto-plasticity problems using the Asymptotic Numerical Method (ANM). Elastoplastic behavior combines two unilateral conditions; the first concerns the transition from the elastic domain to the elastoplastic domain and the second concerns the elastic discharge. In the context of the Asymptotic Numerical Method (ANM), the elastoplastic behaviour law has to be changed to a regular one in order to obtain an analytical representation of every variable. The main point is to find an appropriate relationship between the plastic multiplier and the threshold function of plasticity. The linear problems found are solved using the Finite Element Method (FEM). An example of a 2D plate subjected to a time dependent monotonic loading is considered, and a study on the different regularization parameters is presented. The obtained results by the proposed algorithm are compared to the reference solution (Abaqus code). The comparison shows that this algorithm is effective for solving such problems.

Keywords: High Order Algorithm · Elastoplastic · Large deformation · Regularizations

1 Introduction

To numerically simulate the metal forming process such as lamination, stamping.... we must consider the major changes the metal go through both geometrical and/or material. These changes are usually permanent inelastic deformation, which means that the standard elastic behavior lows are no longer valid, which bring us to use plasticity behavior lows combined with the large strain theory. In previous work that exist in the literature (Assidi et al., 2009), (Hamdaoui et al., 2016) and (Yoshida et al. 2002), the majority of researchers have been showing an interest in developing elasto-plastic models to simulate this type of behaviors. As a matter of fact, the past decades have seen a great upsurge in research in modeling elasto-plastic materials. The first research considers that the plastic deformation is small compared to the elastic strain and there,

L. Azrar et al. (Eds.): CIP 2022, LNME, pp. 195–202, 2023.
https://doi.org/10.1007/978-3-031-23615-0_20

one uses the elastoplastic theory in small deformation (Assidi et al., 2009), (Hamdaoui et al., 2016). The second considers that plastic deformations are important, which is the case in shaping structures (Yoshida et al. 2002), (El Kihal et al., 2022). In order to correctly resolve large deformation plasticity problems, one must take into consideration the objective stress rate, meaning that the mechanical response of a material should not depend on the frame of reference.

In this paper, we propose to solve the large deformation elasto-plastic problems using the Asymptotic Numerical Method (Cochelin et al. 1994) and (Cochelin et al., 2007). The ANM associates three techniques, a perturbation technic, a continuation procedure and a discretization method. In this context, the solution of every step is represented by a power series and forms a family of linear problems. These problems are solved by decomposing only one tangent matrix per step. The length of every step is determined using properties of convergence of the Taylor series which result in naturally adapted step length.

We introduce first the general problem of elasto-plastic structure in large deformation theory, and then we try to regularized the four equations that prevents us from applying the perturbation technic (El Kihal et al., 2022). The elasto-plastic behavior with the elastic unloading exhibits three states, the elastic state, the plastic state and the elastic unloading state. So, to identify a unique regularized constitutive relation, one must regularize the two corresponding unilateral conditions (El Kihal et al., 2022) and (Assidi et al., 2009). The key point to establish this regular problem has already been discussed in previous works of elasto-plasticity in small deformations (Hamdaoui et al., 2016), we take the same regularization technics as in (Assidi et al., 2009), and adopt it to our problem. In other words, determine a regularized relation between the plastic multiplier and the yield function, and add a regularization parameter to the equivalent stress to avoid the singularity point, then add another regularization parameter to the load function for the same reason.

2 Large Deformation Regularized Problem

The variational problem translating the equilibrium of solid in the case of large deformation plasticity is written in the matrix form:

$$\int_{\Omega_0} \langle \delta L \rangle^t \left[A(f) \right] \{\tau\} d\Omega_0 = C(t) \int_{\partial\Omega_0} \langle \delta v \rangle \{F_{ext}\} dS_0 \tag{1}$$

where Ω_0 is the undeformed configuration and F^d is the force applied to the structure, $\left[A(f)\right]$ is the matrix of the inverse of the deformation gradient, $\{\tau\}$ represents the stress Kirchhoff tensor, $\{L\}$ represent the vector of the velocity gradient, and v is the velocity vector. We consider that the strain rate tensor is unambiguously divided into two parts (elastic and plastic) additively as $\{D\} = \{D^p\} + \{D^e\}$. The behavior low that describes such problem is written in the following matrix form:

$$\left\{\tau^J\right\} = \left[C^{elas}\right]\left(\{D\} - \{D^p\}\right) \tag{2}$$

where $\{\tau^J\}$ is the Jaumann stress tensor, $[C^{elas}]$ is the representing matrix of the material elasticity tensor. The plastic behavior part is written as:

$$\{D^p\} = \dot{\lambda}\left\{\frac{\partial f_y}{\partial \tau}\right\} = \dot{\lambda}\{n\} \tag{3}$$

where $\dot{\lambda}$ is the plastic multiplier, $\{n\}$ is the vector of direction flow and f_y is the yield function. The normal vector $\{n\}$ can be written as:

$$\{n\} = \frac{3}{2}\frac{\tau^d}{q} \tag{4}$$

and the yield function is defined as:

$$\begin{cases} f_y = \frac{q-\tau_e}{\tau_e} \\ \tau_e = \tau_y + h\bar{\epsilon}^p \\ \bar{\epsilon}^p = \int \sqrt{\frac{3}{2}D^p : D^p}\, dt = \int \dot{\lambda} \\ q^2 = \frac{3}{2}\tau^d : \tau^d \end{cases} \tag{5}$$

where τ_e represents the effective stress, τ_y is the yield stress, h is the plastic modulus, $\bar{\epsilon}^p$ is the equivalent plastic strain, q is the equivalent stress and τ^d is the deviator of the Kirchhoff stress tensor. The condition of the yield function and the plastic multiplier can be reformulated from the Kuhn-Tucker's condition:

$$\dot{\lambda} \geq 0, f_y \leq 0, \dot{\lambda} f_y = 0 \tag{6}$$

Combining the previous relations with the unilateral relation, one gets the consistency relation:

$$\dot{\lambda} = g(f_y)\langle n : D\rangle \tag{7}$$

The unilateral relation is presented by two non-regularized functions $g(f_y)$ and $\langle . \rangle$, which are written as:

$$\begin{cases} g(f_y) = 0 \text{ } if \text{ } f_y < 0 \\ g(f_y = 0) = \frac{2\mu}{2\mu+h} \end{cases} \qquad \begin{cases} \langle n : D\rangle = 0 \quad if \text{ } n : D < 0 \\ \langle n : D\rangle = n : D \text{ } else \end{cases} \tag{8}$$

μ represents the shear modulus. The two functions are going to be replaced by:

$$\dot{\lambda} = G(f_y)H(D) \tag{9}$$

With

$$\begin{cases} G(f_y) = \dfrac{\eta_3}{\frac{f_y^2}{2\mu}\tau_e + \eta_3\left(\frac{3}{2} + \frac{h}{2\mu}(1+f_y)\right)} \\ H(H - \xi) = \eta_2^2 \dot{\tau}_c^2 \end{cases} \tag{10}$$

The Von Mises equivalent stress, which isn't an analytic equation when the stress tends to zero. So, we add a regularization parameter η_1 to avoid the zero in the square root, and the new expression of the equivalent stress is given by the following formula: $q^2 = \frac{3}{2}\tau^d : \tau^d + \eta_1^2\tau_y^2$. Furthermore, two additional $\hat{F}$ and Den variables are introduced to simplify the lettering of $G(f_y)$, that will be split into three elementary equations:

$$\hat{F} = f_y^2;\ Den = \frac{\hat{F}}{2\mu}\tau_e + \eta_3\left(\frac{3}{2} + \frac{h}{2\mu}(1 + f_y)\right);\ GDen = \eta_3 \tag{11}$$

To reproduce a typical behavior of the elastoplastic law (elastic, plastic, elastic unloading), C(t) is defined by the following hyperbolic equation:

$$\left(C - C_m\frac{t}{T_m}\right)\left(C - C_m\left(2 - \frac{t}{T_m}\right)\right) = \eta_4^2 C_m^2 \tag{12}$$

where C_m and T_m are the load parameters and η_4 is a regularization parameter.

In structural mechanics problems involving elasto-plastic constitutive law, the equations are given by:

$$R(U(t), C(t)) = 0 \tag{13}$$

where R represents the residual vector and $U(t)$ is the vector of unknowns with the following components:
$\left(v, \tau, \tau^j, D, D^p, G, H, \omega, f, F, L, l, R, q, n, \xi, Den, f_y, \hat{F}, \tau_e, \dot{\lambda}\right)$, and $C(t)$ is a loading parameter.

3 Numerical Resolution

The vector unknown U and the loading C are expressed as a Taylor series expansion truncated at order N as:

$$\begin{cases} U(a) = U^j + \sum_{i=1}^{N} a^k U_k \\ C(a) = C^j + \sum_{i=1}^{N} a^k C_k \end{cases} \quad a \in [0, a_{amx}] \tag{14}$$

where $\left(U^j, C^j\right)$ are a known and a given regular solutions corresponding to $a = 0$, a_{max} is the convergence radius of the series. For this study, we chose to work with the parameterization equation used in (El Kihal et al., 2022), written as:

$$a = \alpha_1\left(u - u^j\right)u_1 + \alpha_2\left(t - t^j\right)t_1 + \alpha_3\left(C - C^j\right)C_1 \tag{15}$$

where t is the time, C is the load and u represents the displacement. The coefficients $\alpha_i, fori = 1, 3$ are equals to 0 or 1. By injecting the Taylor series representations (14) into the nonlinear problem (13) we obtain, the linear problems verified by each term of the series:

$$order1 : \int_{\Omega_0} \langle\delta L\rangle\left({}^t\left[A(f_0)\right]\{\tau_1\} + {}^t\left[A(f_1)\right]\{\tau_0\}\right)d\Omega_0 = C_1(t)\{F_{ext}(\delta v)\} \tag{16}$$

order k $(2 \leq k \leq N)$:

$$\int_{\Omega_0} \langle \delta L \rangle \left({}^t[A(f_0)]\{\tau_k\} + {}^t[A(f_k)]\{\tau_0\} \right) d\Omega_0 = C_k(t)\{F_{ext}(\delta v)\} + \left\{ P_k^{nl} \right\} \tag{17}$$

where

$$\{\tau_k\} = L(v_k) + \left\{ \tau_k^{nl} \right\}; \ \{f_k\} = L(v_k) + \left\{ f_k^{nl} \right\} \tag{18}$$

The terms $\{P_k^{nl}\}$, $\{\tau_k^{nl}\}$, $\{f_k^{nl}\}$ are known and calculated previously. Once the problem is discretized, we apply the finite element method to solve the linear problems (16) and (17). The tangent matrix is given by:

$$\begin{aligned} [K_t]\{v_0\} &= C_1\{F^{ext}\} for\, k = 1 \\ [K_t]\{v_{k-1}\} &= C_k\{F^{ext}\} + \{P_k^{nl}\} for\, k \geq 2 \end{aligned} \tag{19}$$

The solution of the problem (13) is computed using a path-following method proposed by (Cochelin et al, 2007), where the solution is obtained branch-by-branch and the final solution point of a branch is the new beginning point of the next branch. The range of validity $a_{\max}$ of the parameter a is evaluated as follow:

$$a_{\max}\left(U^{(i)}\right) = \left(\in \frac{\left\| \left\{ U_1^{(i)} \right\} \right\|}{\left\| \left\{ U_N^{(i)} \right\} \right\|} \right)^{\left(\frac{1}{N-1}\right)} \tag{20}$$

where $\in$ represents the tolerance parameter and $U^{(i)}$ represents the components of the unknown vector U. For this study, we define the validity range $a_{\max}$ as in (El Kihal et al., 2022) and (Assidi et al. 2009). We chose $a_{\max}$ as the smallest of all the components of the vector $U^{(i)}$:

$$a_{\max} = \inf\left(a_{\max}\left(U^{(i)}\right)\right) \tag{21}$$

Once $a_{\max}$ calculated, the starting solution $\left(U^j, C^j\right)$ in (13) to the next branch is evaluated by $U^j = U(a_{\max})$ and $C^j = C(a_{\max})$.

4 Numerical Example

We consider a $2D$ plate of length $L = 100$ mm and width $l = 40$. The plate is submitted to a monotonic loading in pressure $C(t)P$ on the right edge until reaching an elastoplastic deformation of 35%. The loading function parameters are chosen as $C_m = 55$ and $T_m = 55$. For symmetry reasons, we work only with the plate quarter as indicated in Fig. 1.

The Taylor series truncation order and the tolerance parameter are $N = 15$ and $\in = 10^{-3}$. The regularization parameters are $\eta_1 = 10^{-2}$, $\eta_2 = 10^{-4}$, $\eta_3 = 9.10^{-5}$ and $\eta_4 = 5.10^{-3}$. We chose for this comparison to work with the parametrization based on the load parameter and time (PCT). In Fig. 2, we compare our response curve to the one obtained by the industrial code ABAQUS (Abaqus, 2014) at the Pt1 indicated in Fig. 1; we see clearly that the response curve found by the Asymptotic Numerical Method approach the reference curve. The ANM completed the whole curve in 116 steps where ABAQUS needed 110 increments and 140 iterations which means 250 steps. We will discuss later on how to optimize more our algorithm in order to obtain the whole response curve with less steps.

To justify the choice of the parameters of regularization, we present a full study on the four parameters η_1, η_2, η_3 and η_4, in which we discuss the influence of every parameter on the solution obtained and number of steps. To do so, we are going to fix the order at 15 and the tolerance parameter at 10^{-3}, and to simplify the task we vary each parameter while keeping the others fixed to describe the domain validity of each one with respect to the others. To avoid division by zero, the first parameter η_1 is added to the equation of the equivalent stress q given previously. We vary it in an interval of $\eta_1 \in \left[10^{-6}, 10^{-1}\right]$. The response curve starts to converge from a value $\eta_1 = 10^{-2}$ where we get the whole response curve in 116 ANM steps, and in 118, 127 steps for the values $\eta_1 = 10^{-3}$ and $\eta_1 = 10^{-4}$ respectively. And it diverges again for the value of $\eta_1 = 10^{-5}$ and $\eta_1 = 10^{-6}$. So, the interval for the first parameter η_1 will be from 10^{-2} until 10^{-4}. Secondly, we discuss the influence of the η_2 which influences on the H function that describe the elastic discharge (10), and it appears also in the equation of ζ that completes the H function. After testing it in the interval of $\eta_2 \in \left[10^{-6}, 10^{-1}\right]$, the solution start to converge from the value $\eta_2 = 10^{-3}$ with more accurate results at $\eta_2 = 10^{-4}$, where we obtain the whole response curve in 116 ANM steps, and the more it's value decreases the more steps are needed, for $\eta_2 = 10^{-6}$ it needs 128 ANM steps. So, the interval is $\eta_2 \leq 10^{-3}$. The third parameter is used to regularize the function $G\left(f_y\right)$ (10). We do the same test that was previously conducted on the two other parameters, we vary it in the interval $\eta_3 \in \left[10^{-6}, 10^{-1}\right]$, and the solution starts to converge from the value $\eta_3 = 9.10^{-4}$. Therefore, the domain of validity is $\eta_3 \leq 9.10^{-4}$.

For the final regularization parameter η_4 which is linked to the loading equation, which is described by a triangle and as previously done, we fix the other parameters and we vary η_4. The parameter η_4 converges for the values $\eta_4 \leq 5.10^{-3}$.

Table 1 indicates the influence of the number of the order for two given tolerance parameters $CRIT1 = 10^{-3}$ and $CRIT2 = 10^{-4}$ for the time parameterization (PT) and the parametrization based on loading and time (PCT). It is shown that, if the number of the series order increases the number of the ANM steps decreases considerably for both parametrization and for the two of the tolerance criteria. In exact numbers, we find that for PT-CRIT1 we get 42.77% decrease in the number of steps from the order 10 to 30, and 57.32% decrease for PT-CRIT2.

5 Conclusion

We propose in this study a high order algorithm based on the Asymptotic Numerical Method technics to simulate large deformation elasto-plasticity problems. We adopted the regularization used in (El Kihal et al., 2022) for the elasto-plastic behavior in order

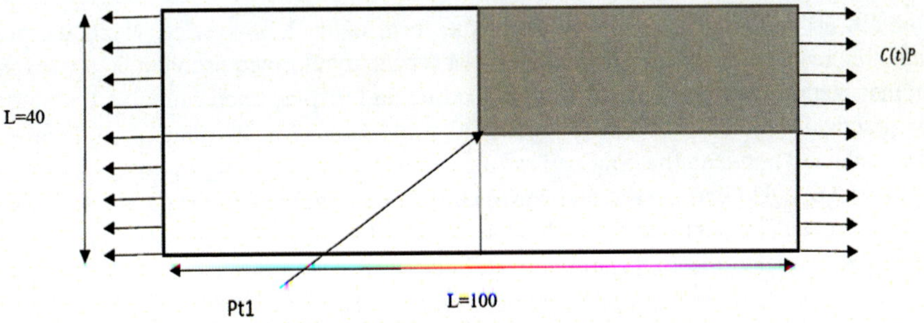

Fig. 1. Tensile test for the 2D plate

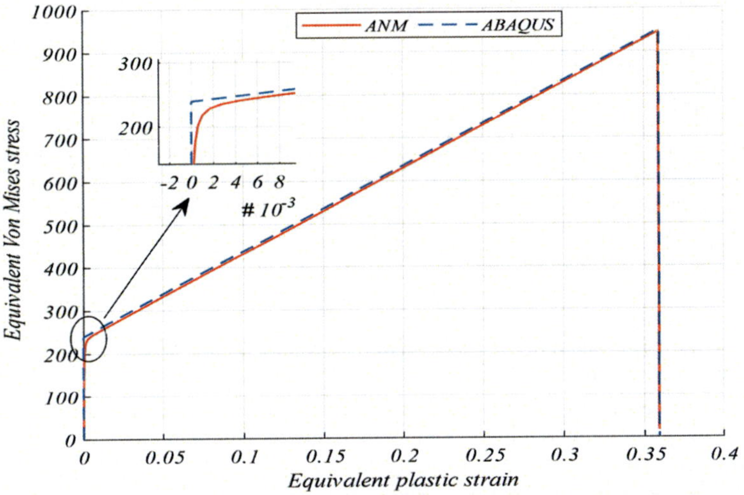

Fig. 2. The equivalent plastic strain versus the equivalent stress. The numerical parameters used are $N = 15$, $\epsilon = 10^{-3}$. The results are compared with the ones of industrial code Abaqus.

Table 1. Number of steps for the two parametrizations (PT, PCT) with two tolerance parameters (CRIT1, CRIT2) for different truncation orders N = 10, N = 15, N = 20, N = 30.

N-order	PT		PCT	
	CRIT1	CRIT2	CRIT1	CRIT2
10	166	239	157	226
15	118	141	116	134
20	104	115	102	114
30	95	102	92	99
ABAQUS	250			

to obtain an analytical representation in the form of a Taylor series. Next, we presented the resolution strategy using the ANM for the regularized problem, and used the parametrization equation based on time, loading and displacement. Finally, the model was practically tested in a tensile numerical simulation of a 2D plate. A comparison was conducted between the obtained results using the proposed algorithm and the ones found by ABAQUS code. Also, an investigation of the influence of the parameterization was presented. The obtained result demonstrates that the algorithm proposed is capable of computing elasto-plasticity problems in finite transformation.

References

El Kihal, C., et al.: Simulation of a 2D Large Transformation Elastoplasticity Problem Using the ANM. In: Abdel Wahab, M. (ed.) NME 2021. LNME, pp. 185–193. Springer, Singapore (2022). https://doi.org/10.1007/978-981-16-8806-5_13

El Kihal, C., et al.: Asymptotic numerical method for finite plasticity. Finite Elem. Anal. Des. **206**, 103759 (2022)

Assidi, M., Zahrouni, H., Damil, N., Potier-Ferry, M.: Regularization and perturbation technique to solve plasticity problems. Int.J. Mater. Form. **2**(1), 1–14 (2009)

Cochelin, B., Damil, N., Potier-Ferry, M.: Méthode asymptotique numérique. Hermes Lavoissier (2007)

Abaqus.: Version. 6.14 Documentation. Dassault Systemes Simulia Corporation **651**, 6.2 (2014)

Cochelin, B.: A path-following technique via an asymptotic-numerical method. Comput. Struct. **53**(5), 1181–1192 (1994)

Crishield, M.A.: A faster modified newton-raphson iteration. Comput. Methods Appl. Mech. Eng. **20**(3), 267–278 (1979)

Hamdaoui, A., Braikat, B., Damil, N.: Solving elastoplasticity problems by the asymptotic numerical method: Influence of the parameterizations. Finite Elem. Anal. Des. **115**, 33–42 (2016)

Yoshida, F., Uemori, T.: A model of large-strain cyclic plasticity describing the Bauschinger effect and workhardening stagnation. Int. J. Plast. **18**(5–6), 661–686 (2002)

A Meshless Collocation Method for Linear Vibration Analysis of Magnetostrictive Actuators

K. Belfallah(✉) and M. Jamal

Laboratoire d'Ingénierie Et Matériaux (LIMAT), Faculté Des Sciences Ben M'Sik, Hassan II University of Casablanca, Sidi Othman, BP 7955 Casablanca, Morocco

kawtar.belfallah-etu@etu.univh2c.ma

Abstract. The main objective of this paper is the modeling and simulation of longitudinal vibrations of a linear magnetostrictive actuator. The mechanical modeling of the actuator is realized by using a one-dimensional beam made of an elastic magnetostrictive material attached to a linear elastic spring. The mathematical model is made based on the Euler Bernoulli assumptions of elastic beams taking into account the coupling between mechanical and magnetic effects under magnetic excitation. The equation governing the longitudinal vibrations of the magnetostrictive actuator is obtained using the Lagrangian Formalism. The equation obtained is a space-time derivative equation with a coefficient that depends on the mechanical, magnetic and magnetostrictive characteristics and also implicitly depends on time. The resolution of this type of equation by analytical way is difficult or even impossible so we resort to numerical methods, for this work we adopt the moving least squares method (MLS) and the finite differences method (FDM). A comparison between the results obtained by the two proposed approaches is presented.

Keywords: Vibration · Magnetostriction · Magnetostrictive material · Actuator · Terfenol-D · Moving Least Squares (MLS) · Finite differences method (FDM)

1 Introduction

Smart or adaptive materials are composed of piezoelectric materials, electrostrictive materials, magnetostrictive materials, and Shape Memory Alloys, they modify their physical properties such as stiffness or viscosity according to an external stimulus such as mechanical stress, electric and magnetic field, or temperature change (Gandhi and Thompson 1992). Each type of material has different qualities which are put to good use, depending on the field of application. The materials we are interested in are magnetostrictive materials that can transform mechanical energy into magnetic energy and vice versa, these materials are used a lot as actuators and sensors (Pasquale 2003). Studies are being carried out on these different materials to optimize them but also to minimize losses by appropriately placing them. For these materials, efficiency and weight must be combined.

L. Azrar et al. (Eds.): CIP 2022, LNME, pp. 203–211, 2023.
https://doi.org/10.1007/978-3-031-23615-0_21

In 1842, the English physicist James Prescott Joule discovered that an iron bar subjected to a longitudinal magnetic field elongated according to this field, while at the same time contracting transversely, as if under the effect of mechanical traction or compression. He called this phenomenon magnetostriction.

Since then, other mechanical effects of the magnetic field have been demonstrated. When subjected to a magnetic field, samples of certain substances not only expand or contract, but also bend, twist, change their density and even change their Young's modulus. Magnetostriction is, in the most general sense of the term, the set of relations linking the mechanical properties of a body to its magnetic properties.

Magnetostrictive materials are ferromagnetic materials that magnetize under applied mechanical stress or undergo shape changes under the application of a magnetic excitation field. The military pioneered the use of magnetostrictive materials during the Second World War for submarine sonar. The two main advantages of these materials are that they can transmit (transducer mode) and receive acoustic waves (sensor mode) with the same material and that, as no physical contact is required between the coil (creating the magnetic field) and the magnetostrictive material, control can be done "remotely". They have potential applications in several sectors, such as aviation and automotive (Shoa et al., 2010). These areas require sensors and actuators that can prevent, detect and repair damage in several critical structural components in real-time (Shoa et al., 2010). The range of applications for this material is vast: active vibration and noise control systems, machine tools, servo valves, hybrid motors, sonar devices, automatic braking systems, etc.

However, the magnetostrictive effect of the materials used during the Second World War (mainly Al-Fe and nickel) was very weak. Around 1950, magnetostrictive sensors and transducers (Jaffe and Berlincourt 1965) were replaced by piezoelectric materials, which had become much more efficient.

In the 1960s, research resumed after the promising study by Mr. Belov in Moscow. He studied the magnetostriction of certain metals of the rare earth family. Then Clark, an American from the Naval Ordnance Laboratory, discovered in 1974 an alloy that, at room temperature and under a weak magnetic field, preserves high magnetostriction, thus Terfenol-D (a dysprosium-terbium-iron alloy) was born. The term "giant magnetostriction" is often given to these new materials to distinguish them from those of the Al-Fe or nickel type whose magnetostrictive effects are less important. The deformation performance is about twice as good as the best piezoelectric materials.

Research on magnetostrictive materials and the dynamic behavior of structures made of these magnetostrictive materials has attracted a lot of interest in recent years (Shang et al., 2008). Modeling tools are therefore essential to help optimize magnetostrictive structures made of magnetostrictive materials. The need for these materials exists in all fields of science and excites researchers to study and model this type of material. We are therefore interested in simulating and modeling the vibratory behavior of a magnetostrictive actuator immersed in a magnetic excitation field using a numerical approach based on meshless methods.

The objective of this work is to perform a modeling and simulation of vibrations of a magnetostrictive actuator. The actuator is modeled by a one-dimensional beam made of a magnetostrictive material of Euler-Bernoulli type attached to a linear elastic spring.

The numerical resolution of the dynamic equation that governs the vibrations of the magnetostrictive actuator is performed by the moving least squares method (MLS) (Liu and Gu 2005) and the finite differences method (FDM) (Leveque 1998). Comparisons of the obtained results are presented.

2 Modeling of the Linear Magnetostrictive Actuator

The considered magnetostrictive actuator is shown in Fig. 1.

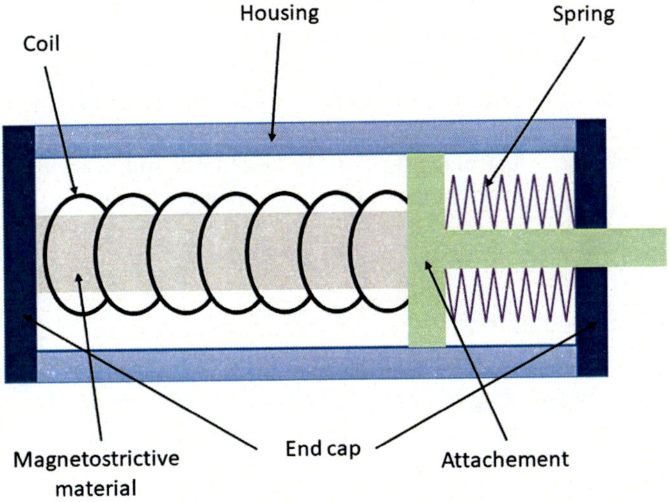

Fig. 1. Experimental system of the linear magnetostrictive actuator

To study the vibratory behavior of the linear magnetostrictive actuator, we model this device by a rod made of an elastic ferromagnetic material (Terfenol-D) of length l, cross section S, mass density ρ, Young's modulus E, magnetoelastic coefficient r and magnetostrictif modulus m, attached to a spring of stiffness k_r. The rod is clamped at x $= 0$ and connected to a spring at $x = l$ and subjected to a longitudinal magnetic field H (see Fig. 2).

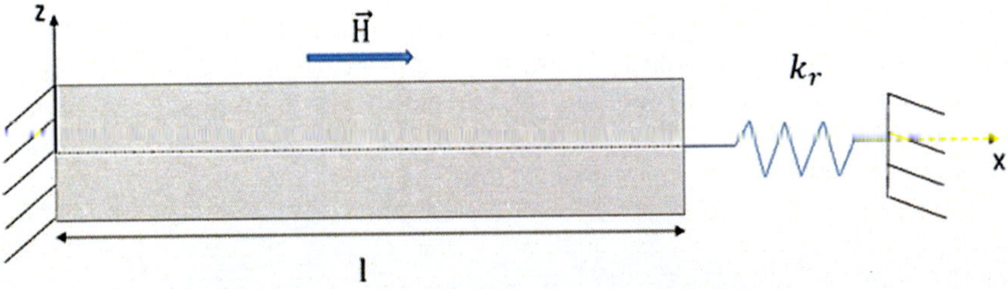

Fig. 2. Unidimensional model of the linear magnetostrictive actuator

The considered unidimensional nonlinear constitutive law of the homogeneous and isotropic magnetostrictive material is given by (Wan et al. 2003):

$$\begin{cases} \sigma = \frac{E\varepsilon - EmH^2}{1+EH^2} \\ B = \mu H + mH\sigma + rH\sigma^2 \end{cases} \tag{1}$$

where σ is the normal stress, B is the magnetic induction, ε is the normal strain and μ is the magnetic permeability.

To find the vibrational equation of the mechanical system, we use the Lagrangian formalism. The Lagrangian of this system is given by:

$$L = T - U = T - U_e + U_m \tag{2}$$

with T, U_e and U_m respectively represents the kinetic energy, the elastic energy and magnetic energy.

By introducing the displacement field of the Terfenol-D rod, we get the following equation for the longitudinal vibration of the studied linear actuator:

$$\begin{cases} \frac{\partial^2 u(x,t)}{\partial t^2} - \frac{E}{\rho} C(t) \frac{\partial^2 u(x,t)}{\partial x^2} = 0 \\ C(t) = \frac{1}{\left(1+rEH^2(t)\right)^2} \end{cases} \tag{3}$$

where $C(t)$ is a function of time t which depends on Young modulus E, magnetoelastic coefficient r and magnetic field H.

The wave propagation Eq. (3), is completed by the following boundary:

$$\left[\frac{\partial \mathrm{u(x,t)}}{\partial \mathrm{x}} + \frac{\mathrm{K_r}}{\mathrm{ES}}\left(1 + \mathrm{rEH}^2(t)\right)^2 \mathrm{u(x,t)}\right]_{\mathrm{x=l}} = \mathrm{mH}^2 \tag{4}$$

And

$$u(x = 0, t) = 0 \tag{5}$$

and by the periodicity conditions:

$$\begin{cases} u(x,\mathrm{t})|_{\mathrm{t=0}} = u(x,T)|_{t=T} \\ \left.\frac{\partial u(x,0)}{\partial t}\right|_{\mathrm{t=0}} = \left.\frac{\partial u(x,T)}{\partial t}\right|_{\mathrm{t=T}} \end{cases} \tag{6}$$

Note that the dynamic Eq. (3) has a variable coefficient with time t and the vibration problem of the linear magnetostrictive actuator is completed by the boundary conditions and the temporal periodicity condition (4), (5) and (6). In the following section, we propose two numerical approaches to solve this problem: the moving least squares (MLS) method and the finite difference method (FDM).

3 Numerical Model

3.1 Space Discretization Procedure Mesh-Free Methods

The displacement field $u(x, t)$ at any point of the rod can be expressed using N displacement values of the scattered particles $(x_1, x_2, \ldots x_N)$, according to the meshless methods (Liu and Gu 2005; Chen et al., 2006), as follows:

$$u(x) = \sum_{i=1}^{n} \phi_i(x) u_i = \phi^T(x) U_S \tag{7}$$

where n is the number of nodes included in the support domain in the axial direction, u_i is the nodal displacement field variable at the ith node, U_S is the vector that gathers all nodal displacement field variables at these n nodes, and $\phi_i(x)$ is the shape function of the ith node.

Inserting (7) in (3) we obtain the following material system:

$$[M]\{\ddot{u}\} + C(t)[K]\{u\} = \{0\} \tag{8}$$

With:

$$[M] = \rho\phi(x); \; [K] = -E\phi_{,xx}(x) \tag{9}$$

$[M]$: is the mass matrix.
$[K]$: is the stiffness matrix.

3.2 Time Discretization Procedure-Newmark Scheme

Using the Newmark implicit time scheme, the time discretized form of the Eq. (8) is expressed by:

$$[M]\{\ddot{u}\}^{n+1} + \boxed{K}\{u\}^{n+1} = \{0\} \tag{10}$$

With:

$$\boxed{K} = C(t)[K] \tag{11}$$

where $\{u\}^{n+1}$ is the nodal displacement, $\{\dot{u}\}^{n+1}$ is the nodal velocity and $\{\ddot{u}\}^{n+1}$ is the nodal acceleration at time $(n+1)\Delta t$, where Δt is the time step and $n = 1, 2, \ldots, N$ given by:

$$\{\dot{u}\}^{n+1} = f_0\{\dot{u}\}^n + f_1\{\ddot{u}\}^n + f_2\{\ddot{u}\}^{n+1} \tag{12}$$

$$\{\ddot{u}\}^{n+1} = g_0\left(\{u\}^{n+1} - \{u\}^n\right) - g_1\{\dot{u}\}^n - g_2\{\ddot{u}\}^n \tag{13}$$

with f_i and g_i (i = 0, 1, 2) are the Newmark coefficients (Newmark 1959).

4 Numerical Application

Consider the vibration of a magnetostrictive rod (Terfenol-D) of length $l = 0.25$ m, section $S = 3.1410^{-4}$ m^2, Young's modulus $E = 26.5$ GPa and mass density $= 9250$ kg m^{-3}. The magnetostrictive modulus $m = 0.0910^{-12}$ $\mathrm{m}^2\mathrm{A}^{-2}$ and the magnetoelastic coefficient $r = -2.7710^{-20}$ $\mathrm{m}^2\mathrm{A}^{-2}\mathrm{Pa}^{-1}$, the stiffness of the spring $k_r = 510^6 Nm^{-1}$. The number of turns in the bobbin $n = 220$ and the AC the excitation current that passes through the solenoid is harmonic: $i(t) = i_{max}\sin(t)$, with $i_{max} = 15$A and the pulsation $= 4.4$ kHz (the frequency $f = 2/ = 700$ Hz).

Figure 3 shows that the frequency of the displacement at the tip is double that of the frequency of the excitation current.

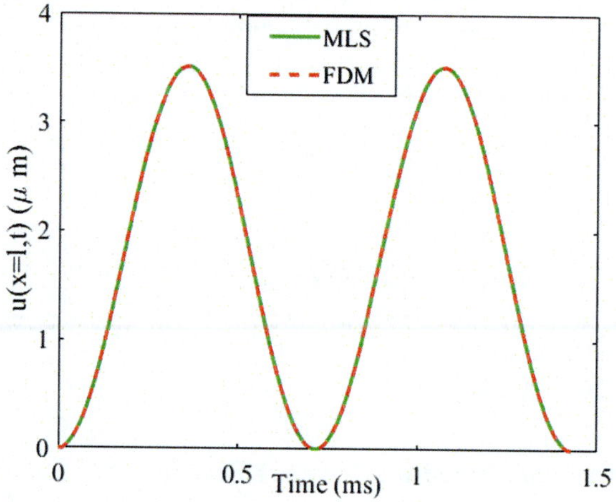

Fig. 3. Displacement versus time at the end of the pusher $u(l, t)$

The Fig. 4 shows that the relationship between deformation (10^{-6}) and the magnetic field is an even function. In a first approach, we can approximate it using a second-order polynomial, this is exactly the behavioral relation used in the modeling of the studied actuator.

The effect of various material parameters (m, r and k_r) on the performance of the linear magnetostrictive actuator is also studied.

1) Impact of the magnetostrictive modulus *m*

The displacement response at the end of the pusher ($x = l$) evaluated for a number of different magnetostrictive modulus values: m $= 0.06 \times 10^{-12}, 0.09 \times 10^{-12}$ and $0.14 \times 10^{-12} m^2 A^{-2}$ for a fixed spring stiffness $k_r = 510^6$ Nm^{-1} and magnetoelastic coefficient $r = -2.77 \times 10^{-20}$ $\mathrm{m}^2\mathrm{A}^{-2}\mathrm{Pa}^{-1}$.

The obtained results show that the magnetostrictive modulus m has an effect on the displacement response, The impact of the magnetostrictive modulus m on the displacement is minimal. If m increases, the displacement decreases (Fig. 5).

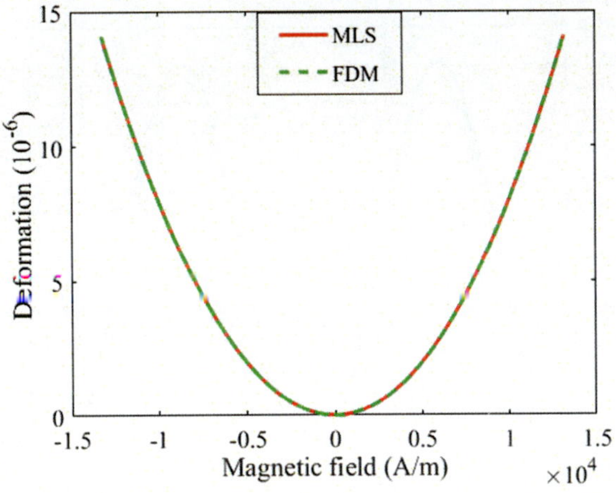

Fig. 4. Magnetostrictive deformation of the Terfenol-D rod

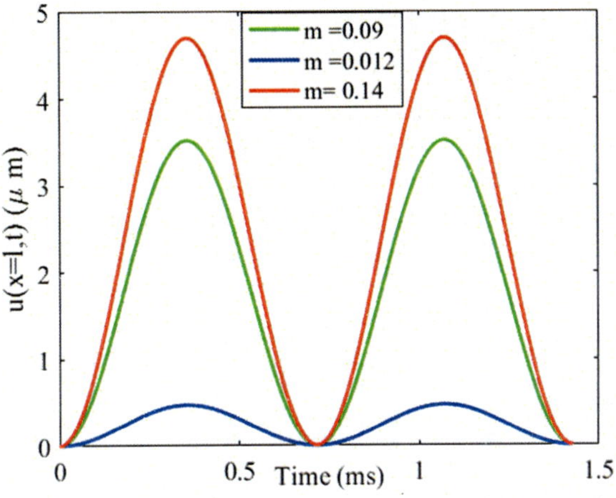

Fig. 5. Influence of magnetostrictive modulus $\boldsymbol{m}$ on the displacement $\boldsymbol{u}(\boldsymbol{l}, \boldsymbol{t})$.

2) **Impact of magnetoelastic coefficient *r***

The displacement at the end of the pusher ($x = l$) is also calculated for three different magneto elastic coefficient values: $r = -2.77 \times 10^{-20}$ $\mathrm{m^2 A^{-2} Pa^{-1}}$, -710^{-20} $\mathrm{m^2A^{-2}Pa^{-1}}$ and 2×10^{-20} $\mathrm{m^2A^{-2}Pa^{-1}}$, while the magnetostrictive modulus is set to $m = 0.06 \times 10^{-12}$ $\mathrm{m^2A^{-2}}$ and $k_r = 510^6$ $\mathrm{Nm^{-1}}$.

The Fig. 6 shows that the change in the magnetoelastic coefficient r does not have a remarkable influence on the displacement response.

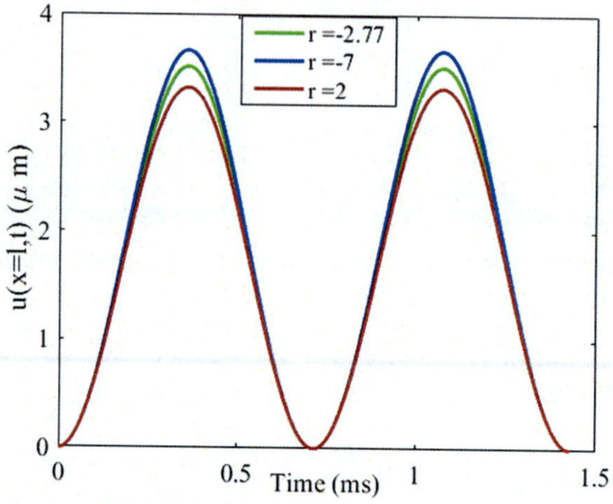

Fig. 6. Influence of magnetoelastic coefficient $\boldsymbol{r}$ on $\boldsymbol{u(l,t)}$.

3) Impact of spring stiffness k_r

The computation of displacement at the end of the pusher ($x = 1$) for different spring stiffness values: $k_r = 510^6$, 2×10^6 and 10×10^6 Nm^{-1} and for $r = -2.7710^{-20}\ \text{m}^2\text{A}^{-2}\text{Pa}^{-1}$ and $m = 0.0910^{-12}\ \text{m}^2\text{A}^{-2}$.

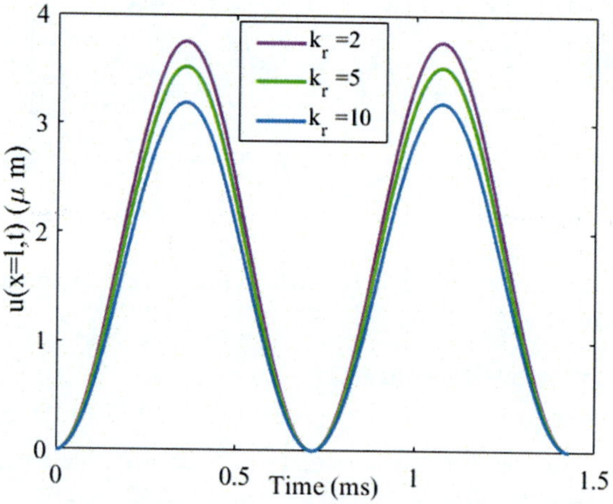

Fig. 7. Effect of the spring stiffness $\mathbf{k_r}$ on $\boldsymbol{u(l,t)}$.

The Fig. 7 show that if the spring stiffness decreases, the amplitude of the displacement increases.

5 Conclusion

In this work, we are interested in modeling the experimental device by a beam-spring model to analyze the longitudinal vibrations of a linear magnetostrictive actuator. The coupling of magnetic and mechanical effects is taken into account in this study. The vibratory response of the actuator is numerically computed by solving an elastic wave propagation equation with coefficients depending on the mechanical, magnetic and magnetostrictive characteristics as well as time. Two numerical approaches combined with the implicit Newmark time scheme. The influence of some magnetic and mechanical parameters is studied. A comparison of the results is also made.

References

Belfallah, K., Laasri, I., Fouaidi, M., Jamal, M.: Analysis of the elasto-magnetic wave propagation in a smart magnetostrictive actuator using a mesh-free method. J. Intell. Mater. Syst. Struct. (2022). https://doi.org/1045389X221128572

Chen, Y., Lee, J., Eskandarian, A.: Meshless methods in solid mechanics. Springer Science & Business Media (2006)

Gandhi, M.V., Thompson, B.D.: Smart materials and structures. Springer Science & Business Media (1992)

Jaffe, H., Berlincourt, D.A.: Piezoelectric transducer materials. Proc. IEEE **53**(10), 1372–1386 (1965)

Leveque, R.J.: Finite difference methods for differential equations. Draft Ver. AMath **585**(6), 112 (1998)

Liu, G.R., Gu, Y.T.: An introduction to meshfree methods and their programing. Springer Science & Business Media (2005)

Newmark, N.M.: A method of computation for structural dynamics. J. Eng. Mech. Div. **85**(3), 67–94 (1959)

Pasquale, M.: Mechanical sensors and actuators. Sens. Actuators, A **106**(1–3), 142–148 (2003)

Shang, X., Pan, E., Qin, L., et al.: Mathematical modeling and numerical computation for the vibration of a magnetostrictive actuator. Smart Mater. Struct. (Print) **17** (2008)

Shoa, T., Madden, J.D.W., Mirfakhrai, T., Alici, G., Spinks, G.M., Wallace, G.G.: Electromechanical coupling in polypyrrole sensors and actuators. Sens. Actuat. A: Phys. **161**(1–2), 127–133 (2010)

Wan, Y.P., Fang, D., Hwang, K.C.: Non-linear constitutive relations for magnetostrictive materials. Int. J. Non-linear Mech. **38**, 1053–1065 (2003)

Check for updates

Numerical Simulation of Non-viscous Multi-species Flows in the Presence of Shocks and Contact Discontinuities

Hind Benakrach(✉), Mohamed Bounouib, Mourad Taha-Janan, and Mohamed Zeriab Essadek

Laboratory of Applied Mechanics and Technologies (LaMAT), ENSAM, STIS Research Center, Mohammed V University, Rabat, Morocco
{hind.benakrach,mohamed.bounouib,m.tahajanan}@um5s.net.ma

Abstract. The use of the Roe flow separation technique for solving the equations governing the multi-species flow of inviscid fluids has been presented in this work. Mathematically, the models are based on Euler's equations, associated with a generalized equation of state. For reasons of stability and accuracy, the finite volume method was applied in this study. Two academic cases of flows in the presence of shocks and contact discontinuities were used to validate the model. The two-dimensional explosion case was chosen for the validation of the model in the mono-species case. The case of instability of Richtmyer-Meshkov was the subject of validation of the case multi-species, this instability is generated by shock waves crossing an interface between two gases of different densities. The results of the two-dimensional numerical modeling obtained, whether for the single or multi-species case, are similar to the results available in the literature, which proves that the model can correctly capture shocks and contact discontinuities.

Keywords: Euler's equation · Finite volume method · Roe solver · Multi-component flows · Explosion problem · Richtmyer-Meshkov instability

1 Introduction

Throughout the years, the diversity of flows and their complicities have attracted the tension of many scientists, particularly complex flows containing shocks and contact discontinuities.

In supernova studies or inertial confinement fusion (ICF), for example, the understanding of the Richtmyer-Meshkov instability phenomenon is not yet fully mastered despite several theoretical, numerical, and experimental studies. This instability is produced when an interface separating two fluids of different densities is subjected to acceleration by a pulse (Richtmyer 1954; Meshkov 1969; Mosedale and Drikakis 2007; Schilling et al. 2007).

The chosen model is based on the resolution of the Euler equations, completed by a generalized equation of state. The latter is based on determining two characteristic parameters for each species present in the flow.

L. Azrar et al. (Eds.): CIP 2022, LNME, pp. 212–220, 2023.
https://doi.org/10.1007/978-3-031-23615-0_22

In order to correctly capture shocks and contact discontinuities without influencing the smooth areas of the flow, the flow separation technique and the gamma model were implemented (Roe 1989; Taha Janan and El Marjani 2007; Benakrach et al. 2019, 2022).

This work aims at treating flows in the presence of shocks and contact discontinuities. Whether the fluid is non-viscous, compressible or incompressible, a single formulation is valid in all cases. Simulation results of single and multi-species flows in a two-dimensional domain are presented in this work.

2 Mathematical Formulation

In a two-dimensional Cartesian reference system, an inviscid fluid is described by the Euler equations in their conservative form:

$$\frac{\partial}{\partial t}\begin{pmatrix} \rho \\ \rho u \\ \rho v \\ \rho E_0 \end{pmatrix} + \frac{\partial}{\partial x}\begin{pmatrix} \rho u \\ \rho u^2 + p \\ \rho u v \\ (\rho E_0 + p)u \end{pmatrix} + \frac{\partial}{\partial y}\begin{pmatrix} \rho v \\ \rho u v \\ \rho v^2 + p \\ (\rho E_0 + p)v \end{pmatrix} = 0 \tag{1}$$

where ρ is the density of the supposed fluid, u and v the components of the velocity vector in the Cartesian coordinate system considered, and p is the static pressure. The total mass-energy E_0 expressed by:

$E_0 = e + \frac{u^2+v^2}{2}$, with e representing the internal energy mass of the fluid.

Associated with these Euler equations, we find the equation of state with two parameters, γ *and* p_∞ and γ *and* p_∞, which allows the closure of the system of equations. This equation is written as follows:

$$\rho e = \frac{p + \gamma p_\infty}{\gamma - 1}$$

Let q represent the conservative variables, E and F represent the inviscid flux:

$$q = \begin{pmatrix} \rho \\ \rho u \\ \rho v \\ \rho E_0 \\ \frac{1}{\gamma-1} \\ \frac{\gamma p_\infty}{\gamma-1} \end{pmatrix}, \quad E = \begin{pmatrix} \rho u \\ \rho u^2 + p \\ \rho u v \\ (\rho E_0 + p)u \\ \frac{u}{\gamma-1} \\ \frac{u\gamma p_\infty}{\gamma-1} \end{pmatrix}, \quad F = \begin{pmatrix} \rho v \\ \rho u v \\ \rho v^2 + p \\ (\rho E_0 + p)v \\ \frac{v}{\gamma-1} \\ \frac{v\gamma p_\infty}{\gamma-1} \end{pmatrix} \tag{2}$$

The global system is written in the following vector and conservative form:

$$\frac{\partial q}{\partial t} + \frac{\partial E}{\partial x} + \frac{\partial F}{\partial y} = 0 \tag{3}$$

3 System Resolution

To solve the global system of equations, the coupled approach is used. This approach consists in solving simultaneously the Euler equations and the species equations. Contrary to the decoupled approach, which starts by solving the Euler equations for a γ and a p_∞ parameter well defined for each species, then solving the species equations (Larrouturou 1991).

In the rest of this section, a one-dimensional study is performed. The same reasoning is valid for the two-dimensional case.

The calculation of convective fluxes is based on the resolution of the Riemann problem applied to each cell. For one dimension, this problem consists in solving the following Cauchy problem:

$$\begin{cases} \frac{\partial W}{\partial t} + \frac{\partial F(W)}{\partial x} = 0; \quad For: \; W \in R^p, \quad p \geq 1 \\ W_0(x) = \begin{cases} W_L, \; x < 0 \\ W_R, \; x \geq 0 \end{cases} \end{cases} \tag{4}$$

The general solution of this problem in (p + 1) constant states, separated by admissible waves, relaxation waves or contact discontinuity waves (Vila 1986).

For a resolution by the finite volume method, the discretization of the vector $F(W)$ is presented by the equation:

$$\oint_{\partial K} F(W).n \, ds = |S|\Phi \tag{5}$$

where:

- S: area of the surface on which the surface integral is evaluated;
- Φ: numerical flow function.

To calculate the numerical flow, it is possible to use either the exact solution (Godunov and Bohachevsky 1959) or the approximate solution of the Riemann problem (Woodward and Colella 1984; van Leer 1979).

Although Godunov's solver is the most efficient, its determination of all the wave structures by an iterative method, so that in the end it is only interested in the state at the interface, consumes an enormous amount of computation time. An approximate solver, such as Roe's, generally gives good results in such a short time. The goal is to approximate the flow at the interfaces based on a linearization of the Jacobian matrix.

$$\frac{\partial W}{\partial t} + \tilde{A}(W_L, W_R)\frac{\partial W}{\partial x} = 0 \tag{6}$$

$\tilde{A}(W_L, W_R)$ is the average matrix of the Jacobian $J(w) = F_W$ or Roe matrix, it must have the following properties:

$$\begin{cases} F(W_L) - F(W_R) = \tilde{A}(W_L, W_R)(W_L, W_R) \\ \tilde{A}(W, W) = J(w) \\ \tilde{A} \; is \; diagonalizable \end{cases}$$

The numerical flow is therefore given by:

$$\begin{cases} \Phi(W_L, W_R) = \frac{1}{2}[F(W_L) + F(W_R) + \Delta F] \\ \Delta F = T^{-1}|\Lambda|T(W_L - W_R) \end{cases} \tag{7}$$

T and T^{-1} are, respectively, the eigenvector matrix of the Roe matrix and its inverse. $|\Lambda|$ is the diagonal matrix of the absolute eigenvectors of the Roe matrix, these matrices are linked by the relation:

$$\tilde{\Lambda} - T^{-1}|\Lambda|T$$

The present development in this section is a discretization of order 1, which introduces a numerical dissipation necessary in the zones of discontinuities to eliminate the oscillations, which influences negatively the sought solution. To remedy this, a discretization up to order three and an introduction of limiters in the numerical flow allows bringing back the accuracy near the shock zone.

4 Validation Tests

4.1 2D Explosion Problem

This Roe solver was tested on a two-dimensional explosion test case. In a square domain of [1.0:1.0], a circle of radius $R = 0.2$ is positioned at the center of the square.

- On the outside of the disc, the fluid is characterized by:

$$(\rho, u, v, p)_{r<r_0} = (1.0, 0.0, 0.0, 1.0)$$

- While inside we have:

$$(\rho, u, v, p)_{r>r_0} = (0.125, 0.0, 0.0, 0.1)$$

Figures 1 and 2 respectively represent the distribution of density and pressure in the (x, y) plane obtained at $t = 0.125$. These results are obtained by [200 × 200] meshes and using a Van Leer limiter.

A circular shock wave moves in the domain due to the pressure difference between the circle and the rest of the square, while an expansion takes place. We find, therefore, a good coincidence with the results presented in the literature (Toro 2009).

4.2 Richtmyer-Meshkov Instability

Taking the case of a classical Richtmyer-Meshkov instability where an air-liquid interface interacts with a shock located in the liquid at $x = 1.325$. The interface is plotted according to the curve described by the following equation:

$$x = x_0 - \varepsilon \cos(2\pi k y), \quad y \in [0, 1]$$

With:

$$x_0 = 1.2, \ \varepsilon = 0.1, \ \text{and } k = 1.$$

In the initial state:

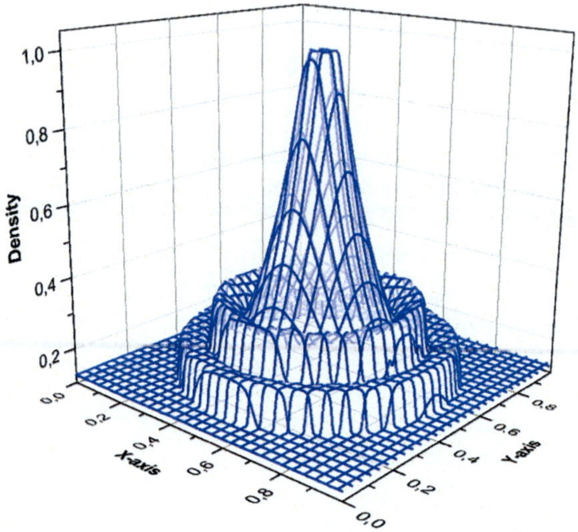

Fig. 1. Density distribution at t = 0.125.

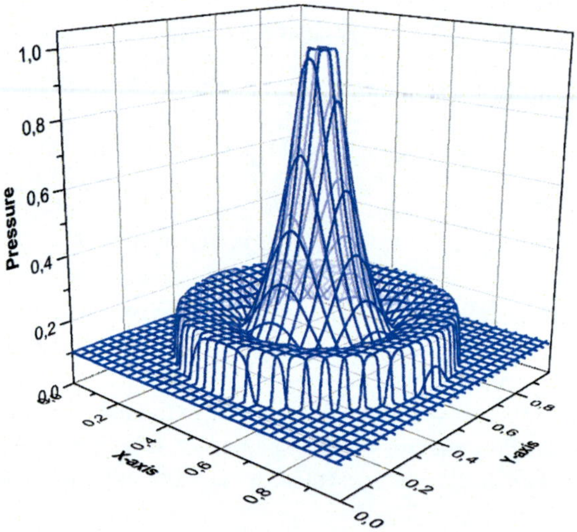

Fig. 2. Pressure distribution at t = 0.125

- In air:

$$(\rho, u, v, p, \gamma, P_\infty)_L = (1.0, 0.0, 0.0, 1.0, 1.4, 0.0)$$

- In the liquid, between the interface and the shock:

$$(\rho, u, v, p, \gamma, P_\infty)_M = (5.0, 0.0, 0.0, 1.0, 4.0, 1.0)$$

- In the liquid, behind the shock:

$$(\rho, u, v, p, \gamma, P_\infty)_R = (7.093, -0.7288, 0.0, 10.0, 4.0, 1.0)$$

For a mesh size of [149 × 71] and a time step of 2×10^{-3}, the results are shown in Figs. 3 and 4.

Figures 3 and 4 represent the evolution of the characteristic variables of the flow along the thx-axis and for $y = 0.5$. These results are obtained at $t = 0.1$.

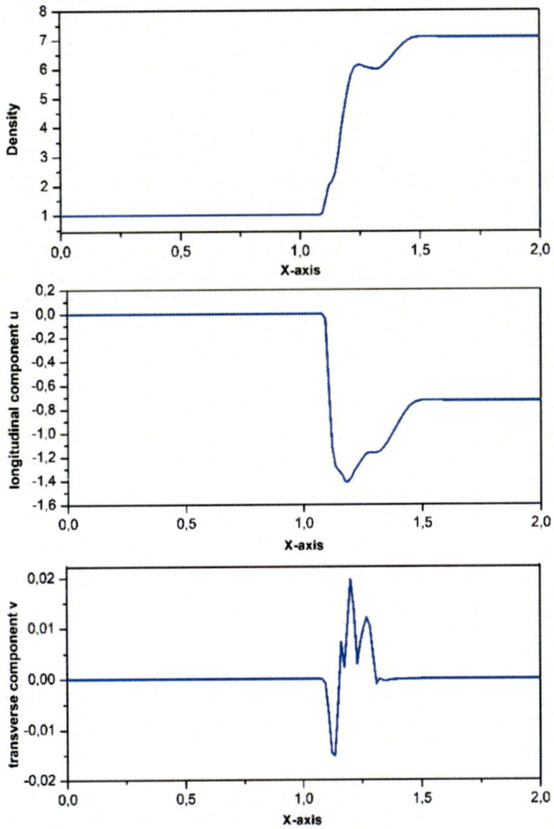

Fig. 3. Characteristics of the Richtmyer-Meshkov instability at t = 0.1 on the x axis: density, longitudinal component-u, transverse component-v.

From Figs. 3 and 4, we can see a very good agreement between the results of the calculation obtained and those presented by Shyue (1998).

In order to better visualize the further evolution of this instability, Figs. 5 and 6 show the distribution of the tooth and pressure in the plane (x, y) at the times 0.3 and 0.5.

Despite the appearance of some small oscillations, the density and pressure distributions, in the (x, y) plane, at $t = 0.3$ and $t = 0.5$, correctly describe the behavior of fluids

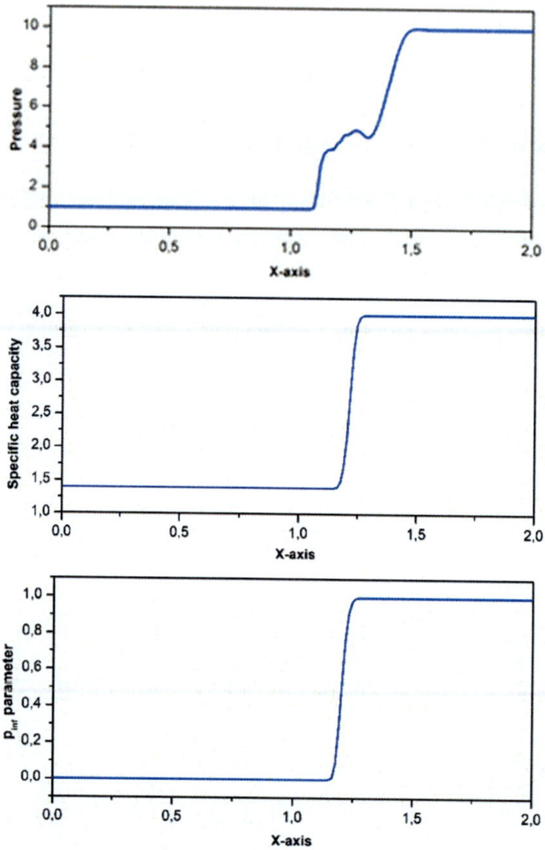

Fig. 4. Characteristics of the Richtmyer-Meshkov instability at $t = 0.1$ on the x-axis: pressure, specific heat capacity, p_{inf} parameter.

are in agreement with the results obtained with Shyue. According to Quirk (1994), these oscillations are caused by the choice of limiter. To remedy this, it is necessary to choose a more dissipative limiter, except that this will affect the sharpness of the shocks.

5 Conclusion

The work presented in this paper shows the ability of the method used to treat multi-species flows, with a generalized equation of state, in a two-dimensional domain. As the results obtained are very encouraging, it is necessary to develop the code for three-dimensional cases so that it can simulate flows with viscous fluids.

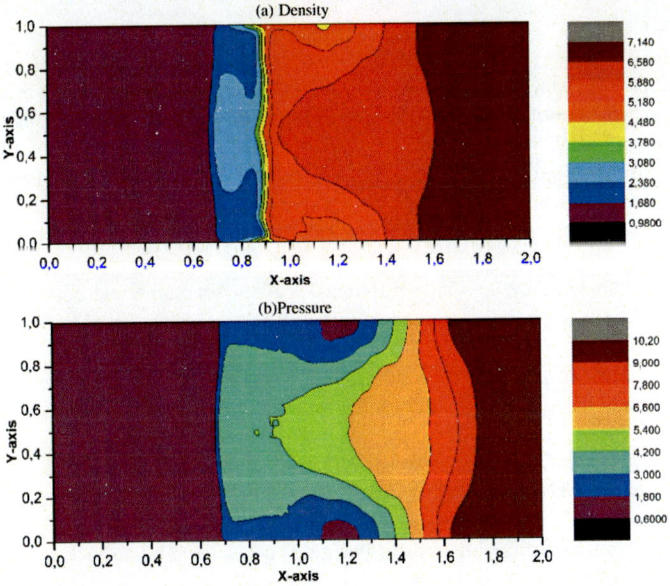

Fig. 5. Pressure and density distribution at $t = 0.3$

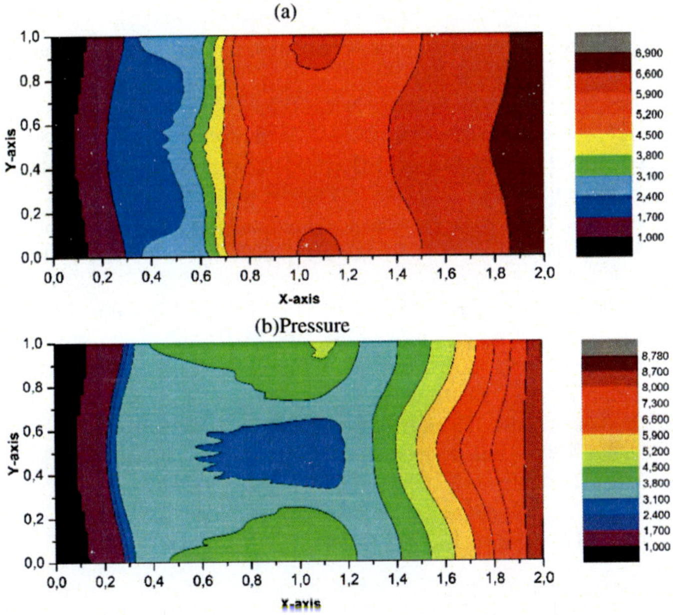

Fig. 6. Pressure and density distribution at $t = 0.5$

References

Benakrach, H., Taha-Janan, M., Es-Sadek, M.Z.: Simulation of compressible and incompressible flows in the presence of shocks. MATEC Web Conf. **286**, 07018 (2019). https://doi.org/10.1051/matecconf/201928607018

Benakrach, H., Bounouib, M., Taha-Janan, M., Essadek, M.Z.: Shocks and discontinuities treatment in multi-species flows using a Riemann solver. In: Benhala, B., Mansouri, K., Raihani, A., Qbadou, M. (eds.). E3S Web Conf. **336**, 00075 (2022). https://doi.org/10.1051/e3sconf/202233600075

Godunov, S.K., Bohachevsky, I.: Finite difference method for numerical computation of discontinuous solutions of the equations of fluid dynamics. Mat. Sb. **47**(89) (3), 271–306 (1959)

Larrouturou, B.: How to preserve the mass fractions positivity when computing compressible multi-component flows. J. Comput. Phys. **95**(1), 59–84 (1991). https://doi.org/10.1016/0021-9991(91)90253-H

van Leer, B.: Towards the ultimate conservative difference scheme. V. A second-order sequel to Godunov's method. J. Comput. Phys. **32**(1), 101–136 (1979). https://doi.org/10.1016/0021-9991(79)90145-1

Meshkov, E.E.: Instability of the interface of two gases accelerated by a shock wave. Fluid Dyn. **4**(5), 101–104 (1969)

Mosedale, A., Drikakis, D.: Assessment of very high order of accuracy in implicit LES models. J. Fluids Eng. **129**(12), 1497–1503 (2007). https://doi.org/10.1115/1.2801374

Quirk, A.: Contribution to the great Riemann solver debate. Int. J. Numer. Methods Fluids **18**, 555 (1994)

Richtmyer, R.D.: Taylor Instability in Shock Acceleration of Compressible Fluids. Los Alamos Scientific Laboratory, New Mexico (1954)

Roe, P.L.: The use of the Riemann problem in finite difference schemes. In: Reynolds, W.C., MacCormack, R.W. (eds.) Seventh International Conference on Numerical Methods in Fluid Dynamics. Lecture Notes in Physics, vol. 141, pp. 354–359. Springer Berlin Heidelberg, Berlin, Heidelberg (1989). https://doi.org/10.1007/3-540-10694-4_54

Schilling, O., Latini, M., Don, W.S.: Physics of reshock and mixing in single-mode Richtmyer-Meshkov instability. Phys. Rev. E **76**(2), 026319 (2007). https://doi.org/10.1103/PhysRevE.76.026319

Shyue, K.-M.: An efficient shock-capturing algorithm for compressible multicomponent problems. J. Comput. Phys. **142**(1), 208–242 (1998). https://doi.org/10.1006/jcph.1998.5930

Taha Janan, M., El Marjani, A.: A flow solver for the Euler and Navier-Stokes equations for multiphase flows with a stiffened gas equation of state. Int. J. Numer. Methods Heat Fluid Flow **17**(8), 823–835 (2007). https://doi.org/10.1108/09615530710825792

Toro, E.F.: Riemann Solvers and Numerical Methods for Fluid Dynamics. Springer Berlin Heidelberg, Berlin, Heidelberg (2009). https://doi.org/10.1007/b79761

Woodward, P., Colella, P.: The numerical simulation of two-dimensional fluid flow with strong shocks. J. Comput. Phys. **54**(1), 115–173 (1984)

Geometrically Nonlinear Free Vibration of Fully Clamped Beam: Comparison of the Experimental with Analytical Results

Afafe Boufnichel[1,2](✉), El Bekkaye Merrimi[1,2], and Atman Jbari[1,2]

[1] ENSAM, Mohammed V University in Rabat, B.P. 6207, Rabat, Morocco
boufnichel.afafe@gmail.com, {elbekkaye.merrimi, atman.jbari}@ensam.um5.ac.ma
[2] Rabat Instituts, Rabat, Morocco

Abstract. Having a high-performance structure is continuously one of the most prominent concerns of mechanical engineering. To develop the dynamic behavior of a structure many studies have been carried out, it is in this context that the objective of our paper fits. The paper's aim is to examine the non-linear dynamic response of a clamped beam at both ends. A theoretical model has been built based on Hamilton's principle, energetic approach and spectral analysis, in order to obtain a nonlinear algebraic equation system that describes the nonlinear motion response of the beam. The objective of this paper is to compare the nonlinear frequency response of a fully clamped beam by using two methods: a numerical method using a subroutine called NS01A from Harwell library and an experimental method using a bench of vibration composed of a beam excited by a magnetic field and an infrared sensor measuring the beam's vibration displacement.

Keywords: Non-linear free vibration · Vibration sensor · Resonance frequency

1 Introduction

For structural safety, improvement and optimization of the dynamic response of a structure widely remain an objective of engineering studies. A very common structural problem has not been treated by many researches in spite of his great importance in optimizing the dynamical behavior is described by non-linear equations.

Innumerable papers were carried out about the non-linear vibrations using approximate analytical [1] and numerical methods. Pillai and Rao [2] investigated the non-linear free vibrations of simply supported beams, they focused on the relation between the large amplitude free vibration and the frequency response of the system using different methods like Galerkin method and the harmonic balance method which been also used by Tseng and Dugundji [3] to examine the dynamical behavior of a buckled beam with fixed ends excited by a harmonic motion of its supporting base. Foda [4] analyzed the nonlinear free vibrations of a beam with pinned ends by reducing the equation of motion to a duffing-type equation with a small perturbation parameter and then he applied the

L. Azrar et al. (Eds.): CIP 2022, LNME, pp. 221–229, 2023.
https://doi.org/10.1007/978-3-031-23615-0_23

method of multiple scales to determine a higher third order perturbation. Yan et al. [5] has conducted a study about the nonlinear dynamical behavior of a parametrically excited clamped beam made of functionally graded materials by using Timoshenko beam theory, Von Karman type geometric nonlinearity, and rotational spring model. In fact, it is not easy to find exact solutions for the nonlinear behavior of beams. Hence, Ahmadian and Mojahedi [6] in order to analyze the nonlinear free vibrations of a beam with pinned ends, he employed modified Lindstedt-Poincare methods and homotopy analysis method it is one of the powerful resolution techniques of nonlinear problems which help us to predict precisely the nonlinear response of the structure.

A structure is mostly threatened by mechanical vibrations, it has a critical impact on its functioning and lifespan. In fact, those mechanical vibrations are mainly caused by dynamic excitations which come either from internal or external environment and as a result the structure presents a linear or nonlinear behavior. Actually, the non-linearity may be referred to geometrical, inertial, or material reasons. Material nonlinearity typically becomes a problem when the system's material starts to deform permanently and goes beyond the yield strength. Geometric nonlinearity occurs in the structure when the system goes through a large deformation which means the strain-displacement relationship is nonlinear, this type of nonlinearities [7–9] is the most frequent category treated over the last decade.

A considerable number of researches has shown some advancement in the analysis of different aspects of non-linear dynamic behavior. The aim of this work is to study the nonlinear vibrations of a clamped beam at both ends in case of flexion. A theoretical model based on Hamilton's principle, energetic approach, and spectral analysis has been used to reduce the problem of nonlinear free vibrations to nonlinear algebraic equations which depends on the stiffness matrix, the mass matrix and a tensor of order four due to the nonlinearity, in case we neglect this tensor of order four we will obtain the eigenvalue problem of classical linear theory. This theoretical model has been developed, in order to calculate the first nonlinear mode shape for different values of vibration displacement amplitude. We will begin by determining the nonlinear motion equation and then we will transform it into a non-dimensional system of equations to make a parameterized study of the actual problem. The nonlinear system of equations obtained has been treated numerically using a FORTRAN's program NS01A in this Ref. [10]. This paper is an attempt to validate an experimental approach for analyzing the dynamic response of a CC beam, which is based on a vibration bench composed of a beam excited by a magnetic field and an infrared sensor measures the amplitude of the beam's vibration. The objective is to compare the experimental data obtained from the non-dimensional nonlinear free vibration of the fully clamped beam and the numerical results corresponding to the same non-dimensional vibration obtained from FORTRAN's program.

2 Theoretical Formulation

In order to calculate the first nonlinear mode shape of a beam with fixed ends corresponding to various vibration amplitude, we should determine the nonlinear motion equation based on energetic approach, spectral analysis and Hamilton's principle.

Consider a homogeneous beam having geometrical and material characteristics (S, E, H, I, L, ρ) clamped at both ends. The transverse vibrations of the beam are given by its deflection W(x, t) by the longitudinal neutral axis (x) and at time (t).

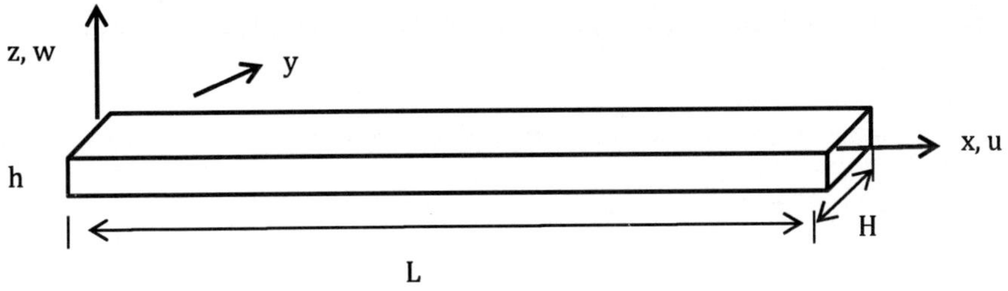

We develop the transverse displacement of beam into a finite series of functions, the bending strain energy V_f, the axial strain energy V_a and the kinetic energy T.

$$V_a = \frac{ES}{8L}\int_0^L \left(\left(\frac{\partial w(x,t)}{\partial x}\right)^2 dx\right)^2, \quad (1)$$

$$V_f = \frac{1}{2}\int_0^L E.I_z.\left(\frac{\partial^2 w(x,t)}{\partial x^2}\right)^2 \quad (2)$$

$$T = \frac{1}{2}\rho.S\int_0^L \left(\frac{\partial w(x,t)}{\partial t}\right)^2 dx \quad (3)$$

where: w(x, t) is the beam's transverse displacement.

We assume that the beam has a harmonic motion, in this case the transverse displacement can be written as:

$$w(\mathrm{x,\,t}) = a_i.W_i(x).\sin(wt) \quad (4)$$

where:

$$a_i(t) = a_i.\sin(wt) \quad (5)$$

And: $W_i(x)$ represent the linear mode shapes.

We replace $w(x, t)$ in the expressions of V_f, V_a, T and rearranging leads to:

$$V_a = \frac{1}{2}a_i.a_j.a_k.a_l.B_{ijkl}.sin^1(wt) \quad (6)$$

where:

$$B_{ijkl} = \frac{ES}{4L}\left(\int_0^L \frac{\partial W_i(x)}{\partial x}.\frac{\partial W_j(x)}{\partial x}dx\right)\left(\int_0^L \frac{\partial W_k(x)}{\partial x}.\frac{\partial W_l(x)}{\partial x}dx\right) \quad (7)$$

B_{ijkl} is the nonlinear stiffness tensor.

$$V_f = \frac{1}{2} a_i . a_j . K_{ij} . sin^2(wt) \tag{8}$$

where:

$$K_{ij} = E.I \int_0^L \left(\frac{\partial^2 W_i(x)}{\partial x^2} \right) \left(\frac{\partial^2 W_j(x)}{\partial x^2} \right) dx \tag{9}$$

K_{ij} is thc lincar rigidity tcnsor.

$$T = \frac{1}{2} a_i . a_j . M_{ij} . W^2 . cos^2(w.t) \tag{10}$$

where:

$$\mathrm{M_{ij}} = \rho s \int_0^L \left(W_i(x) W_j(\mathrm{x}) \right) \mathrm{dx}; \quad \mathrm{M_{ij}} \text{ is a mass tensor} \tag{11}$$

The nonlinear motion equation may be obtained by the application of Hamilton's principle for a conservative system which leads to:

$$2a_i . K_{ir} + 3a_i a_j a_k B_{ijkr} - 2w^2 . a_i . M_{ir} = 0; \quad \mathrm{r} = 1 \ldots \mathrm{n} \tag{12}$$

In order to obtain a non-dimensional system of equations, we will put a non-dimensional formulation:

$$\begin{cases} x^* = \frac{x}{L} \\ W_i^*(x) = \frac{W_i(x)}{H} \end{cases} \tag{13}$$

By replacing (13) in the tensor's expressions, we will obtain a non-dimensional motion equation which can be written in a matrix form as the following:

$$2[K^*]\{\mathrm{A}\} + 3[B^*(\mathrm{A})]\{\mathrm{A}\} = 2W^2[M^*]\{\mathrm{A}\}; \; \{A\}^T = \{a_1, a_2, \ldots, a_n\} \tag{14}$$

It's a system of nonlinear algebraic equations where {A} is a column vector of the basic functions contribution coefficients.

$$2a_i K_{ir}^* + 3a_i a_j a_k B_{ijkr}^* - 2w^{*2} a_i M_{ir}^* = 0, \quad \mathrm{r} = 1, \ldots, \mathrm{n} \tag{15}$$

Simplifying by 2, the equation becomes:

$$a_i K_{ir}^* + \frac{3}{2} a_i a_j a_k B_{ijkr}^* - w^{*2} a_i M_{ir}^* = 0, \quad \mathrm{r} = 1 \ldots \mathrm{n} \tag{16}$$

From which, we obtain the nonlinear frequency ω^{*2}:

$$w^{*2} = \left(a_i K_{ir}^* + \frac{3}{2} a_i a_j a_k B_{ijkr}^* \right) / a_i M_{ir}^*, \quad \mathrm{r} = 1 \ldots \mathrm{n} \tag{17}$$

We put:

$$K_{nl}^{*} = \frac{3}{2} a_j a_k B_{ijkl}^{*} \tag{18}$$

Equation (15) corresponds to:

$$\left([K^{*}] + [K^{*}]\right)\{A\} - \omega^2 [M^{*}]\{A\} = \{0\} \tag{19}$$

where $[M^{*}]$ is the non-dimensional mass tensor, $[K^{*}]$ is the non-dimensional linear stiffness tensor and $[K_{nl}{}^{*}]$ is the non-dimensional, nonlinear stiffness tensor.

Each term of the matrix $[K_{nl}{}^{*}]$ depends on the components of the column vector $\{A\}$ which represents the contribution coefficients of the basic functions.

3 Numerical Results

The numerical results obtained in Ref. [10] for the nonlinear fundamental mode of a totally clamped beam showed that the contribution of the coefficient a_1 of the first basic function remains predominant for the large ranges of vibration's amplitudes [11] compared with the contribution of the coefficients functions of higher bases, for that reason we will consider just the first mode shape to determine the vibratory behavior of the beam.

After determining the nonlinear motion equation of the clamped beam at both ends, we will use a routine called NS01A from the Harwell library to obtain the numerical solutions of this equation in the neighborhood of the first resonance, for which the contribution of basic functions $\{A\}$ are already chosen and the tensors $[M^{*}]$, $[K^{*}]$, and $[K^{*}]_{nl}$ are calculated using a routine called PREP.

We have determined the nonlinear frequency and the corresponding maximum displacement and we have represented the nonlinear pulsation according to the maximum displacement (Fig. 1).

4 Experimental Results and Discussions

In order to confirm an experimental method used for studying the vibratory response of a mechanical structure, we will use a vibratory bench and compare the experimental measurements with the theoretical results. A 1.5 × 26 × 1200 mm rectangular steel beam clamped at both ends was excited in a harmonic manner by an electromagnetic exciter which did not impose any additional rigidity or mass to the beam. An infrared sensor was used to measure the dynamic displacement across the beam. The excitation point was chosen at the middle of the beam in order to have a symmetric distribution of all the contribution functions in either the length or width directions. The experimental instrumentation used in this work is illustrated in Fig. 2.

As shown in Fig. 3 the beam was excited in the middle harmonically by an electromagnetic coil which was excited by a current generator. An IR sensor sends the vibratory response of the beam to an Arduino card which will be treated on LabVIEW interface as illustrated in Fig. 3.

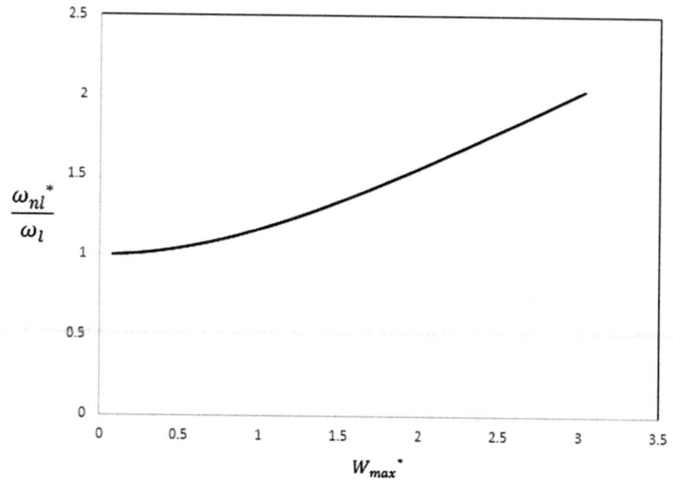

Fig. 1. Backbone curve of clamped-clamped beam, in the vicinity of the first mode.

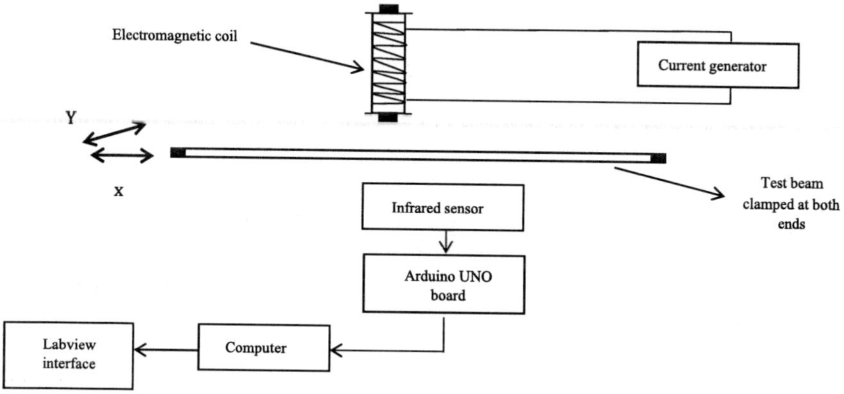

Fig. 2. Experimental layout for beam's displacement measurements.

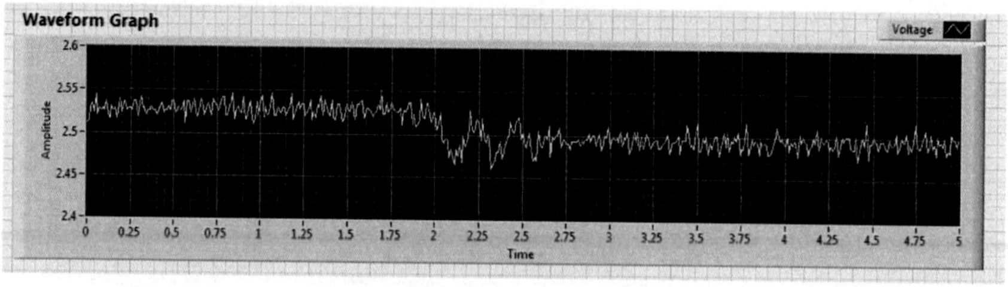

Fig. 3. Nonlinear frequency response of CC beam following 8 V excitation voltage.

Vibration processing:

Digital filtering: Digital Low-Pass Filter, Order = 4, Fc = 6 Hz Acquired vibration (Fig. 4).

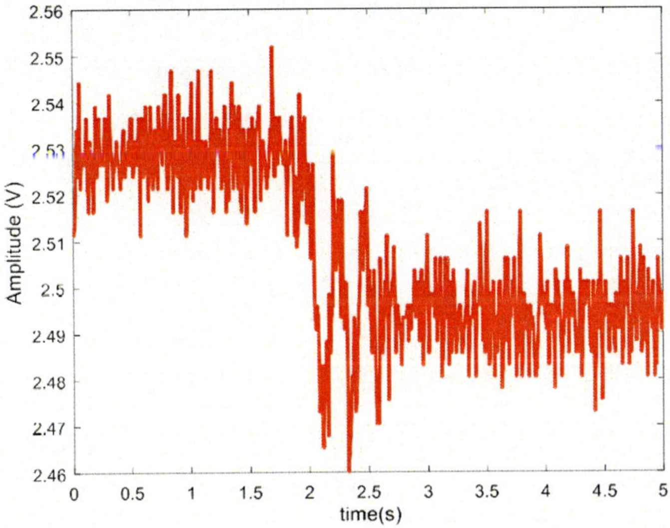

Fig. 4. The vibrational response of a CC beam following 8 V excitation voltage.

And then we obtain a filtered vibration (Fig. 5; Table 1).

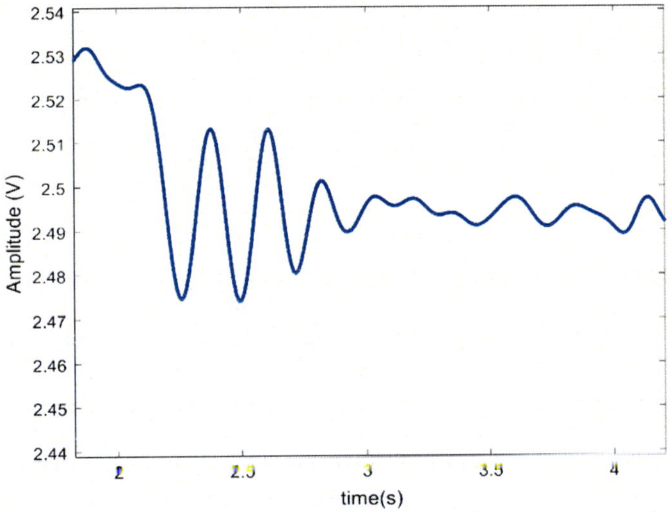

Fig. 5. Filtered vibration.

Table 1. Frequency ratio and amplitude vibration corresponding to 8 V excitation voltage.

Excitation	$\omega_{nl}^{*}/\omega_{l}$	W_{max}^{*}
8 V	1.57	2.53

By using an excel interpolation of the graph line illustrated in Fig. 6 we could determine the value corresponding to 2.53 maximal displacement, we found that the frequency ratio is equal 1.77.

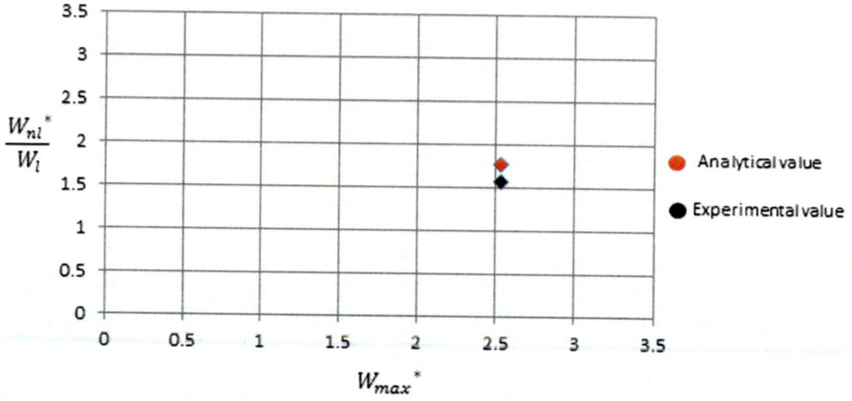

Fig. 6. Comparison of frequency ratio corresponding to analytical and experimental approaches of CC beam.

Results Discussion:

As follows from Fig. 6, it has been found that the experimental frequency ratio corresponding to the embedded beam at both ends is approximately equal the analytical value with incertitude of 0.2.

5 Conclusion

This paper is a contribution to the latest researches about the nonlinear free vibrations of an embedded beam at both ends. The main purpose of this work was to compare two different approaches: Analytical approach based on nonlinear motion equation resolved numerically using FORTRAN's program, and an experimental approach based on a bench of vibration, after comparing the results it is proved that this bench of vibration still an effective method to study the nonlinear frequency response of the beam. In the future, we are willing to extend our research to other structures with different boundary conditions.

References

1. Azrar, L., Benamar, R., White, R.G.: A semi-analytical approach to the non-linear dynamic response problem of S-S and C-C beams at large vibration amplitudes part I: general theory and application to the single mode approach to free and forced vibration analysis. J. Sound Vib. **224**, 183207 (1999)
2. Pillai, S.R.R., Rao, B.N.: On nonlinear free vibrations of simply supported uniform beams. J. Sound Vib. **159**, 527–531 (1992)
3. Tseng, W.Y., Dugundji, J.: Nonlinear vibrations of a buckled beam under harmonic excitation. J. Appl. Mech. **37**(1970), 292–299 (1970)
4. Foda, M.A.: On non-linear free vibrations of a beam with pinned ends. J. King Saud Univ. Eng. Sci. **7**(1), 93–107 (A.H. 1415/1995) (1995)
5. Yan, T., Yang, J., Kitipornchai, S.: Nonlinear dynamic response of an edge-cracked functionally graded Timoshenko beam under parametric excitation. Nonlinear Dyn. **67**(1), 527–540 (2012)
6. Ahmadian, M.T., Mojahedi, M.: Free vibration analysis of a nonlinear beam using homotopy and modified Lindstedt-Poincare methods. J. Solid Mech. **2**(1), 29–36 (2009)
7. El Bikri, K., Benamar, R., Bennouna, M.M.: Geometrically non-linear free vibrations of clamped–clamped beams with an edge crack. Comput. Struct. **84**(7), 485–502 (2006)
8. Chajdi, M., Merrimi, E., El Bikri, K.: Geometrically nonlinear free and forced vibration of clamped-clamped functionally graded beam with discontinuities
9. Chajdi, M., Merrimi, E., El Bikri, K.: Geometrically nonlinear free vibration of composite materials: clamped-clamped functionally graded beam with an edge crack using homogenization method (2017)
10. Benamar, R.: Nonlinear dynamic behavior of a fully clamped beams and rectangular isotropic and laminated plates. Ph.D. thesis, University of Southampton (1990)
11. El Kadiri, M., Benamar, R., White, R.G.: Improvement of the semi-analytical method, for determining the geometrically non-linear response of thin straight structures. Part 1: application to clamped-clamped and simply supported – clamped beams. J. Sound Vib. (2002)

Dynamic Behaviour Study of Inter-component Vehicle in Frontal Crashes

Echarkaoui Somman[1](✉), Doha Souhail[1], Bouchra Rzine[2], Mohammed Radouani[1], and Benaissa Elfahime[1]

[1] Moulay Ismail University, Meknès, Morocco
e.somman@edu.umi.ac.ma, {m.radouani,b.elfahime}@ensam-umi.ac.ma
[2] Faculté des Sciences et Techniques de Fès, Fez, Morocco

Abstract. The growth of the global automotive market is pushing car manufacturers to move towards fast and efficient methods to minimize costs and withstand strong competition. The design of a light vehicle involves the choice of mechanical, electrical and structural components while respecting technical, functional requirements and, above all, active and passive passenger safety standards [1]. Indeed, manufacturers are making considerable efforts to reduce the rate of road accident victims by investing in crashworthiness and crash energy dissipation to ensure user safety.

In this paper, we present a study, by using the finite element method, of the component vehicle dynamic behaviour in a frontal crash against a deformable impactor [2, 4]. In general, the interaction between components during the impact creates an energy flow that propagates by transmission or dissipation through the vehicle structure elements. The damping of the force flow depends on the morphology of the component and the physical properties of the material and the mechanical assemblies used [8].

The objective of this study is to develop a solution to improve the ability of the cradle subsystem to absorb kinetic energy and achieve a very precise sinking [3]. This condition allows the impact kinematics to be controlled in such a way as to prevent the powertrain from entering the passenger compartment and therefore protect the front occupants of the vehicle [5].

Keywords: Vehicle · Crashworthiness · Frontal crash · Simulation · Finite elements method · Energy absorption

1 Introduction

Passive Safety in the automotive vehicle is based on the one hand on the ability of the structure to absorb energy throughout the impact by programmed deformation and on the other hand on additional equipment or devices designed to protect the vehicle occupants from injury. Passive Safety, in its absorption aspect, considers the capacity of the structure to transform its kinetic energy into plastic deformation energy [8]. The design constraints of the structure are directly dictated by accident analyses and the resulting design standards [6].

L. Azrar et al. (Eds.): CIP 2022, LNME, pp. 230–237, 2023.
https://doi.org/10.1007/978-3-031-23615-0_24

The level of crashworthiness must be capped (possibly in the form of a force-depression curve, then referred to as programmed resistance) so as to limit the decelerations of the vehicle and therefore the occupants. The non-deformability of the passenger cabin must be guaranteed in order to prevent intrusion and to ensure that the doors can be opened (maintaining the vehicle's evacuation capacity) [2].

In the automotive sector, frontal impact accounts for a very large proportion of accidents. As an indication, the classic chronology of deformation of the structure can be summarised in the form of Table 1 to which a deceleration curve can be associated.

Table 1. Sequences of a crash scenario.

Time (ms)	Crushing distance (mm)	Observation
0	0	Contact Crushing absorbers "DANNER"
5	250	Radiator crushed against the powertrain
42		Failure of the powertrain connections
44		Bracing of engine, gearbox and steering column against the dashboard High stresses on the body of the vehicle Intrusion of the pedal and steering column
44–100	600	Deformation of the tube, the under-seat and the dashboard

2 FE Study of a Cradle Sub-system During a Frontal Crash

Accurate modelling of the mechanical phenomena involved in very large deformations requires precise knowledge not only of the geometry and its boundary conditions, but also of the intrinsic characteristics of the materials and the dynamic behaviour of the assemblies of structural elements [7].

The predictive simulation of dynamic behaviour requires the consideration of "Nonlinearities":

- Material non-linearities.
- Geometrical non-linearities.
- Contact non-linearities.

The cradle is a critical mechanism in the functioning of passive safety. In a frontal crash, the amount of energy adsorbed by the programmed deformation through these components maps out a crash scenario so as to prevent the impact of the vehicle's front components (engine, radiator and suspension) against the occupants (Fig. 1).

In the present case, the cradle contributes significantly to the realisation of such a crash scenario. The role of the cradle is to prevent the intrusion of the front components

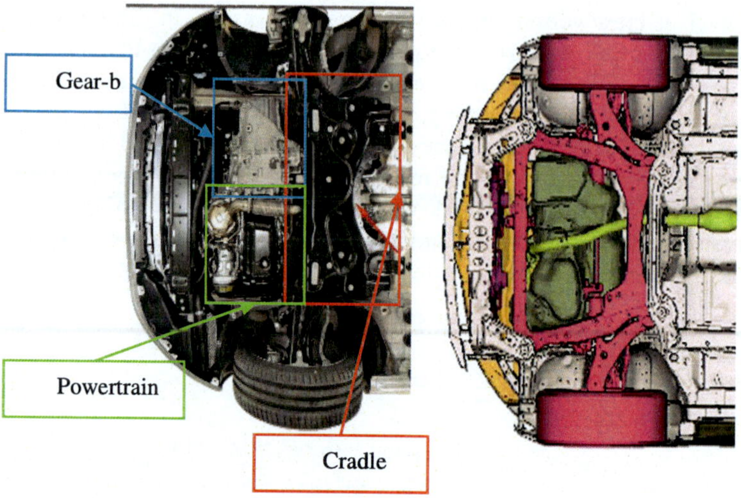

Fig. 1. An overview of the vehicle's components

(Powertrain, Radiator, …), hypothesis infinitely rigid compared to the components of the cabin, to prevent damage to the occupants [7, 9].

Controlling the flow of forces through the cradle automatically implies control of the kinematics of the Powertrain and other components. Figure 2 illustrates the contact points of the cradle with the vehicle's external components. The impact scenario requires that the cradle fixings be maintained in the correct scenario to meet the crash requirements.

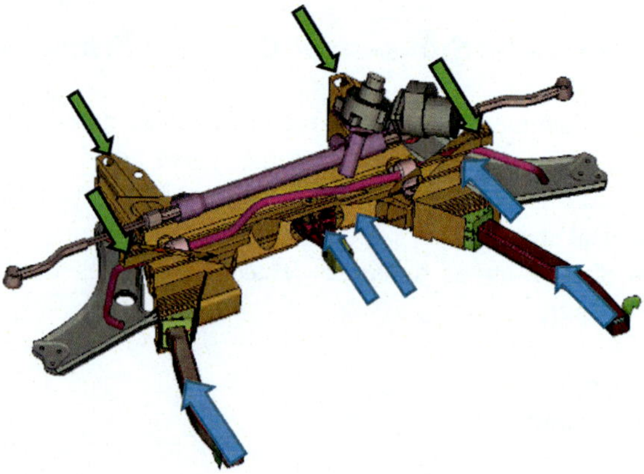

Fig. 2. Overview of incoming/outgoing cradle efforts

The modelling of the whole system requires a significant amount of computing resources. Therefore, partial modelling of the geometric model with the same input hypotheses minimises the computational cost (Fig. 3).

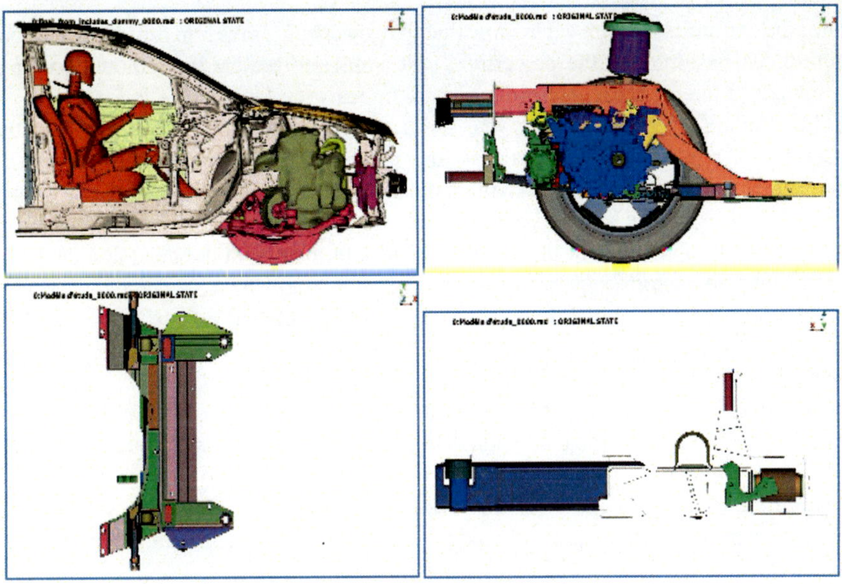

Fig. 3. Simplified model with the main characteristics

This stage of the study represents a simplification of the FEA model in order to reduce the computing time. The impactor models the entire powertrain with a point mass assigned to the centre of the impactor of about 60 kg.

The design of the geometric model of the counter-impactor is based on the iterative method in order to meet the objective of the study, we have kept the basic structure of the cradle and act on the insertion of a component in order to increase its resistance in this point. The FE model is built by CAD in CATIA V5, and the pre-processing part is done in ANSA (Fig. 4).

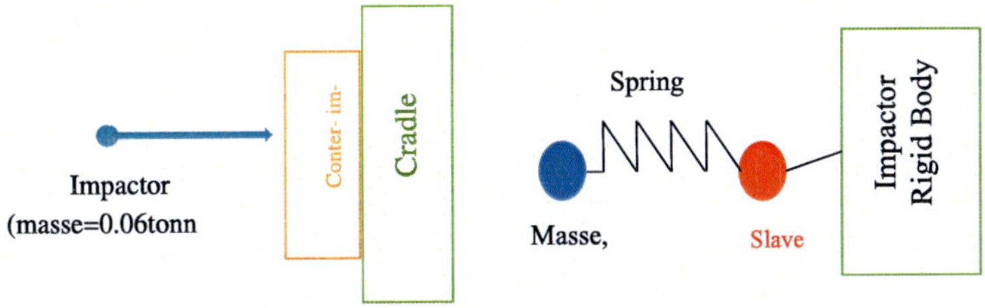

Fig. 4. Physical modelling of loading

At the input phase, and as defined in the previous paragraph. Assign the correct physical properties of the components (materials, thicknesses, mathematical formulations of the mesh elements, …). Then, define the initial conditions (loads and contacts) and the boundary conditions (cradle fixation).

For the extreme scenario, we assume that the powertrain sinks into the cradle without damping. We assume that the powertrain unit sinks into the cradle without damping, which leads to the measurement of the maximum resisting load before failure.

This method of loading allows the measurement of the total force resisting the impactor's intrusion into the depth of the cradle. The application of an imposed displacement explains the fact of having a constant kinetic energy along the crash.

For this case, we aim is to maximise the resistant force of the cradle while respecting the 20 mm holding depth. For this study, we will optimise both the head and the body of the counter-impactor.

The results presented are based on tree models developed by acting on the contribution of the counter-impactor to increase the force during its interaction with the Powertrain unit.

In the post-processing level, the verification of the output files is essential to validate the stability conditions of the calculation. In addition, certain physical quantities, such as the conservation of kinetic and internal energy, the added mass on the elements and the stability of the time step, determine the conformity of the calculation with certain tolerances depending on the precision of the study and the criticality of the application field.

Figures 5, 6 and 7 shows three configurations of the cradle. The first is the basic configuration without a counter-impactor, the second and third are the two configurations in which we have integrated the counter-impactor to maximise the effort required in order to achieve the correct crash scenario.

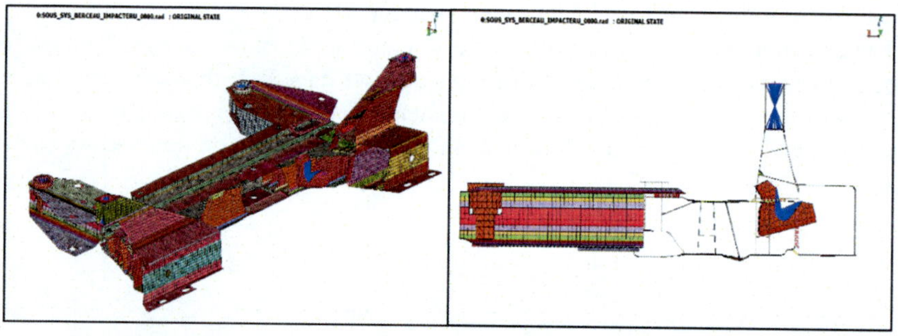

Fig. 5. Comparison between the tree cradles: old version (Configuration_1)

This design guarantees a management of energy dissipation through the counter-impactor and then the cradle. The 3rd configuration allows to increase the contact force between powertrain and the designed counter-impactor. This solution allows the strength of the intermediate cradle plates to be taken advantage of. For the implementation of the solution, some changes to the cradle are required. In the first instance. Creation of openings in the intermediate cradle plates. Next, inserting a part that points to the whole of the sheets. The contact area between the counter-impactor and the sheets significantly increases the cradle's resistance strength.

The analysis of the kinematics of the model reveals the coherence of the phase of the data setting at the level of contact management (absence of penetrations between

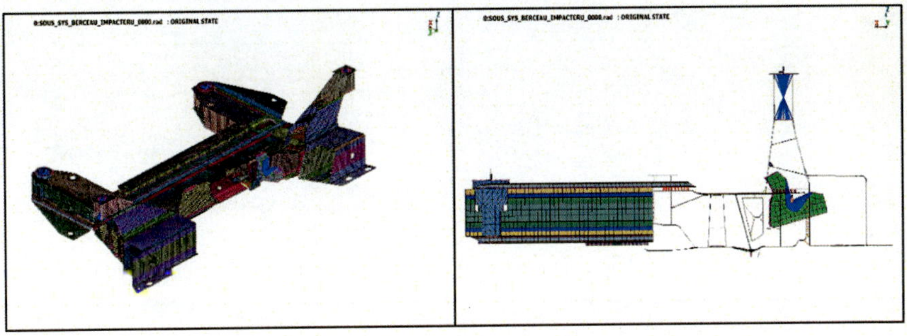

Fig. 6. Comparison between tree subframes: first proposition (Configuration_2)

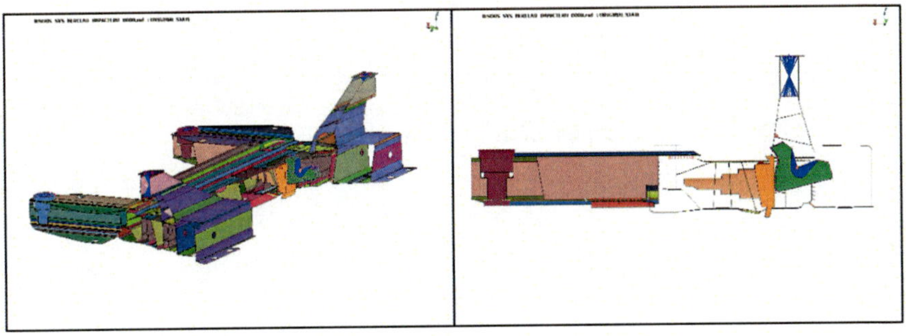

Fig. 7. Comparison between tree subframes: final proposition (Configuration_3)

the parts) and the mesh rupture. The physical deformation of the structure reflects the absorption of kinetic energy and its dissipation along the cradle until failure.

3 Results and Discussion

The results of the curves where the horizontal axis represents crash distance in mm. This is the equivalence of the crash of the part in the cradle. While the vertical axis describes the interfacing force between the parts in contact (Slave/Master). The aim is to adapt the geometry of the counter-impactor body in such a way as to have simultaneous contact. To quantify the resulting load, it is simply a matter of measuring the load computed by the tension spring as defined in the model Fig. 4.

Figure 8 describes the capacity of the counter-impactor to maintain a minimum force of 160 kN for a duration of 3.2 ms equivalent to 20 mm of minimum crushing distance of powertrain in the cradle. Both parameters are acceptance criteria for qualifying the cradle subsystem in crash tests (Fig. 9).

The design of the cradle structure and its interaction with the surrounding vehicle components contributes greatly to the management of energy dissipation during a frontal crash. It is designed to absorb the impact energy or kinetic energy generated by a vehicle deceleration. According to the 3 configurations presented above. The ability to absorb energy is clearly shown in Fig. 9. A significant increase in internal energy as measured

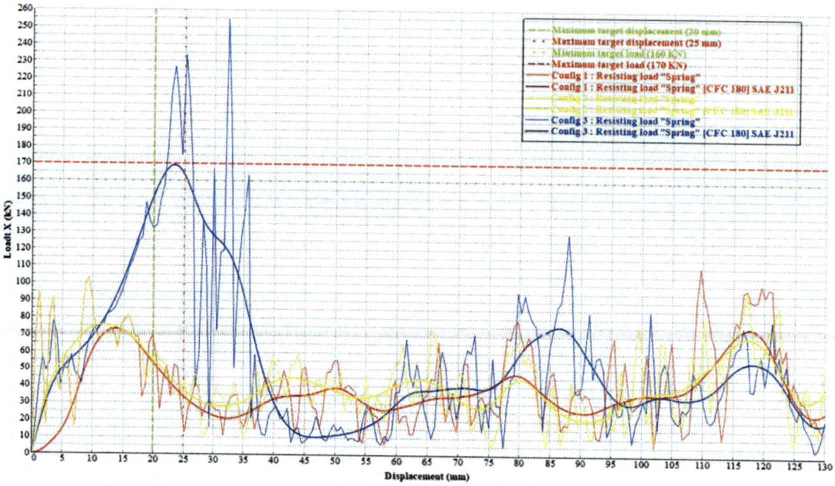

Fig. 8. The evolution of the force on the computation spring

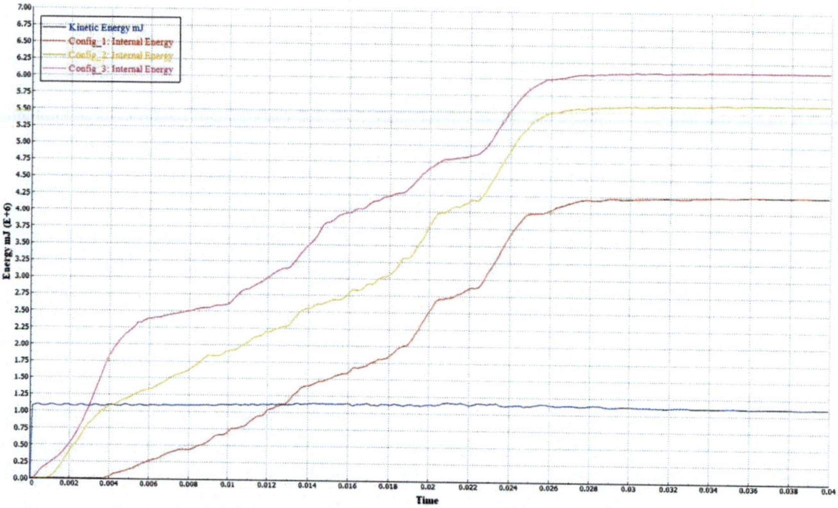

Fig. 9. Energy graph of the three configurations.

on the graph reflects the absorption of the impact by the structure and therefore reduces its transmission to the occupants.

4 Conclusion

One of the sensitive parameters in this process is optimising the weight of the structure. Several modifications have been taken into consideration. One of these modifications is the change of material. Previously the vehicle models were made of steel. However, in view of the improvements and weight optimisations, it is necessary to think about other materials. Therefore, the manufacturer has chosen to replace steel with aluminium alloy due to its lightness and elasto-plastic behaviour.

The present study is part of the dimensioning of the undercarriage parts of a new electric vehicle. After a numerical validation of a cradle subsystem, describing the frontal offset crash test scenario. It consists in designing a counter-impactor which resists a force from the powertrain precisely from the gearbox impactor. Then guaranteeing its failure at specific load and displacement.

The cradle is considered the key sub-assembly in passive safety, whose role is firstly to absorb energy from the external environment and secondly to absorb the forces generated by the powertrain. This ensures correct kinematics of the powertrain in the event of impact.

To control the kinematics of the powertrain in the front end of the vehicle, and to ensure the efficient transmission of forces to the vehicle structure and chassis. In other words, the powertrain must slide into the cradle at a specific position. Secondly, the connection between the powertrain and its external environment must be ruptured. This scenario allows us to avoid the impact of the powertrain, which is assumed be infinitely rigid in comparison to the structure, with the passenger compartment in order to protect the occupants.

References

1. Akrout, S.E., Mhaiti, N., Radouani, M., El Fahime, B.: Modeling and simulation of a vehicle crash test. Adv. Sci. Technol. Eng. Syst. J. **6**(1), 937–946 (2021)
2. Güler, M.A., Cerit, M.E., Mert, S.K., Acar, E.: Experimental and numerical study on the crashworthiness evaluation of an intercity coach under frontal impact conditions. Proc. Inst. Mech. Eng. Part D J. Automob. Eng. 095440702092764 (2020). https://doi.org/10.1177/0954407020927644
3. Hou, L., et al.: An integrated multi-objective optimization method with application to train crashworthiness design. Struct. Multidiscip. Optim. **63**(3), 1513–1532 (2020). https://doi.org/10.1007/s00158-020-02758-2
4. Kim, S., Cho, H.: A study on the stiffness change of a passenger car's front frame body before and after a collision accident. Int. J. Mech. Eng. Robot. Res. **10**(5) (2021). https://doi.org/10.18178/ijmerr.10.5.270-275
5. Li, Q.Q., Li, E., Chen, T., Wu, L., Wang, G.Q., He, Z.C.: Improve the frontal crashworthiness of vehicle through the design of front rail. Thin-Walled Struct. **162**, 107588 (2021). ISSN 0263-8231
6. Mei, T.X., Goodall, R.M.: Recent development in active steering of railway vehicles. Veh. Syst. Dyn. **39**(6), 415–436 (2003). https://doi.org/10.1076/vesd.39.6.415.14594
7. Singh, H., Ganesan, V., Davies, J., Paramasuwom, M., Gradischnig, L.: Vehicle interior and restraints modeling development of full vehicle finite element model including vehicle interior and occupant restraints systems for occupant safety analysis using THOR dummies. Report No. DOT HS 812 545. National Highway Traffic Safety Administration, Washington, DC (2018)
8. Xiaoming, C., Hikmat, M., David, A.W., Baccouche, M.R.: Aluminum subframe design for crash energy management. In: 2004 SAE World Congress, Detroit, Michigan, 8–11 Mar 2004
9. Eichmueller, G., Meywerk, M.: On computer simulation of components for automotive crashworthiness, validation and uncertainty quantification for simple load cases. Int. J. Crashworthiness (2019). https://doi.org/10.1080/13588265.2019.1582577

Numerical Modelling of Nonlinear Vibrations of Mechanical Structures

Hafida Chekkouchi(✉) and El Bekkaye Merrimi

Mechanical Engineering Department, ENSAM, Royal Armed Forces, 10100 Rabat, Morocco
hafida_chekkouchi@um5.ac.ma, elbekkaye.merrimi@ensam.um5.ac.ma

Abstract. Non-linear vibration problems are challenging to solve which makes this type of research crucial for the scientific community. Indeed, most non-linear systems cannot have an exact solution, that is why many solutions developed in the literature are approximations. This paper will present the first part of a study concerning the semi-analytical modeling of geometric nonlinear vibrations of thin structures, which we have focused on beams and plates. We have chosen the study of a beam because it presents the ideal case by its simplicity as a continuous system to project our equations in order to validate the proposed model. The method used is a programmed one based on FORTRAN to obtain semi-analytical results presenting the non-linear dynamic response of the beam. The various assumptions adopted for the study of the non-linear geometric free vibration problem of a beam are given. Then, the general formulation and numerical details of the nonlinear problem are generated, including expressions for the energies and the equation of motion. Then, these expressions are discretized, using Hamilton's principle, and finally the discretized expressions are given in dimensionless form. We rely on the parameters and the dynamic properties of the structure in the dynamic analysis which consists in executing the program with these data in order to verify the behavior of the studied system.

Keywords: Vibration · Non-linear · Semi-analytical · Thin structures

1 Introduction

Even though the study of vibrations is not a new field of research, it knows in recent years a surge of interest to provide solutions to the nonlinear dynamic mechanics problems, that is possible to set up the prerequisite for the stability of oscillations of elastic structural elements [1, 2]. This article deals with the modal analysis of geometric nonlinear vibrations of thin structures, through the development of a semi-analytical model. The vibration is always caused by some kind of exciting forces; also the vibrating object is entirely determined by the strength of the excitation, its direction, its maximum amplitude and its frequency. It is

El Bekkaye Merrimi—This author contributed equally to this work.

L. Azrar et al. (Eds.): CIP 2022, LNME, pp. 238–247, 2023.
https://doi.org/10.1007/978-3-031-23615-0_25

for this reason that vibration analysis makes it possible to identify the dynamic properties of the structure. Most mechanical systems are more complex than a simple structure, and they will not necessarily move as a whole under the influence of a force. Here we will not dwell on them in detail, because they are complex enough, however, to understand the essence of vibration analysis of structures, it is useful to examine the non-linear dynamic response of beams and plates. In the last few years, studies on the static and dynamic behavior of various mechanical structures, such as beams, plates, shells, etc., have drawn the attention of a large number of researchers.

By delving through the literature [3–6] (which does not represent an exhaustive list), we find that there are two types of vibration: linear and non-linear. If the basic components of a vibrating system behave in a linear way, the resulting vibration is a linear vibration. Nevertheless, if one of the fundamental parts of a vibrating system exhibits non-linear behavior, the resultant vibration is a non-linear vibration. A number of factors, including geometrical effects, the nonlinear characteristics of materials, and nonlinear boundary conditions, frequently combine to cause nonlinearity. The system is usually linearized and these effects are omitted for the sake of simplicity.

In contrast to the equation regulating linear vibration, the equation of motion governing non-linear vibration will be a non-linear differential equation. Vibrations are frequently a concern, especially for components of flexible systems. To study vibrations in mechanical systems, mathematical equations are first required, as they can be used to examine both forced and free vibration problems.

The Equations of Movement are the equations that explain the kinematics and physics of systems. There are numerous methods designed to solve these equations, and among them, we mention Newton's Second Law [7–10], the principle of Hamilton and the energy conservation, etc.

The features of vibrational phenomena include eigenfrequencies, eigenvalues, eigenvectors, and an eigenmode. They're known as modal characteristics.

To maximize the search for the best design and to most accurately predict the behavior of a mechanical system, it is essential to implement accurate and representative models and simulations of the entire system. Through simulations, this model is utilized to validate the preconception phase's requirements. These simulations are necessary to ensure that the system is operating in line with the different criteria.

Generally speaking, A system's behavior (performance, operation, etc.) is typically controlled by physical effects (mechanical, thermal, etc.). The performance of the behavioral simulations is therefore crucial. The use by the designer of multi-physics models at this stage actually allows him to numerically validate the product, for which it is necessary to represent certain physical phenomena.

Dynamical system simulation techniques are usually used to resolve problems with the model's physical behavior. A mathematical expression based on a physical constitutive law attempts to model a system's behavior using empirical laws. The selection of starting ideas is improved, the design cycle time is shortened, and the cost of optimization is decreased when designers have access to a methodology that enables them to generate models with high precision. Analyzing the modal damping and resonant pulsation damping qualities is the key to solving a vibration problem. To build simplified differential equations, the earliest methods are analytical and entirely predicated on a variety of assumptions [11–13].

The extraction of damping properties is therefore possible and these can be solved quickly utilizing simple techniques. But in actual applications, we deal with structures that have different boundary conditions and geometric shapes. Such formulations result in complicated equations that are challenging to resolve analytically [6].

2 The Graphic Presentation

The beam and the plate are characterized by (Fig. 1):

L: The length,
b: The width,
h: The thickness,
S: The surface of the cross section,
I: The moment of inertia of the cross-sectional area,
E: The modulus of elasticity (the module of young),
ρ: The density of the material.

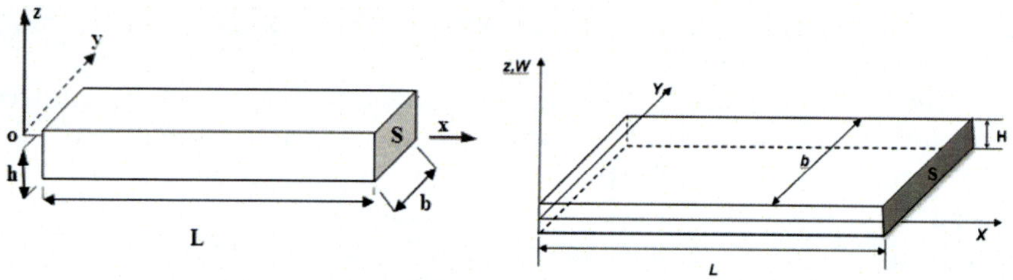

Fig. 1. Free rectangular plate and beam in the coordinate system (o, x, y, z)

3 Formulation of Energies for the Beam and Plate

3.1 Formulation of Energies for the Beam

– bending strain energy

$$V_b = \frac{1}{2}\int_0^L EI(\frac{\partial^2 W}{\partial x^2})^2 dx \tag{1}$$

– non-linear strain energy

$$V_a = \frac{ES}{8L}(\int_0^L (\frac{\partial W}{\partial x})^2 dx)^2 \tag{2}$$

– kinetic energy

$$T = \frac{1}{2}\rho S \int_0^L (\frac{\partial W}{\partial t})^2 dx \tag{3}$$

3.2 Formulation of Energies for the Plate

– bending strain energy

$$V_b = \frac{1}{2}\int_S D(\frac{\partial^2 W}{\partial x^2} + \frac{\partial^2 W}{\partial y^2})^2 dS \tag{4}$$

– non-linear strain energy

$$V_a = \frac{3D}{2H^2}\int_S [(\frac{\partial W}{\partial x})^2 + (\frac{\partial W}{\partial y})^2]^2 dS \tag{5}$$

– kinetic energy

$$T = \frac{1}{2}\rho H \int_S (\frac{\partial W}{\partial t})^2 dS \tag{6}$$

4 Expression of the Movement Equation for the Beam

A mechanical problem's solution reflects the level of stress and strain the system has endured over time as a result of mechanical stressors. It is necessary to establish a formalism for the study of a moving body in order to analyze the deformation of the system at every given time. One of two ways exists to describe movements:

- The Lagrangian approach consists of following the evolution of each material point solid over time. The mechanics of continuous media uses very often this approach to describe the deformation and displacement point by point of the solid.
- The Eulerian approach consists in following the evolution of an environment during the time by focusing on a precise point in space which will see several material points pass chronologically. This approach is widely used in fluid mechanics to study the transport of fluid along a domain delimited.

4.1 Hypotheses

Beam Hypothesis The separation of variables utilized to solve equation (4). In this section, we'll look at an analytical approach that makes use of the harmonic scaling method and is expressed as follows to apply a transverse harmonic load to the beam:

$$W(x,t) = a_i(t)w_i(x) \tag{7}$$

and $w_i(x)$ is called normal mode or characteristic beam function.

Plate Hypothesis The applied methodology is based on harmonic scales. The plate is subjected to a transverse harmonic load, which is expressed as shown below:

$$W(x,y,t) = a_i(t)w_i(x,y) \tag{8}$$

4.2 Hamilton's Principle

The Hamiltonian principle states that the motion of an elastic system with conservative behavior occurs between two instants t_0 and t_1,

$$H = \int_{t_0}^{t_1} (T - V)dt \tag{9}$$

is stationary :

$$\partial H = \int_{t_0}^{t_1} (\partial T - \partial V)dt = 0 \tag{10}$$

5 Non-dimensional Formulation

5.1 Dimensionless Beam Parameters

In order to conduct a parameterized analysis of the problem under consideration, this formulation seeks to identify equations that are independent of both beam properties and dimensions.

$$\text{we pose:} \qquad \begin{cases} x* = \frac{x}{L} \\ w*_i\,(x) = \frac{w_i(x)}{H} \end{cases} \tag{11}$$

H: is the thickness of the beam in the case of a beam of rectangular section.

By replacing these equations in the expressions of the tensors, we obtain:

- The linear stiffness tensor:

$$K_{ij} = \int_0^L (\frac{\partial^2 W_i}{\partial x^2})(\frac{\partial^2 W_j}{\partial x^2})dx \tag{12}$$

- The mass tensor:

$$M_{ij} = \int_0^L (W_i(x)W_j(x))dx \tag{13}$$

- The Tensor of Nonlinear Stiffness:

$$B_{ijkl} = \frac{ES}{4L} \int_0^L (\frac{\partial W_i}{\partial x})(\frac{\partial W_j}{\partial x})dx) \int_0^L (\frac{\partial W_k}{\partial x})(\frac{\partial W_l}{\partial x})dx \tag{14}$$

The dynamic equations were written via the Lagrange method. A matrix equation can be used to express the discrete motion equations of the beam:

$$[K]\{A\} + \frac{3}{2}[B(A)]\{A\} = \omega^2[M]\{A\}, \tag{15}$$

$$\{A\} = \{a_1, a_2, \ldots, a_n\}$$

It is an algebraic system of non-linear equations whose unknowns are the contribution coefficients of the basis functions written as parts of the vector {a}.

5.2 Dimensionless Plate Parameters

The expressions of the tensors are:

- The linear stiffness tensor:

$$K_{ij} = \int_S D(\frac{\partial^2 W_i}{\partial x^2} + \frac{\partial^2 W_i}{\partial y^2})(\frac{\partial^2 W_j}{\partial x^2} + \frac{\partial^2 W_j}{\partial y^2})dxdy \tag{16}$$

- The mass tensor:

$$M_{ij} = \rho H \int_S W_i(x,y)W_j(x,y)dxdy \tag{17}$$

- The Tensor of Nonlinear Stiffness:

$$B_{ijkl} = \frac{3D}{H^2} \int_S ((\frac{\partial W_i}{\partial x})(\frac{\partial W_j}{\partial x}) + (\frac{\partial W_i}{\partial y})(\frac{\partial W_j}{\partial y}))((\frac{\partial W_k}{\partial x})(\frac{\partial W_l}{\partial x}) + (\frac{\partial W_k}{\partial y})(\frac{\partial W_l}{\partial y}))dxdy \tag{18}$$

The matrix equation can be written as follows:

$$([K] - \omega^2[M]\{A\} + 3/2[B(A)]\{A\} = \{0\} \tag{19}$$

6 Results and Discussion

A program developed in FORTRAN , to obtain the numerical resolution. The non-linear pulsation was plotted against the maximum displacement, as shown in Fig. 2. After creating a loop to fix a_1, for each value of a_1, we determined the non-linear frequency and the corresponding maximum displacement (Fig. 3).

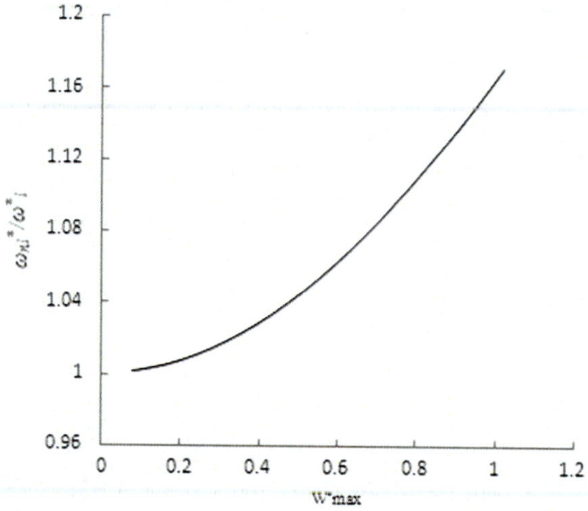

Fig. 2. Backbone curve of the free beam in the vicinity of the first mode

The behavior of the beam and the plate is non-linear. The linearity is clear in the resulting curves.

The results obtained for the different tensors of the beam are presented below.

- Results of the mass tensor calculation:

$$[M_{ij}] = \begin{bmatrix} 1 & & & & & \\ & 1 & & & & \\ & & 1 & & & \\ & & & 1 & & \\ & & & & 1 & \\ & & & & & 1 \end{bmatrix}$$

- Results of the linear stiffness tensor calculation:

$$[K_{ij}] = \begin{bmatrix} 500.5639 & & & & & \\ & 14617.6301 & & & & \\ & & 89135.4080 & & & \\ & & & 308208.4575 & & \\ & & & & 793408.0075 & \\ & & & & & 1686361.3257 \end{bmatrix}$$

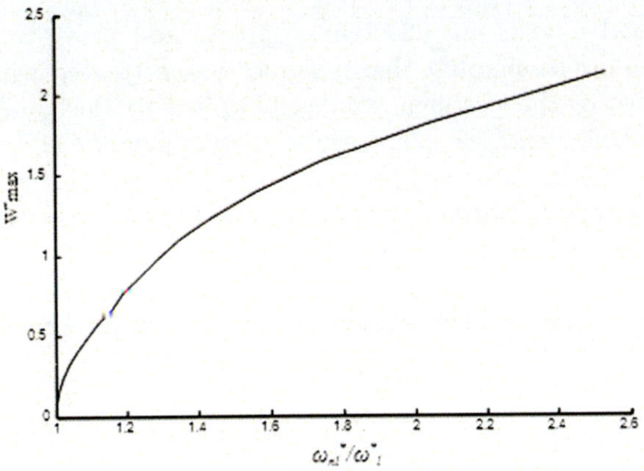

Fig. 3. Backbone curve of the free plate in the vicinity of the first mode

– Results of the nonlinear stiffness tensor calculation:

$$[B_{ijkl}] = \begin{bmatrix} 12.30 & -9.73 & -7.62 & -6.11 & -5.07 & -4.26 \\ -9.73 & 98.90 & -24.35 & -22.98 & -20.85 & -18.53 \\ -7.62 & -24.35 & 264.00 & -38.03 & -37.96 & -35.86 \\ -6.11 & -22.98 & -38.03 & 508.04 & -51.22 & -51.39 \\ -5.07 & -20.85 & -37.96 & -51.22 & 831.06 & -63.14 \\ -4.26 & -18.52 & -35.86 & -51.38 & -63.13 & 1234.32 \end{bmatrix}$$

The comparison of the analytical results in reference [8] (marked with R as in the case below) with those obtained with the present modeling (noted by P) shows that the method used gives good results.

for example :

$M(_{ij})_R \approx M(_{ij})_P \approx 1$, (i, j= 1, 1),
$K(_{ij})_R \approx K(_{ij})_P \approx 500.56$, (i, j= 1, 1) and
$B(_{ijkl})_R \approx B(_{ijkl})_P \approx 12.30$, (i, j, k, l= 1, 1, 1, 1).

The values are valid as long as they comply with the references.

7 Conclusion

This paper presents an approach to modeling the nonlinear geometric vibrations of the beam and the plate. It is used to obtain semi-analytical results that show the dynamic responsiveness of these structures. NSO1A subroutine was used to numerically solve the set of nonlinear algebraic equations. The resulting curve is definitely convergent, which indicates that the proposed model is validated to be used with different structures. That demonstrated to us that the pulsation and typical bending stress in the nonlinear case depend on the vibratory amplitude.

Semi-analytical models are effectively applied and prove to be a valuable method. They allow to simplify the dynamics to be treated whilst preserving the particularities of the problem studied. Applied to the nonlinear dynamic response of the beam and the plate, these developments could, in the future, considerably reduce the computation times. The same nonlinear vibration analysis study can also be performed with prestressed beams, non-isotropic beams, composite beams and cracked beams.

Finally, it should be noted that vibrations are a very open field of research and that current prediction methods still need many improvements.

References

1. Merrimi, E.B., El Bikri, K., Benamar, R.: A homogenisation procedure for the steady state periodic forced response of laminated composite beams at large vibration amplitudes. In: Advanced Materials Research, vol. 535, pp. 1811–1814. Trans Tech Publications (2012)
2. Nicoletti, R.: On the natural frequencies of simply supported beams curved in mode shapes. J. Sound Vibr. **485**, 115597 (2020). https://doi.org/10.1016/j.jsv.2020.115597
3. Froio, D., Rizzi, E., Simões, F.M.F., Costa, A.P.D.: Critical velocities of a beam on nonlinear elastic foundation under harmonic moving load. Procedia Eng. **199**, 2585–2590 (2017). https://doi.org/10.1016/J.PROENG.2017.09.348
4. Bekkaye Merrimi, E., Bikri, K.E., Benamar, R.: Geometrically non-linear steady state periodic forced response of a clamped-clamped beam with an edge open crack. Comptes Rendus Mecanique **339** (2011). https://doi.org/10.1016/j.crme.2011.07.008
5. Chajdi, M., Merrimi, E.B., Bikri, K.E.: Geometrically nonlinear free vibration of composite materials: clamped-clamped functionally graded beam with an edge crack using homogenisation method, vol. 730. KEM (2017). https://doi.org/10.4028/www.scientific.net/KEM.730.521
6. Durnin, J., Miceli, J., Eberly, J.H.: Diffraction-free beams. Phys. Rev. Lett. **58**, 1499–1501 (1987). https://doi.org/10.1103/PhysRevLett.58.1499
7. El Kadiri, M., Benamar, R.: Improvement of the semi-analytical method, for determining the geometrically non-linear response of thin straight structures: part ii-first and second non-linear mode shapes of fully clamped rectangular plates. J. Sound Vibr. **257**(1), 19–62 (2002)
8. El Kadiri, M., Benamar, R., White, R.G.: Improvement of the semi-analytical method, for determining the geometrically non-linear response of thin straight structures. Part I: application to clamped-clamped and simply supported-clamped beams. J. Sound Vibr. **249**(2), 263–305 (2002). https://doi.org/10.1006/jsvi.2001.3808
9. Froio, D., Rizzi, E.: Analytical solution for the elastic bending of beams lying on a variable winkler support. Acta Mech. **227**, 1157–1179 (2016). https://doi.org/10.1007/S00707-015-1508-Y
10. Merrimi, E.B., Bikri, K.E., Benamar, R.: A homogenisation procedure for the steady state periodic forced response of laminated composite beams at large vibration amplitudes, vol. 535, pp. 1811–1814 (2012)

11. Wu, K., Zheng, G.: A comprehensive static modeling methodology via beam theory for compliant mechanisms. Mech. Mach. Theory **169** (2022). https://doi.org/10.1016/j.mechmachtheory.2021.104598
12. Shi, Z., Li, H.: Multi-degree-of-freedom beam rotation based on multi-beam vector superposition. Opt. Commun. **502** (2022). https://doi.org/10.1016/j.optcom.2021.127402
13. Fallah, A., Aghdam, M.M.: Thermo-mechanical buckling and nonlinear free vibration analysis of functionally graded beams on nonlinear elastic foundation (2011). https://doi.org/10.1016/j.compositesb.2011.08.041

Numerical Investigation on Particle Deposition in Curved Turbulent Pipes with Thermophoresis

Fatima Zahrae Erraghroughi[1,2](✉), Kawtar Feddi[2,3], Anas El Maakoul[2], Abdellah Bah[1], Abdellatif Ben Abdellah[2,3], and Fouad Dimane[4]

[1] Energy Research Center, Thermal and Energy Research Team-ERTE-ENSAM, Mohammed V University in Rabat, 10100 Rabat, Morocco
fatimazahrae_erraghroughi@um5.ac.ma, a.bah@um5s.net.ma

[2] Renewable Energy and Advanced Materials Laboratory, International University of Rabat, 11 100 Sala Al Jadida, Morocco
kawtar.feddi@uir.ac.ma

[3] College of Science and Technology of Tangier, Abdelmalek Esaadi University, Tangier, Morocco

[4] Laboratoire des Sciences Appliquées (LSA), ENSA, Al Hoceima, Université Abdelmalek Essaadi, Tangier, Morocco

Abstract. Particulate fouling in pipelines and bends is a serious problem since it can diminish the efficiency and imbalance the operation of various industrial equipment. An accurate description of the deposition process will contribute to defining mitigation strategies and methods to prevent or reduce this problem. In this study, particle deposition is analyzed numerically in a 3D curved pipe. The particulate turbulent flow in a 90° bend was predicted by the Reynolds stress model (RSM) along with the Lagrangian DPM model using ANSYS Fluent© *code*. The model credibility was evaluated by an experimental study and the effect of particle Stokes number and thermophoresis (hot and cold bend walls) on deposition efficiency was carefully explored. The results demonstrated that deposition efficiency depends on St number, an increasing trend was recorded as the particle Stokes number increased. Furthermore, it was found that small particles were notably more sensitive to thermophoretic force compared to large ones and the deposition efficiency can be enhanced or reduced depending on whether the wall is cold or hot.

Keywords: Turbulent bend flow · DPM model · Deposition efficiency · Thermophoresis

1 Introduction

Upon their benefits in offering an enhancement of the thermal and hydraulic performances, the curved pipes or bends are a required component in various applications. Including for example the transportation pipelines of gas and oil, the heat exchangers, the combustion engines, the electronic components, etc.

L. Azrar et al. (Eds.): CIP 2022, LNME, pp. 248–256, 2023.
https://doi.org/10.1007/978-3-031-23615-0_26

One of the most challenges in reducing their efficiency is particle deposition. Due to the significant damage caused by this problem, extensive research has been addressed to investigate the underlying mechanisms experimentally and numerically.

Pui et al. [1] developed an empirical correlation to compute particle deposition efficiency in 90° bend pipe based on an experimental study. In 1993, Brockmann [2] extended Pui's correlation to consider the impact of bend angle. Later, McFarland et al. [3] established a new model including the effect of Stokes number, bend angle, as well as bend curvature. Peters and Leith [4] demonstrated that bend angle had no measurable impact on particle deposition.

Large Eddy Simulation (LES) was used to analyze numerically particle deposition in 90° bends [5, 6]. Zhang et al. [7] comprehensively demonstrated that the Reynolds Stress turbulence Model (RSM) along with the Enhanced Wall treatment (EWT) predicts the particle deposition in bends with high accuracy. Arsalanloo and Abbasalizadeh [8] used the same numerical model of Zhang to study the impact of swirling flow on the deposition efficiency in bend pipe.

Using numerical analysis, Inthavong [9] studied particle deposition in laminar flow. Guo et al. [10] studied particle deposition in various square bend walls. Seyfi et al. [11] investigated the black powder deposition in a bend of a gas transport network based on a numerical model.

The present work investigates the particle deposition efficiency in a 90° curved pipe. The investigation was done at different Stokes number and temperature differences.

2 Numerical Model

Particle deposition efficiency is predicted in a circular 90° bend pipe by means of the CFD code Ansys FLUENT version 19. The geometric dimensions are selected based on Pui's work [1] and listed in Table 1. Properties of air are: $T_{in} = 288°$ K, density $\rho = 1.225\,\text{kg/m}^3$ and dynamic viscosity $\mu = 1.7894e - 5$ Pa.s. Intending to avoid the impact of inlet and outlet conditions on flow development, an extended horizontal and vertical pipes of lengths 15D and 4D are employed before and after the elbow section.

Table 1. Geometric and physical properties

Parameter	
Pipe diameter D	8.51 mm
Bend curvature ratio δ	5.6
Re	10000
ρ_p	895 kg/m^3
dp	0.4–11.9 μm
St	0.0025–1.62

Figure 1 depicts the 3D view of the bend along with the mesh used in this study. 25000 dilute particles are injected in a square surface located at 2D away from the bend entrance. The boundary conditions are velocity inlet at bend inlet while pressure outlet is used at the bend outlet. The one-way coupling is used and particles are considered trapped when they contact a wall boundary.

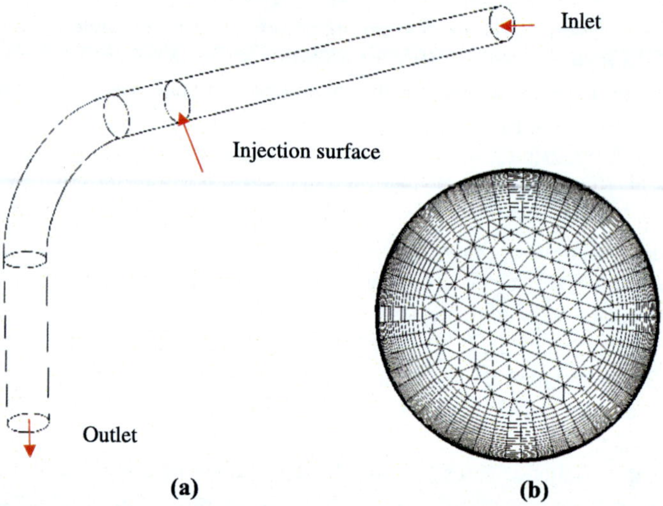

Fig. 1. (a) 90° bend tube geometry, and (b) mesh at bend cross section

The governing equations for isothermal and incompressible air flow are written as [12]:

$$\frac{\partial(\overline{u_i})}{\partial x_i} = 0 \tag{1}$$

$$\overline{u}_j \frac{\partial \overline{u_i}}{\partial x_j} = g_i - \frac{1}{\rho}\frac{\partial \overline{P}}{\partial x_i} + \frac{\partial}{\partial x_j}\left(2\upsilon S_{ij} - \overline{u_j^{'} u_i^{'}}\right) \tag{2}$$

$$\frac{\partial(\rho \overline{u}_i T)}{\partial x_i} = \frac{\partial}{\partial x_i}\left[\frac{\lambda}{c_p}\frac{\partial T}{\partial x_i}\right] \tag{3}$$

where $\overline{u}$ and $\overline{P}$ are the average velocity and pressure, respectively. g is the gravity vector, S_{ij} is the mean rate of the strain tensor, ρ is the air density, υ is the kinematic viscosity, T is the air temperature, λ and c_p are air conductivity and air specific heat capacity, respectively. The RSM turbulence model is used to model the Reynolds stress term. The Enhanced Wall Treatment is adopted to deal with the zone near the wall [13, 14].

The particle trajectory is investigated by the integration of the following equation [15]:

$$\frac{dU_{p,i}}{dt} = \sum F_i \tag{4}$$

where: F_i presents all the forces impacting the particle movement. In this work drag, buoyancy, gravity, Brownian, Saffman lift, and thermophoretic forces are considered, respectively. Equation (4) becomes:

$$\frac{du_{p,i}}{dt} = F_D(u_i - u_{p,i}) + F_G + F_B + F_S + F_{Th} \tag{5}$$

The drag force is considered as:

$$F_D = \frac{18\mu}{\rho_p d_p^2} \frac{C_d \mathrm{Re}_r}{24} \tag{6}$$

where: Re_r is the relative Reynolds number, d_p is the diameter of particles, ρ_p is particle density, and C_d presents the drag coefficient expressed as [16]:

$$C_d = a_1 + \frac{a_2}{\mathrm{Re}_r} + \frac{a_3}{\mathrm{Re}_r^2} \tag{7}$$

a_1, a_2, and a_3 are constants that employ within various Re ranges.

The gravity and buoyancy forces are expressed as a one force:

$$F_G = \frac{g_i(\rho_p - \rho)}{\rho_p} \tag{8}$$

F_B is the Brownian force defined as a Gaussian white noise process [17]:

$$F_B = \zeta_i \sqrt{\frac{\pi s_0}{\Delta t}} \tag{9}$$

where ζ_i is a Gaussian random number, s_0 presents the spectral intensity of the noise, and Δt is the time step magnitude.

The Saffman lift force is expressed by [18]:

$$F_s = \frac{2\rho K_c \upsilon^{0.5} S_{ij}}{\rho_p d_p (S_{lk} S_{kl})^{1/4}} (u_i - u_{p,i}) \tag{10}$$

where $K_c = 2.594$ and S_{ij} presents the deformation tensor.

F_T is the thermophoretic force is:

$$F_{Th} = -D_{T,p} \frac{1}{m_p T} \frac{dT}{dx} \tag{11}$$

where: $D_{T,p}$ is the thermophoretic coefficient [19], and m_p is the particle mass.

The Discrete Random Walk Model (DRWM) [20] is used to incorporate the effect of flow fluctuations on particle deposition. The DRWM uses a stochastic method to generate the local fluctuating velocities and the turbulent dispersion of particles is predicted through the integration of the trajectory equation using the instantaneous air velocity which is a combination of the average and the fluctuating velocity components.

The momentum, energy, and turbulence transport equations are spatially discretized by the second-order upwind scheme. PRESTO! scheme is used for pressure. 10^{-5} is selected as a convergence residual criterion. Moreover, the area-weighted average total outlet pressure and the air mass flow rate variation are applied as monitors for convergence. The Runge-Kutta [21] is used to discretize the discrete phase equation.

Particle deposition efficiency is described as the ratio between N_{bend} the number of particles trapped at the bend, and N_{total} the total number of injected particles.

$$\eta = \frac{N_{bend}}{N_{total}} \tag{12}$$

It is described as a function of *St*:

$$St = \frac{\tau_D}{\tau_s} = \frac{C_c \rho_p d_p^2 / 18\mu}{(D/2)/U_0} = \frac{C_c \rho_p d_p^2 U_0}{18\mu(D/2)} \tag{13}$$

$$C_c = 1 + \frac{\lambda}{d_p}\left(2.514 + 0.8\exp\left(-0.55\frac{d_p}{\lambda}\right)\right) \tag{14}$$

where τ_D and τ_s are the particle are the system response time respectively. U_0 is the mean air velocity, C_c is the Cunningham coefficient and λ presents the air mean free path. Before performing any computations, a mesh sensitivity test is conducted. Three grids are investigated and Fig. 2 presents the obtained results.

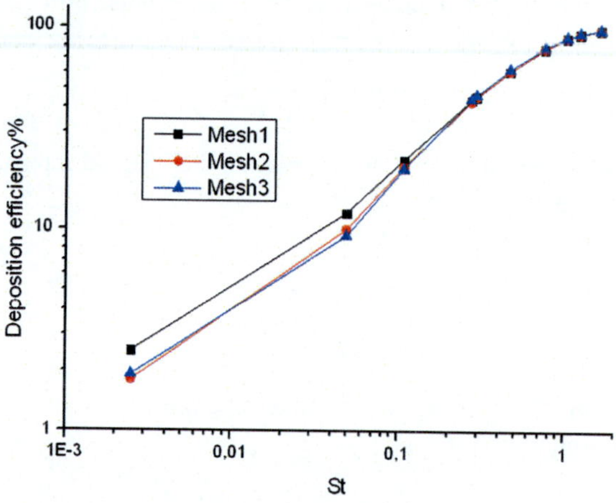

Fig. 2. Mesh sensitivity test

The figure demonstrates that the solution remains the same when refining the mesh between Mesh 2 and Mesh 3. Moreover, by considering the mesh quality as well as the CPU time, Mesh 2 was selected to conduct the simulations.

3 Results and Discussion

The experimental analysis of Sudo et al. [22] and the LES modeling of Yan et al. [23] were selected to validate the present model. Comparison of the numerical predictions and the

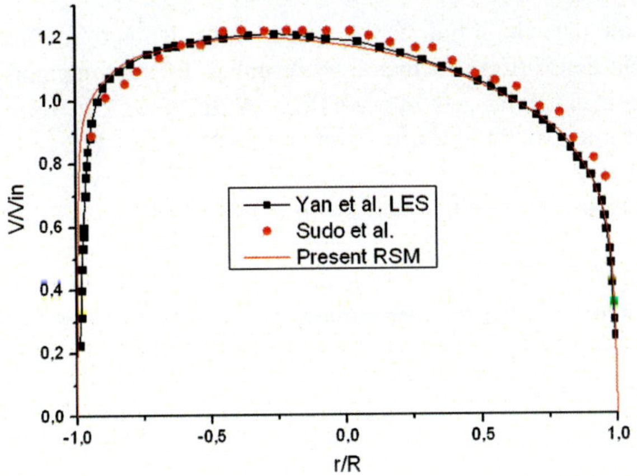

Fig. 3. The comparison of normalized velocity at the bend inlet with Sudo et al. [22] and Yan et al. [23]

experimental normalized velocity at bend inlet is presented in Fig. 3. It is demonstrated that the obtained results are in accordance with both the studies.

Pui et al. [1] established a correlation to evaluate the particle deposition efficiency in 90° curved pipe:

$$\eta = \left(1 - 10^{-0.963St}\right) \times 100\% \tag{15}$$

Figure 4 compares the predicted particle deposition efficiency with literature results [1].

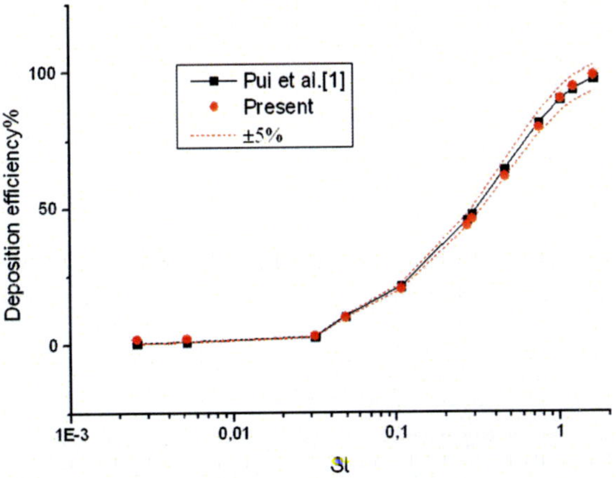

Fig. 4. Particle deposition efficiency versus Stokes number, comparison with Pui's work [1]

It is shown that the current investigation of deposition efficiency agreed well with the experimental measurements. Which shows the reliability of the current model. It is observed that as the particle St number is increased the deposition efficiency increased. Generally, particle deposition is affected by turbulent fluctuations and particle inertia. Small particles are characterized by their low inertia, they can follow the flow, get captured and entrained by the turbulent eddies thus they mostly exit the bend, a small number of them get deposited due to the secondary flow. However, the capture capacity of the eddies is weakened upon the increase of particle St number. Owing to centrifugal force, high inertia particles cross the streamlines and accumulate on the outer bend.

In order to explore the thermophoretic effect, the flow velocity and temperature are maintained at 17.2 m/s and 288 °K, respectively. However, two values of wall temperature are considered to study the effect of cold and hot walls ($T_W = 228$ °K and $T_W = 388$ °K). The results are depicted in Fig. 5.

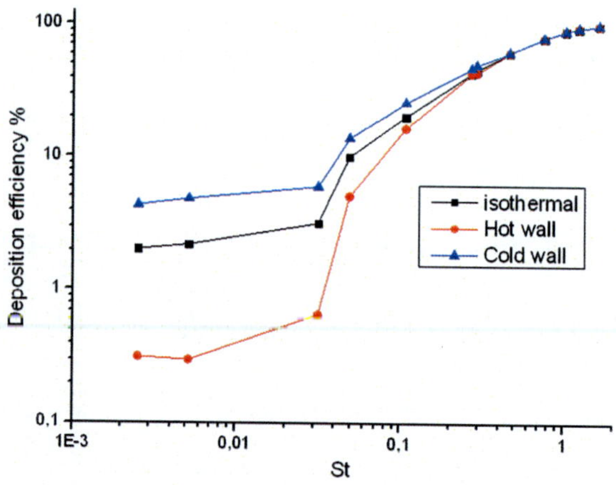

Fig. 5. Effect of thermophoresis on particle deposition efficiency in isothermal, cold and hot walls

It is shown that thermophoresis influences the small particles only (St $<$ 0.29), for particles with St $\geq$ 0.29, the isothermal, cold wall, and hot wall cases have the same deposition trend. Furthermore, it is found that thermophoretic force prohibits particles from depositing on the hot wall, but an enhancement of deposition was recorded for the cold wall.

This is associated with the fact that the thermophoretic force is directly impacted by temperature gradient and particles move through the fluid in the opposite direction of that gradient. Therefore, when the wall is heated, particles close to the wall are exposed to the thermophoretic force and move toward the core region of the fluid. As a result, particles are prevented from accumulating on the wall.

While in case of decreasing wall temperature, the particles in the fluid experience the thermophoresis effect, they tend to migrate toward the wall. Thus, the thermophoretic force drives particles to adhere to the pipe wall.

4 Conclusion

The anisotropic RMS turbulence model along with the DPM model were used to investigate particle deposition in turbulent bend pipe flow.

Particle deposition efficiency, as well as the normalized velocity are validated against experimental data available in the literature. The accuracy of the present model has been proven and the effect of particle diameter and thermophoresis on deposition efficiency was investigated. It was found that particle deposition efficiency depends on particle Stokes number. Under the same inlet temperature, it was shown that the thermophoresis influence only small particles (St < 0.29). However, for the large Stokes number, the impact was not significant.

References

1. Pui, D.Y.H., Romay-Novas, F., Liu, B.Y.H.: Experimental study of particle deposition in bends of circular cross section. Aerosol Sci. Technol. **7**, 301–315 (1987)
2. Brockmann, J.E.: Sampling and transport of aerosols. In: Aerosol Measurement: Principles, Techniques, and Applications, vol. 2, pp. 143–195 (2001)
3. McFarland, R.A., Gong, H., Muyshondt, A., Wente, W.B., Anand, N.K.: Aerosol deposition in bends with turbulent flow. Environ. Sci. Technol. **31**, 3371–3377 (1997)
4. Peters, T.M., Leith, D.: Particle deposition in industrial duct bends. Ann. Occup. Hyg. **48**(5), 483–490 (2004)
5. Breuer, M., Baytekin, H.T., Matida, E.A.: Estimation of aerosol deposition in bends using LES and an efficient Lagrangian tracking method. J. Aerosol Sci. **37**, 1407–1428 (2006)
6. Berrouk, A.B., Laurence, D.: Stochastic modeling of aerosol deposition for LES of 90° bend turbulent flow. Int. J. Heat Fluid Flow **29**, 1010–1028 (2008)
7. Zhang, P., Roberts, R.M., Bénard, A.: Computational guidelines and an empirical model for particle deposition in curved pipes using an Eulerian-Lagrangian approach. J. Aerosol Sci. **53**, 1–20 (2012)
8. Arsalanloo, A., Abbasalizadeh, M.: Numerical study on deposition of particles in a 90° bend in the presence of swirling flow using Eulerian-Lagrangian method. Powder Technol. **320**, 285–294 (2017)
9. Inthavong, K.: A unifying correlation for laminar particle deposition in 90-degree pipe bends. Powder Technol. **345**, 99–110 (2019)
10. Guo, J., Chen, Z., Shen, B., Wang, J., Yang, L.: Numerical study on characteristics of particle deposition efficiency on different walls of 90° square bend. Powder Technol. **364**, 572–583 (2020)
11. Seyfi, S., Mirzayi, B., Seyyedbagheri, H.: CFD modeling of black powder particles deposition in 3D 90-degree bend of natural gas pipelines. J. Nat. Gas Sci. Eng. **78**, 103330 (2020)
12. Versteeg, H.K., Malalasekera, W.: An Introduction to Computational Fluid Dynamics: The Finite Volume Method. Pearson Education (2007)
13. Chen, C.J., Jaw, S.Y.: Fundamentals of Turbulence Modeling. Taylor and Francis, Washington (1998)
14. Wolfshtein, M.: The velocity and temperature distribution in one-dimensional flow with turbulence augmentation and pressure gradient. Int. J. Heat Mass Transf. **12**(3), 301–318 (1969)
15. Maxey, M.R., Riley, J.J.: Equation of motion for a small rigid sphere in a nonuniform flow. Phys. Fluids **26**(4), 883–889 (1983)

16. Morsi, S.A.J., Alexander, A.J.: An investigation of particle trajectories in two-phase flow systems. J. Fluid Mech. **55**(2), 193–208 (1972)
17. Li, A., Ahmadi, G.: Dispersion and deposition of spherical particles from point sources in a turbulent channel flow. Aerosol Sci. Technol. **16**(4), 209–226 (1992)
18. Saffman, P.G.: The lift on a small sphere in a slow shear flow. J. Fluid Mech. **22**(2), 385–400 (1965)
19. Talbot, L.R.K.R.W.D.R., Cheng, R.K., Schefer, R.W., Willis, D.R.: Thermophoresis of particles in a heated boundary layer. J. Fluid Mech. **101**(4), 737–758 (1980)
20. Jang, D.S., Acharya, S.: Calculation of particle dispersion due to turbulence in elliptic flows. AIChE J. **34**(3), 514–518 (1988)
21. Cash, J.R., Karp, A.H.: A variable order Runge-Kutta method for initial value problems with rapidly varying right-hand sides. ACM Trans. Math. Softw. (TOMS) **16**(3), 201–222 (1990)
22. Sudo, K., Sumida, M., Hibara, H.: Experimental investigation on turbulent flow in a circular-sectioned 90-degree bend. Exp. Fluids **25**(1), 42–49 (1998)
23. Yan, Y., Zhao, Y., Yao, J., Wang, C.H.: Investigation of particle transport by a turbulent flow through a 90° bend pipe with electrostatic effects. Powder Technol. **394**, 547–561 (2021)

Improved Analytical/Statistical Modelling of the Shock Wave-Laminar Boundary Layer Around a Thin Airfoil: Standard Atmosphere Case

Nasser Eddegdag[1](✉), Omar El-Aajine[1], Aze-eddine Naamane[2], and Mohammed Radouani[1]

[1] Moulay Ismail University, Meknes, Morocco
{n.eddegdag,o.elaajine}@edu.umi.ac.ma, m.radouani@ensam-umi.ac.ma

[2] Royal Moroccan Air School, Marrakech, Morocco

Abstract. This present work aims to develop an improved and more precised analytical modelling of a steady irrotational laminar Shock Wave-Boundary Layer Interaction for weak shockwaves around a thin airfoil at a low incidence in the standard atmosphere. This study adapts the asymptotical modelling of our problem treated by our research team previously [1], and improves the analytical resolution process by integrating the empirical parameter *m*. Then, confrontation of our analytical model to experimental results obtained through experimentation in Supersonic Wind Tunnel AF300. And finally, a CFD numerical simulation in ANSYS Fluent R13 was conducted in order to validate our model, followed up by a statistical study in SPSS taking into consideration analytical and numerical results in order to establish the exact analytical expression of *m* for each airfoil depending on upstream Mach numbers. The comparison of results obtained from the numerical, experimental and analytical confrontations of our improved and previous models showed promising results and greatly decreased the approximation errors obtained in our previous studies.

Keywords: Shock Wave-Boundary Layer Interaction · Laminar · Standard atmosphere · Empirical parameter · CFD · Statistical study · Supersonic Wind Tunnel

1 Introduction

Because of their ubiquitous presence in high-speed flights and their impact on vehicles and components performances Shock Wave-Boundary Layer Interaction "SWBLI" has been studied for more than 50 years [2]. Despite remarkable progress in computational and measurements capabilities, there are more important quantities that cannot be predicted very accurately. In high Reynolds number flows, the no-slip condition at solid surfaces leads to the formation of a boundary layer in which the velocity and the temperature adjust rapidly to meet the local freestream conditions. Hence, a shockwave is a

L. Azrar et al. (Eds.): CIP 2022, LNME, pp. 257–272, 2023.
https://doi.org/10.1007/978-3-031-23615-0_27

localized pressure increase and is classified as an adverse pressure gradient. The effects of a moderate adverse pressure gradient are a thickening of a boundary layer [3]. But in extreme cases, where the shockwave is strong enough and large enough, it causes boundary layer separation, which then has a global effect on the structure of the flow and hence the solid. The study, knowledge and understanding harvested from SWBLI studies is vital given the omnipresent nature of SWBLI in high-speed flights. Consider the example of two missiles flying in close proximity to each other or that of the side intake with central bodies of a supersonic aircraft, Wing-Body interactions and static props in the presence of a shockwave are all applications in which SWBLI plays the main role [4].

In our previous studies, we were interested in the modelling of laminar steady irrotational SWBLI around a thin airfoil located in the standard atmosphere. The modelling process initiated from Navier-Stokes equations adapted to our assumptions, to which we applied the asymptotic methods, adimensionalisation and linearization, in order to obtain the equation governing our problem [1].

Afterwards, we applied an asymptotical expansion on the governing equation following β and Re^{-1} individually at $o(1)$. With $\overline{U}_{\infty}$, characterizing the dimensionless upstream velocity, which in our case in the standard atmosphere, for which the solution is considered simple since we consider that the upstream flow is slightly disturbed by the thin airfoil, irrotational and varies linearly following $\overline{\eta}$ which is considered the only variable characterizing the upstream velocity. In addition, we applied the singular perturbations methods in respect to the principle of least degeneracy, in order to find expression forms of perturbation velocity potential φ near the airfoil body far from the leading edge, both inside the boundary layer $\hat{\varphi}_{01}(\overline{s}, \hat{\eta})$ and in the undisturbed flow outside the boundary layer $\overline{\varphi}_{00}(\overline{s}, \overline{\eta})$ [5, 6].

The resolution process of our past work consisted of adopting the same technics (Triple Deck Technic) used in this present work [1], however the linkage methods between the different decks of the boundary layer, especially the linkage between the sub-viscous deck and the main deck where the logic adopted was to connect the two layers based on the continuity of the relative thickness that tends in each deck to meet an intermediate variable, neglecting the effect of connectivity in velocity components are what caused slight lack of precision in our previous model [7]. Mach numbers in each layer tend to be continuous and converge in each layer to an intermediate variable m which is the local incompressible connection Mach number that characterizes the continuity of velocity (continuity of variables and their derivatives) and the unique incompressible aspect of the sub-viscous deck. The neglection of this empirical parameter has effects on the formulation of the velocity field expressions inside and outside the boundary layer since the resolution methods to find exact expressions of constants link all the expressions together, hence inserting and taking into consideration the effect of that parameter will minimize errors in Mach number expressions in the different study parts of our problem.

2 Resolution Process

The resolution process of our problem focalizes on finding exact expressions of constants in the formulas of $\overline{\varphi}_{00}$ and $\overline{\varphi}_{01}$, in order to establish exact formulas of Mach numbers inside and outside the boundary layer near the wall far from the leading edge of the thin airfoil. The technic adopted in our process is the "triple-layer" model, which was proposed in 1969, independently and almost simultaneously, by Neiland [8] and Stewartson and Williams [9], makes it possible to account for the behavior of the boundary layer (here it will be of a laminar boundary layer) in various circumstances where it undergoes "an accident" in the vicinity of a line drawn on an obstacle. Naturally, we are interested in the local structure of the flow when $Re \to \infty$ and we are then led to locally distinguish three layers (triple deck) (Fig. 1): ***Upper deck***, ***Main deck*** and ***Lower deck***.

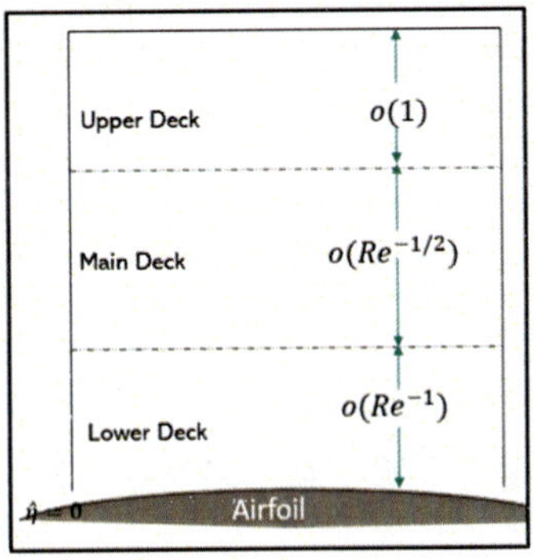

Fig. 1. The triple deck model of NEILAND and STEWARTSON-WILLIAMS

2.1 The Connection Between the Main Deck and the Lower Deck (Link Between the Incompressible and the Compressible)

Near the wall away from the leading edge when $\overline{\eta} \to 0$, the steady flow is considered that of a low viscosity and homogeneous incompressible fluid. Thus, we have the equation of the velocity field for a low viscous incompressible flow, from which we have the equation of the velocity field in the viscous sub-layer [5] of thickness $\tilde{\eta}$ relative to:

$$\begin{cases} \tilde{U}^{LD}(\overline{s}, \hat{\eta}) = \frac{\tilde{U}_{\infty}(0)\, cos(\alpha)}{3} \tilde{\eta}^2 \\ \tilde{V}^{LD}(\overline{s}, \tilde{\eta}) = 0 \end{cases} \quad (1)$$

where α is the angle of incidence, $\overline{\eta}$ and $\overline{s}$ are normal and tangential coordinate in the Frenet coordinate system $\left(P, \vec{e}, \vec{n}, \vec{b}\right)$, $\tilde{U}^{LD}$ and $\tilde{V}^{LD}$ are the dimensionless longitudinal and transverse velocity in the Lower deck, with $\hat{\eta}$ is the normal coordinate (dimensionless) in the vicinity of the wall in the Main deck and $\tilde{\eta}$ is the normal coordinate (dimensionless) in the vicinity of the wall in the Lower deck.

Thus, we are interested in the area of connection between the Lower deck and the Main deck (Fig. 2) when $\tilde{\eta} \to \tilde{\eta}^*$ and $\hat{\eta} \to \hat{\eta}^*$ with $\tilde{\eta} = \frac{\hat{\eta}}{Re^{-p}}$, with $\left(Re^{-1}\right)^p$ a gauge function a priori unknown and they must be determined so that the model in triple layer, set up is asymptotically, when $Re^{-1} \to 0$, consistent [10]. As well as the Local number of Mach tends towards an incompressible Mach, hence $M_{local} \to m$ with $0 \prec m \prec 0.3$.

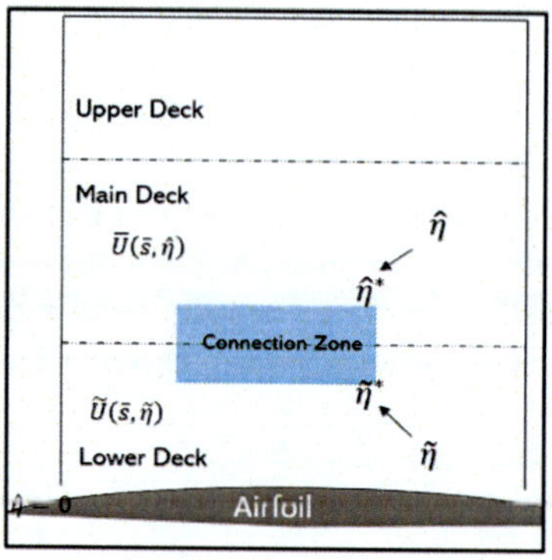

Fig. 2. Diagram showing the connection between the Lower deck and the Main deck

Thus, by taking the normal component of the connection between the Main deck and Lower, we have:

$$\hat{V}_{MD}\left(\bar{s},\hat{\eta}\right)=\tilde{V}^{LD}(\bar{s},\tilde{\eta})\Rightarrow A=B\frac{2\tan\left(\frac{m^2}{2}\pi\right)}{\sqrt{\frac{m^4}{1-m^2}}} \tag{2}$$

And by taking the tangential component of the connection we have: $\hat{U}_{MD}\left(\bar{s},\hat{\eta}\right)=\tilde{U}^{LD}(\bar{s},\tilde{\eta})$, and so we get

$$A=\frac{1.7783m\cos(\alpha)}{\cos\left(\frac{m^2}{2}\pi\right)\left(\frac{m^5}{4}+\frac{m^3}{2}\right)\left(\frac{m^4}{1-m^2}\right)}Re^{2p+1} \tag{3}$$

With $p=-1$, since the constants of the perturbed potentials must be of order $o(1)$, hence we will have a condition on p which will manage the link between $\hat{\eta}^*$ and $\tilde{\eta}^*$. And so, the connection zone is of thickness l_0Re^{-1} (where l_0 is the length of the profile wall) [11]. But, since A is of the order $o(1)$, thus, we find that for $1\leq A\leq 5$ we must have $0.153\leq m\leq 0.2$. Hence, we have the exact expression of non-disturbed viscous non-dimensional velocity potential in the Main deck:

$$\hat{\varphi}_{01}\left(\bar{s},\hat{\eta}\right)=\left[\begin{array}{c}\dfrac{0.77\left(\frac{M_\infty^5}{4}+\frac{M_\infty^3}{2}\right)}{\sqrt{\frac{M_\infty^4}{M_\infty^2-1}}\frac{m\cos(\alpha)}{\cos\left(\frac{m^2}{2}\pi\right)\left(\frac{m^5}{4}+\frac{m^3}{2}\right)\left(\frac{m^4}{1-m^2}\right)}Re^{-2}}\\ \left(\begin{array}{c}\cos\left(\frac{M_\infty^2}{2}\pi\right)\sin\left(\frac{1}{2}\sqrt{\frac{M_\infty^4}{M_\infty^2-1}}\bar{s}\right)\\ -\frac{\sin\left(\frac{M_\infty^2}{2}\pi\right)}{2\tan\left(\frac{m^2}{2}\pi\right)}\sqrt{\frac{m^4}{1-m^2}}\cos\left(\frac{1}{2}\sqrt{\frac{M_\infty^4}{M_\infty^2-1}}\bar{s}\right)\end{array}\right)\end{array}\right]\hat{\eta}^2+D_3 \tag{4}$$

With D_3 being a constant.

2.2 The Connection Between the Main Deck and the Upper Deck (Link Between Boundary Layer and Exterior Supersonic Flow)

We have the connection between the Upper layer and the Main layer is characterized by the sonic line which characterizes the passage from the subsonic flow to a supersonic flow in $\overline{\varphi}_{00}(\overline{s}, \overline{\eta})$ (Fig. 3), thus the connection zone is characterized by a transonic flow in which $M_\infty^2 = 1 \Rightarrow \overline{U}^2 + \overline{V}^2 = 1$.

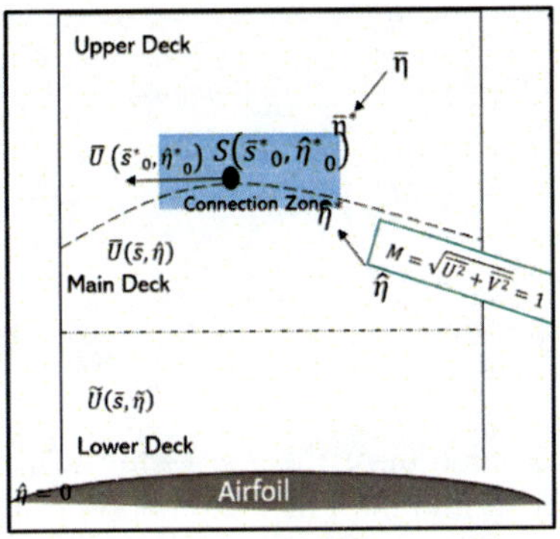

Fig. 3. Diagram showing the connection between the Main deck and the Upper deck

Thus, in this zone where;

$$\begin{pmatrix} \hat{\eta} \to \hat{\eta}^* \text{ et } \overline{\eta} \to \overline{\eta}^* \text{ with } \hat{\eta}^* = \frac{\overline{\eta}^*}{Re^{-q}} \\ \text{also } \overline{s} \to \overline{s}^* \end{pmatrix} \tag{5}$$

We suppose that the curve of the sonic line is characterized by a vertex $S(\overline{s}_0^*, \hat{\eta}_0^*)$ [10, 12] in which:

$$\begin{cases} \overline{V}_{MD}(\overline{s}_0^*, \hat{\eta}_0^*) = 0 \\ \overline{V}_{ext}(\overline{s}_0^*, \overline{\eta}_0^*) = 0 \end{cases} \text{ and } \begin{cases} \overline{U}_{MD}(\overline{s}_0^*, \hat{\eta}_0^*) = \delta_0 \\ \overline{U}_{ext}(\overline{s}_0^*, \overline{\eta}_0^*) = \delta_0 \end{cases} \text{ with } \delta_0 = \pm 1 \tag{6}$$

So, to find the coordinates of the vertex, we solve the system:

$$\begin{cases} \overline{V}_{MD}(\overline{s}_0^*, \hat{\eta}_0^*) = 0 \\ \overline{U}_{MD}(\overline{s}_0^*, \hat{\eta}_0^*) = \delta_0 \end{cases} \tag{7}$$

Hence, we get

$$\overline{s}_0^* = \frac{2\,arctan\left(\frac{tan\left(\frac{M_\infty^2}{2}\pi\right)}{2\,tan\left(\frac{m^2}{2}\pi\right)}\sqrt{\frac{m^4}{1-m^2}}\right)}{\sqrt{\frac{M_\infty^4}{M_\infty^2-1}}} \tag{8}$$

Although, since $0 \leq \overline{s}_0^* \leq 1$, by plotting $\overline{s}_0^*$ as a function of M_∞, we notice that our model is valid for a supersonic Mach as $1.41 \leq M_\infty \leq 1.6864$. Also, we have

$$\overline{U}_{MD}\left(\overline{s}_0^*, \hat{\eta}_0^*\right) = \frac{1}{1-\hat{\eta}_0^*}\frac{\partial\hat{\varphi}_{01}\left(\overline{s}, \hat{\eta}\right)}{\partial\overline{s}} = \delta_0 = \pm 1 \tag{9}$$

Hence

$$\hat{\eta}_0^* = \frac{1-\sqrt{1-4\tau}}{2\tau} \text{ with}$$

$$\tau = \left[\begin{array}{c}\frac{0.385m\,cos(\alpha)}{cos\left(\frac{m^2}{2}\pi\right)\left(\frac{m^5}{4}+\frac{m^3}{2}\right)\left(\frac{m^4}{1-m^2}\right)}Re^{-1}\left(\frac{M_\infty^5}{4}+\frac{M_\infty^3}{2}\right)\left(\frac{M_\infty^4}{M_\infty^2-1}\right)\\ \left[\begin{array}{c}cos\left(\frac{M_\infty^2}{2}\pi\right)cos\left(\frac{1}{2}\sqrt{\frac{M_\infty^4}{M_\infty^2-1}}\overline{s}_0^*\right)\\ +\frac{sin\left(\frac{M_\infty^2}{2}\pi\right)}{2\,tan\left(\frac{m^2}{2}\pi\right)}\sqrt{\frac{m^4}{1-m^2}}sin\left(\frac{1}{2}\sqrt{\frac{M_\infty^4}{M_\infty^2-1}}\overline{s}_0^*\right)\end{array}\right]\end{array}\right] \tag{10}$$

And so, after finding the coordinates for the vertex $S\left(\overline{s}_0, \hat{\eta}_0\right)$, we resolve the following system (11) in order to find the exact expression of the undistributed inviscid dimensionless velocity potential in the Upper deck and hence find the values of the constants in $\overline{\varphi}_{00}(\overline{s}, \overline{\eta})$, we have:

$$\overline{\varphi}_{00}(\overline{s}, \overline{\eta}) = (1-\overline{\eta})^{\frac{-M_\infty^2}{2}}\left[\begin{array}{c}cos\left(\frac{M_\infty^2}{2}\pi\right)(C_1+C_2)\,cos\left(\frac{1}{2}\sqrt{\frac{M_\infty^4}{M_\infty^2-1}}\overline{s}\right)\\ +sin\left(\frac{M_\infty^2}{2}\pi\right)(C_1-C_2)\,sin\left(\frac{1}{2}\sqrt{\frac{M_\infty^4}{M_\infty^2-1}}\overline{s}\right)\end{array}\right]D \tag{11}$$

We put $\begin{cases} Z = C_1 + C_2 \\ X = C_1 - C_2 \end{cases}$.

In the connection zone between the Upper and Main deck we have: $\overline{\eta}_0^* = Re^{-q}\hat{\eta}_0^*$, although since $0 \leq \overline{\eta}_0^* \leq 1$, then we conclude that $\overline{\eta}_0^* = \hat{\eta}_0^*$.

To find the values of constants Z and X, we resolve the following equations:

$$\overline{V}_{ext}\left(\overline{s}_0^*, \overline{\eta}_0^*\right) = 0 \Rightarrow Z = -X\frac{tan^2\left(\frac{M_\infty^2}{2}\pi\right)}{2\,tan\left(\frac{m^2}{2}\pi\right)}\sqrt{\frac{m^4}{1-m^2}} \tag{12}$$

$$\overline{U}_{ext}\left(\overline{s}_0^*, \overline{\eta}_0^*\right) = \overline{U}_{MD}\left(\overline{s}_0^*, \overline{\eta}_0^*\right) = -1 \Rightarrow DX = -\frac{2}{\left(1-\overline{\eta}_0^*\right)^{\frac{-M_\infty^2}{2}}\sqrt{\frac{M_\infty^4}{M_\infty^2-1}}\sin\left(\frac{M_\infty^2}{2}\pi\right)\chi} \tag{13}$$

with

$$\chi = \begin{bmatrix} \frac{tan\left(\frac{M_\infty^2}{2}\pi\right)}{2\,tan\left(\frac{m^2}{2}\pi\right)}\sqrt{\frac{m^4}{1-m^2}}sin\left(arctan\left(\frac{tan\left(\frac{M_\infty^2}{2}\pi\right)}{2\,tan\left(\frac{m^2}{2}\pi\right)}\sqrt{\frac{m^4}{1-m^2}}\right)\right) \\ +cos\left(arctan\left(\frac{tan\left(\frac{M_\infty^2}{2}\pi\right)}{2\,tan\left(\frac{m^2}{2}\pi\right)}\sqrt{\frac{m^4}{1-m^2}}\right)\right) \end{bmatrix} \quad (14)$$

Accordingly, we have:

$$\overline{\varphi}_{00}(\overline{s},\overline{\eta}) = \frac{2(1-\overline{\eta})^{\frac{-M_\infty^2}{2}}}{\left(1-\overline{\eta}_0^*\right)^{\frac{-M_\infty^2}{2}}\sqrt{\frac{M_\infty^4}{M_\infty^2-1}}\chi}\begin{bmatrix} \frac{tan\left(\frac{M_\infty^2}{2}\pi\right)}{2\,tan\left(\frac{m^2}{2}\pi\right)}\sqrt{\frac{m^4}{1-m^2}}cos\left(\frac{1}{2}\sqrt{\frac{M_\infty^4}{M_\infty^2-1}}\overline{s}\right) \\ -sin\left(\frac{1}{2}\sqrt{\frac{M_\infty^4}{M_\infty^2-1}}\overline{s}\right) \end{bmatrix} \quad (15)$$

3 Results for Disturbance Potential and Mach Number

The application of the asymptotic method in the equations, both in the Navier-Stokes model and in the boundary conditions, gave two systems predicted by the asymptotic expansion of $\overline{\varphi}_0$. According to the generalized development of $\overline{\varphi}_0$ with respect to Re^{-1} [1], we have:

$$\overline{\varphi}_0(\overline{s},\overline{\eta}) = \overline{\varphi}_{00}(\overline{s},\overline{\eta}) + Re^{-1}\overline{\varphi}_{01}(\overline{s},\overline{\eta}) \quad (16)$$

However, thanks to the singular perturbations' methods, the triple deck technique and the boundary conditions, the expressions $\overline{\varphi}_{00}(\overline{s},\overline{\eta})$ have been obtained, so we will have the tangential and normal velocity components outside the boundary layer. From result of the velocity perturbation potential which was obtained in our calculations, we can deduce the expression of the velocity field on the wing profile. According to the linearization theorem or Hartman–Grobman theorem [1], we have:

$$\overline{U}_{ext}(\overline{s},\overline{\eta}) = \overline{U}_\infty(\overline{s},\overline{\eta}) + \frac{\beta}{1-\overline{\eta}}\frac{\partial\overline{\varphi}_{00}(\overline{s},\overline{\eta})}{\partial\overline{s}}$$
$$\text{and } \overline{V}_{ext}(\overline{s},\overline{\eta}) = \beta\frac{\partial\overline{\varphi}_{00}(\overline{s},\overline{\eta})}{\partial\overline{\eta}} \quad (17)$$

With β the relative thickness of the airfoil, $\overline{U}_\infty(\overline{s},\overline{\eta})$ is the speed of the dimensionless standard atmosphere. Hence, we get:

$$\overline{U}_{ext}(\overline{s},\overline{\eta}) = -1.155M_\infty(\overline{\eta}-1) + \frac{\beta(1-\overline{\eta})^{\frac{-M_\infty^2}{2}-1}}{\left(1-\overline{\eta}_0^*\right)^{\frac{-M_\infty^2}{2}}\chi}K_1$$
$$\text{and } K_1 = \frac{tan\left(\frac{M_\infty^2}{2}\pi\right)}{2\,tan\left(\frac{m^2}{2}\pi\right)}\sqrt{\frac{m^4}{1-m^2}}sin\left(\frac{1}{2}\sqrt{\frac{M_\infty^4}{M_\infty^2-1}}\overline{s}\right) - cos\left(\frac{1}{2}\sqrt{\frac{M_\infty^4}{M_\infty^2-1}}\overline{s}\right) \quad (18)$$

And

$$\overline{V}_{ext}(\overline{s},\overline{\eta}) = \beta \frac{M_{\infty}^{2}(1-\overline{\eta})^{\frac{-M_{\infty}^{2}}{2}-1}}{\left(1-\overline{\eta}_{0}^{*}\right)^{\frac{-M_{\infty}^{2}}{2}}\sqrt{\frac{M_{\infty}^{4}}{M_{\infty}^{2}-1}}\chi} K_{2}$$

$$\text{and } K_{2} = \frac{tan\left(\frac{M_{\infty}^{2}}{2}\pi\right)}{2\,tan\left(\frac{m^{2}}{2}\pi\right)}\sqrt{\frac{m^{4}}{1-m^{2}}}cos\left(\frac{1}{2}\sqrt{\frac{M_{\infty}^{4}}{M_{\infty}^{2}-1}}\overline{s}\right) - sin\left(\frac{1}{2}\sqrt{\frac{M_{\infty}^{4}}{M_{\infty}^{2}-1}}\overline{s}\right) \quad (19)$$

We obtained thanks to the methods of the singular perturbations and the boundary conditions the expression $\hat{\varphi}_{01}(\overline{s},\hat{\eta})$, so we will have the tangential and normal velocity components inside the boundary layer. From the result of the speed disturbance potential that was obtained in our calculations, we can deduce the expression of the speed field on the wing profile. According to the linearization theorem or Hartman–Grobman theorem, we have:

$$\overline{U}_{BL}(\overline{s},\hat{\eta}) = \beta\frac{\partial\hat{\varphi}_{01}(\overline{s},\hat{\eta})}{\partial\overline{s}} \text{ and } \overline{V}_{BL}(\overline{s},\hat{\eta}) = \beta\frac{\partial\hat{\varphi}_{01}(\overline{s},\hat{\eta})}{\partial\overline{\eta}} \quad (20)$$

Hence,

$$\overline{U}_{BL}(\overline{s},\hat{\eta}) = \beta\left[\frac{0.385\left(\frac{M_{\infty}^{5}}{4}+\frac{M_{\infty}^{3}}{2}\right)\left(\frac{M_{\infty}^{4}}{M_{\infty}^{2}-1}\right) m\,cos(\alpha)}{cos\left(\frac{m^{2}}{2}\pi\right)\left(\frac{m^{5}}{4}+\frac{m^{3}}{2}\right)\left(\frac{m^{4}}{1-m^{2}}\right)}Re^{-2}\left[\begin{array}{l} cos\left(\frac{M_{\infty}^{2}}{2}\pi\right)cos\left(\frac{1}{2}\sqrt{\frac{M_{\infty}^{4}}{M_{\infty}^{2}-1}}\overline{s}\right) \\ +\frac{sin\left(\frac{M_{\infty}^{2}}{2}\pi\right)}{2\,tan\left(\frac{m^{2}}{2}\pi\right)}\sqrt{\frac{m^{4}}{1-m^{2}}}sin\left(\frac{1}{2}\sqrt{\frac{M_{\infty}^{4}}{M_{\infty}^{2}-1}}\overline{s}\right)\end{array}\right]\right]\hat{\eta}^{2} \quad (21)$$

And

$$\overline{V}_{BL}(\overline{s},\hat{\eta}) = \beta\left[\frac{1.54\left(\frac{M_{\infty}^{5}}{4}+\frac{M_{\infty}^{3}}{2}\right)\sqrt{\frac{M_{\infty}^{4}}{M_{\infty}^{2}-1}}\, m\,cos(\alpha)}{cos\left(\frac{m^{2}}{2}\pi\right)\left(\frac{m^{5}}{4}+\frac{m^{3}}{2}\right)\left(\frac{m^{4}}{1-m^{2}}\right)}Re^{-2}\left[\begin{array}{l} cos\left(\frac{M_{\infty}^{2}}{2}\pi\right)sin\left(\frac{1}{2}\sqrt{\frac{M_{\infty}^{4}}{M_{\infty}^{2}-1}}\overline{s}\right) \\ -\frac{sin\left(\frac{M_{\infty}^{2}}{2}\pi\right)}{2\,tan\left(\frac{m^{2}}{2}\pi\right)}\sqrt{\frac{m^{4}}{1-m^{2}}}cos\left(\frac{1}{2}\sqrt{\frac{M_{\infty}^{4}}{M_{\infty}^{2}-1}}\overline{s}\right)\end{array}\right]\right]\hat{\eta} \quad (22)$$

Our main goal is to find the local Mach number on the wall far from the leading edge of the wing airfoil. From this number, we can deduce all the other physical quantities: pressure, density, temperature, flow speed … a formula was useful for us to carry out our calculations:

$$M_{local} = \sqrt{\left[\overline{U}(\overline{s},\overline{\eta})\right]^{2} + \left[\overline{V}(\overline{s},\overline{\eta})\right]^{2}} \quad (23)$$

With $1.41 \leq M_\infty \leq 1.6864$.

Hence, we obtain the core result of our study which is the Local Mach Number around the airfoil, which characterizes all the governing parameters that play a substantial factor in the control and definition of the behavior of the supersonic flow of a laminar viscous fluid around a thin airfoil.

4 Validation by Numerical Simulation

In this part, the numerical approach for the study of the problem will be approached, under the same assumptions as in the theoretical approach. Indeed, it will make it possible to obtain the precise behaviour of a viscous laminar supersonic airflow around the NACA 43013 airfoil after setting the simulation software Ansys Fluent R19.3.

4.1 Modelling of the Problem in Ansys Fluent R19.3

The modelling of the problem on the Ansys fluent software requires a specific study and a great precision in the different stages of the numerical simulation, the study geometry is airfoil NACA 43013, the calculation domain and meshing were done similar to pervious work by our research team [6], taking into account a hyper fine and precise mesh in the vicinity of the edge of the airfoil in order to be able to model the different sub-layers of the boundary layer. Indeed, we have achieving *104777* nodes and *103477* elements stacking in the direction normal to the boundary using a feature called "Inflation" (Fig. 4).

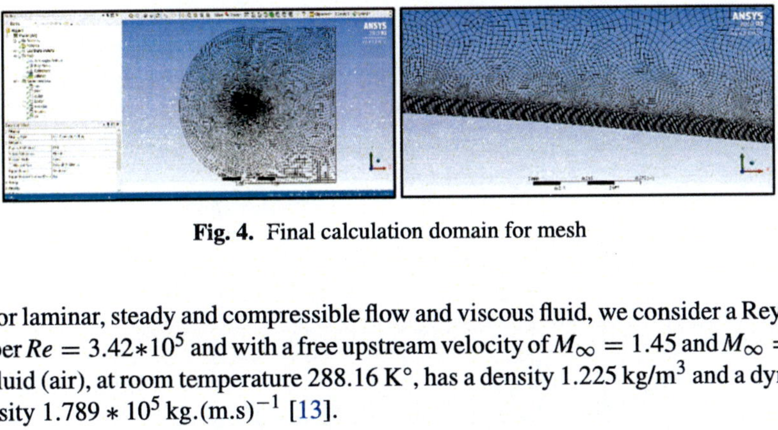

Fig. 4. Final calculation domain for mesh

For laminar, steady and compressible flow and viscous fluid, we consider a Reynolds number $Re = 3.42 * 10^5$ and with a free upstream velocity of $M_\infty = 1.45$ and $M_\infty = 1.5$. The fluid (air), at room temperature 288.16 K°, has a density 1.225 kg/m^3 and a dynamic viscosity $1.789 * 10^5$ kg.(m.s)$^{-1}$ [13].

4.2 Visualization and Analysis of Results for NACA 43013

For $M_\infty = 1.45$, Fig. 5 is obtained. First of all, it allows to visualize the variation of the Mach number from the leading edge and all along the airfoil. In particular, it is observed that on the upper surface the Mach number keeps increasing from the leading edge to the trailing edge where ${M_{local}}^l = 1.666$. While on the lower surface, it grows faster from the leading edge to the trailing edge where ${M_{local}}^2 = 1.616$. Secondly, the figure clearly depicts the curved detached shock wave on the leading edge, with the presence of a blue-green stagnation sonic bubble where the flow is locally rotational unsteady and incompressible and $M < 1.00$, we can also observe at the end of the extrados leading edge where the shape of the airfoil is more of a convex corner the formation of infinite Mach lines that are the result of Prandtl-Meyer expansion waves, same phenomena occurs in the middle of the intrados part of the airfoil near the war, and finally we have the weak oblique SW formed near the intrados part of the leading edge where the airfoil takes the form of a concave corner [14].

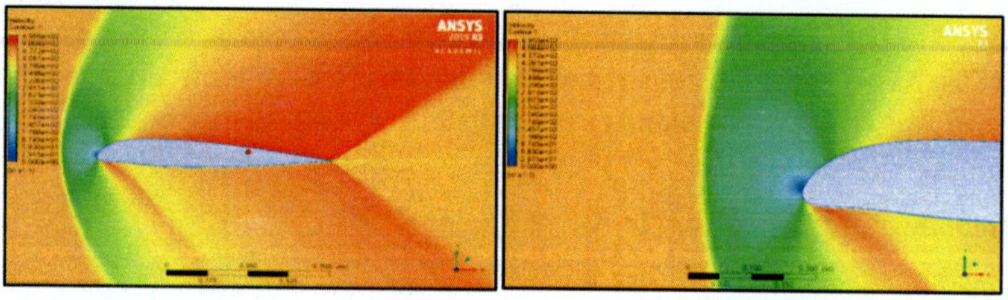

Fig. 5. Velocity evolution around the NACA 43013 wing airfoil, simulation for $M_\infty = 1.45$

Finally, the third observed phenomenon is the SWBLI, which is the result of viscous supersonic flow. According to Fig. 6, we note the existence of a boundary layer all along the wall of the thickness profile $l_0 = 0.4076$ mm where the Mach is subsonic and the speed tends towards a zero-value approaching the surface of the profile, at the end of the boundary layer all around the airfoil we have Mach lines [*expansion waves-weak SW*] that are incident shocks towards the boundary layer, taking into consideration for our case their weak nature causing the SWs to go through the boundary layer without changing its global aspect and to slightly locally separate, all over the extrados, its Lower Deck while overall maintaining its same total width. [*The width of the different decks changes imperceptibly in many points of the airfoil, in a pseudo-periodic nature, while maintaining the same total width of the boundary layer*].

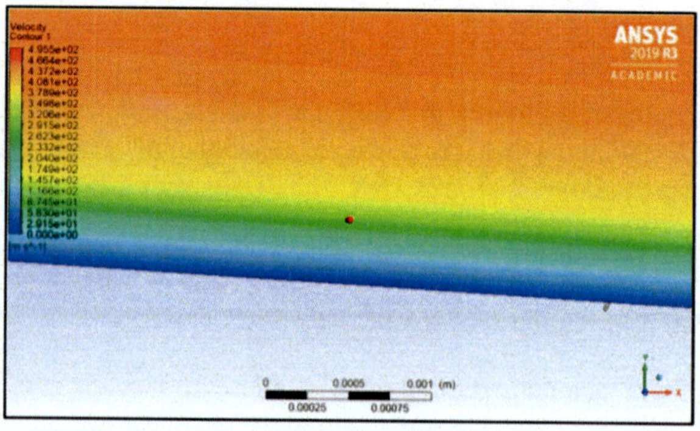

Fig. 6. Visualization of the boundary layer around the profile by simulation on Ansys Fluent-in dark blue Lower deck, in light blue Main deck, in light green Upper deck

5 Confrontation of Numerical-Analytical Results

The various theoretical studies carried out since the advent of the supersonic cover large areas and make it possible to compare several parameters. Our focus on our study is to compare the velocity evolution by comparing numerical and analytical results of our model of local Mach numbers near airfoil wall.

5.1 Comparison of Numerical-Analytical Results

We took results in the wall far from the leading edge; hence we took results from an x_0 which characterizes the end of the leading edge, then for the incompressible Mach m which characterizes the connection between the viscous underlayer and the intermediate layer, we decided to fix its value thanks to the numerical results to have a better precision and a weak error. Local Mach extrados and intrados located in the wall far from the leading edge, thus the position of the maximum relative camber characterizes the end of the leading-edge zone. We perform our calculations and extract our results for x greater than *0.15* m. Similarly, we have according to the theoretical model that $0 \leq \overline{\eta} \leq 0.13257143$, from which we have y limited in 0 and 0.19 for x included between 0.15 and 0.208866, as well as y limited by 0 and 0.132 for x between 0.208866 and 1. It can be seen from Fig. 7 that for x in the first half of the extrados, up to 40% of the chord, the error is very minimal, almost $\varepsilon^1_{ectrados} = 2.283977\%$, while approaching the trailing edge the error increases to a value of $\varepsilon^2_{ectrados} = 10.10676\%$, since the region of $x \in [0.2088, 1]$ is an instability region with [0.2088, 0.32] a transition region which is the response of the oscillation that perturbs the surrounding flow.

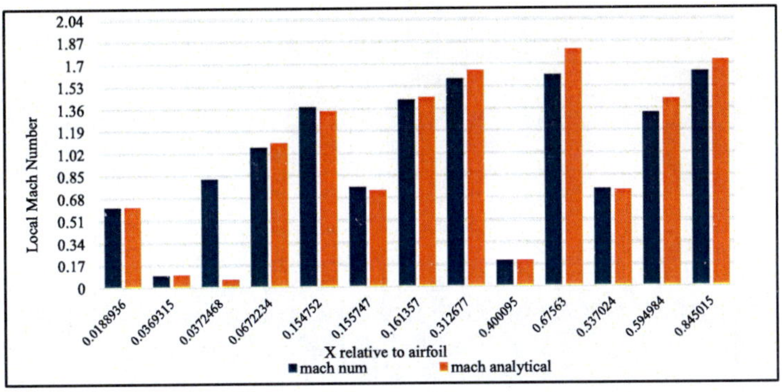

Fig. 7. Numerical and analytical Local Mach Number around the NACA 43013 wing airfoil, for Upstream Mach Number $M_\infty = 1.45$

5.2 Statistical/Analytical Formulation of Empirical Parameter *m*

We have the incompressible Mach *m* that varies; according to x and y coordinates taken in different regions of the wall of the NACA 43013 wing profile as well as in different positions of $\overline{\eta}$ (characterizing the lower, intermediate and upper deck of the boundary layer), for a fixed upstream infinite Mach number $M_{\infty} = 1.45$; according to a well-defined variable curve which is not linear (see Fig. 8).

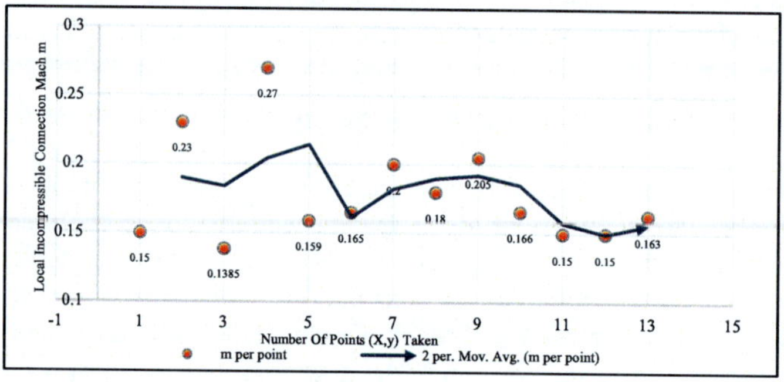

Fig. 8. Evolution of empirical parameter m for the points (x, y) taken

Thus, by plotting the incompressible Mach number *m* in a 3D Wireframe contour on EXCEL according to x and y coordinates for all chosen points, it's clear that *m* varies in quasi-linear way according to each parameter x and y individually (see Fig. 9) with *m* varying approximately between 0.1 and 0.2 (*maximum density in orange*) (which is consistent with our analytical results $0.152 < m < 0.2$).

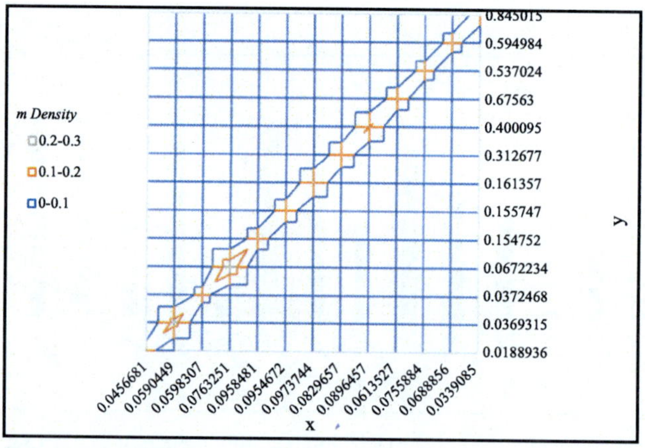

Fig. 9. Wireframe contour graph of the evolution of *m* density following (x, y) for all points taken for $M_{\infty} = 1.45$ for NACA 43013

And so, for a first determination of the value of m, we plot the data of the incompressible Mach *m* according to the coordinate points x and y at $M_{\infty} = 1.45$ on the SPSS software, then we plot its boxplot, and we note that *m* is approximately equal to $m = 0.164912$ (the value of the median of the curve of *m* boxplot) (see Fig. 10).

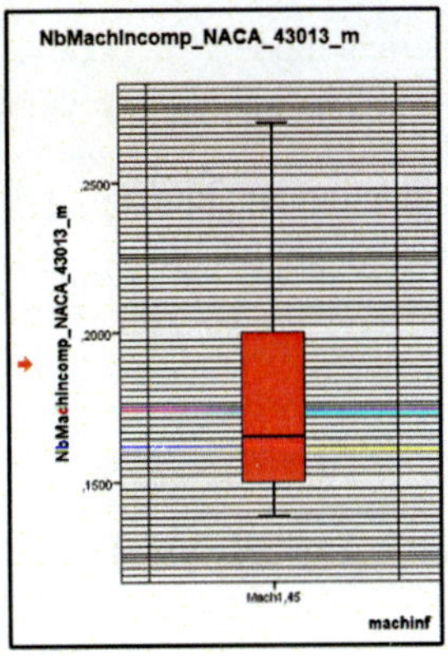

Fig. 10. Boxplot graph on SPSS software for m as a function following points (x, y) for $M_\infty =$ 1.45 for NACA 43013

Then for more precision for the m, we note thanks to the numerical simulation on Ansys Fluent that m depends mainly on three parameters for each specific airfoil $(\overline{y}, \overline{x}, M_\infty)$, from which, for more precision and to have exact results of our theoretical model, we assume that m varies for NACA 43013 according to these three parameters according to an empirical distribution function with:

$$m(\overline{y}, \overline{x}, M_\infty) = \alpha_1\overline{y} + \alpha_2\overline{x} + \alpha_3 M_\infty \tag{24}$$

where α_1, α_2 and α_3 are constants.

For $M_\infty = 1.45$ and NACA 43013, we combine the calculated data into a single simple system of three equations:

$$\begin{cases} 0.0188936\alpha_2 + 0.0456681\alpha_1 + 1.45\alpha_3 = 0.15 \\ 0.0369315\alpha_2 + 0.0590449\alpha_1 + 1.45\alpha_3 = 0.23 \\ 0.0372468\alpha_2 + 0.0598307\alpha_1 + 1.45\alpha_3 = 0.1385 \end{cases} \tag{25}$$

In order to deduce, with more precision, the equation of m as a function of $(\overline{y}, \overline{x}, M_\infty)$, we solve the system (33) on the LINGO Software, we obtain the equation below which is the approximative statistical-analytical formulation of the local incompressible deck-connection Mach number m for $M_\infty = 1.45$ for the NACA 43013 airfoil:

$$m_{NACA\,43013}(\overline{y}, \overline{x}, M_\infty = 1.45) = -168.301688\overline{y} + 129.2466\overline{x} + 5.39408395 \tag{26}$$

Same for $M_\infty = 1.55$, following the same statistical/analytical method we obtain an approximative formulation for empirical parameter m for NACA 43013 as following:

$$m_{NACA\,43013}(\overline{y}, \overline{x}, M_\infty = 1.5) = -130.33688\overline{y} + 39.387003\overline{x} + 6.486107077 \tag{27}$$

Consequently, all these comparisons on various parameters of the flow finally prove that the numerical simulation transcribes in a coherent and precise manner the flow around the profile in accordance with the theoretical approach. Indeed, the differences between the theoretical and numerical models are excessively small while being precise. Finally, the numerical approach leads to very fine concrete results. Thus, numerical simulation is a powerful, precise and reliable tool whose implementation is, moreover, much less cumbersome than the theoretical approach. However, a numerical simulation remains a series of theoretical calculations whose goal is to get as close as possible to reality, without ever reaching it. Therefore, it remains to be determined a method that will allow the phenomena to be truly observed in real or similar conditions. The experimental approach seems to meet these criteria.

6 Confrontation of Experimental-Analytical-Numerical Results and Discussion

The experimental aspect of our research was already carried out in the previous studies of our research team for NACA 43013 airfoil with two pressure taps in the extrados at 31% and 69% of the chord [1]. To confront and validate our analytical model via experimentation we compared the different results for two upstream Mach numbers $M_{Upstream} = 1.55$ and $M_{Upstream} = 1.45$. For our analytical model, we took into consideration the statistical/analytical formulas of the empirical parameter m for each upstream Mach number, to enhance even more our model precision, results are shown in Table 1.

Table 1. Experimental-numerical-analytical local Mach numbers results for NACA 43013 at 31%c and 69%c for $M_\infty = 1.55$ and $M_\infty = 1.45$

Mach upstream	X	Y	Exp local Mach	Numerical local Mach	Analytical local Mach
1.45	0.31	0.094	1.6	1.58245	1.59057
1.45	0.69	0.045	2.21	1.8625	2.0125
1.55	0.31	0.094	1.61	1.62156	1.60822
1.55	0.69	0.045	2.39	1.9524	2.1103

By comparing the experimental and analytical results for local Mach number near the wall far from the leading edge for a supersonic irrotational laminar steady flow of a viscous compressible fluid around thin airfoil NACA 43013, we can conclude to two main results:

- First, our improved statistical/analytical model is validated and proven to be accurate and exacte comparing to numerical simulation and experimentation.
- Second, the injection of the empirical parameter m gave more exactitude to our analytical model comparing to reality {experimentation}, and decreased the relative error comparing to our previous model, as following:
  - For $M_\infty = 1.45$, for our previous model the error at 31%c was 1.25% and for our improved model it decreased to 0.5929%, same at 69%c the error decreased to

8.9367% comparing to 16.289% for the previous model, minimizing the relative error by more than half.

- For $M_\infty = 1.55$, for our previous model the error at 31%c was 1.242% and for our improved model it decreased to 0.1106%, same at 69%c the error decreased to 11.7029% comparing to 20.502% for the previous model, minimizing the relative error by more than half.

The large difference between the experimental, numerical and analytical results in the second pressure tap which is located at 69% of the chord (near the trailing edge), can be explained by the fact that the flow in reality is turbulent, thus in the part of the airfoil close to the trailing edge is a part of instability, contrary to what we presumed in our assumptions that the flow is laminar and irrotational.

7 Conclusion

The benefit of the SWBLI modelling is highly significant, since it doesn't only describe with high precision and takes into consideration a multitude of parameters that influence the flow behavior around a large variety of NACA airfoils with relative curvature $\lambda = 1$ in supersonic field with upstream Mach number $1.41 \leq M_\infty \leq 1.6864$, but it also is the fundamental base for any future passive flow control that could lead into the amelioration of the different factors governing the reliability and the airworthiness of the aircraft's wings such as the aerodynamic coefficients. The integration of the empirical parameter m (Local Incompressible Connection Mach Number) added more value and precision to our analytical model, both comparing to numerical simulation where our model was validated, and experimentation results confrontation that proved the improvement of our model's precision by elaborating an exact formula for the empirical parameter based on statistical/analytical methods. As perspectives for our study, our research team aims to furthermore elaborate an exact fully analytical model for the empirical parameter m taking into consideration the parameters and variables that influence its evolution $m(\overline{x}, \overline{y}, M_\infty, \beta)$, introducing the parameter β that refers to the airfoil type.

Acknowledgements. The authors would like to express profound gratitude and respects to the late Professor Mohammed Hasnaoui, who passed away due to Covid 19 during this study, in December 2020. His outstanding morality, astonishing expertise in the field of fluid mechanics and asymptotical modelling, as well as supervising role were very enriching; we, as a team, are honoured to have been his pupils.

References

1. El-Aajine, O., Eddegdag, N., Naamane, A., Radouani, M., El Fahime, B.: Asymptotic modeling of a viscous laminar flow around thin airfoils: resolution and experimental treatment in case of supersonic flow (2022)
2. Dolling, D.S.: Fifty years of shock-wave/boundary-layer interaction research: what next? AIAA J. **39**(8), 1517–1531 (2001)

3. Ben Dor, G., Igra, O., Elperin, T.: Handbook of Shock Waves, Vol. 2 Shock Wave Interactions and Propagation (2001)
4. Ligrani, P.M., McNabb, E.S., Collopy, H., Anderson, M., Marko, S.M.: Recent investigations of shock wave effects and interactions. Adv. Aerodyn. **2**(1), 1–23 (2020)
5. Hasnaoui, M., Naamane, A., Akhmari, H.: Asymptotic modeling the aerodynamic coefficients of the NACA airfoil. Model. Meas. Control B **88**(2–4), 58–66 (2019)
6. Naamane, A., Hasnaoui, M.: Supersonic flow around a dihedral airfoil: modeling and experimentation investigation. Int. J. Aerosp. Mech. Eng. **13**(6), 413–417 (2019)
7. Zeytounian, R.K.: Singular coupling and the triple-deck model. In: Asymptotic Modelling of Fluid Flow Phenomena, pp. 471–525 (2002)
8. Neiland, V.Y.: Theory of laminar boundary layer separation in supersonic flow. Fluid Dyn. **4**(4), 33–35 (1969)
9. Stewartson, K., Williams, P.G.: Self-induced separation. Proc. R. Soc. Lond. A Math. Phys. Sci. **312**(1509), 181–206 (1969)
10. Zeytounian, R.K. (Ed.): Les Modèles Asymptotiques de la Mécanique des Fluides 1, pp. 158–171. Springer Berlin Heidelberg, Berlin, Heidelberg (1986)
11. Zeytounian, R.K.: Mécanique des fluides fondamentale, vol. 4, pp. 207–2026. Springer Science & Business Media (1991)
12. Smith, F.T., Brighton, P.W.M., Jackson, P.S., Hunt, J.C.R.: On boundary-layer flow past two-dimensional obstacles. J. Fluid Mech. **113**, 123–152 (1981)
13. Matak, L., Nikolić, K.K.: CFD analysis of F-16 wing airfoil aerodynamics in supersonic flow. In: The Science and Development of Transport—ZIRP 2021, pp. 197–210. Springer, Cham (2022)
14. Billig, F.S.: Shock-wave shapes around spherical- and cylindrical-nosed bodies. J. Spacccr. Rockets **4**(6), 822–823 (1967)

Transmission and Reflection and Coefficients of the Periodic Thermoelastic Multilayer Structures with Effect of Interface Using the Pseudo-Stroh Formalism

J. Houyouk[1](✉), J. A. Manyo Manyo[1], G. E. Ntamack[1], and L. Azrar[2,3]

[1] Groupe de Mécanique, Matériaux et Acoustique, Département de Physique, Université de Ngaoundéré, Ngaoundéré, Cameroon
jacobhouyouk@yahoo.com

[2] Department of Applied Mathematics and Informatics, Research Center STIS, Team M2CS, ENSAM, Mohamed V University, Rabat, Morocco
l.azrar@um5s.net.ma

[3] Department of Mechanical Engineering, Faculty of Engineering, King Abdulaziz University, Jeddah, Saudi Arabia

Abstract. Transmission and reflection coefficients of plane waves for multilayered systems of thermoelastic materials are investigated in this paper. The Pseudo-Stroh formalism is employed to solve the elastic displacements, temperature and heat flux. The general solutions of simply supported plates are used with the dual variable and position method in the corresponding layered structures to determine the transmission and reflection coefficients. Two examples of application are proposed to highlight on the effects of the temperature and the imperfect interface on the acoustic behavior of the thermoelastic multilayer structures. To model the isothermal conditions at the interface, the linear spring-type model is used and it is assumed a continuity of traction and a discontinuity of displacement field throughout the interface. The interface is assumed to be mechanically compliant and weakly conducting thermally. One of the multilayers used is the perfectly bonded composed of aluminum and polyethylene. The other multilayers system is composed of the same materials with imperfect interface. Numerical results are presented to illustrate the variation of acoustic transmission and reflection coefficients with imperfect interface depending on temperature and frequency.

Keywords: Transmission and reflection coefficients · Imperfect interface · Temperature · Thermoelastic material · Pseudo-Stroh formalism

1 Introduction

In high-temperature environments, structures subject to vibration and thermomechanical resonance phenomena have a significantly reduced service life (Vattré and Pan 2020). In the multilayered media, wave propagation is of particular interest such as the analysis of composite structures and smart materials that are used in submarine acoustics and

L. Azrar et al. (Eds.): CIP 2022, LNME, pp. 273–284, 2023.
https://doi.org/10.1007/978-3-031-23615-0_28

even in geophysics. The dynamic and static performances of structural systems are usually improved by multilayered structures using distributed or patched active layers (Oh 2006). The study of isothermal elastic problems in a layer has been approached using the transfer matrix method (TMM) (Bufler 1971). To relate the states variables from the lower to the upper surfaces of the multilayered structure, the global transfer matrix is established (Verma 2011), coupling the temperature and deformation fields are coupled. The propagation of generalized thermoelastic free waves in multilayered plates of isotropic media has also been studied and extended to staggering anisotropes (Hawwa and Nayfeh 1995). The dynamic distribution of displacements and thermal stresses has later been investigated in multilayered media or the generalized thermoelasticity (Verma and Hasebe 2002). This study has dealt with problems of wave propagation in transversely isotropic and anisotropic plates, and the interaction of free harmonic waves with multilayered media, combining the linear transformation and TMM approach. It has been demonstrated that taking into account the interface effect in multilayered structures improves the service life of these structures, offering applications in several fields such as aeronautics, automotive and especially in civil engineering (Zaki et al. 2014). They presented the modeling and numerical simulation of static and dynamic behaviors of multilayer beams and plates with different types of interfaces, applied to thick, bimorph and viscoelastic structures with imperfect interface.

The present work carries out a comprehensive analysis on the thermoelastic multilayered plate. Since the propagator matrix is unsuitable because of its instabilities in higher frequencies (Manyo Manyo et al. 2021), the dual variable and position (DVP) method required for a good analysis in all frequency range is used (Pan 2019).

Considering the temperature, a combination within the Pseudo-Stroh Formalism (PSF) and the DVP method is elaborated. The transmission and reflection influences on the elastic solids with thermal effects are investigated. The DVP method combined to the Pseudo-Stroh formalism is used to determine the acoustic transmission and reflection coefficients for multilayered with imperfect interface, under thermomechanical loading.

2 Basic Equations

The constitutive equations of heat conduction for thermoelastic material are given by (Wang and Pan 2007; Vattré and Pan 2020):

$$\sigma_{ij} = c_{ijkl}\gamma_{kl} - \beta_{ij}\theta \tag{1}$$

$$q_i = -k_{ij}\theta_{,j} \tag{2}$$

where the stress σ_{ij}, γ_{kl} and q_i are the components of the stress tensor, engineering strain tensor and the heat flux, respectively. β_{ij} is the stress temperature. k_{ij} is the thermal conductivity coefficient and θ is the temperature change. c_{ijkl} is the stiffness tensor. The comma followed by index $j = \{1, 2, 3\}$ in Eq. (2) denotes partial differentiation relatively to the x_j axis.

The strains and displacements u_i can be expressed in terms of the displacements, with $i = \{1, 2, 3\}$ are related by:

$$\gamma_{ij} = \frac{1}{2}\left(u_{i,j} + u_{j,i}\right) \tag{3}$$

The motion equations without the body forces and heat equilibrium equation are given by (Vattré and Pan 2020):

$$\begin{cases} \sigma_{ij,j} = \rho \frac{\partial^2 u_i}{\partial t^2} \\ q_{i,i} = 0 \end{cases} \tag{4}$$

with ρ the mass density.

2.1 Heat Conduction

The change of temperature for simply supported boundary conditions can be assumed as follows (Wang and Pan 2007):

$$\theta = f e^{\eta z} \sin(px) \sin(qy) \tag{5}$$

where $p = \frac{m\pi}{L_x}$, $q = \frac{n\pi}{L_y}$, m and n two positives integers, f and η are unknowns constants to be determined.

Combining Eqs. (2) and (5), the heat flux components can be obtained as

$$\begin{aligned} q_1 &= -k_{11} p f e^{\eta z} \cos(px) \sin(qy),\ q_2 = -k_{22} q f e^{\eta z} \sin(px) \cos(qy), \\ q_3 &= +k_{33} \eta f e^{\eta z} \sin(px) \sin(qy) \end{aligned} \tag{6}$$

Combining Eqs. (2), (4) and (5), one obtains the following relationship:

$$\left(k_{33}\eta^2 - k_{11}p^2 - k_{22}q^2\right) f = 0 \tag{7}$$

Hence the following eigenrelations is resulted:

$$M \begin{bmatrix} f \\ g \end{bmatrix} = \eta \begin{bmatrix} f \\ g \end{bmatrix} \tag{8}$$

where:

$$M = -\begin{bmatrix} 0 & 1/k_{33} \\ k_{11}p^2 + k_{22}q^2 & 0 \end{bmatrix} \tag{9}$$

The general solution for the thermal field of the plate, obtained by superposing Eqs. (5) and (7), is giving as follow:

$$\begin{bmatrix} \theta \\ q_3 \end{bmatrix} = \begin{bmatrix} f_1 & f_2 \\ g_1 & g_2 \end{bmatrix} \begin{bmatrix} e^{\eta_1 z} & 0 \\ 0 & e^{-\eta_1 z} \end{bmatrix} \begin{bmatrix} \chi_1 \\ \chi_2 \end{bmatrix} \tag{10}$$

where χ_1 and χ_2 will be defined later. $\begin{bmatrix} f_1 & g_1 \end{bmatrix}^T$ and $\begin{bmatrix} f_2 & g_2 \end{bmatrix}^T$ are eigenvectors of Eq. (8) that are assumed to be independent.

2.2 General Solutions

The simply supported plates admit as general solutions expressed as relationship (Chen et al. 2017):

$$u = \begin{pmatrix} u_1 \\ u_2 \\ u_3 \end{pmatrix} = e^{(sz+i\omega t)} \begin{pmatrix} a_1 \cos(px) \sin(qy) \\ a_2 \sin(px) \cos(qy) \\ a_3 \sin(px) \sin(qy) \end{pmatrix} + f e^{(\eta z+i\omega t)} \begin{pmatrix} c_1 \cos(px) \sin(qy) \\ c_2 \sin(px) \cos(qy) \\ c_3 \sin(px) \sin(qy) \end{pmatrix} \tag{11}$$

By substituting Eq. (11) into Eq. (3), and subsequently into Eq. (1) the traction vector is obtained as follow:

$$\sigma = \begin{pmatrix} \sigma_{13} \\ \sigma_{23} \\ \sigma_{33} \end{pmatrix} = e^{(sz+i\omega t)} \begin{pmatrix} b_1 \cos(px) \sin(qy) \\ b_2 \sin(px) \cos(qy) \\ b_3 \sin(px) \sin(qy) \end{pmatrix} + f e^{(\eta z+i\omega t)} \begin{pmatrix} d_1 \cos(px) \sin(qy) \\ d_2 \sin(px) \cos(qy) \\ d_3 \sin(px) \sin(qy) \end{pmatrix} \tag{12}$$

where $p = m\pi/L_x$ and $q = n\pi/L_y$, $\{m, n\}$ are integers, $\{L_x, L_y\}$ are the dimensions of the plate, ω the pulsation and t the time.

2.3 Pseudo-Stroh Fundamental Matrix

It has been shown that the eigenvectors b and d are respectively related to a and c as follows (Chen et al. 2017):

$$b = \left(sT_0 - R_0^t\right)a = -\frac{1}{s}(Q_0 + sR_0)a \tag{13}$$

$$d = \left(\eta T_0 - R_0^t\right)c - \gamma_2 = \frac{1}{\eta}(Q_0 + \eta R_0)c + \frac{1}{\eta}\gamma_1 \tag{14}$$

where the 3×3 matrices Q_0, R_0, T_0 are defined as:

$$\begin{cases} Q_0 = -\begin{bmatrix} (p^2c_{11} + q^2c_{66}) & pq(c_{66} + c_{12}) & 0 \\ pq(c_{66} + c_{12}) & (p^2c_{66} + q^2c_{22}) & 0 \\ 0 & 0 & (p^2c_{55} + q^2c_{44}) \end{bmatrix} + \rho\omega^2 I_3 \\ R_0 = \begin{bmatrix} 0 & 0 & pc_{13} \\ 0 & 0 & qc_{23} \\ -pc_{55} & -qc_{44} & 0 \end{bmatrix} \\ T_0 = \begin{bmatrix} c_{55} & 0 & 0 \\ 0 & c_{44} & 0 \\ 0 & 0 & c_{33} \end{bmatrix} \end{cases} \tag{15}$$

By substituting Eq. (11) into Eq. (3), then into Eq. (1), and finally into Eq. (4), we obtain the following relations:

$$\begin{cases} \left[Q_0 + s\left(R_0 - R_0^t\right) + s^2T_0\right]a = 0 \\ \left[Q_0 + \eta\left(R_0 - R_0^t\right) + \eta^2T_0\right]c = \gamma_1 + \eta\gamma_2 \end{cases} \tag{16}$$

Using Eqs. (13) and (14), Eq. (16) can rewritten as follows:

$$N\begin{bmatrix} a \\ b \end{bmatrix} = s\begin{bmatrix} a \\ b \end{bmatrix} \tag{17}$$

$$N\begin{bmatrix} c \\ d \end{bmatrix} = \eta\begin{bmatrix} c \\ d \end{bmatrix} + \gamma \tag{18}$$

N is the fundamental Stroh matrix given by:

$$N = \begin{bmatrix} T_0^{-1}R_0^t & T_0^{-1} \\ -Q_0 - R_0T_0^{-1}R_0^t & -R_0T_0^{-1} \end{bmatrix},\quad \gamma = -\begin{bmatrix} 0 & T_0^{-1} \\ I_3 & -R_0T_0^{-1} \end{bmatrix}\begin{bmatrix} \gamma_1 \\ \gamma_2 \end{bmatrix} \tag{19}$$

Equation (18) is an eigenvalue problem. Its resolution in mechanic has been investigated by several authors (Pan and Han 2005; Ewolo Ngak et al. 2019). In the present study, we focus on the acoustics transmission and reflection coefficients of the multilayered by imperfect interface.

2.4 Propagation Within a Layer

Assuming different eigenvalues, the general solutions in the layer j for the thermoelastic field of the plate obtained by superposing Eqs. (6) and (11) are given as follow (Chen et al. 2017):

$$\begin{bmatrix} u \\ \sigma \end{bmatrix}_j = \begin{bmatrix} A_1 & A_2 \\ B_1 & B_2 \end{bmatrix}_j \begin{bmatrix} e^{S_1 z} & 0 \\ 0 & e^{S_2 z} \end{bmatrix}_j \begin{bmatrix} K_1 \\ K_2 \end{bmatrix} + \begin{bmatrix} C_1 & C_2 \\ D_1 & D_2 \end{bmatrix}_j \begin{bmatrix} e^{\eta_1 z} & 0 \\ 0 & e^{\eta_2 z} \end{bmatrix}_j \begin{bmatrix} f_1 & 0 \\ 0 & f_2 \end{bmatrix}_j \begin{bmatrix} \chi_1 \\ \chi_2 \end{bmatrix} \tag{20}$$

where K_1 and K_2 are two vectors each with 3 constants to be determined.

By using the conditions at the interfaces combined with the general solutions of each layer giving in Eq. (20), the general solutions of the multilayered plate can be obtained (Manyo Manyo et al. 2021).

From Eq. (10), the temperature and the heat flux can be expressed for the jth layer delimited by the abscissa surfaces z_j and z_{j+1} as follows, with h_j being the thickness of the layer:

$$\begin{bmatrix} \theta \\ q_3 \end{bmatrix}_{z=z_{j+1}} = \begin{bmatrix} f_1 & f_2 \\ g_1 & g_2 \end{bmatrix} \begin{bmatrix} e^{\eta_1 z_{j+1}} & 0 \\ 0 & e^{\eta_2 z_{j+1}} \end{bmatrix} \begin{bmatrix} \chi_1 \\ \chi_2 \end{bmatrix}_j \tag{21}$$

More over, if we locate ourselves at the lower surface $z = z_j$, Eq. (10) leads to:

$$\begin{bmatrix} \chi_1 \\ \chi_2 \end{bmatrix}_j = \begin{bmatrix} e^{\eta_1 z_j} & 0 \\ 0 & e^{\eta_2 z_j} \end{bmatrix}^{-1} \begin{bmatrix} f_1 & f_2 \\ g_1 & g_2 \end{bmatrix}^{-1} \begin{bmatrix} \theta \\ q_3 \end{bmatrix}_{z=z_j} \tag{22}$$

Equations (21) and (22) relate the thermal variables of the upper and lower surfaces of the layer j as follows:

$$\begin{bmatrix} \theta \\ q_3 \end{bmatrix}_{z=z_{j+1}} = \begin{bmatrix} f_1 & f_2 \\ g_1 & g_2 \end{bmatrix} \begin{bmatrix} e^{\eta_1 h_j} & 0 \\ 0 & e^{\eta_2 h_j} \end{bmatrix} \begin{bmatrix} f_1 & f_2 \\ g_1 & g_2 \end{bmatrix}^{-1} \begin{bmatrix} \theta \\ q_3 \end{bmatrix}_{z=z_j} \tag{23}$$

Moreover, if we replace the unknown constant vectors χ_1 and χ_2 of Eq. (22) in Eq. (20), the vectors K_1 and K_2 are rewritten as follows:

$$\begin{bmatrix} K_1 \\ K_2 \end{bmatrix}_j = \begin{bmatrix} e^{\eta_1 z_j} & 0 \\ 0 & e^{-\eta_1 z_j} \end{bmatrix}^{-1} \begin{bmatrix} A_1 & A_2 \\ B_1 & B_2 \end{bmatrix}^{-1} \begin{bmatrix} u \\ \sigma \end{bmatrix}_{z=z_j} - \begin{bmatrix} A_1 & A_2 \\ B_1 & B_2 \end{bmatrix}^{-1} \begin{bmatrix} c_1 & c_2 \\ d_1 & d_2 \end{bmatrix} \begin{bmatrix} f_1 & 0 \\ 0 & f_2 \end{bmatrix} \begin{bmatrix} e^{-\eta_1 z_j} & 0 \\ 0 & e^{\eta_1 z_j} \end{bmatrix} \begin{bmatrix} f_1 & f_2 \\ g_1 & g_2 \end{bmatrix}^{-1} \begin{bmatrix} \theta \\ q_3 \end{bmatrix}_{z=z_{j+1}} \quad (24)$$

Using Eqs. (20–24), one obtains the propagation relation between the unknowns of the upper and the lower surfaces:

$$\begin{bmatrix} u \\ \sigma \\ \theta \\ q_3 \end{bmatrix}_{z=z_{j+1}} = \begin{bmatrix} \begin{bmatrix} E_{11} & E_{12} \\ E_{21} & E_{22} \end{bmatrix} & \begin{bmatrix} G_{11} & G_{12} \\ G_{21} & G_{22} \end{bmatrix} \\ 0_{2\times 6} & \begin{bmatrix} M_{11} & M_{12} \\ M_{21} & M_{22} \end{bmatrix} \end{bmatrix}_j \begin{bmatrix} u \\ \sigma \\ \theta \\ q_3 \end{bmatrix}_{z=z_j} = P_j(h_j) \begin{bmatrix} u \\ \sigma \\ \theta \\ q_3 \end{bmatrix}_{z=z_j} \quad (25)$$

where $P_j(h_j)$ is the propagator within the layer j of thickness h_j, and its components are calculated as follows:

$$\begin{cases} \begin{bmatrix} E_{11} & E_{12} \\ E_{21} & E_{22} \end{bmatrix} = \begin{bmatrix} A_1 & A_2 \\ B_1 & B_2 \end{bmatrix} \begin{bmatrix} e^{S_1 h_j} & 0 \\ 0 & e^{S_2 h_j} \end{bmatrix} \begin{bmatrix} A_1 & A_2 \\ B_1 & B_2 \end{bmatrix}^{-1} \\ \begin{bmatrix} G_{11} & G_{12} \\ G_{21} & G_{22} \end{bmatrix} = \begin{bmatrix} C_1 & C_2 \\ D_1 & D_2 \end{bmatrix} \begin{bmatrix} f_1 g_2 e^{\eta_1 h_j} & -f_1 f_2 e^{\eta_1 h_j} \\ -f_2 g_1 e^{-\eta_1 h_j} & f_1 f_2 e_2^{-\eta_1 h_j} \end{bmatrix} \cdots \\ \qquad + \begin{bmatrix} E_{11} & E_{12} \\ E_{21} & E_{22} \end{bmatrix} \begin{bmatrix} C_1 & C_2 \\ D_1 & D_2 \end{bmatrix} \begin{bmatrix} -f_1 g_2 & f_1 f_2 \\ f_2 g_1 & -f_1 f_2 \end{bmatrix} \\ \begin{bmatrix} M_{11} & M_{12} \\ M_{21} & M_{22} \end{bmatrix} = \begin{bmatrix} f_1 & f_2 \\ g_1 & g_2 \end{bmatrix} \begin{bmatrix} e^{\eta_1 h_j} & 0 \\ 0 & e^{-\eta_1 h_j} \end{bmatrix} \begin{bmatrix} f_1 & f_2 \\ g_1 & g_2 \end{bmatrix}^{-1} \end{cases} \quad (26)$$

2.5 Imperfect Interface

The considered structure is a multilayer rectangular plate with isotropic layers poled in the z-direction (Fig. 1). The conditions at the linear spring-type model interface are taken as follows under isothermal conditions (Fan and Wang 2003; Vattré and Pan 2020): continuity for traction vector and the heat flux, discontinuity of displacement field and temperature for weakly conducting interface. The sliding displacement are assumed to be linearly proportional to the interfacial traction components. For highly conducting interfaces, we consider continuity for the temperature and discontinuity for the heat flux throughout the interface (Vattré and Pan 2020).

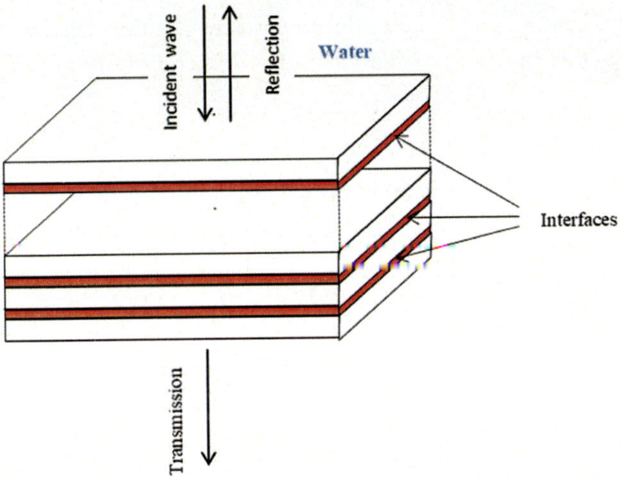

Fig. 1. Geometry of the problem.

The following relationship are considered at the interfaces (Wang and Pan 2007):

$$\begin{cases} \sigma_{xz}\left(x, y, z^{+}\right) = \sigma_{xz}\left(x, y, z^{-}\right) \\ \sigma_{yz}\left(x, y, z^{+}\right) = \sigma_{yz}\left(x, y, z^{-}\right) \\ \sigma_{zz}\left(x, y, z^{+}\right) = \sigma_{zz}\left(x, y, z^{-}\right) \\ u_x\left(x, y, z^{+}\right) - u_x\left(x, y, z^{-}\right) = \alpha_1^{(j)}\sigma_{xz}\left(x, y, z^{-}\right) \\ u_y\left(x, y, z^{+}\right) - u_y\left(x, y, z^{-}\right) = \alpha_2^{(j)}\sigma_{yz}\left(x, y, z^{-}\right) \\ u_z\left(x, y, z^{+}\right) - u_z\left(x, y, z^{-}\right) = \alpha_3^{(j)}\sigma_{zz}\left(x, y, z^{-}\right) \end{cases} \tag{27}$$

$\{\alpha_1^{(j)}, \alpha_2^{(j)}, \alpha_3^{(j)}\}$ are the positive constants. The perfect interface case is observed for $\{\alpha_1^{(j)}, \alpha_2^{(j)}, \alpha_3^{(j)}\} = 0$.

2.5.1 Imperfect Interface in Heat Conduction

The following relationship models the interfacial thermal resistance for weakly conducting interfaces (Wang and Pan 2007):

$$\begin{cases} q_3\left(x, y, z_j^{+}\right) = q_3\left(x, y, z_j^{-}\right) \\ \theta\left(x, y, z_j^{+}\right) - \theta\left(x, y, z_j^{-}\right) = -\beta_\theta^{(j)} q_3\left(x, y, z_j^{-}\right) \end{cases} \tag{28}$$

where $\beta_\theta^{(j)}$ is a positive constant. The thermally perfect interface corresponds to $\beta_\theta^{(j)} = 0$. And the thermally highly conducting corresponds to the following relationships:

$$\begin{cases} \theta\left(x, y, z_j^{+}\right) = \theta\left(x, y, z_j^{-}\right) \\ q_3\left(x, y, z_j^{+}\right) - q_3\left(x, y, z_j^{-}\right) = \gamma_\theta^{(j)} \Delta_s \theta\left(x, y, z_j^{-}\right) \end{cases} \tag{29}$$

where: $\Delta_s = \frac{\partial^2}{\partial x^2} + \frac{\partial^2}{\partial y^2}$, $\gamma_\theta^{(j)}$ is a positive constant. The thermally perfect interface corresponds to $\gamma_\theta^{(j)} = 0$.

2.5.2 Thermomechanical Interface Matrix

In this part, we consider that the interface is mechanically compliant and thermomechanically conducting. Combining Eqs. (27) and (28) or (29) the following relation is obtained:

$$\begin{bmatrix} u \\ \sigma \\ \theta \\ q_3 \end{bmatrix}_{z=z_j^-} = \Omega_{j-1} \begin{bmatrix} u \\ \sigma \\ \theta \\ q_3 \end{bmatrix}_{z=z_{j-1}^+} = \begin{bmatrix} \Omega_{11} & 0_{6\times 2} \\ 0_{2\times 6} & \Omega_{22} \end{bmatrix} \begin{bmatrix} u \\ \sigma \\ \theta \\ q_3 \end{bmatrix}_{z=z_{j-1}^+} \tag{30}$$

The top and bottom of the jth interface j are represented by z_j^+ and z_j^- respectively. For a lower thermal conductive interface, the matrices Ω_{11} and Ω_{22} are respectively used to relate the mechanical and thermal properties. They are given by:

$$\Omega_{11} = \begin{bmatrix} I_3 & diag\left(\alpha_1^{(k)}, \alpha_2^{(k)}, \alpha_3^{(k)}\right) \\ 0_3 & I_3 \end{bmatrix} \tag{31}$$

$$\Omega_{22} = \begin{bmatrix} 1 & \beta_\theta^{(k)} \\ 0 & 1 \end{bmatrix} \tag{32}$$

However, for a thermal highly conducting interface, we have:

$$\Omega_{22} = \begin{bmatrix} 1 & 0 \\ -\gamma_\theta^{(k)}\left(p^2 + q^2\right) & 1 \end{bmatrix} \tag{33}$$

The state variables of the upper and the lower surfaces of any arbitrary multilayer with imperfect interface are related as follows:

$$\begin{bmatrix} u \\ \sigma \\ \theta \\ q_3 \end{bmatrix}_{z=H} = \left(\prod_{j=N_b}^{2} P_j(h_j)\Omega_{j-1} \right) P_1(h_1) \begin{bmatrix} u \\ \sigma \\ \theta \\ q_3 \end{bmatrix}_{z=0} \tag{34}$$

N_b is the number of layers of the plates.

Due to the appearance of higher frequencies, this traditional propagation matrix method is unstable. The dual variable and position method is thus arranged with the PSF (Vattré and Pan 2020). Equation (34) can be recast in following form:

$$\begin{bmatrix} u \\ \sigma \\ \theta \\ q_3 \end{bmatrix}_{z=z_{j+1}} = P(h_j) = \begin{bmatrix} P_{11} & P_{12} \\ P_{21} & P_{22} \end{bmatrix} \begin{bmatrix} u \\ \sigma \\ \theta \\ q_3 \end{bmatrix}_{z=z_j} \tag{35}$$

Moreover, for the reason of clarity, we introduce $\overline{u}_\theta$ and $\overline{t}_q$ and Eq. (35) becomes:

$$\begin{bmatrix} \overline{u}_\theta(z) \\ \overline{t}_q(z) \end{bmatrix}_{z=z_{j+1}} = \begin{bmatrix} P_{11} & P_{12} \\ P_{21} & P_{22} \end{bmatrix} \begin{bmatrix} \overline{u}_\theta(z) \\ \overline{t}_q(z) \end{bmatrix}_{z=z_j} \tag{36}$$

With:

$$\begin{cases} \overline{u}_\theta(z) = \left[u(z)\ \theta(z)\right]^t \\ \overline{t}_q(z) = \left[\sigma(z)\ q_3(z)\right]^t \end{cases} \tag{37}$$

Thus, we can write:

$$V_1^j = Ph_j V_0^j \tag{38}$$

where: $V_1^j = \begin{bmatrix} \overline{u}_\theta \\ \overline{t}_q \end{bmatrix}_{z_{j+1}}$ and $V_0^j = \begin{bmatrix} \overline{u}_\theta \\ \overline{t}_q \end{bmatrix}_{z_j}$ denote unknown vectors of the upper and lower surfaces of the jth layer respectively. The bottom and the top of the layer j are indicate by the indices 0 and 1 respectively.

From Eq. (36), one establishes the following relation between the displacements and thermal unknowns:

$$\begin{bmatrix} \overline{u}_\theta(z_j) \\ \overline{u}_\theta(z_{j+1}) \end{bmatrix} = \begin{bmatrix} -P_{21}^{-1}P_{22} & P_{21}^{-1} \\ P_{12} - P_{21}^{-1}P_{11}P_{22} & P_{21}^{-1}P_{11} \end{bmatrix}_j \begin{bmatrix} \overline{t}_q(z_j) \\ \overline{t}_q(z_{j+1}) \end{bmatrix} = \begin{bmatrix} S_{11} & S_{12} \\ S_{21} & S_{22} \end{bmatrix} \begin{bmatrix} \overline{t}_q(z_j) \\ \overline{t}_q(z_{j+1}) \end{bmatrix} \tag{39}$$

where $[S]$ is the compliance matrix.

2.6 Transmission and Reflection Coefficients

Using Eq. (39), we can write the direct and reverse reflection coefficients (R_d and R_r, respectively) and the transmission coefficient (T_c) parameters related to the compliance matrix. The acoustic transmission and reflection coefficients are determined by the following relationships (Chen et al. 2017):

$$\begin{cases} R_d = \dfrac{(s_{11} - y_f)(s_{22} - y_f) + (s_{12})^2}{(s_{11} + y_f)(s_{22} - y_f) + (s_{12})^2} \\ R_r = \dfrac{(s_{11} + y_f)(s_{22} + y_f) + (s_{12})^2}{(s_{11} + y_f)(s_{22} - y_f) + (s_{12})^2} \\ T_c = \dfrac{2y_f s_{12}}{(s_{11} + y_f)(s_{22} - y_f) + (s_{12})^2} \end{cases} \tag{40}$$

where: $s_{ij} = S_{ij}^N(3,3)$, $y_f = \cos\phi/(j\omega Z_f)$, with ϕ the incidence angle and $Z_f = \rho_f c_f$ the acoustical impedance of the surrounding fluid.

3 Numerical Results

In this section, the formulation is applied to analyze the transmission and reflection coefficients of the periodic multilayered plate. We first present in this part a comparison of the results obtained in the case with perfect interface in order to validate the presented methodology and the present results for N = 1 period. Later on we consider the interface effects and temperature in this structure. The multilayer structure is immerged in water having a density $\rho = 1000\,\text{kg.m}^{-3}$ and velocity $c = 1480\,\text{m.s}^{-1}$ (Siryabe et al. 2013). Table 1 give the material properties. Numerical results are computed in the [0.75–3.5] MHz frequency range.

Table 1. Materials properties

Material	$c_{11} = c_{22} = c_{33}$ (GPa)	$c_{12} = c_{13} = c_{23}$ (GPa)	$c_{44} = c_{55} = c_{66}$ (GPa)	d (mm)	ρ (kg/m^3)
AL	113.9	60.15	26.9	2	2800
PE	5.28	2.57	1.35	1	940

c_{ij}: elastic modulus, d: thickness, ρ: density

Our period is similar to Marechal and collaborators ones (Maréchal et al. 2014). Two layers respectively made of aluminium (AL) and polyethylene (PE) form one period. Acoustic transmission and reflection are shown depending on the insonification side of the multilayered structure.

In this subsection, the theoretical transmission and reflection coefficients spectra of plane wave in multilayer structures are reminded taking into account interface effect and temperature. For the configuration AL/PE and PE/AL, one observes a difference of spectra in Fig. 2(a) and (b) respectively. In the reverse case, the undulations related to the vibration of the polyethylene plate are obtained and we notice that the acoustic reflection coefficient decreases when the frequency increases. The minima of the acoustic reflection are close to zero. In the two configurations (AL/PE and PE/AL), the amplitude of the reflection coefficients reaches the maximum when the temperature increases. However, in both cases, we have two minima spaced and tend towards zero. Concerning the transmission coefficients, we observe a reduction of its amplitude when the frequency increases. This coefficient does not depend on the insonation side. When the parameter of interface increases at a fix temperature, his amplitude increases at the high frequencies. It is also observed that the temperature and the interface effect cause the decrease in the amplitude of the acoustic transmission coefficient.

4 Conclusion

The aim of this study was to analysed the transmission and reflection coefficients for plane waves at normal incidence on a thermoelastic structure with interface effect. We used the DVP method to solve the propagation equation. Numerical examples are presented both

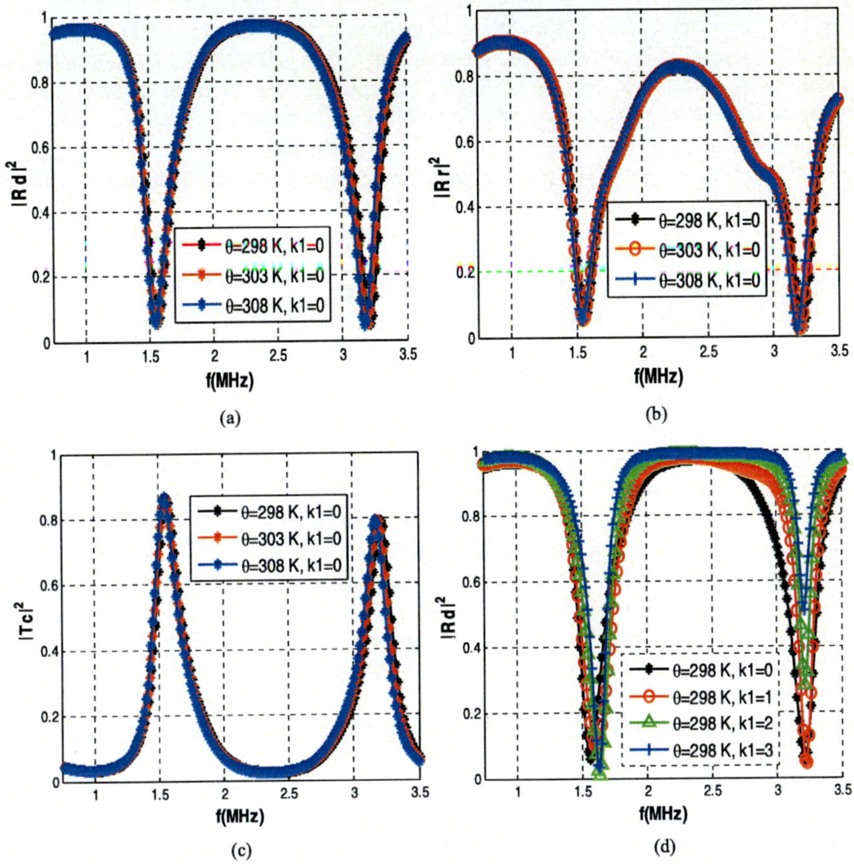

Fig. 2. Transmission and reflection coefficients for N = 1 of (a) AL/PE with θ, (b) PE/AL with θ, (c) AL/PE with θ and (d) AL/PE or PE/AL with k1 with interface effect at $\theta = 328$ K and $\theta_{ref} = 298$ K.

with interface and without interface. It is observe that the insonation side of the plate considerably influences the studied variables. For the perfect interface case, it is clearly seen that the temperature influences the transmission and reflection coefficients. The forbidden bands are bigger as the temperature increases. The imperfect interface effect is studied and we observe that the transmission and reflection coefficients considerably depend on the interface parameter. The proposed method is efficient and stable for arbitrary number of layers at high frequency and at any temperature.

References

Vattré, A., Pan, E.: Thermoelasticity of multilayered plates with imperfect interfaces. Int. J. Eng. Sci. **158**(103409), 1–30 (2020)

Oh, I.K.: Piezoelectric suppression of thermoelastic snap-through in active piezolaminated curved shells. Smart Mater. Struct. **15**, 1616–1626 (2006)

Bufler, H.: Theory of elasticity of a multilayered medium. J. Elast. **19**, 125 (1971)
Verma, K.L.: The general problem of thermoelastic wave propagation in multilayered anisotropic media with application to periodic media. Int. J. Appl. Eng. Res. Dindigul **1**, 908–922 (2011)
Hawwa, M.A., Nayfeh, A.H.: The general problem of thermoelastic waves in anisotropic periodically laminated composites. Compos. Eng. **51**, 499–517 (1995)
Verma, K.L., Hasebe, N.: Wave propagation in transversely isotropic plates in generalized thermoelasticity. Arch. Appl. Mech. **72**, 470–482 (2002)
Zaki, S., Azrar, L., Mouchtachi, A.: Modeling and numerical simulation of static and dynamic behavior of multilayered plates with interface effects. MATEC Web Conf. **11**(02006), 1–6 (2014). https://doi.org/10.1051/matecconf/20141102006
Manyo Manyo, J.A., Ntamack, G.E., Azrar, L.: 3D-dynamic modeling of cross-ply magneto-electro-elastic laminates based on the pseudo-Stroh formalism. Mech. Adv. Mater. Struct. **28**(13), 1337–1354 (2021). https://doi.org/10.1080/15376494.2019.1668094
Pan, E.: Green's functions for geophysics: a review. Rep. Prog. Phys. **82**(10), 106801 (2019). https://doi.org/10.1088/1361-6633/ab1877
Wang, X., Pan, E.: Exact solutions for simply supported and multilayered piezothermo-elastic plates with imperfect interfaces. Open Mech. J. **1**, 1–10 (2007)
Chen, J., Guo, J., Pan, E.: Reflection and transmission of plane wave in multilayered nonlocal magneto-electro-elastic plates immersed in liquid. Compos. Struct. **162**, 401–410 (2017)
Pan, E., Han, F.: Exact solution for functionally graded and layered magneto-electro-elastic plates. Int. J. Eng. Sci. 321–339 (2005)
Ewolo Ngak, F.P., Ntamack, G.E., Azrar, L.: Dynamic analysis of multilayered magnetoelectroelastic plates based on the pseudo-Stroh formalism and Lagrange polynomials. J. Intell. Mater. Syst. Struct. **30**, 939–962
Fan, H., Wang, G.: Screw dislocation interacting with imperfect interface. Mech. Mater. **35**, 943–953 (2003)
Siryabe, E., Ntamack, G.E., Marechal, P.: Holes effects in plane periodic multilayered viscoelastic media. Open J. Acoust. **3**(3), 80–87 (2013). https://doi.org/10.4236/oja.2013.33013
Maréchal, P., Lenoir, O., Khaled, A., ElKettani, M.E.C., Chenouni, D.: Viscoelasticity effect on a periodic plane medium immersed in water. Acta Acust. Acust. **100**, 1036–1043 (2014)

Dynamic Analysis of Axially Functionally Graded Beams on Nonlinear Foundation Subjected to Multiple Harmonic Moving Loads

Anas Ouzizi[1(✉)], Farah Abdoun[1], and Lahcen Azrar[1,2]

[1] Research Center STIS, M2CS, ENSAM Rabat, Mohammed V University, Rabat, Morocco
anas.ouzizi@gmail.com, farah.abdoun@ensam.um5.ac.ma, l.azrar@um5r.ac.ma
[2] Department of Mechanical Engineering, Faculty of Engineering, King Abdulaziz University, Jeddah, Saudi Arabia

Abstract. In this work, an extensive investigation is effectuated on the behaviour of axially functionally graded Euler-Bernoulli beam on Winkler type non-linear viscoelastic foundation and submitted to numerous harmonic moving loads. The non-linear Winkler foundation is assumed to be cubic. The nonlinear dynamic problem is formulated and Hamilton's principle is applied to generate the governing variable coefficients equations. In this study, the numerical results are obtained using the Galekrin's scheme and fourth Order Runge-Kutta method. The transverse dynamic response of AFG beam is obtained for different material gradation models, moving load velocities and different values of load magnitude and frequency. The supported pinned-pinned supports boundary condition and three successive harmonic loads are considered in the analysis. Results indicate that the above-mentioned parameters play a very important role on the transverse dynamic response of the AFG beam under successive harmonic moving loads. There must be a large number of mode shapes for convergence. Furthermore, the obtained results are compared and evaluated with existing approaches in the literature.

Keywords: Functionally graded beam · Nonlinear foundation · Dynamic analysis · Harmonic moving loads

1 Introduction

In engineering practice, systems exposed to moving loads or moving masses are generally used in highway industry, railway technologies, highway industries and the optimization design of tracks such as railway engineering equipment, bridges, pavements for highways and airports, machine parts, and so on. A critical problem that has captured the interest of many researchers over the last few decades is

Farah Abdoun, Lahcen Azrar are contributed equally to this work.

L. Azrar et al. (Eds.): CIP 2022, LNME, pp. 285–293, 2023.
https://doi.org/10.1007/978-3-031-23615-0_29

the dynamic investigation of various systems such as beams under moving loads or moving masses. A book by Fryba[1] presents and discusses a thorough review of analytical and semi-analytical techniques in order to treat these kinds of issues.

One of the most fundamental advanced generations of composition material are functionally graded materials (FGMs), where the material characteristics are changing continuously along spatial directions. These materials are classified as non-homogeneous composites since they are made of a combination of two different elements, often metal and ceramic. The FGM technology concept was first implemented in Japan in 1984 for original material fabrication by Sendai Group. Moreover, structures fabricated from FGM have a great practical use in different applications, such as aerospace parts, space vehicles, automotive industries and other engineering structures [2–4]. Given that the FGMs have a wide material variations and applications, Mastering their static and dynamic behaviors is of big interest. Therefore, a detailed study has been done on the dynamic response of FGMs beams by Huang and Li. A new method for non-forced vibration of beams formed of non-uniform axially functionally graded materials was established by [5]. Utilizing Lagrange's equations to derive the equation that describe the behavior of a the physical system, Simsek et al. [6] investigated the response of an (AFG) beam of type Euler-Bernoulli with moving harmonic load. Using the finite element approach (FEM), Shahba et al. [7] explored the stability analysis of axially FG tapered Timoshenko beams for free vibration.

The foundations also have a significant practical role in engineering applications. For simplicity, most available researchers take the foundations as linear elastic ones. Calim [8] used the complementary functions method to analyze the behaviour of FGM beams lying on visco-elastic foundations. For more practical models. It is more necessary to take into account non-linear foundations, such as cubic types [9]. Rahman et al. [10] applied a novel technique to perform non-linear vibration analysis of a beam on a non-linear elastic foundation. In [11] a study of FG beams on non-linear elastic foundations was established utilizing an analytical methodology via variational iteration method.

Based on a review of the research literature , A few papers on beams lying on nonlinear viscoelastic foundations have also been published, however the majority of the existing research is on linear foundations and homogeneous beams. The performance of vibration modes using the Galerkin's approach for analyzing the dynamic behaviour of beams under a moving force have been studied by Ding et al. [12]. Ebrahimi et al. [13] conducted a frequency response analysis of curved magneto-electro-viscoelastic FG nano-beams on a visco-elastic foundation.

According to review of the literature stated above, Despite the availability of published articles on the behaviour of beams on several kind of foundations, research on AFG beams is exceedingly rare. Thus, it is critical to deeply analyze the dynamic behavior of AFGM beams supported by a non-linear visco-elastic foundation and exposed to various moving forces with harmonic excitation. Moreover, analyzing the impact of all system characteristics on the dynamic response of the considered AFG beam can deliver important knowledge in this study, which is the goal of this research.

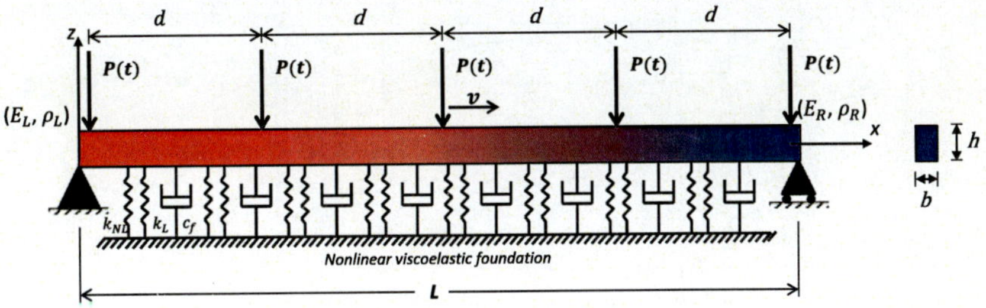

Fig. 1. The beam and the foundation models.

2 AFG Beam-Foundation Model

Considering an Euler-Bernoulli beam with AFG material. The length is L with a constant cross-section A. The considered AFG beam is supported by a viscoelastic Winkler foundation with cubic non-linearity. Multiple harmonic forces P with speed v are applied to the beam, as indicated in Fig. 1. The foundation used in this study written as follow [9]:

$$P_f(x,t) = k_L w(x,t) + c_f \frac{\partial w(x,t)}{\partial t} + k_{NL} w^3(x,t) \tag{1}$$

where cf, k_{NL} and k_L are the viscous damping parameter, the cubic non-linear stiffness and the linear stiffness of the foundation, respectively. And $P_f(x,t)$ represent for the reaction force of the foundation. It should be noted that the material parameters of the investigated AFG beam, namely mass density(ρ) Young's modulus of elasticity(E), are taken as follows

$$E(x) = E_R + (E_L - E_R)\left(1 - \frac{x}{L}\right)^k, \quad \rho(x) = \rho_R + (\rho_L - \rho_R)\left(1 - \frac{x}{L}\right)^k \tag{2}$$

where k denotes the power-law index that governs the variation profile of the considered material over the beam's length , the subscripts L and R denote the left and right ends of the beam, respectively. Figure 2 indicates the influence of the power-law index k on volume fraction variation. The governing differential equation is expressed as follows:

$$\begin{aligned} &\rho(x)A\frac{\partial^2 w(x,t)}{\partial t^2} + \frac{\partial^2}{\partial x^2}\left[E(x)I\frac{\partial^2 w(x,t)}{\partial x^2}\right] + k_L w(x,t) + c_f\frac{\partial w(x,t)}{\partial t} \\ &+ k_{NL} w^3(x,t) = \sum_{n=1}^{J} P_0 \sin(\Omega t)\delta\left(x + nd - vt\right), \end{aligned} \tag{3}$$

where $w(x,t)$ denotes the transverse displacement, I, d and n are the moment of inertial, distance between moving forces and number of moving forces, respectively. The moving forces are treated using the Dirac delta function $\delta(.)$, P_0 is the magnitude of the force, Ω is the moving harmonic load's excitation frequency, and v is the moving load's speed.

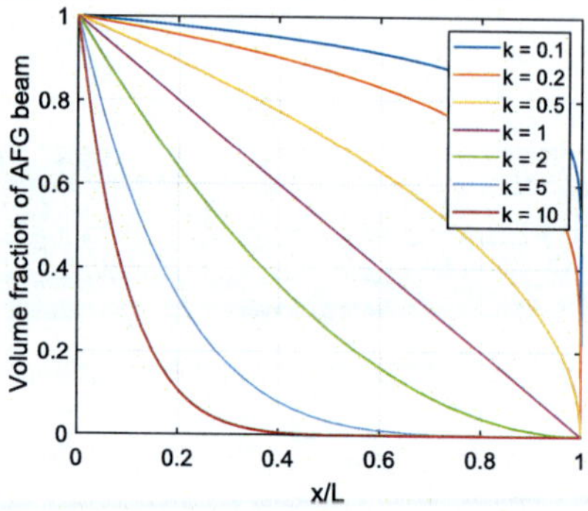

Fig. 2. Variation of the volume fraction through the AFG material ratio (x/L) in terms of the power-law index k

3 Galerkin Discritisation

In order to use the Galerkin's approach, the deflection $w(x,t)$ can be written as follows [14]:

$$w(x,t) = \sum_{i=1}^{N} q_i(t)\phi_n(x) \qquad i = 1,2,3,\ldots,N \tag{4}$$

where $q_i(t)$ denote the unknown time-dependent coefficients, N is the number of modes shapes and $\phi_i(x)$ are the basis functions which are declared for a simply supported beam as

$$\phi_i(x) = \sin\left(\frac{i\pi x}{L}\right) \qquad i = 1,2,3,\ldots,N \tag{5}$$

Since the beam is supposed to be pinned-pinned, the initial conditions and boundary conditions are given as follow:

$$w(x,t)|_{0,L} = 0, \qquad \frac{\partial^2 w(x,t)}{\partial x^2}|_{0,L} = 0, \tag{6}$$

$$w(x,t)|_{t=0} = 0, \qquad \frac{\partial w(x,t)}{\partial t}|_{t=0} = 0, \tag{7}$$

Substituting Eq. (4) into Eq. (3), and multiplying both sides of Eq. (3) by the same functions $\phi_j(x)$ and integrating over an interval from 0 to L, the non-linear time dependent amplitude equations of an AFG beam with a nonlinear viscoelastic Winkler foundation system are written as follows:

$$[M]\{\ddot{q}\} + [C]\{\dot{q}\} + [K]\{q\} + \{Q(q)\} = F(t), \tag{8}$$

where,

$$M_{ij} = A\int_0^L [\rho(x)\phi_i(x)\phi_j(x)]dx \tag{9a}$$

$$C_{ij} = c_f \int_0^L [\phi_i(x)\phi_j(x)]dx \tag{9b}$$

$$K_{ij} = I \int_0^L [E(x)\phi_i''(x)\phi_j''(x)]dx + k_L \int_0^L [\phi_i(x)\phi_j(x)]dx \tag{9c}$$

$$Q_j = k_{NL} \int_0^L \left\{ \sum_{i=1}^{N} [q_i(t)\phi_i(x)] \right\}^3 \phi_j(x)dx \tag{9d}$$

$$F_j = P_0 \sum_{n=1}^{J} \sin(\Omega t) \sin\left(\frac{j\pi vt}{L} - \frac{j\pi nd}{L}\right), \quad for \ 0 \leqslant t \leqslant L/v \tag{9e}$$

In Eq. (8), $[K]$,$[M]$ are stiffness and mass matrices, respectively that are computed by an elaborated numerical integration procedure. $[C]$ is the damping matrix and $\{Q\}$ is a colon vector which stands for the impact of the foundation's nonlinearity. $F(t)$ is the load vector which depend on time and produced by the moving forces and n is the number of moving load to be considered. $\{q\} = [q_1, q_2, q_3, \cdots, q_N]^\top$ is the displacement vector. Due to property variations along the beam length, the stiffness and mass matrices for this AFG beams are not symmetric, and the vector $Q(q, t)$ is responsible for non-linearity. The equation (8) is used for investigating the dynamic behaviour of the an AFG beam on nonlinear Winkler foundation under multiple harmonic moving loads and is solved by the fourth order Range-Kutta algorithm using a script code in MATLAB.

4 Results and Discussion

4.1 Validation Model

In this subsection, an existing case is first examined to determine the validity of the proposed approach. It concerns a pinned-pinned AFG beam subjected to moving force [6]. The AFG beam's and moving load's properties are derived from [6]. The vertical displacement for the AFG beam versus the normalized time is depicted in Fig. 3; for the power-law exponent ($k = 1$) and the dynamic response is normalized by the static response $w_0 = P_0L^3/48E_RI$ and various speed v. The vertical displacement at the middle of the AFG beam's is determined for $\Omega = 0$ (Fig. 3(a)) and $\Omega = 45 \ rad/s$ (Fig. 3(b)), it is noted that the current procedure's outcomes are well compared and have good agreements. It should be noted that the one mode analysis is not accurate for such a problem and a hight number of eigenmodes is required as already demonstrated in [15].

4.2 Application

The Material properties of the AFG Beam, mechanical parameters of the nonlinear visco-elastic foundation and the harmonic moving loads are listed in Table 1.

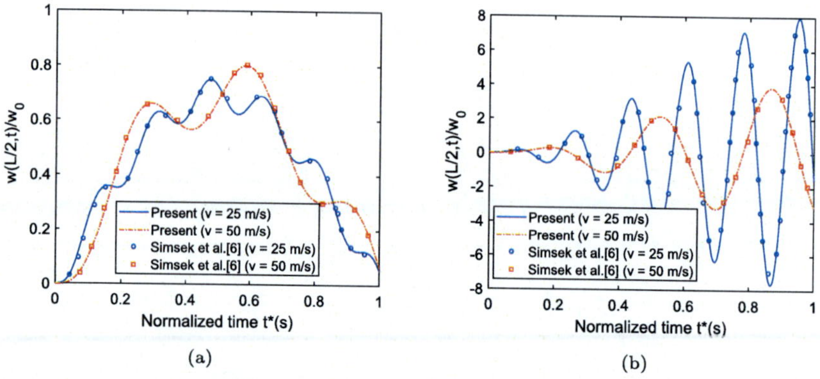

Fig. 3. The vertical deflection at the middle of the beam for $N = 1$ and $k = 1$: (a) $\Omega = 0$ and (b) for $\Omega = 45rad/s$

Table 1. Properties of the AFG beam, foundation and moving loads

Item	Notation	Value
Beam		
Young's modulus (steel)	E_R	$210\ GPa$
Young's modulus (ceramic)	E_L	$390\ GPa$
Mass density Right	ρ_R	$7850\ kg/m^3$
Mass density Left	ρ_L	$3960\ kg/m^3$
Cross-sectional area	A	$7.69 \times 10^{-3}\ m^2$
Second moment of area	I	$3.055 \times 10^{-5}\ m^4$
Length	L	20
Winkler foundation		
Linear stiffness	k_L	$3.503 \times 10^7\ N/m^2$
Nonlinear stiffness	k_{NL}	$4 \times 10^{14}\ N/m^4$
Viscous damping	c_f	$1732.5 \times 10^3\ Ns/m^2$
Harmonic moving load		
Load	P_0	$65\ kN$
Initial velocity	v	$25\ m/s$
Loads separation distance	d	$3\ m$

In Fig. 4 the deflection $w(L/2, t)$ of the AFG beam's midpoint under two successive harmonic loads for ($k = 1$, $v = 25$m/s and $\Omega = 20$rad/s) is presented. From the observation one notes the effect of modes N on the solution's convergence. One can also observe two peaks, which correspond to the two moving loads. Moreover The transverse deflection two moving loads is larger than one moving harmonic load.

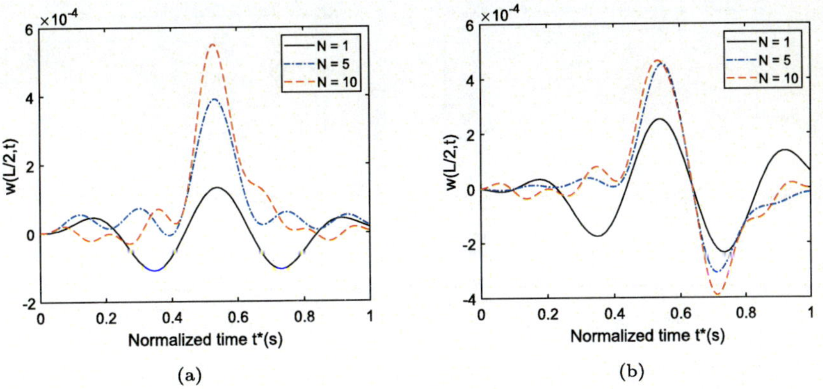

Fig. 4. Vertical displacement at mid span with $k = 1$: for (a) One moving load and (b) Two successive moving loads.

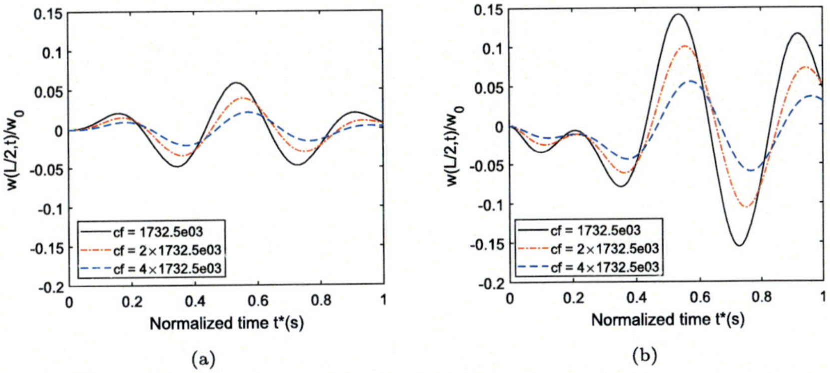

Fig. 5. Vertical displacement at mid span with $k = 1$ and $\Omega = 20$rad/s for (a) One moving load and (b) Three successive moving loads

Figure 5 depicts the dynamic deflection with the power-law index $k = 1$ and for selected values of the viscous damping ($c_f = 1732.5 \times 10^3, 2 \times 1732.5 \times 10^3, 4 \times 1732.5 \times 10^3 \;\; Ns/m^2$) for one and three successive moving loads. It can be observed that while the viscous damping coefficient c_f increases, the vertical displacements decrease for the both number of moving loads. It is apparent that more large displacements are obtained for $n = 3$ (Three successive moving load)(Fig. 5(b)) than those obtained for $n = 1$ (one moving load)(Fig. 5(a)).

Figure 6 depicts the maximum value of the variation of the normalized transverse deflections at the middle of the considered AFG beam versus k for $\Omega = 20, 40 rad/s$ and diverse values of the moving harmonic force speed. It is clear from Fig. 6, that the vertical displacements increases, when the power-law index k increases for both frequency values. As shown in Fig. 6, when $\Omega = 20 rad/s$ a large deflections are obtained than for $\Omega = 40 rad/s$. It is also clear that rising

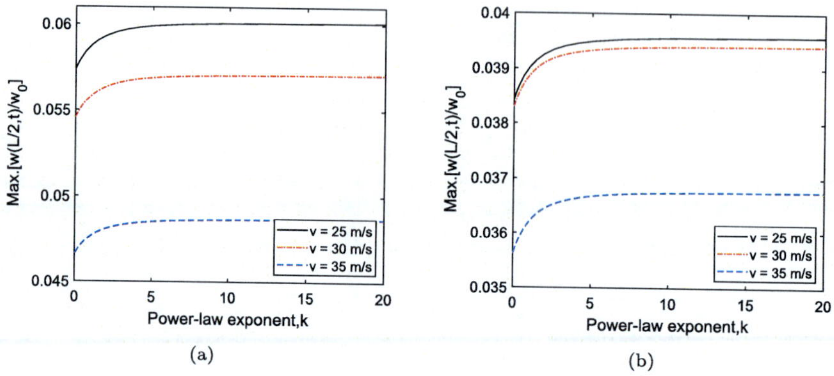

Fig. 6. The maximum vertical displacement at mid span with different k for (a) $\Omega = 20rad/s$ and (b) $\Omega = 40rad/s$

the speed value of the harmonic moving force results in a reduction in beam deflections. Moreover, from Fig. 6 it is concluded that the moving load speed and the power-law exponent k have a considerable impact on the AFG beam's dynamic response.

5 Conclusion

The investigation of the dynamic behaviour of an AFG beam lying on non-linear Winkler foundation is presented. A harmonic moving force model were considered, in which a number of moving loads are used. To get the governing equations, the Euler-Bernoulli beam theory was applied, and the dynamic equation was stated using Hamilton's principle. According to the comprehensive analysis of the results, various parameters such as foundation viscous damping and its variation, the power-law exponent, and moving force velocities have a major impact on the vertical displacements of the AFG beam. To verify the accuracy of the results and the current formulation, the vertical displacements are obtained for different parameter's values, and a comparison study shows a good agreement with the published results. All the previously described effects have a considerable impact on the dynamic response of the AFG beam on a non-linear visco-elastic foundation.

Acknowledgments. The authors would like to acknowledge the financial support of the CNRST and the Moroccan Ministry of Higher Education and Scientific Research with the project PPR2/06/2016.

References

1. Frýba, L.: Vibration of solids and structures under moving loads. Thomas Telford (1999)
2. Duc, N.D.: Nonlinear static and dynamic stability of functionally graded plates and shells. Vietnam Natl Univ Press, Vietnam. Google Scholar (2014)
3. Asghari, M., Rahaeifard, M., Kahrobaiyan, M.H., Ahmadian, M.T.: The modified couple stress functionally graded Timoshenko beam formulation. Mater. Des. **32**(3), 1435–1443 (2011)
4. Sankar, B.V.: An elasticity solution for functionally graded beams. Compos. Sci. Technol. **61**(5), 689–696 (2001)
5. Huang, Y., Li, X.F.: A new approach for free vibration of axially functionally graded beams with non-uniform cross-section. J. Sound Vib. **329**(11), 2291–2303 (2010)
6. Şimşek, M., Kocatürk, T., Akbaş, ŞD.: Dynamic behavior of an axially functionally graded beam under action of a moving harmonic load. Compos. Struct. **94**(8), 2358–2364 (2012)
7. Shahba, A., Attarnejad, R., Marvi, M.T., Hajilar, S.: Free vibration and stability analysis of axially functionally graded tapered Timoshenko beams with classical and non-classical boundary conditions. Compos. Part B: Eng. **42**(4), 801–808 (2011)
8. Calim, F.F.: Free and forced vibration analysis of axially functionally graded Timoshenko beams on two-parameter viscoelastic foundation. Compos. Part B: Eng. **103**, 98–112 (2016)
9. Kanaka Raju, K., Venkateswara Rao, G.: Effect of a non-linear elastic foundation on the mode shapes in stability and vibration problems of uniform columns/beams. J. Sound Vib. **160**(2), 369–371 (1993)
10. Rahman, M., Hasan, A.S., Yeasmin, I.A.: Modified multi-level residue harmonic balance method for solving nonlinear vibration problem of beam resting on nonlinear elastic foundation. J. Appl. Comput. Mech. **5**(4), 627–638 (2019)
11. Yaghoobi, H., Torabi, M.: An analytical approach to large amplitude vibration and post-buckling of functionally graded beams rest on non-linear elastic foundation. Journal of Theoretical and Applied Mechanics **51**(1), 39–52 (2013)
12. Ding, H., Chen, L.Q., Yang, S.P.: Convergence of Galerkin truncation for dynamic response of finite beams on nonlinear foundations under a moving load. J. Sound Vib. **331**(10), 2426–2442 (2012)
13. Ebrahimi, F., Fardshad, R.E., Mahesh, V.: Frequency response analysis of curved embedded magneto-electro-viscoelastic functionally graded nanobeams. Adv. Nano Res. **7**(6), 391–403 (2019)
14. Fletcher, C.A.: Computational galerkin methods. In Computational Galerkin Methods (pp. 72–85). Springer, Berlin, Heidelberg (1984)
15. Ouzizi, A., Abdoun, F., Azrar, L.: Nonlinear dynamics of beams on nonlinear fractional viscoelastic foundation subjected to moving load with variable speed. J. Sound Vib. 116730 (2022)

Numerical Analyze of the Flow Around an Airfoil Excited by Ultrasonic Radiation Force

Abdessamad Mehdari[1(✉)] and Mohamed Agouzoul[2]

[1] Engineering and Health Sciences and Technologies Research Center Laboratory of Applied Mechanics and Technologies (LAMAT), National School of Arts and Crafts (ENSAM), Mohammed V University, Rabat, Morocco
abdessamad-mehdari@um5.ac.ma

[2] Mechanical and Energetic Engineering Research Team: Modelling and Experimentation ERG2 (ME), Mohammadia School of Engineering (EMI), Mohammed V University, Rabat, Morocco

Abstract. In this study, we are interested to improve the aerodynamics performance by analyzing numerically 2D fluid flow around an aircraft wing airfoil. The active method of fluid flow control based an external energy source has been used and representing by an ultrasonic excitation. The waves propagation origin at an acoustic streaming field which is obviously influence the boundary layer under the radiation force located in the far field. The simulation has been performed by using the CFD software ANSYS fluent using User-Defined Function (UDF) code for adding an ultrasonic radiation force term to the Navier Stokes equations for an incompressible fluid flow at low Reynolds number. The results have revealed that adding acoustic power enhances the aerodynamics coefficients of the NACA0012 compared to a normal flow. The lift coefficient has been slight outstandingly increased although the drag coefficient has also been stepped up to a low rate. Likewise, the present work provides us to design an embedded system could improve the aerodynamics coefficients.

Keywords: Incompressible flow · NACA0012 airfoil · Acoustic excitation

1 Introduction

In the last decade, developing a control technique to enhance a fluid flow parameter has been widely adopted (Agbaglah and Mavriplis 2017; Alziadeh and Mohany 2018). We generally distinguish two classes of methods: passive methods and active methods. The passive methods are generated by a slight modification of the geometry, which is carried out to improve the considered phenomenon. The active methods are based by adding energy to the flow, like the injection unsteady parietal or external excitation. The both methods can be represented in advance (open loop control) or time's coordinated with actual flow itself (closed loop control or feedback control).

Fluid-excited by an acoustic energy has attracted our attention for the active methods. In drilling application, acoustic excitation can generate a several advantage despites to the complexity of the flow-sound interaction mechanism. These advantages can be

L. Azrar et al. (Eds.): CIP 2022, LNME, pp. 294–302, 2023.
https://doi.org/10.1007/978-3-031-23615-0_30

presented with a simple design concern material and equipment to provide a higher levels parameter. Also, the acoustic excitation leads to a forceful coupling between the flow and sound fields (Arafa and Mohany 2019). The study of acoustic streaming generated by ultrasonic waves propagation in a fluid has been the subject of many studies (Eckart 1948; Lightill 1978; Sedaghatkish et al. 2020). In fact, when a viscous fluid is exciting by an ultrasound, a mass flow can be caused as a derivative of the fluid-acoustic coupling which is known as acoustic streaming. The acoustic streaming has enormous applications in biology and in microfluidic devices.

Moreover, acoustic excitation is widely used in the aeronautical field and in particular for the flow around a wing. It reveals that fluid-excited with acoustic energy is able to affect the laminar boundary layer instabilities which is indirectly affect the tonal noise emission (Friend and Yeo 2011).

In this paper, we present a numerical investigation to analyze a dynamic boundary layer over an airfoil under an external energy. This energy is represented by an ultrasonic beam generating an acoustic radiation force. Analyzing its effect on computational model provide us dynamic aspect of the fluid flow excited. The numerical study has been implemented through ANSYS Fluent 17.0 using User-Defined Function (UDF) code.

The results allow us to evaluate the dynamic behavior of the fluid flow, which is illustrated by the aerodynamics coefficients variations for different angle of attack. The results are matched to numerical studies and to experimental researches.

2 Computational Model

2.1 Geometry

In this work, we simulate a turbulent flow past to NACA0012 airfoil, which is characterized by a symmetric profile, a closed trailing edge and 12% thickness ratio to 1 m chord length. The two-dimensional (2D) computational domain adopted for the present analysis is illustrated in Fig. 1. The acoustic excitation source is assumed spherical wave located at a distance 'a' from airfoil. 'L' and 'H' respectively present the length and the thickness of the domain. R is the radius for the inlet domain.

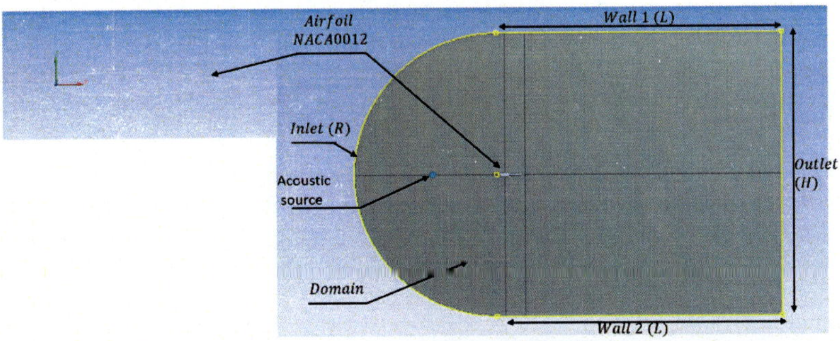

Fig. 1. Computational domain

2.2 Mathematical Model

Ultrasonic waves propagating in a viscous fluid generate a physical forces of the sound waves called *"acoustic streaming or Eckart Streaming"*. Under the acoustic radiation force $\overrightarrow{f^{ac}}$ (Nyborg 1965), the flow is modelled as a turbulent and incompressible flow which is governed by the 2D Navier-Stokes and continuity equations are as follow:

$$\rho\frac{\partial}{\partial x_i}(u_i u_j) = -\frac{\partial p}{\partial x_i} + \mu\frac{\partial}{\partial x_i}\left(\frac{\partial u_i}{\partial x_j} - u_i' u_j'\right) + f_i^{ac} \quad (1)$$

$$\frac{\partial u_i}{\partial x_j} = 0 \quad (2)$$

where, μ is the dynamic viscosity, ρ is the density, p is the pressure and u_i is the velocity component. The represents the turbulent fluctuations incorporated in the Reynolds Stress. According the Boussinesq approximation form of the incompressible flow (Alfonsi 2009), the present stresses are writing as:

$$u_i' u_j' = -\nu_T\left(\frac{\partial u_i}{\partial x_j} + \frac{\partial u_j}{\partial x_i}\right) + \frac{2}{3}k\delta_{ij} \quad (3)$$

As it concerns boundaries conditions, the non-slip hypothesis describe the fluid flow velocities components on over to the NACA boundary (the fluid particles adhere to the surface of the airfoil). Also, we assume that the acoustic effect in the model's inlet is negligible.

2.3 Effect of Acoustic Excitation on Computational Model

Effect of acoustic radiation from a source is characterized by a field emission decomposed into two regions: The region located near the source which is called the *Fresnel zone (or near field)*. It is the region in which the acoustic field has practically the shape of the source (Cylindrical, spherical, …). In this zone the sound pressure varies strongly and rapidly in space along the propagation's axis. The acoustic intensity has been faded on moving away from the source. The length of this zone called Fresnel length is defined by the position of the last maximum of the acoustic pressure. It is given by (Blackstock 2000).

$$L_F = \frac{D^2}{4\lambda} \quad (4)$$

where D is the diameter of the source and λ is the wavelength.

On the other hand, the area located beyond the near field is called the *far field or Fraunhofer area.* It is a region in which the spatial variations impacts have weakly acted the acoustic intensity and we assume that (Rossing 2013):

$$(k.r)_{\min} = \frac{F_r}{4.c}a \gg 1 \quad (5)$$

where F_r is acoustic frequency, c is the sound speed, k is wave number and r is the spatial position of acoustic propagation.

In addition, it is useful to note that the waves can have a very strong curvature nearby its source. However, further away from the source, the curvature diminishes significantly and generating essentially plane wave.

Consider that the acoustic waves propagate from the source at $F_r = 2\,\text{MHz}$, with $a = 0.25\,\text{m}$ and $c = 344\,[\text{m/s}]$. The present model characterizes by:

$$(k.r)_{\min} = \frac{f}{4.c}a \simeq 23250 \gg 1 \tag{6}$$

According to (6), the acoustic radiation force is written in the following form (Palmeri et al. 2005):

$$\overrightarrow{f^{ac}} = Q.\frac{\alpha_{ac}}{c}\frac{P_{ac}}{2\pi r^2}.\overrightarrow{n} \tag{7}$$

$\overrightarrow{n}$ normal vector of the acoustic wave-front.

Where P_{ac} [W] is the temporal average power and Q is the directivity factor. The α_{ac} [Np/m] (Np: Neper) is the acoustic absorption coefficient which indicates the capacity to absorb the energy of a sound wave. The larger it is, the more absorbent the material.

In 1972, Evans et al. published much data for absorption of sound depending to four variables: air pressure; temperature; relative humidity and frequency (Bass et al. 1990). After that improved analytical expressions are developed (Palmeri et al. 2005). Actually, ISO 9613-1 presents a calculator for any model assumes uniform meteorological conditions.

The α_{ac} for the present computational model is given in Table 1.

Table 1. Acoustic absorption coefficient

Atm. pressure: 101.325 kPa; temperature = 20 °C; relative humidity = 40%		
Fr [MHz]	α_{ac} [dB(decibel)/m]	α_{ac} [Np/m]
2	640.71185	73.76467774

Also, we assume that the acoustic waves have propagated in all directions: $Q = 1$.

2.4 Numerical Model

The domain has decomposed to multi-zonal blocks, where C-type structured grid has been used. The mesh is globally characterized by good quality parameters. The average of orthogonality and skewness are in excellent quality (0.9676; 5.375E−2). For the average aspect ratio is 1.3427 and y+ wall less than 4.315 (Fig. 2).

The Pressure-based solver was selected due to the flow incompressibility. The viscous model is considered and the standard k-epsilon model (2) is adopted, which provides good predictive capability for flows separations and improves the calculations of the boundary layer.

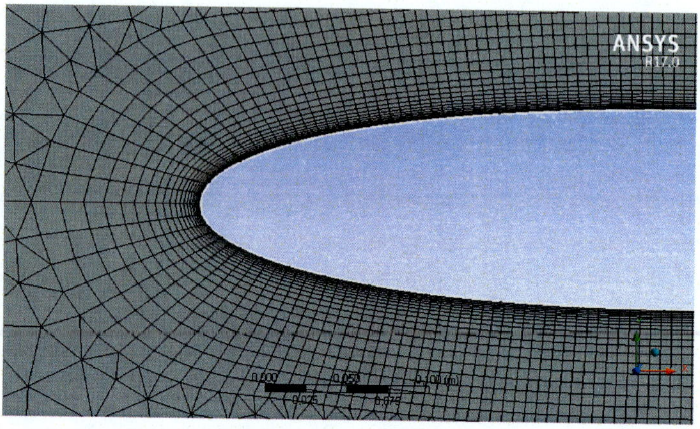

Fig. 2. Meshing domain

The SIMPLE algorithm was applied for pressure–velocity coupling. Also, the first and second order upwind were employed respectively to discretize the turbulence equations and the convective terms in the momentum. The under relaxations factors for solution were stabilized at the 0.8 for the turbulence equations and 0.3 for pressure.

The inlet (left), wall 1 and wall 2 boundaries were fixed with a uniform inlet velocity. The outlet (right) boundary condition was free-stream boundary.

To analyze computation fluid dynamic behavior over the airfoil under ultrasonic excited, UDF code was written by DEFINE_PROFILE macros in order to define a cell zone condition for our domain.

In addition, in the aforementioned data, the geometrical and numerical parameters are listed in Table 2.

Table 2. Geometrical and numerical parameters

Fluid		Geometry	
Inlet velocity [m/s]	7.55	Length 'L' [m]	10
Density [kg/m^3]	1.204	High 'H' [m]	10
Cinematic viscosity [m^2/s]	1.511E−05	Distance 'a' [m]	0.25

According the above parameters, our model was computed under the Reynolds number 5.E5.

3 Results and Discussions

As part of validating our model, a simulation without excitation took place. The convergence criterion for resulting system of equations was solved until a O(6) which all residuals were satisfied. The present results were matched with the numerical simulation data of Kianoosh et al. (2013) and Huang et al. (2004). Both simulations investigated the flow behavior aver a NACA0012 airfoil under a Reynolds number of 5.E5 and an Angle Of Attack (AOA) varied from 0° degree up to 18°. Other hand, the current computation was matched to the experimental results of Jacobs et al. (1937).

The present work was compared with the above-mentioned studies in Fig. 3. The figure indicates that the present numerical results show good agreement with the simulation of Kianoosh and Huang and the experimental data of Jacobs et al.

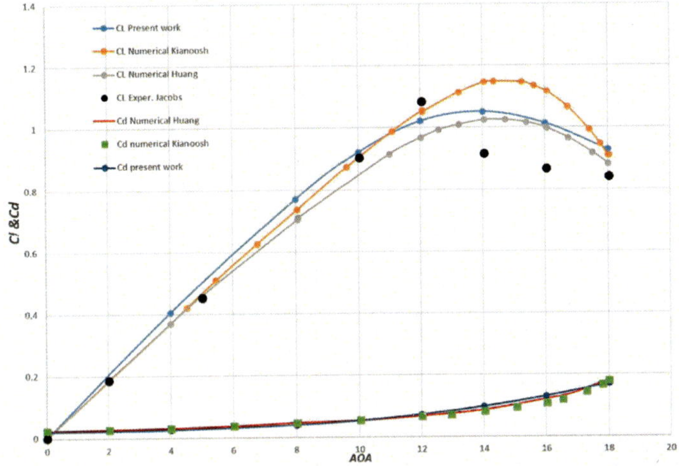

Fig. 3. Comparison among the present model results, the numerical results of Kianoosh et al. (2013) and Huang et al. (2004), and the experimental results of Jacobs et al (1937).

The effects of acoustic excitation on lift and drag coefficients are illustrated in Fig. 4 (a) and (b). An excitation on the 80 W of the power was applied at different angle of attack. The figure indicates that acoustic's effect is more important in increasing the angle of attack. Increasing an angle of attack enfeebles the velocity intensity near the airfoil which makes an acoustic effect more important.

Also, the present result shows that the lift coefficient is roughly impacted, when stall angle is not affected and the phenomena occurs at AOA of 14°. With any changing impact of the stall angle. Although, the drag coefficient is increased but the increasing lift rate is too high to increasing drag rate.

After analyzing the acoustic excitation effects on the lift and drag coefficient, varying the excitation power amplitude was also investigated. Table 3 presents the power's effects on the aerodynamics coefficients under an angle of attack 16°. It is clearly that increasing the power amplitude, the rate lift is increasing despite of the slight drag increasing too. The rate increases with a high ultrasonic power. From 80 W to 200 W acoustic powers, it's clearly that the rates coefficients have increased. The rate of lift coefficient passes to 7.05% to 18.09%. However, The rate of the drag pass to 4.65% to 11.62%.

Generally, the acoustic excitation positively affects the lift coefficient with a marginal effect on the drag coefficient.

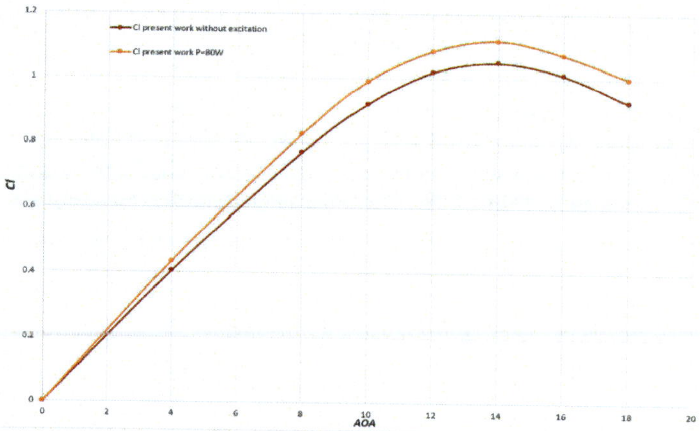

(a). Lift coefficient vs Angle Of Attack

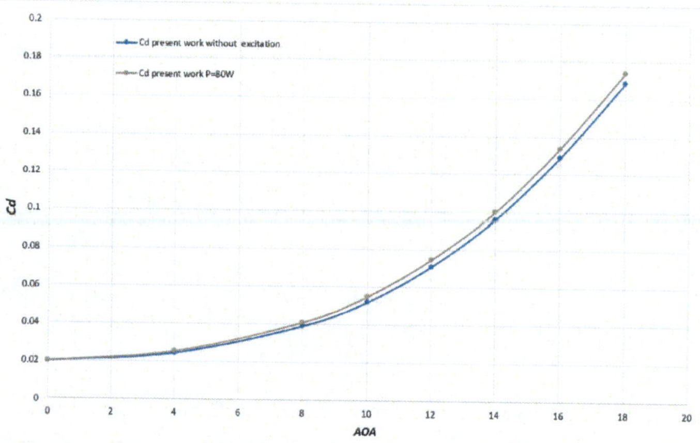

(b). Drag coefficient vs Angle Of Attack

Fig. 4. Effects of acoustic excitation on aerodynamics coefficients. Pac = 80 W

Table 3. Aerodynamics coefficients under different acoustic powers on AOA 16°

Acoustic power (MW)	Aerodynamics coefficient		Rate	
	Cd	Cl	Cd	Cl
0	0.129	1.012	–	–
80	0.135	1.0832	4.65%	7.05%
200	0.144	1.195	11.62%	18.09%

According to Table 4, the aerodynamics coefficients were investigated an angle of attack 10 for different acoustic powers amplitudes. Figure 5 shows that lift coefficient is linearly increasing as acoustic amplitude higher. This increase is accompanied by a slight undesirable effect in terms of increased drag. However, the rate of increase in lift coefficient (order E−3) is largely compared to the rate of increase in drag (3E−6).

Table 4. Aerodynamics coefficients under different acoustic powers on AOA 10°

Aerodynamics coefficients	Acoustic power (W)						
	0	60	80	90	100	140	200
Cd	0.052	0.05469	0.05525	0.05551	0.0559	0.05709	0.05894
Cl	0.9199	0.973	0.9906	0.9996	1.008	1.0429	1.0952

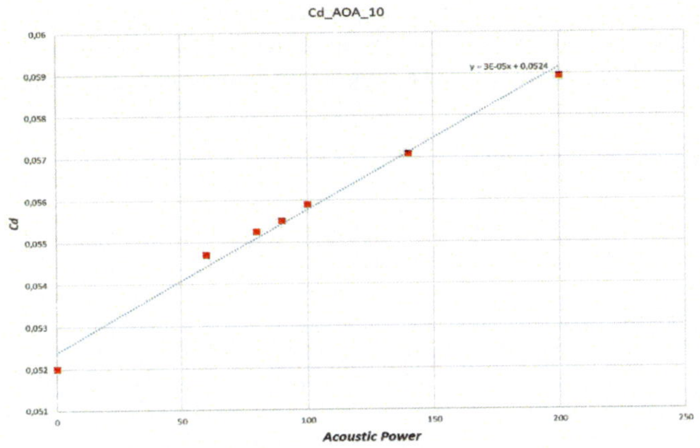

(a). Drag coefficient vs Acoustic power for AOA=10^0

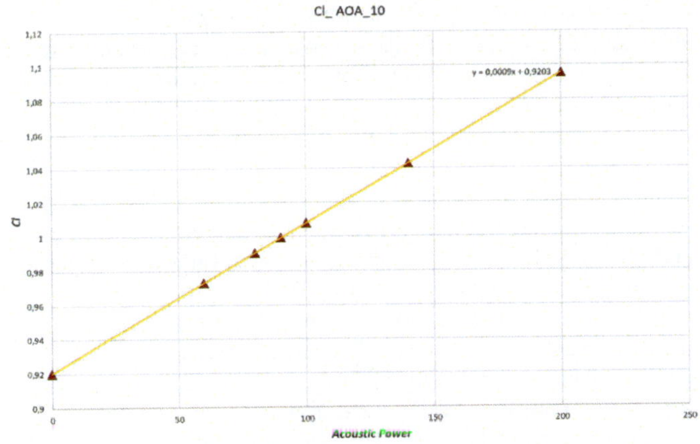

(b). Lift coefficient vs Acoustic power for AOA=10^0

Fig. 5. Effects of acoustic excitations on aerodynamics coefficients on AOA = 10°

This linear behavior, we allow us to design an embedded system to improve aerodynamic performance. In particular, the system can provide a gradient of the forces to overcome some critical situations.

4 Conclusion

The present paper investigated the fluid flow behavior excited by ultrasonic radiation force over the NACA0012 airfoil. In order to analyze this work, ANSYS FLUENT had been used using the UDF code to apply acoustic force in the domain. The results obtained show the importance of the external acoustic excitation to enhance the aerodynamics coefficients.

The lift coefficient obviously increases although the drag coefficient also slightly increases. Subsequently, these results will be used to design an embedded system in order to improve the aircraft performance and offer opportunities of a safe control in critical situation. The system should be being an active control and flow mastering tools.

References

Agbaglah, S., Mavriplis, G.: Computational analysis of physical mechanisms at the onset of three-dimensionality in the wake of a square cylinder. J. Fluid Mech. **833**, 631–647 (2017)

Alfonsi, G.: Reynolds-averaged Navier-Stokes equations for turbulence modeling. Appl. Mech. Rev. **62**, 040802 (2009)

Alziadeh, M., Mohany, A.: Near wake characteristics and acoustic resonance excitation of crimped spirally finned cylinders in cross-flow. ASME J. Press. Vessel Technol. **140**, 051301 (2018)

Arafa, N., Mohany, A.: Wake structures and acoustic resonance excitation of a single finned cylinder in cross-flow. J. Fluids Struct. **86**, 70–93 (2019)

Bass, H.E., et al.: Atmospheric absorption of sound. J. Acoust. Soc. Am. **88**, 2019–2021 (1990)

Blackstock, D.T.: Fundamentals of Physical Acoustics. Wiley (2000)

Eckart, C.: Vortices and streams caused by sound waves. Phys. Rev. **73**, 68 (1948)

Friend, J., Yeo, L.Y.: Microscale acoustofluidics: microfluidics driven via acoustics and ultrasonics. Rev. Mod. Phys. **83**, 647 (2011)

Jacobs, E., Sherman, A.: Airfoil section characteristics as affected by variations of the Reynolds number. NACA Report No. 586-231 (1937)

Huang, et al.: Numerical study of blowing and suction control mechanism on NACA0012 airfoil. J. Aircraft **41**, AI368E-05/C10301(A) (2004)

Kianoosh, Y., et al.: Numerical study of blowing and suction slot geometry optimization on NACA 0012 airfoil. J. Mech. Sci. Technol. **28**, 1–14 (2013)

Lightill, S.J.: Acoustic streaming. J. Sound Vib. **61**, 391–418 (1978)

Nyborg, W.: Acoustic streaming. Phys. Acoust. **2**, 265–331 (1965)

Palmeri, M.L., et al.: A finite-element method model of soft tissue response to impulsive acoustic radiation force. IEEE Trans. Ultrason. Ferroelectr. Freq. Control **52**, 1699 (2005)

Rossing, T.D.: Handbook of Acoustics, 2nd edn. Spring Science and Business Media (2013)

Sedaghatkish, A., et al.: Acoustic streaming and thermosensitive liposomes for drug delivery into hepatocellular carcinoma tumor adjacent to major hepatic veins; an acoustics–thermal–fluid-mass transport coupling model. Int. J. Therm. Sci. **158**, 106540 (2020)

Process and Integrated Manufacturing

Digital Twin-Driven Approach for the Rapid Reconfiguration of Manufacturing Systems

Mohammed Abadi[1(✉)], Chaimae Abadi[1], Asmae Abadi[2], and Hussain Ben-Azza[1]

[1] Laboratory of Mechanics, Mechatronics and Command Modeling, Information Processing and Control of Systems (MTICS) Team, Ecole Nationale Supérieure d'Arts et Métiers (ENSAM Meknes), Moulay Ismail University, Meknes, Morocco
abadi.s.mohammed@gmail.com

[2] Euromed Research Center, Euromed University of Fez, Fez, Morocco

Abstract. In order to respond to the increasing demands on products individualization, nowadays manufacturing systems must allow a big flexibility, reactivity and reconfigure ability. We propose in this work to deal with these complexities through the proposition and development of a new digital twin-driven approach to assist manufacturers in the achievement of the rapid reconfiguration of their manufacturing systems. Our aim is to allow manufacturers to produce multiple variations of customized products at a very near price of standardized mass products. To do so, we propose to gather the main two families of reconfiguration methods, namely, knowledge-based reasoning methods using inference ontologies and artificial intelligence-based optimization methods in one integrated approach that takes into consideration all the critical requirements of nowadays RMS particularly in the industry 4.0 context and to ensure the concurrent reconfiguration of both the digital and hardware components of the manufacturing systems. A case of study implementing the proposed approach in the Digital Twin-based reconfirmation of the Flexible machining cell of our engineering school "ENSAM" is developed in the last section.

Keywords: Smart manufacturing · Digital twin · Ontology · Industrialization · Artificial intelligence · Flexible manufacturing system

1 Introduction

Nowadays, products and costumers needs are becoming more and more complex and highly personalized. To meet these requirements, manufacturing systems must be flexible and reconfigurable enough to allow the production of multiple variations of production sequences in short cycle times with low reconfiguration and production costs (Alsafi et al. 2010). To deal with these complexities, new paradigms have emerged in nowadays industrial companies, namely smart manufacturing (Tao et al. 2019), Digital Twin and Industry 4.0 technologies. In fact, it is envisioned that rapid commissioning and decommissioning of system elements are enabled to save significant time and cost (Wang et al. 2017). Thus, multiple advanced manufacturing technologies and strategies have been

L. Azrar et al. (Eds.): CIP 2022, LNME, pp. 305–314, 2023.
https://doi.org/10.1007/978-3-031-23615-0_31

developed to boost the automatization of manufacturing system to achieve smartness (Jiewua et al. 2020).

The reconfiguration ability throughout modularization of industrial manufacturing system is a relevant strategy to ensure highly-frequent changeover however this flexibility in reconfiguration should be coupled with cost efficiency as well as ability for fast digital and hardware reconfiguration and high reliability (Bruccoleri et al. 2006).

Reconfiguration of the manufacturing system calls for strong coupling of multidimensional optimization variables (process plans, process parameters, typological parameters, buffer configuration, delivery date…). Thus, the identification and the modeling of manufacturing systems reconfiguration are becoming more and more complex. Digital twin technologies provide a potent online simulation tool for supporting reconfiguration of manufacturing system (Tao et al. 2018).

A new digital twin-driven approach is proposed in this work to assist manufacturers in the achievement of the rapid reconfiguration of their manufacturing systems. The methodology integrates the artificial intelligence-based optimization methods (Lu et al. 2017) with the use of inference ontologies (Abadi et al. 2018) in one integrated approach that takes into consideration all the critical requirements of nowadays RMS particularly in the industry 4.0 context. In fact, our idea is to model and automate the execution of Group technology and engineering rules using the Semantic Web Rule Language (SWRL) and the inference abilities of inference ontologies.

To do so, we review in Sect. 2 the main related works concerning Digital Twin and Reconfigurable Manufacturing Systems. We propose then in Sect. 3 our new Digital twin-driven approach for the rapid reconfiguration of manufacturing systems. Then, we develop in Sect. 4 a case of study implementing the proposed approach in the Digital Twin-based reconfirmation of the Flexible machining cell of our engineering school "ENSAM". Finally, in Sect. 5, we present conclusions and perspectives of this research.

2 Related Work

We review in this section the two main concepts combined in this work which are Digital Twin and Reconfigurable Manufacturing Systems paradigms.

2.1 Digital Twin Concept

The digital twin has become one of the most widely used tools for treating problems faced by cyber-physical systems. First, the digital twin is a virtual representation of a physical entity containing all the necessary data and information for its complete description. The term digital twin was proposed by Vickers of Nasa, as well as its concept was used at a conference on project life cycle management in 2003.

Subsequently, the concept has been used recently by Michael Grieves and John Vickers who conceived of it in their white paper on the origins of the digital twin (Grieves et al. 2014). In fact, A standard and general structure for the Digital Twin model, containing three dimensions, has been first established by Grieves. Indeed, these three dimensions are (Jones et al. 2019):

- Physical artefact: It contains several sensors that collect the states of the subsystems and the working conditions of the devices;
- Virtual representation: It models with high accuracy the different properties of the physical entity, namely geometries, mechanical properties, attitudes and regulations.
- Data connections: The different technologies, in particular network communication, IoT and network security maintain the connections between the different elements of the DT model (Abadi et al. 2022)

Then, this architecture proposed by Grieves was kept by (Tao et al. 2018), and added another dimension notably:

- Services: They consist of a functional service that ensures the conversion of data and algorithms into support services for the proper functioning of the DT, as well as a business service that presents the standard inputs and outputs in a simpli fied form through software interfaces.

2.2 Reconfigurable Manufacturing Systems

As its name indicates, a Reconfigurable Manufacturing System (RMS) is a system designed from the outset to rapidly evolve its structure, as well as its hardware and software components, in order to quickly adjust its production capacity and functionality within a family of parts in response to rapid market changes or intrinsic system changes (Chalfoun 2014).

The RMS must be designed initially using hardware and software components that can be software components that can be integrated quickly and reliably.

Typically, according to authors (Koren et al. 1999; Mehrabi et al. 2000; Dashchenko 2006; ElMaraghy et al. 2006), each manufacturing system must have the following main characteristics to be reconfigurable: Modularity of all components of the RMS system, integrability, personalization, scalability, convertibility, diagnosability and mobility.

3 Digital Twin-Driven Approach for the Rapid Reconfiguration of Manufacturing Systems

3.1 Motivations

We propose in this section a new digital twin-driven approach to assist manufacturers in the achievement of the rapid reconfiguration of their manufacturing systems. The aim of the proposed approach is to allow manufacturers to produce multiple variations of customized products at a very near price, or ideally the same price, of standardized mass products. To do so, we propose to gather the main two families of reconfiguration methods, namely, knowledge-based reasoning methods (Lee et al. 2015; Wang et al. 2016), and artificial intelligence-based optimization methods (Lu et al. 2019; Wang et al. 2017) in one integrated approach that takes into consideration all the critical requirements of nowadays RMS particularly in the industry 4.0 context.

In fact, these previous reconfiguration methods, taken each one alone, present many limitations. The majority of the first ones, knowledge-based reasoning methods, lack the consideration on both the software and hardware implementation while the second ones, artificial intelligence-based optimization methods, suffers from substantial computation complexity, which hinders the realization of rapid and/or online reconfiguration of the manufacturing system (Tao et al. 2019). Our aim is to combine the potentialities of each of them with the relevant emerging technologies of digital twins to respond to the rapid reconfiguration of nowadays complex manufacturing systems.

3.2 The Proposed Approach

We consider the industrialization phase of a new product. We place our approach in the context of the reconfiguration of the existing manufacturing systems and not in the context of designing a new one.

The idea is to perform in a concurrent manner the reconfiguration steps on the cyber and physical systems of the RMS digital twin in order to build a real-time synchronization between them, since the beginning of the reconfiguration process, in order to take advantage from the enabling technologies that we propose to incorporate in the RMS digital twin.

To do so, we complete in this paper our previous contribution (Abadi et al. 2021) where we proposed a physical reconfiguration framework for RMS by developing in this paper its digital twin to allow the automation of the approach and to achieve the scalability, convertibility and intelligence needed in nowadays RMS.

Figure 1 represents the flowchart of the digital-twin based reconfiguration approach. The product architecture is the first input in our approach. Once it is developed, the analysis of the product modules is initiated.

To do so, the digital mock-up of the product is analyzed in order to decompose the product into a set of functional modules. Antecedent similar modules are then searched in the RMS core ontology that we propose to develop and which constitutes one of the backbones of the cyber system of the RMS digital twin. Our aim here is first to identify whether the deductive or the generative approach must be implemented to perform the reconfiguration, then second to enhance the use of two of relevant techniques in our RMS which are the open-architecture machine tools (OAMT) (Jiewu et al. 2020) and the modularization of machine tools. In fact, after identifying the different functional modules of the product, we propose to use the inference abilities of the core RMS ontology to determinate if similar modules have been already developed by the team. If they do exist, a deductive approach is implemented and the previous programs, and configurations are readopted. To do so, we propose to enrich the RMS core ontology with a set of inference rules formalizing the different techniques of Group Technology (GT) (Kashkoush et al. 2014). In fact, each configuration of the manufacturing system is relative to the production of an individualized product order that belongs to one specific product family. GT permits to determinate and gather, into families, different parts in order to take advantage of their similarities in design and production (Kashkoush et al. 2014). Using GT we will deduce the reference process plan to be adopted to the new product specifications and using OAMT we will integrate standardized plug-and-play platform and modules such as reconfigurable modular fixture, extensible processing channel, adjustable intermittent buffer and plug-and-play circulation device (Tao et al. 2019). Our idea is to use the inference abilities of the RMS ontology to rapidly create

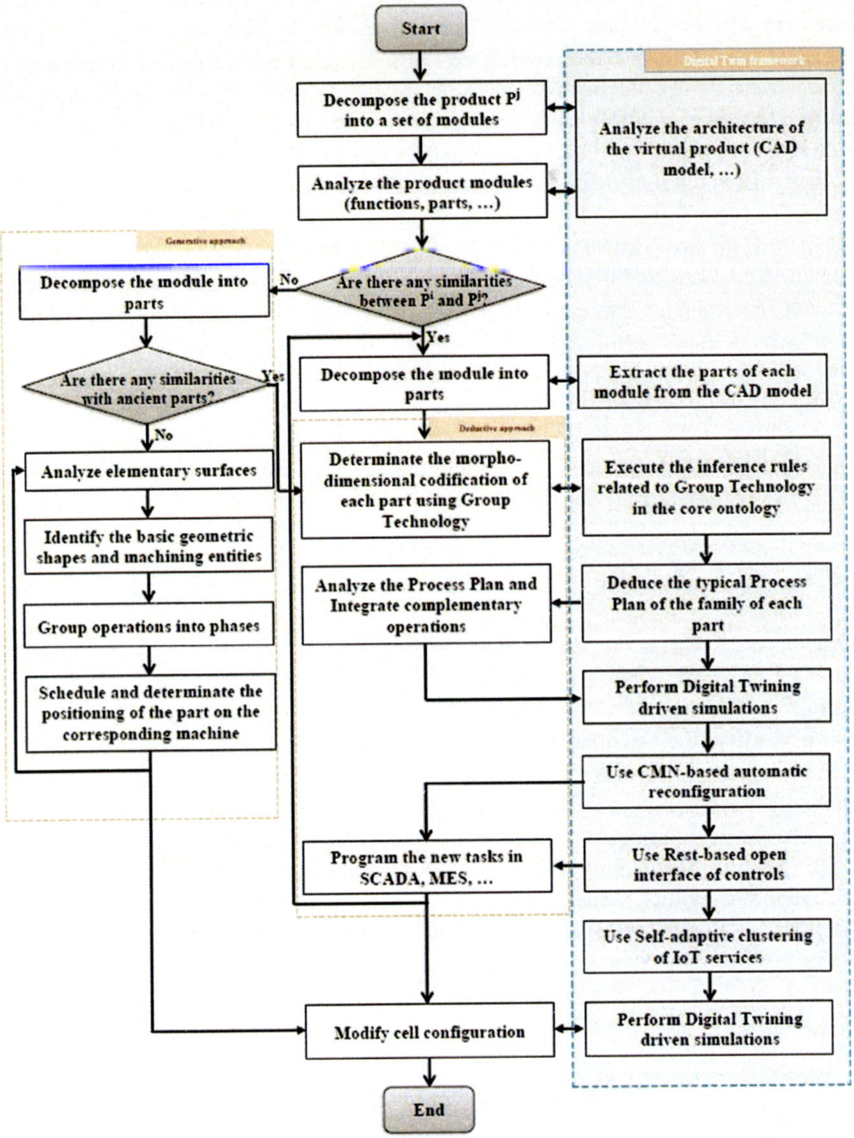

Fig. 1. The proposed Digital twin-driven approach for the rapid reconfiguration of manufacturing systems

a reconfigurable digital model of the physical OAMT using Group Technology and its reference models modelled previously in the RMS ontology.

Once the new reconfiguration of the digital twin is identified, the next step in our proposed approach is to program the new tasks in the production monitoring information systems and to perform the reconfiguration of the IoT system. To do so, we propose to use the Representational StateTransfer (REST) (Battle et al. 2008) architectural style where the position of all operational objects is identified by a uniform resource identifier (i.e., URI) and Uniform Interface (e.g., HTTP). The connection of the resources in the cluster to the network will be then performed by building a REST-based embedded Web server on the smart gateway as the cluster head node (Li et al. 2019). The final step of the deductive approach will be to perform the digital twin driven simulations.

As shown in the flowchart, the reconfiguration steps of the digital twin are the same even in the case of the implementation of the generative approach. This approach is implemented if no similar module has been studied or developed by the manufacturers before. In this case the architectural decomposition then the analysis of elementary surfaces of each part are implemented. The basic geometric shapes and machining entities and phases are then identified. Finally, the scheduling and the positioning of each part on its corresponding machines are developed. The reconfiguration of the digital twin in this case will be performed exactly the same way as in the deductive approach, using the chosen industry 4.0 technologies, namely the open-architecture machine tools (OAMT), REST-enabled reconfigurable IIoT system, the CMN-based automatic reconfiguration and the Digital twining-driven simulations.

4 Case of Study

In order to demonstrate the applicability of the proposed approach, we develop in this section a case of study concerning the digital twin-based reconfiguration of the flexible machining cell of ENSAM. Our purpose is to follow the different steps of our approach in order to allow the rapid reconfiguration of both the digital and physical entities of the cell's Digital Twin in order to enhance its reactivity for the development of new industrial products. Figure 2 represents the production monitoring information system and the hardware components of the flexible cell (Abadi et al. 2021).

The flexible machining cell is designed to allow the machining of the different component of desk sets, namely penholders and baseplates. The initial configuration of the cell allows also the assembly of different categories of Desk sets.

The new reconfiguration of the ENSAM cell must allow the production of a new variant of the Desk sets containing a new instrument which is a desk watch. To do so, we implement the first steps in our approach to initially position our self either in the generative or deductive product development approach.

To do so, the proposed RMS ontology will constitute the core element of the inference system. It will also be used to ensure the interoperability between the other digital entities of the Digital Twin of the ENSAM flexible cell.

Thus, as explained before, we started by enriching our RMS core ontology with a set of inference rules (Table 1) that model the Group technology rules in order to implement the first steps of our approach and to check if any similar products or modules have been already developed in the flexible cell and if it is the case, to reproduce and readjust the previous process plans and configurations. For instance, rule 1 models the deferent specification that must be satisfied by a product to be classified in 'Desk Set' product family. Rule 2 gives an example of the automatic generation of product components according the family of product to each it belongs. For example, any variant of desk sets which must satisfy the principal function 'hold pen' must automatically has as

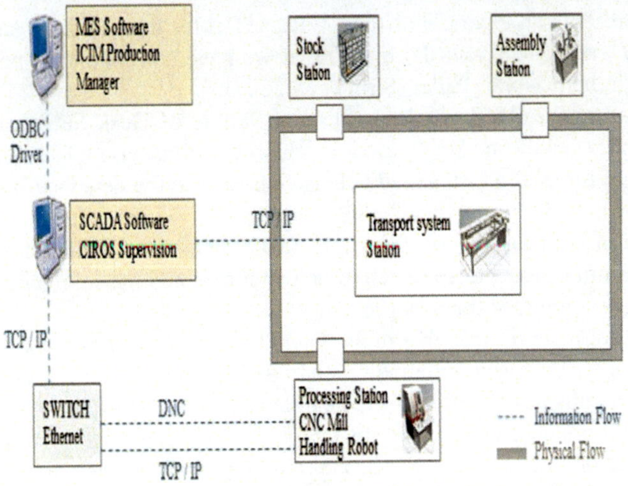

Fig. 2. The architecture of ENSAM flexible machining cell studied.

a component a 'Pen Holder' and so on. Rule 4 illustrate the use Group Technology techniques in affecting a reference manufacturing Process Plan to each new product according to its specifications and product family.

Table 1. Sample of the SWRL GT rules encoded in the RMS core ontology.

Rule number	SWRL rule	Rule category
1	Product(?p) ^ hasPrincipalFunction(?p, "HoldInstruments") ^ hasAreaOfUse(?p, "Desk") - > BelongsTo_ProductFamily(?p, "DeskSet")	Product family affectation
2	Product(?p) ^ BelongsTo_ProductFamily(?p, "DeskSet") ^ hasPrincipalFunction(?p, "HoldPen") - > hasComponent(?p, "PenHolder")	Product components generation
3	Product(?p) ^ HasSimilarProduct(?p, 1) - > useApproch(?p, "DeductiveApproch")	Product development approach determination
4	Product(?p) ^ BelongsTo_ProductFamily(?p, "DeskSet") - > hasReferenceProcessPlan(?p, "DS_ProcessPlan")	Reference process plan affectation

Table 1 presents an example of the SWRL GT rules that we have encoded in the SWRL tab of Protégé 5.0 and the results that we have obtained after performing the reasoning with the reasoner Pellet (Sirin et al. 2007).

As inferred in the RMS ontology, the new variant of Desk Sets that we want to develop is composed of one single module which constitutes on a base plate with four holes instead of three (the firth added hole is dedicated to the new instrument, a watch, required in the new desk set). Since similar products have been already configured and manufactured in the flexible cell, the implementation of the SWRL rules of the RMS ontology have infer to as the approach to be conducted and the reference process plan to be readopted to produce the new product.

The open-architecture machine tools (OAMT), REST-enabled reconfigurable IIoT system and the CMN-based automatic reconfiguration are then used to ensure reconfiguration of the physical entities of the flexible cell. Then, since the RMS ontology is a formal OWL domain ontology and the Ontology Web Language used to encode it is based on Description Logic and XML, the introduction of the new base plate characteristics and information to the MES software of the flexible cell will be automatic due to the interoperability between the XML files and the OWL RMS ontology. The new base plate is an aluminum one with an additional hole. The reference Process Plan of the Base Plates family is then readjusted by incorporation a new operation of drilling to it. The same reconfiguration has been done on the SCADA software of the cell, CIROS Production (Festo 2010) and the CMN-based automatic reconfiguration is performed to integrate it into the milling program of the CNC machine too. The geometric and dimensional characteristics of the new base plate have been also taken into consideration in the modification of the "Pick and Place" program of the handling robot. For the pen holder (Abadi et al. 2021), predefined programs are encoded previously in the MES software ICIM Manager Database and allow its manufacturing.

5 Conclusion and Perspectives

Throughout this paper, we propose a new digital twin-driven approach to assist manufacturers in the reconfiguration of their manufacturing systems. The novel proposed approach combines the potentialities of Industry 4.0 techniques with those of inference ontologies in order to allow the rapid reconfiguration of the physical and digital entities of the manufacturing systems. In fact, our proposed methodology takes into account the two main cases of generative and deductive industrial products development approaches. We based the reconfiguration process in our approach on the use of the open-architecture machine tools (OAMT), REST-enabled reconfigurable IIoT system, the CMN-based automatic reconfiguration and the Digital twining-driven simulations. Then we used the inference abilities of a core RMS proposed ontology to model and automate the execution of Group Technology rules and to deduce rapidly the new reconfigurations to perform within the RMS Digital Twin.

As perspectives, we intend to incorporate more artificial intelligence methods into the reconfiguration methodology in order to enhance more its reactivity and performance. We will also try to combine several artificial intelligence tools to obtain a better reconfiguration of industrial systems. And finally we will integrate other aspects in the reconfiguration of these industrial systems, such as the degradation of the equipment, the external factors impacting the process, the security of the equipment and the personnel, etc.

References

Tao, F., Tang, Y., Zou, X., Qi, Q.: A field programmable gate array implemented fibre channel switch for big data communication towards smart manufacturing, Robot. Comput.- Integr. Manuf. **57**, 166–181 (2019)

Alsafi, Y., Vyatkin, V.: Ontology-based reconfiguration agent for intelligent mechatronic systems in flexible manufacturing. Robot. Comput. Integr. Manuf. **26**, 381–391 (2010)

Wang, X.V., Wang, L.: A cloud based production system for information and service integration an internet of things case study on waste electronics, Enterp. Inf. Syst. **11**, 952–968 (2017)

Lu, Y., Xu, X.: Cloud-based manufacturing equipment and big data analytics to enable on-demand manufacturing services. Robot. Comput. Integr. Manuf. **57**, 92–102 (2019)

Bruccoleri, M., Pasek, Z.J., Koren, Y.: Operation management in reconfigurable manufacturing systems: reconfiguration for error handling. Int. J. Prod. Econ. **100**, 87–100 (2006)

Jones, D.E., Snider, C., Kent, L., Hicks, B.: Early stage digital twins for early stage engineering design. In: International Conference on Engineering Design, ICED19 5–8 August 2019, delft, The Netherlands

Tao, F., Cheng, J., Qi, Q., Zhang, M., Zhang, H., Sui, F.: Digital twin-driven product design, manufacturing and service with big data. Int. J. Adv. Manufact. Technol. **94**(9–12), 3563–3576 (2018)

Chalfoun, I.: Conception et déploiement des Systèmes de Production Reconfigurables et Agiles (SPRA). Micro et nanotechnologies/Microélectronique. Université Blaise Pascal Clermont-Ferrand II, 2014. Français

Koren, Y., Heisel, U., Jovane, F., Moriwaki, T., Pritschow, G., Ulsoy, G., Van Brussel, H.: Reconfigurable manufacturing systems. Ann. CIRP **48**(2), 527–540 (1999)

Mehrabi, M.G., Ulsoy, A.G., Koren, Y.: Reconfigurable manufacturing systems: key to future manufacturing. J. Intell. Manufact. **11**(4), 403–419 (2000)

Dashchenko, A.I.: Reconfigurable Manufacturing Systems and Transformable Factories. Editions Springer, Netherlands (2006)

ElMaraghy, H.A.: Flexible and reconfigurable manufacturing systems paradigms. Int. J. Flex. Manuf. Syst. **2006**, 261–276 (2006)

Lee, J., Bagheri, B., Kao, H.: A cyber-physical systems architecture for Industry 4.0- based manufacturing systems. Manuf. Lett. **3**, 18–23 (2015)

Wang, S., Wan, J., Zhang, D., Li, D., Zhang, C.: Towards the smart factory for industrie 4.0: a self-organized multi-agent system assisted with big data based feedback and coordination. Comput. Netw. **101**, 158–168 (2016)

Battle, R., Benson, E.: Bridging the semantic web and Web 2.0 with representational state transfer (REST). Web semantics: science, services and agents on the World Wide Web. **6**, 61–69 (2008)

Li, M., Xu, G., Lin, P., Huang, G.Q.: Cloud-based mobile gateway operation system for industrial wearables. Robot. Comput.-Integr. Manuf. **58**, 43–54 (2019)

Jicwu, L., et al.: Digital twin-driven rapid reconfiguration of the automated manufacturing system via an open architecture model. Robot. Comput.-Integr. Manufact. **63**, 101895 (2020)

Abadi, A., Ben-Azza, H., Sekkat, S.: Improving integrated product design using SWRL rules expression and ontology-based reasoning. Procedia Comput. Sci. **127**, 416–425 (2018)

Abadi, C., Manssouri, I., Abadi, A.: Reconfiguration of flexible manufacturing systems considering product morpho-dimensional characteristics and modular design. In: International Conference on Integrated Design and Production (pp. 559–565). Springer, Cham (2019)

Kashkoush, M., ElMaraghy, H.: Product family formation for reconfigurable assembly systems. Procedia CIRP **17**, 302–307 (2014)

Festo: Festo Didactic, CIROS Supervisions, User Manual (2010)

Sirin, E., Parsia, B., Grau, B.C., Kalyanpur, A., Katz, Y.: Pellet: a practical owldl reasoned. Web semantics: science, services and agents on the World Wide Web. **5**, 51–53 (2007)

Abadi, M., Chaimae, A., Asmae, A., Hussain, B.A.: A smart decision making system for the selection of production parameters using digital twin and ontologies. Int. J. Adv. Comput. Sci. Appl. **13**(2), (2022)

Orthogonal Cutting of UD-CFRP Using Micromechanical Modeling

Amira. Hassouna[1](✉), Slah. Mzali[1], Farhat. Zemzemi[2], and Salah. Mezlini[1]

[1] Ecole Nationale d'Ingénieurs de Monastir, Laboratoire de Génie Mécanique (LGM), Université de Monastir, Rue Ibn Eljazzar 5019, Sousse, Tunisie
Hassounaamira0@gmail.com

[2] Ecole Nationale d'Ingénieurs de Sousse, Laboratoire de Mécanique de Sousse (LMS), Université de Sousse, BP 264 Cité Erriadh 4023, Sousse, Tunisie

Abstract. The specific properties of composite materials have attracted the attention of researchers and manufacturers in several fields. Understanding the machining of composites required the analysis of mechanisms occurring during this process. The objective of this study is to investigate the effect of fiber orientation angle (θ) on the fiber failure mechanisms, cutting force and lateral damage during orthogonal cutting of a Unidirectional Carbon Fiber Reinforced Polymer (UD-CFRP). Therefore, a three-dimensional micromechanical model was developed upon Abaqus/Explicit. It is found that the chip formation mechanisms, cutting force and lateral damage are highly affected by the fiber orientation angle. For fiber orientation angle of zero degree, the chip separation occurs after the rupture of fiber by buckling. For fiber orientation angle of ninety degrees, the chip is formed by the bending of fibers and then their breaking perpendicular to their axes. It is also concluded that the cutting force and the lateral damage for fiber orientation angle of ninety degrees are higher than those for fiber orientation angle of zero degree.

Keywords: Micromechanical model · Fiber orientation angle · Chip formation mechanisms · Lateral damage · Cutting force

1 Introduction

Compared with the metal machining, the composition of Carbon Fiber Reinforced Polymer (CFRP) makes the cutting process difficult to achieve and generates the appearance of defects. To control the damage and ensure a good surface quality, the mechanisms of chip formation need to be analyzed in detail. Therefore, several researchers make effort to study the effect of fiber orientation and cutting parameters on cutting forces and chip formation mechanisms using finite elements analysis (Calzada et al., 2012; Hassouna et al., 2020; Nayak et al., 2005). Particularly, A two dimensional(2D) finite element model was developed by Nayak et al. (2005) to study the effect of the fiber orientation on the cutting and thrust forces. The workpiece was composed of a single fiber surrounded by a matrix. These compounds were assumed as a brittle isotropic material

L. Azrar et al. (Eds.): CIP 2022, LNME, pp. 315–322, 2023.
https://doi.org/10.1007/978-3-031-23615-0_32

with an elastic behavior. The fiber matrix interface was modeled with nodal debonding. It was found that the fiber orientation has a significant effect on both forces. The cutting force increased with the variation of the fiber orientation from 0 to 90°. Another work was conducted by Gopala et al. (2007) using a 2D micromechanical model based on the plane strain assumption. The workpiece was composed of two domains: the first one was modeled with an Equivalent Homogenous Material (EHM) and the second one was formed of fibers, interfaces and matrix. The matrix was supposed to be an elastoplastic material and the fiber has elastic behavior. The interface was modeled with cohesive elements. Two types of composites were tested; unidirectional carbon fiber reinforced polymer (UD-CFRP) and unidirectional glass fiber reinforced polymer (UD-GFRP). The results revealed the sensitivity of the chip formation mechanisms to the fiber orientation and the type of composite. For a fiber orientation of 90°, the combination of bending and crushing mechanisms were observed during the machining of UD-CFRP. However, for UD-CFRP, the fracture of fiber was caused by the bending mechanism.

To take into account the out-of-plane failure, several researches are focused on three dimensional (3D) micromechanical model (Gao et al. 2015; Su, 2019). Particularly, Geo et al. (2015) investigated the effect of the fiber orientation and the cutting speed on the cutting force and the surface roughness. The carbon fiber was supposed to be an anisotropic and elastic material. The matrix was assumed to be an isotropic and elastoplastic material. It was concluded that the maximum cutting force was observed for 90° fiber orientation and the high surface roughness was detected for 135° fiber orientation. The increase of the cutting speed results in an increase of the cutting force and a decrease in the surface roughness.

Despite the scientific research on the orthogonal cutting of composite, few woks were focused on 3D micromechanical modeling. Therefore, a 3D finite element model was developed using the micromechanical analysis. The effect of the fiber orientation on chip formation mechanisms, cutting force and lateral damage was investigated.

2 Finite Element Modelling

2.1 Description of the FE Model

A 3D micromechanical model is developed using Abaqus/Explicit for simulating the orthogonal cutting of UD-CFRP composites. Figure 1 presents the geometry, the boundary conditions and the mesh of the finite element model. The workpiece dimensions were 850 * 300 * 150 μm^3. The workpiece is divided into two parts; (i) the first one is modeled with fiber, matrix and interface, separately and (ii) the second one is modeled with Equivalent Homogenous Material (EHM). The workpiece is composed of 15 fibers and 15 interfaces. The cutting tool is supposed as a rigid body having a 0° rake angle, a 5° clearance angle and a 0.01 mm nose radius. A cutting speed of 33 mm/min and a depth of cut of 15 μm are applied. The bottom surface of the workpiece is clamped, and the right side is constrained along the x axis. The workpiece side parallel to the tool is constrained along the z axis.

In this model, different types of mesh are used. The matrix and the fibers are meshed with continuum three-dimensional 8-node linear brick elements of type C3D8R. An 8-node three-dimensional cohesive element of type COH3D8 is applied to the interface.

Finally, the EHM is meshed with quadrilateral continuum shell elements (SC8R). A fine mesh is considered in the vicinity of the cutting tool and a coarse mesh is applied elsewhere. This model is composed of 563,086 elements.

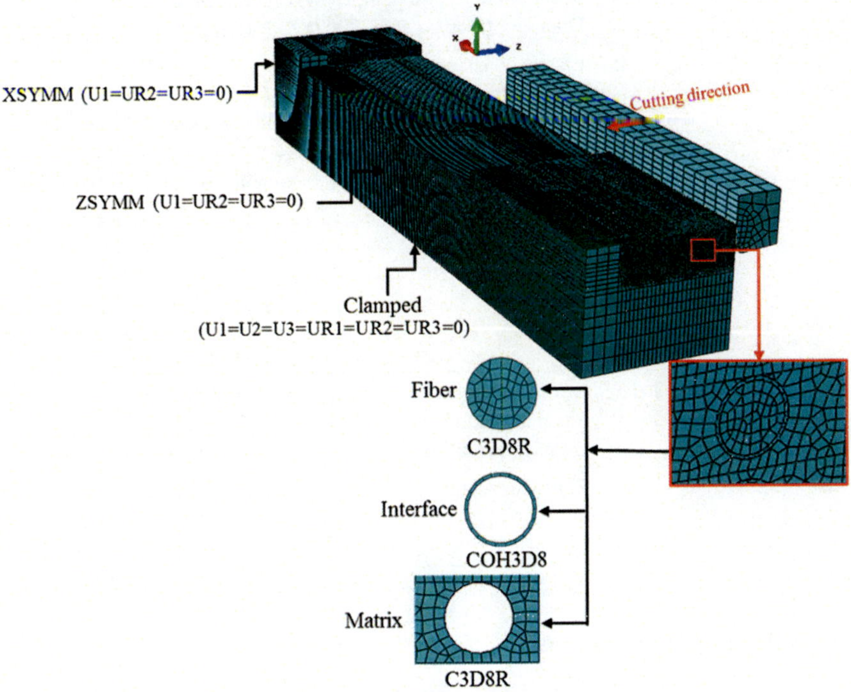

Fig. 1. Geometry, boundary conditions and mesh of the finite element model.

The "GENERAL CONTACT" algorithm available in Abaqus/Explicit is applied for modelling the contact between tool/matrix, tool/fiber, tool/interface, fiber/fiber and fiber/matrix. For assembling the interface with both the fiber and matrix, the Tie constraint is picked. The Coulomb friction model with a constant friction coefficient of 0.5 is considered between the tool and the different constituents of the workpiece.

2.2 Material Properties

In this study, the matrix is assumed isotropic with elastoplastic behavior. The progressive damage algorithm is used for modeling the damage behavior of matrix. The carbon fiber exhibits isotropic elastic behavior. The 'Brittle Cracking' constitutive model available in Abaqus/Explicit is applied for modelling the damage of fibers. The fiber-matrix interface is defined using the traction-separation concept. The damage initiation is based on the nominal quadratic stress "QUADS". Concerning the damage evolution, the mixed-mode fracture energy is adopted.

The properties of the different constituents of the composite are presented in Table. 1.

Table 1. Properties of fiber, matrix, interface and EHM.

<table>
<tr><td>Fiber (carbon)</td><td>Elastic constants (Xu et al. 2014)
Tensile strength (Xu et al. 2014)
Shear strength (Xu et al. 2014)
Diameter (Gopala et al. 2007)</td><td colspan="2">E = 230 GPa, υ = 0.2
2000 MPa
380 MPa
10 mm</td></tr>
<tr><td rowspan="7">Matrix (epoxy) (Song et al. 2016)</td><td rowspan="7">Elastic constants
Plastic constants
Ductile damage</td><td colspan="2">$E = 3.1$ GPa $\upsilon = 0.33$</td></tr>
<tr><td>Yield stress</td><td>Plastic strain</td></tr>
<tr><td>24</td><td>0</td></tr>
<tr><td>58</td><td>0.03</td></tr>
<tr><td>66</td><td>0.04</td></tr>
<tr><td>70</td><td>0.05</td></tr>
<tr><td colspan="2">Fracture strain = 0.033
stress triaxiality = 1, strain rate = 0.1</td></tr>
<tr><td>Interface (Song et al., 2016)</td><td>Normal strength
Shear strength
Damage evolution</td><td colspan="2">167.5 MPa
25 MPa
50 N/m</td></tr>
<tr><td>EHM</td><td>Elastic constants</td><td colspan="2">$E_{11} = 126$ GPa; $E_{22} = 11$ GPa; $G_{12} = 6.6$ GPa; $\upsilon = 0.28$</td></tr>
</table>

3 Results and Discussion

3.1 Effect of Fiber Orientation Angle on Chip Formation Mechanisms

Figure 2 shows the chip formation mechanisms for two fiber orientation angles 0 and 90°. It is found that the fiber orientation angle highly influences the chip formation mechanisms.

- For a fiber orientation angle of 90° (Fig. 2 (a)), as the tool attacks the workpiece, it causes the matrix disintegration due to the compression force. Then, the fibers undertaken a bending deformation perpendicular to their axis until breaking.
- For a fiber orientation angle of 0° (Fig. 2(b)), the advance of the cutting tool generates the separation between fibers and matrix as a result of interfacial failure. The fibers ahead the tool undergo an opening in mode I combined with a fiber buckling mechanism. Subsequently, a fiber crushing mechanism is observed resulting in a fracture in mode II.

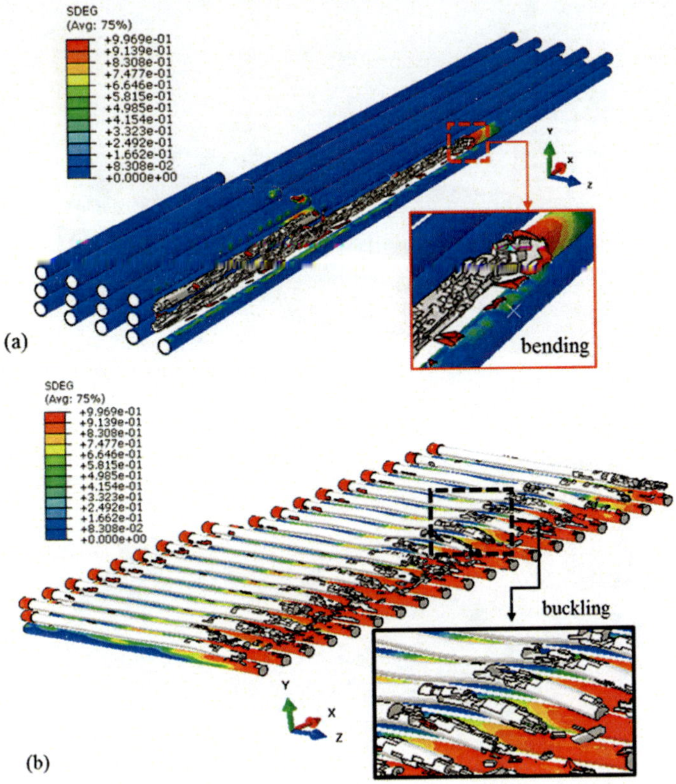

Fig. 2. Chip formation mechanisms for: (a) $\theta = 90°$ and (b) $\theta = 0°$.

3.2 Effect of Fiber Orientation Angle on Lateral Damage

Figure 3 shows the lateral damage in the workpiece (matrix damage) for two fibers orientation angles 0 and 90°. It is observed that machining perpendicular to the fibers (fiber orientation angle of 90°) causes more damage in the workpiece than that obtained for the fiber orientation angle of 0°. The damage zone for the fiber orientation angle of 90° is 65 μm (Fig. 3 (a)) compared to that observed for the fiber orientation angle of 0°, which is 8.92 μm (Fig. 3(b)). This is explained by the bending of fibers before breaking takes place perpendicular to the direction of fibers for a fiber orientation angle of 90°. In this case, the damage in the matrix is accentuated in order to inhibit the bending of fibers.

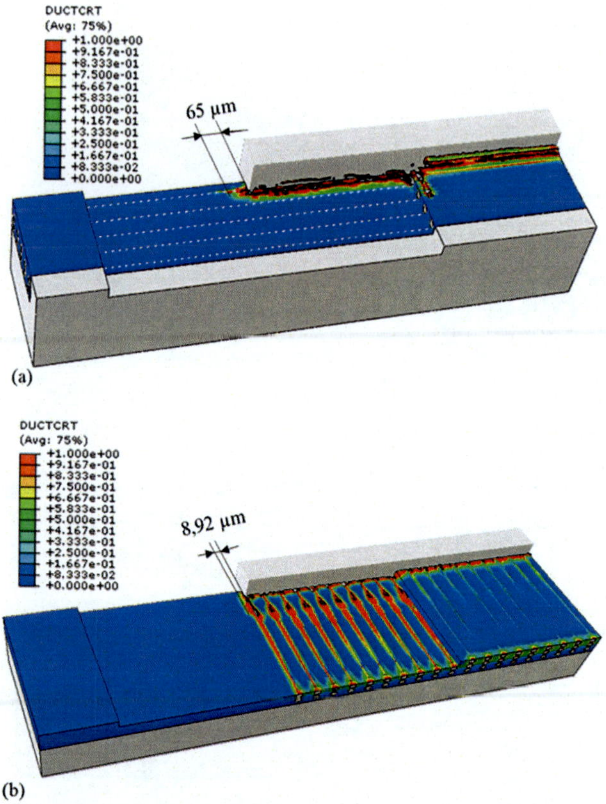

Fig. 3. Lateral damage as a function of fiber orientation angle: (a) $\theta = 90°$ and (b) $\theta = 0°$.

3.3 Effect of Fiber Orientation Angle on Cutting Force

Figure 4 illustrates the evolution of cutting force as a function of time for two fiber orientation angles 0 and 90°. It can be noticed that the fiber orientation angle has a significant effect on the cutting force. In fact, for the two fiber orientation angles, forces evolutions show three zones:

Zone 1: In this zone, the cutting force value is low and the fiber breaking has not yet taken place. The tool has been just cracking the matrix.
Zone 2: This begins when the tool comes into contact with the fibers. This zone is accompanied by an increase in the cutting force.
Zone 3: In this zone, the separation between matrix and fibers continues, accompanied by bending of the fibers leading to the removal of material.

Furthermore, it should be noted that the cutting force for a fiber orientation angle of 90° is greater than that observed for a fiber orientation angle of 0°. This can be attributed to the chip formation mechanisms observed previously. In fact, the resistance of the fibers in the direction perpendicular to their axes causes the increase of cutting force.

This result is in agreement with the literature work (Yan et al. 2019). In addition, it can be seen that the extent of zone 2 for a fiber orientation of 0° is wider than that of 90°. This is explained by the high number of fibers in front of the tool, which takes longer to break.

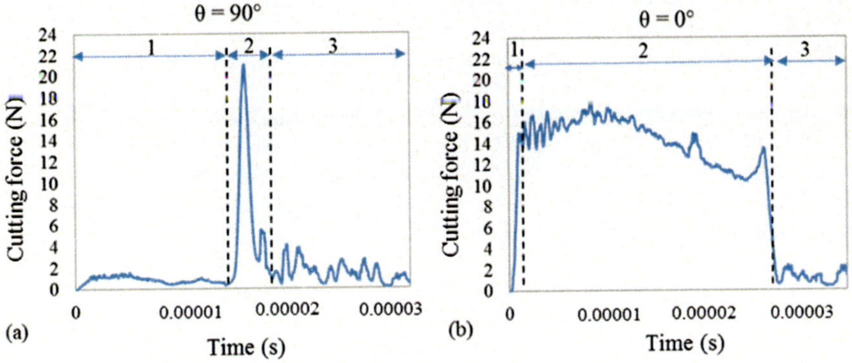

Fig. 4. Evolution of cutting force versus time for: (a) $\theta = 90°$ and (b) $\theta = 0°$.

4 Conclusion

This work aims to investigate the effect of the fiber orientation angle on the chip formation mechanisms, the cutting force and the lateral damage using micromechanical analysis. The results revealed that the fiber orientation angle strongly influences these machinability outputs. For $\theta = 90°$, the chip separation is formed by the bending of fibers, however, for $\theta = 0°$, the buckling mechanism is responsible for fiber breaking. It is found also that the maximum cutting force and lateral damage are observed for $\theta = 90°$.

References

Calzada, K.A., Kapoor, S.G., Devor, R.E., Samuel, J., Srivastava, A.K.: Modeling and interpretation of fiber orientation-based failure mechanisms in machining of carbon fiber-reinforced polymer composites. J. Manuf. Process. **14**(2), 141–149 (2012). https://doi.org/10.1016/j.jmapro.2011.09.005

Gao, C., Xiao, J., Xu, J., Ke, Y.: Factor analysis of machining parameters of fiber-reinforced polymer composites based on finite element simulation with experimental investigation. Int. J. Adv. Manufact. Technol. **83**(5–8), 1113–1125 (2015). https://doi.org/10.1007/s00170-015-7592-2

Gopala, G.V., Mahajan, P., Bhatnagar, N.: Micro-mechanical modeling of machining of FRP composites—cutting force analysis. Compos. Sci. Technol. **67**, 579–593 (2007). https://doi.org/10.1016/j.compscitech.2006.08.010

Hassouna, A., Mzali, S., Zemzemi, F.: Orthogonal cutting of UD-CFRP using multiscale analysis : finite element modeling. J. Compos. Mater. 1–14 (2020).https://doi.org/10.1177/0021998319899129

Nayak, D., Bhatnagar, N., Mahajan, P.: Machining studies of ud-frp composites part 2: finite element analysis. Mach. Sci. Technol. **9**(4), 503–528 (2005). https://doi.org/10.1080/10910340500398183

Song, D., Li, Y., Zhang, K., Cheng, H., Liu, P., Hu, J.: Micromechanical analysis for microscopic damage initiation in fiber/epoxy composite during interference- fit pin installation. Mater. Des. **89**, 36–49 (2016). https://doi.org/10.1016/j.matdes.2015.09.118

Su, Y.: Effect of the cutting speed on the cutting mechanism in machining CFRP. Compos. Struct. **220**(April), 662–676 (2019). https://doi.org/10.1016/j.compstruct.2019.04.052

Xu, W., Zhang, L.C., Wu, Y.: Elliptic vibration-assisted cutting of fibre-reinforced polymer composites : understanding the material removal mechanisms Metal elastic body. Compos. Sci. Technol. **92**, 103–111 (2014). https://doi.org/10.1016/j.compscitech.2013.12.011

Yan, X., Reiner, J., Bacca, M., Altintas, Y., Vaziri, R.: A study of energy dissipating mechanisms in orthogonal cutting of UD-CFRP composites. Compos. Struct. **220**, 460–472 (2019). https://doi.org/10.1016/j.compstruct.2019.03.090

Evaluation of AutoML Tools for Manufacturing Applications

Meryem Chaabi[1(✉)], Mohamed Hamlich[1], and Moncef Garouani[1,2]

[1] CCPS Laboratory, ENSAM, University of Hassan II, Casablanca, Morocco
meryem.chaabi-etu@etu.univh2c.ma

[2] UR 4491, LISIC, Laboratoire d'Informatique Signal et Image de la Cote d'Opale, Univ. Littoral Cote d'Opale, 62100 Calais, France

Abstract. In today's industrial environment, the increased availability of real time data offers a great opportunity to perform data-driven decision making. And so, improving the manufacturing performance. Machine learning algorithms provide the ability to learn effectively from data. However, building an accurate machine learning model requires data scientists knowledge. Hence, the task of training an effective learning model becomes costly, time consuming, and laborious. Unfortunately, in several companies, industrial practitioners lack machine learning knowledge. In this context, we investigated to clarify the potential of Automated machine learning. AutoML is a set of tools that enable non-ML-expert to create automatically a successful ML model. Our study examines whether AutoML methods could achieve convincing results in various manufacturing applications such as quality control and predictive maintenance. We evaluated four AutoML tools: AMLBID, Autoweka, AutoSklearn, and TPOT on 7 industrial datasets. The experimental results have proved that some AutoML tools provided better performance than classical learning models with configuration performed by non-ML experts.

Keywords: AutoML · Machine learning · Data analysis · Quality control · Predictive maintenance

1 Introduction

In the era of industry 4.0, manufacturing companies are witnessing a significant increase in available data [1]. Sensing technologies have made it possible to extract an enormous amount of real time data from the production process. However, the fundamental challenge for these companies remains to use the full potential of available data in order to boost manufacturing operational performance [2].

Machine learning (ML) algorithms have emerged as a powerful tool for converting the available data into valuable knowledge to empower decision making by creating models in terms of states or trends. ML approaches were investigated for use in various manufacturing applications, such as Remaining Useful

L. Azrar et al. (Eds.): CIP 2022, LNME, pp. 323–330, 2023.
https://doi.org/10.1007/978-3-031-23615-0_33

Lifetime (RUL) prediction, predictive quality, and defect detection [3]. Several ML methods have yielded satisfactory results, however, adopting these methods in manufacturing systems is still challenging.

Building effective learning models rely heavily on data scientists' skills [4], which constitutes a time-consuming task. It can be challenging for a non-machine learning expert to determine the proper ML algorithm and its optimal hyperparameters for a specific application. However, collaboration with a data scientist for training and evaluation of ML solutions to different problems could prove to be costly, especially in industrial environments where changes occur rapidly.

Recently, automated machine learning (AutoML) had emerged as a promising solution for a non-machine learning expert to build accurate learning models [5]. In this context, this contribution investigates the potential use of automated machine learning for manufacturing applications as a viable alternative to traditional machine learning methods.

The remainder of the paper is organized as follows. Section 2 provides an overview of the related works. Section 3 presents AutoML tools, datasets, and experimental results. The paper is concluded in Sect. 4.

2 Related Works

Several machine learning methods have been explored for manufacturing applications. Among these, the quality management and predictive maintenance use cases have received growing interest.

2.1 Quality Management

Quality management encompasses defect detection, process monitoring, predictive quality, and process parameter optimization [6].

Jurkovic et al. [7] investigated the performance of Polynomial regression, SVR, and ANN algorithms to predict cutting parameters in a high speed turning process. Process parameter optimization entails selecting optimal process parameters, leading to a high quality product. Schmitt et al. [8] proposed a predictive model based quality inspection using machine learning. The model was trained on historic data sets in the cloud. The proposed system seeks early knowledge of the expected product quality. The study findings have proved that the proposed model had significantly reduced inspection volumes. The paper [9] developed a multi-target regression technique to perform quality prediction in a mining process. Prediction of product quality contributes to better control of the manufacturing process. The study estimated the amount of silica and iron concentrates in the ore at the end of the process. Peres et al. [10] Evaluated different machine learning to predict dimensional defects in a real automotive multistage assembly line. XGBoost and Random Forests have yielded the best results in this non-linear and high dimensional data use case. Jiang et al. [11] and Mujeeb et al. [12] proposed defect detection solutions. Reference [11] developed a deep learning approach, while [12] used autoencoders to classify whether a product is defect-free or defective.

2.2 Predictive Maintenance

Predictive maintenance is a support process based on regular monitoring of the system in order to examine the system's health state. Predictive maintenance aims to forecast the failure time of the system components to avoid unplanned downtimes and economic losses.

Kolokas et al. [13] used LSTM networks to predict equipment faults. The proposed method system had managed to detect changes of sensorial features 5-10 minutes before a breakdown, which is not a sufficient interval to prevent the faults. Paolanti et al. [14] developed a predictive maintenance system based on Random Forest. This system had achieved 95% of overall accuracy. Ayvaz et al. [15] developed a predictive maintenance system by utilizing the real-time data extracted from production lines. The suggested solution aims to determine potential failures before they occur by using machine learning methods. The results of the study showed that Random Forest and XGBoost had yielded satisfactory results. Benkedjouh et al. [16] proposed a method to predict Remaining Useful Life of cutting tools using support vector regression. RUL represents the expected length of time left for the system before it falls down. Their method have shown promising results. Zhu et al. [17] proposed a deep learning-based method for RUL prediction of rotating components.

To summarize, a plethora of manufacturing applications have been investigated based on machine learning algorithms. Overall, these methods achieved promising results. However, the complex task of selection of models and configuration of hyperparameters prevents industrial practitioners from benefiting from the increased available data [1].

3 Methodology

3.1 Automated Machine Learning

Machine learning pipelines design consists of several demanding tasks [18] such as feature engineering, model selection, and hyperparameters optimization as shown in Fig. 1. Automl is a set of tools that enable non-ML-experts to build accurate machine learning models. It aims to facilitate the use of machine learning and reduce repetitive tasks in ML pipelines [19,20].

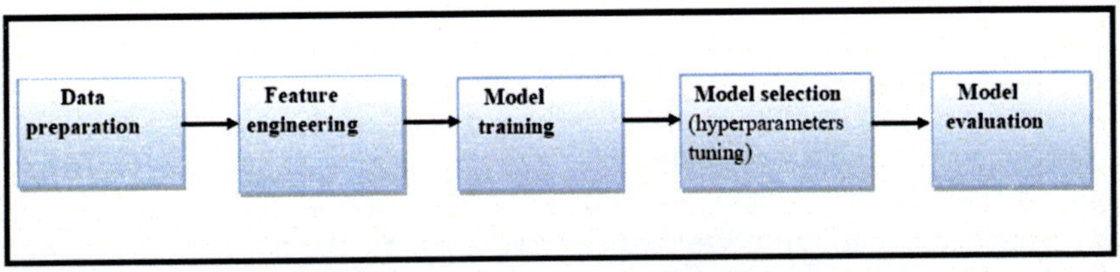

Fig. 1. Machine learning pipeline

In the literature, several AutoML tools with different hyperparameter optimization approaches have been proposed. AutoML tools considered in this paper are as follows:

- AutoSklearn [21]: uses Bayesian optimization and meta-learning to find automatically optimal ML model. It takes into account the past best combinations of ML models and their respective hyperparameters.
- AMLBID [22]: has emerged as a meta-learning [2] based decision support system for the selection and optimization of ML algorithms intended for Big Industrial data mining. It consists of a knowledge base of + 4 million evaluated ML pipeline. At its core, AMLBID has a search space of 8 Scikitlearn classifiers and a range of 8000 hyperparameters configuration of the supported algorithms [1].
- Autoweka [23]: AutomL tool based on WEKA. It uses Bayesian optimization to determine and tune the ML algorithm.
- TPOT [24]: Automated machine learning tool that designs and optimize ML pipelines via genetic programming.

3.2 Datasets

We considered 7 manufacturing datasets. The use cases are related to: quality management and predictive maintenance. Table 1 presents the collected datasets.

Table 1. The datasets details

Dataset	Task	No. of classes	No. of instances
[16]	RUL prediction	2	61,000
[25]	Quality control	7	1941
[26]	Predictive quality	3	2000
[27]	APS failure prediction	2	60,000
[28]	Failure risk analysis	4	959
uci_AI4I	Predictive maintenance	5	10,000
Tool wear	Predictive maintenance	2	7586

3.3 Empirical Results

For our experiments, we evaluated the effectiveness of 4 AutomL tools for 7 manufacturing applications. We compared Automl tools performance against classical machine learning performances. We adopted the accuracy to measure the performance of the AutoML tools. Table 2 and Fig. 2 present the results.

In 4 out of 7 cases, AMLBID provided the best results (97.06%, 99.41%, 99.71%, 93.74%). TPOT, AutoSklearn, Autoweka produced the best accuracies

Table 2. Performances of selected AutoML frameworks on the benchmark datasets

Dataset	AutoML tools accuracy				Original paper result
	AutoWeka	TPOT	Auto-sklearn	AMLBID	
[16]	85.39	99.07	97.82	99.41	98.95
[25]	99.56	97.21	98.79	99.54	80.74
[26]	92.61	95.17	96.32	97.06	95
[27]	94.76	99.33	97.16	99.10	99.02
[28]	86.04	91.20	82.15	93.74	85
uci_AI4I	69.15	72.6	72.9	72.6	–
Tool wear	99.15	99.28	99.64	99.71	–

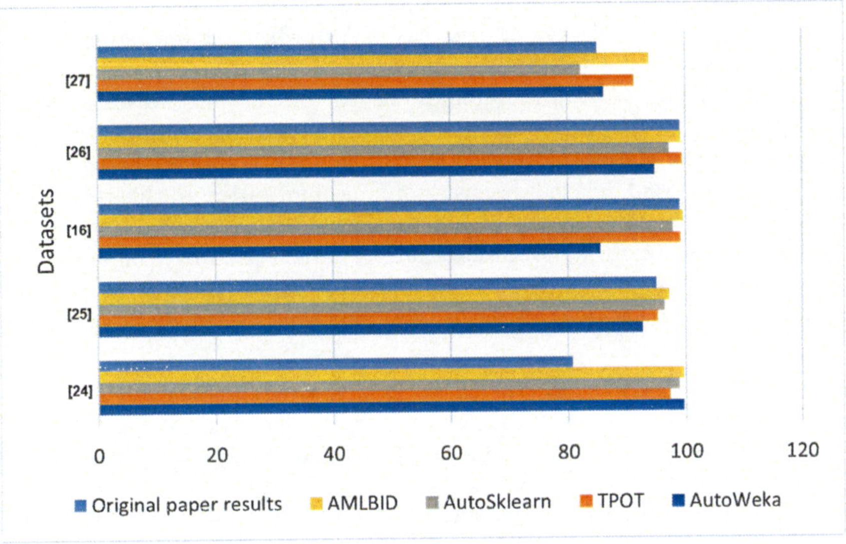

Fig. 2. Comparative results of the 4 AutoML tools and original papers

for the 5, 6, 4 datasets respectively. Overall, Automl methods have yielded satisfactory results whether for binary or multiclass classification. Moreover, AMLBID and TPOT have achieved extremely close results. Their performance was better than learning models with configuration performed by non-ML experts. Whereas, Autoweka and AutoSklearn have failed in some applications to outperform the results from the related papers.

There is no AutomL tool that consistently outperformed the other AutomL tools. However, we noticed that AMLBID performed slightly better in most use cases. Those results demonstrated a certain degree of effectiveness of the AutoML methods in handling manufacturing applications. Thus, AutoML provided the opportunity for non-ML experts in the manufacturing environment to exploit available data.

4 Conclusion

In this paper, we have explored AutoML tools for important manufacturing applications. Our study aimed to compare performance results between 4 open source AutoML tools, and learning models with a configuration that is set by industrial researchers.

Our work has showed that in most cases, AutoML tools outperformed classical machine learning algorithm for quality management and predictive maintenance applications. AutoMl simplifies the use of machine learning for non-MLexperts. This work revealed the great opportunities that AutoML could offer to manufacturers. They could create predictive quality and predictive maintenance models to avoid delayed discovery of nonconformities. In this perspective, AutoML opens new horizons for ML related manufacturing applications.

Given the nature of industrial data often characterized by complex characteristics, we are planning in the future to explore more AutoML tools on vast and large industrial datasets.

References

1. Garouani, M., Ahmad, A., Bouneffa, M., Hamlich, M., Bourguin, G., Lewandowski, A.: Towards big industrial data mining through explainable automated machine learning. Int. J. Adv. Manuf. Technol. (2022). https://doi.org/10.1007/s00170-022-08761-9
2. Garouani, M., Ahmad, A., Bouneffa, M., Lewandowski, A., Bourguin, G., Hamlich, M.: Towards the automation of industrial data science: a meta-learning based approach. In: Proceedings of the 23rd International Conference on Enterprise Information Systems. SCITEPRESS—Science and Technology Publications (2021). https://doi.org/10.5220/0010457107090716
3. Chaabi, M., Hamlich, M.: A sight on defect detection methods for imbalanced industrial data. In: ITM Web of Conferences, vol. 43, p. 01012. EDP Sciences (2022). https://doi.org/10.1051/itmconf/20224301012
4. Hamlich, M., Ramdani, M.: Fuzzy ant miner. In: IADIS International Conference Intelligent Systems and Agents 2012, ISA 2012, IADIS European Conference on Data Mining 2012, ECDM 2012, EID: 2-s2.0-84887463736, ISBN: 978-972893969-4
5. Huttler, F., Kotthoff, L., Vanschoren (eds.): Automated Machine Learning. Springer, Cham (2019). https://doi.org/10.1007/978-3-030-05318-5_10
6. Çınar, Z.M., Nuhu, A.A., Zeeshan, Q., Korhan, O., Asmael, M., Safaei, B.: Machine learning in predictive maintenance towards sustainable smart manufacturing in industry 4.0. Sustainability **12**(19), 8211 (2020). https://doi.org/10.3390/su12198211

7. Jurkovic, Z., Cukor, G., Brezocnik, M., Brajkovic, T.: A comparison of machine learning methods for cutting parameters prediction in high speed turning process. J. Intell. Manuf. **29**(8), 1683–1693 (2016). https://doi.org/10.1007/s10845-016-1206-1
8. Schmitt, J., Bönig, J., Borggräfe, T., Beitinger, G., Deuse, J.: Predictive model-based quality inspection using machine learning and edge cloud computing. Adv. Eng. Inform. **45**, 101101 (2020)
9. Dogan, A., Birant, D., Kut, A.: Multi-target regression for quality prediction in a mining process. In: 2019 4th International Conference on Computer Science and Engineering (UBMK). IEEE (2019). https://doi.org/10.1109/ubmk.2019.8907120
10. Peres, R.S., Barata, J.. Leitao, P., Garcia, G.: Multistage quality control using machine learning in the automotive industry. IEEE Access **7**, 79908–79916 (2019). https://doi.org/10.1109/access.2019.2923405
11. Jiang, J., Cao, P., Zichen, L., Lou, W., Yang, Y.: Surface defect detection for mobile phone back glass based on symmetric convolutional neural network deep learning. Appl. Sci. **10**(10), 3621 (2020)
12. Mujeeb, A., Dai, W., Erdt, M., Sourin, A.: One class based feature learning approach for defect detection using deep autoencoders. Adv. Eng. Inform. **42**, 100933 (2019)
13. Kolokas, N., Vafeiadis, T., Ioannidis, D., Tzovaras, D.: Forecasting faults of industrial equipment using machine learning classifiers. In: Innovations in Intelligent Systems and Applications (INISTA). IEEE (2018)
14. Paolanti, M., Romeo, L., Felicetti, A., Mancini, A., Frontoni, E., Loncarski, J.: Machine learning approach for predictive maintenance in industry 4.0. In: 2018 14th IEEE/ASME International Conference on Mechatronic and Embedded Systems and Applications (MESA). IEEE (2018). https://doi.org/10.1109/mesa.2018.8449150
15. Ayvaz, S., Alpay, K.: Predictive maintenance system for production lines in manufacturing: a machine learning approach using IoT data in real-time. Expert Syst. Appl. **173**, 114598 (2021)
16. Benkedjouh, T., Medjaher, K., Zerhouni, N., Rechak, S.: Health assessment and life prediction of cutting tools based on support vector regression. J. Intell. Manuf. **26**(2), 213–223 (2013). https://doi.org/10.1007/s10845-013-0774-6
17. Zhu, J., Chen, N., Peng, W.: Estimation of bearing remaining useful life based on multiscale convolutional neural network. IEEE Trans. Ind. Electron. **66**(4), 3208–3216 (2019)
18. Hamed, O., Hamlich, M.: Improvised multi-robot cooperation strategy for hunting a dynamic target. In: International Symposium on Advanced Electrical and Communication Technologies (ISAECT). IEEE (2020)
19. Garouani, M., Ahmad, A., Bouneffa, M., Hamlich, M., Bourguin, G., Lewandowski, A.: Towards meta-learning based data analytics to better assist the domain experts in industry 4.0. In: Lecture Notes on Data Engineering and Communications Technologies. Springer, Singapore (2022). https://doi.org/10.1007/978-3-030-97610-1_22
20. Hamed, O., Hamlich, M., Ennaji, M.: Hunting strategy for multi-robot based on wolf swarm algorithm and artificial potential field. Indones. J. Electr. Eng. Comput. Sci. **25**(1), 159 (2022). https://doi.org/10.11591/ijeecs.v25.i1.pp159-171
21. Feurer, M., Klein, A., Eggensperger, K., Springenberg, J.T., Blum, M., Hutter, F.: Auto-sklearn: efficient and robust automated machine learning. In: Hutter, F., Kotthoff, L., Vanschoren, J. (eds.) Automated Machine Learning. TSSCML, pp. 113–134. Springer, Cham (2019). https://doi.org/10.1007/978-3-030-05318-5_6

22. Garouani, M., Ahmad, A., Bouneffa, M., Hamlich, M.: AMLBID: an auto-explained automated machine learning tool for big industrial data. SoftwareX **17**, 100919 (2022)
23. Kotthoff, L., Thornton, C., Hoos, H.H., Hutter, F., Leyton-Brown, K.: Auto-WEKA: automatic model selection and hyperparameter optimization in WEKA. In: Hutter, F., Kotthoff, L., Vanschoren, J. (eds.) Automated Machine Learning. TSSCML, pp. 81–95. Springer, Cham (2019). https://doi.org/10.1007/978-3-030-05318-5_4
24. Olson, R.S., Moore, J.H.: TPOT: a tree-based pipeline optimization tool for automating machine learning. In: Hutter, F., Kotthoff, L., Vanschoren, J. (eds.) Automated Machine Learning. TSSCML, pp. 151–160. Springer, Cham (2019). https://doi.org/10.1007/978-3-030-05318-5_8
25. Tian, Y., Mengyu, F., Fang, W.: Steel plates fault diagnosis on the basis of support vector machines. Neurocomputing **151**, 296–303 (2015)
26. Saravanamurugan, S., Thiyagu, S., Sakthivel, N.R., Nair, B.B.: Chatter prediction in boring process using machine learning technique. Int. J. Manuf. Res. **12**(4), 405 (2017). https://doi.org/10.1504/ijmr.2017.088399
27. Rafsunjani, S., Safa, R.S., Al Imran, A., Rahim, M.S., Nandi, D.: An empirical comparison of missing value imputation techniques on APS failure prediction. Int. J. Inf. Technol. Comput. Sci. **11**(2), 21–29. ISSN: 20749007, 20749015. https://doi.org/10.5815/ijitcs.2019.02.03
28. Mazumder, R.K., Salman, A.M., Li, Y.: Failure risk analysis of pipelines using data-driven machine learning algorithms. Struct. Saf. **89**, 102047 (2021)

An Integrated Preliminary Approach Elaboration for the Analysis of a Blended Wing Body Aerostructure Concept

Mohamed Hakim(✉) and Saad Choukri

Mechanical Engineering and Energetics Research Team: Modeling and Experimentation, Mohammadia School of Engineers, Mohammed V University, Agdal, Rabat 10090, Morocco
mohamed_hakim@um5.ac.ma, choukri@emi.ac.ma

Abstract. Growing interest in UAVs applications pushed the researchers to study new conceptual designs that should meet specific requirements and aerodynamic performances. In this context, the aim of our present paper is the elaboration of a preliminary approach for a blended wing body (BWB) concept analysis and validation. For this goal, reference is made to the general aircraft design cycle and then focused on the main first phases; mainly the conceptual and the preliminary designs and their associated requirements. In this proposed approach, design tasks and tools are integrated. Our interest was then put on static longitudinal stability as an advanced requirement involved in preliminary analysis. We showed that this advanced analysis can be carried out early in the preliminary phase. Using XFLR5 software, which is a medium fidelity numerical tool for aerodynamic plane analysis, a parametric study is conducted on planform geometry and wing twist. The findings are BWB aerostructure that fit the cruise flight requirements.

Keywords: Aeronautics · UAV · Design cycle · Blended wing body · Numerical testing

1 Introduction

The use of UAVs (Unmanned Aerial Vehicle) for civil and tactical applications is gaining interest with the emergence of new aeronautical project developers [1]. We already distinguish two categories of UAVs. The first category is that of fixed-wing UAVs, where applications cover fields such as observation and surveillance, mapping, precision agriculture, and all missions requiring to cover long distances. We qualify this category; of mini UAV, where the altitude ceiling is around 5000 m with an autonomy beyond 1 h. In addition to this, there are the medium and high altitude UAVs reserved mainly for military or scientific applications. The second category is that of rotary-wing UAVs, such as multi-rotors,

Mohamed Hakim and Saad Choukri contributed equally to this work.

L. Azrar et al. (Eds.): CIP 2022, LNME, pp. 331–340, 2023.
https://doi.org/10.1007/978-3-031-23615-0_34

and is dedicated to the inspection of engineering or industrial structures and the taking of aerial photos or videos, where the range is relatively limited. This classification by type of missions leads us to adopt, during the development of mini UAVs, the design cycle of conventional aircraft with, certainly, different constraints in terms of regulations and safety, but a greater freedom of choice of new concepts; Fig. 1. For example, the absence of the pilot and passengers does not require a cockpit, nor a tubular fuselage and reduces the space management to the payload, propulsion and control modules. But, contrary to the conventional construction of light civil aircraft, freighters or airliners, where the references of comparisons exist, the "Baselines", and whose performances are known [2], the mini UAVs are characterized by a variety of quasi unique missions. In this case, the engineering of new concepts requires intensive studies and analyses, necessitating a multitude of tests and iterations. Our approach is to adopt the design cycle methodology and its deliverables, and to integrate the numerical tools adapted to the preliminary analysis. It is in this sense that we undertook this work where we based our preliminary design on the XFLR5 software, and analyzed the stability in level flight (cruise flight) for different configurations of an aerostructure with integrated fuselage.

Fig. 1. Various unmanned aerial vehicles

2 Basis for a Preliminary Approach

The preliminary approach consists, in general, to validate the concept and to establish the configuration of the aerostructure responding the various requirements, such as; the mission, the dynamic and aeroelastic stability, the strength, etc.

The elaborated methodology is based on the aircraft design cycle and the different associated tasks, as presented in [2], where we integrate, Table 1, known numerical methods and adapted to each phase of the cycle [3]. For an aircraft of conventional type and mission, where many requirements are generally known or estimated from a similar reference and baselines, the validation phase of concept only concerns the application of DATCOM (Data Compendium) procedures, based on analytical methods (Lifting Line Theory; LLT) [4], for the estimation of weight or drag among others. This is not the case for UAVs where the mission requirements are unique and without aeroelastic or mass distribution similarities, for example. In this case, it is necessary to iterate on the shapes, aerodynamic profiles, mass distribution... This is not easy with the DATCOM tools, even digitized, and justifies our choice to introduce numerical tests and analysis of candidate configurations very early in the cycle. The numerical methods adopted in these simulations are mainly based on the equations of potential flows (Laplace, Glauert-Prandtl for compressibility effect) and discretized by $2D$ (VLM) or $3D$ panels [5–7]. They allow at least to capture the geometrical and functional complexity of the aircraft with a low computational cost, but with limitations inherent to the modeling assumptions. These methods cover a good part of the preliminary design with a transition to high fidelity methods, based on Navier-Stokes/finite volumes, generally used in the phase of detailed design or optimization, but require large amount of modeling efforts and high computational costs. It should be noted that this cycle also concerns structural design and its numerical tools, from beam theory to non-linear finite elements, but they are not concerned in our present application.

Table 1. Major design phases and methodologies in UAV development

		Phase 1 Conceptual design	Phase II Preliminary design	PHASE III Detail design
	Known	Basic mission requirements Range Altitude Speed Basic material properties	Aeroelastic requirements Fatigue requirements Flutter requirements Overall strength requirements	Local strength requirements Producibility Functional requirements
Numerical testes	Candidate configuration	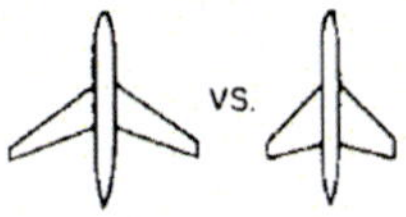	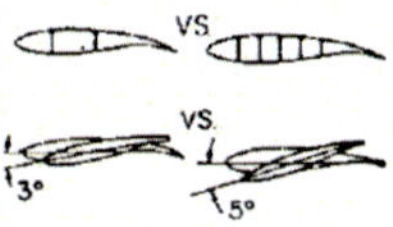	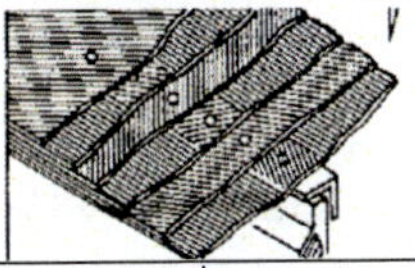
Numerical testes	Analysis	DATCOM, Lifting line, VLM 2D, 3D panels	DLM, RANS	URANS, FEM
	Results	Geometry and design objectives : Airfoil type, AR, λ Drag level, Weight goals, Coast goals	Basic internal arrangement Complete external configuration Local flow problems solver	Detail design Joints, fitting, attachments Design refinements as results of test and operation

3 Application to the Blended Wing Body

3.1 Definition

As shown in Fig. 1, the fuselage is integrated in the wing for the Parrot, AgEagle and ATyges UAVs, this configuration called the Blended Wing Body [8–10]. This concept was developed to encounter various economic and environmental constraints. In fact, it offers a reduction in structural weight, drag with a higher lift-to-drag ratio and therefore a decrease in fuel consumption compared to the existing generation. Also, a notable reduction in greenhouse gas emissions and noise as well as an increase in carrying capacity. On the other way, there're also some disadvantages for the BWB, one of them is the lack of the traditional horizontal tail and vertical fin which lead to a difficulty in longitudinal control and stability [11,12] subsection. The longitudinal stability is discussed in Sect. 3.2.

In this study, we are interested in the mini UAV which Davis and McMaster had classified them in their works [13–15]. We can see clearly from their classifications that mini UAVs had mass (aerostructure + payload) values of 10–100 kg, flight speed of 10–100 m/s, wingspan < 10 m, with flight altitudes up to 5000 m and Reynolds number around 10^5. The latter has been studied in previous works [15,16], which showed that the performance of most conventional airfoils decreases significantly at this critical Reynolds number range. Thus, the determination of an aerodynamic airfoil that circumvents this constraint is essential, which is the subject of our study [17], where from thirty-two airfoils existing in the literature we were able to select five. Among these, the NACA63(3) - 018 airfoil was chosen for the comparison between the configurations developed in the following. Those configurations will have a maximum speed of $V_{max} < 35$ m/s, a dimension of $3 * 2$ m^2 and a maximum weight of 15 kg evaluated by a study on the structure.

3.2 Longitudinal Stability and Cruise Flight

The longitudinal stability is the quality that stabilizes an aircraft about its lateral axis. It involves the pitching motion as the nose of the aircraft rises and falls in flight. In analyzing stability, it is important to remember that a body that is free to rotate will always rotate about its center of gravity. To achieve the static longitudinal stability, the wing moment must be such that, if the airplane is suddenly pitched up, the wing moment will change so that will provide unbalanced but restorative moment which, in turn, will bring the nose down. Likewise, if the aircraft nose drops, the resulting change in moment will bring the nose up Fig. 2. Equation 1 that describes the static longitudinal stability of wing only is defined as:

$$C_{m_{cg}} = C_{m_{ac}} + C_{L_0}\left(\frac{x_{cg}}{\bar{c}} - \frac{x_{ac}}{\bar{c}}\right) + C_{L_\alpha}\left(\frac{x_{cg}}{\bar{c}} - \frac{x_{ac}}{\bar{c}}\right)\alpha \qquad (1)$$

So for the BWB to be statically stable, Fig. 2 give us the idea that the aerodynamic center must be aft of the center of gravity and the restorative moment must counteract the disturbance which leads to $C_{m_\alpha} = C_{L_\alpha}\left(\frac{x_{cg}}{\bar{c}} - \frac{x_{ac}}{\bar{c}}\right) < 0$ and $C_{m_0} = C_{m_{ac}} + C_{L_0}\left(\frac{x_{cg}}{\bar{c}} - \frac{x_{ac}}{\bar{c}}\right) \geq 0$. Where $C_{m_{cg}}$, $C_{m_{ac}}$,C_{L_0}, C_{L_α}, x_{cg}, x_{ac}, α and $\bar{c}$ are respectively the pitching moment coefficient at center of gravity, the pitching moment coefficient at aerodynamic center, lift coefficient at zero angle of attack, the lift curve slope, position of gravity center, position of the aerodynamic center (neutral point), angle of attack and the mean chord.

3.3 Numerical Testing Tool: XFLR5

XFLR5 [18] is an open-source software that couples the $2D$ airfoil analysis of XFOIL [19] with a $3D$ solver to simulate the performance of an aircraft configuration at any time during the design process. It includes:

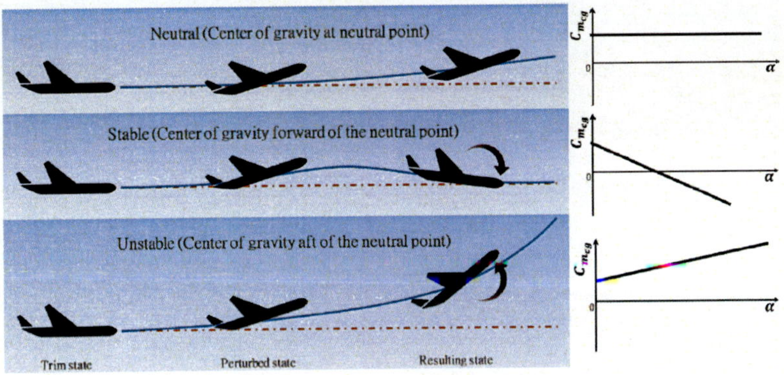

Fig. 2. Pitching moment coefficient $C_{m_{cg}}$ versus angle of attack α for different center of gravity positions

- XFOIL which is an interactive program for the design and analysis of subsonic isolated airfoils. The XFOIL code is a combination of a panel method and an integral boundary layer formulation for analyzing the potential flow around airfoils. The code was developed to quickly predict airfoil performance at low Reynolds numbers and its convergence is achieved by iterating between outer and inner flow solutions on the boundary layer displacement over the thickness of the boundary layer displacement. Thus, the XFOIL code calculates the viscous pressure distribution and captures the influence of limited trailing edge separation and laminar separation bubbles. It had shown an excellent result comparing to other tools [20,21].
- Wing design and related analysis based on tree numerical methods, Fig. 3, the non linear Lifting Line Theory (LLT), the Vortex Lattice Method (VLM), and a $3D$ Panel Method.

The XFLR5 software is intensively used [22,23], and thus has been verified and validated within the potential methods validity. In our case of flying wing where the wing is everything, the software is suitable for our intended use to get trends, and to understand sensitivity to design parameters. It is also noted that increasing panel density may overcomes some limitations of constant strength singularities giving reliable results.

3.4 Design and Solution Validation Process

Many requirements are involved in the preliminary design. These are fulfilled by evaluating corresponding performance. Some of the important requirements are of flight stability concern. This may be the case of the basics as take off, landing, level flight, turn manoeuvre... Every one of these should follow a requirement fitting process as described in Fig. 4. In this study, our primary requirement, is cruise flight or trimmed level flight.

Three candidate configurations, Fig. 5, had been developed and discussed in terms of performances and stability, Fig. 6, to select the best configuration for

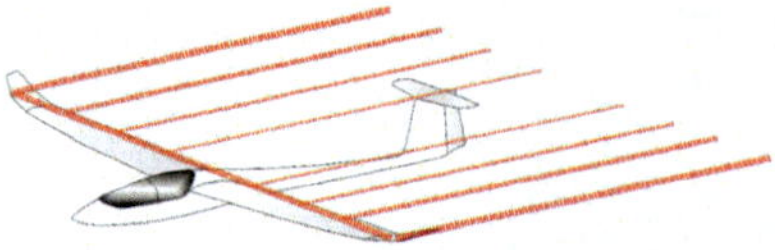

(a) Lifting line horshoe

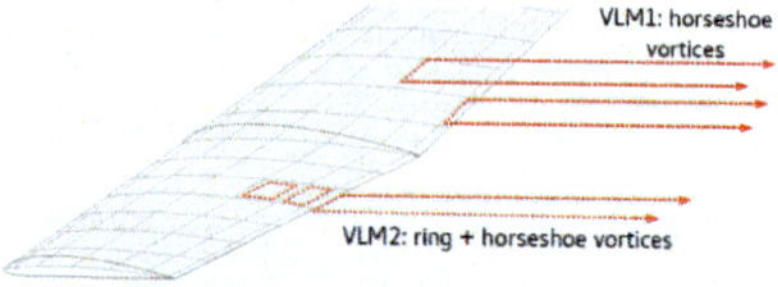

(b) VLM panels

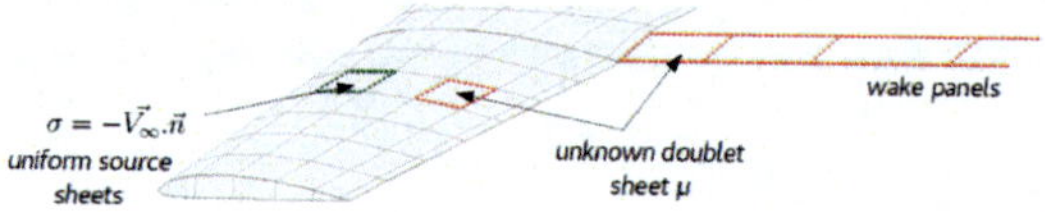

(c) 3D panel

Fig. 3. Numerical methods in XFLR5

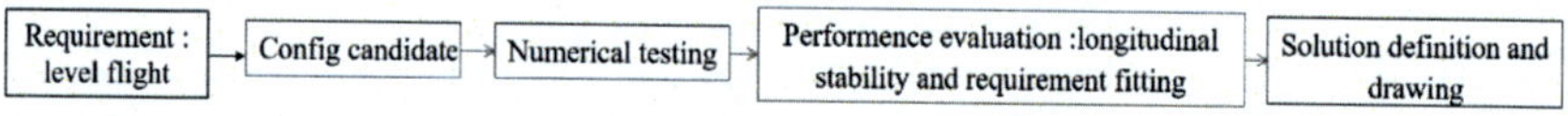

Fig. 4. Requirement fitting process

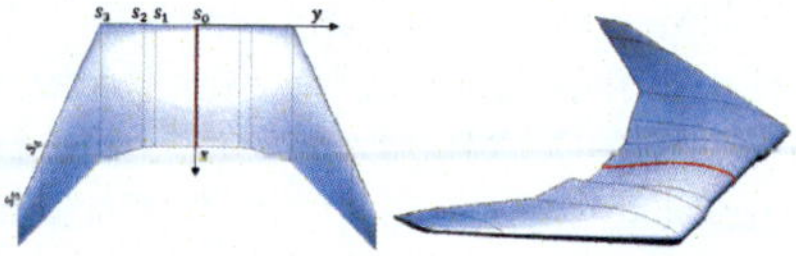

(a) Configuration 1 ($-4° <$ twist $< 10°$)

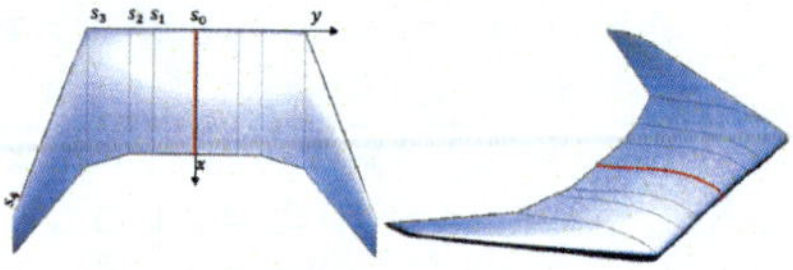

(b) Configuration 2 ($0° <$ twist $< 4°$)

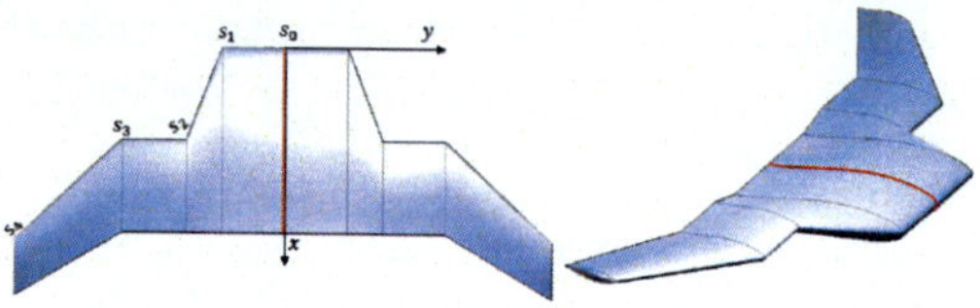

(c) Configuration 3 ($0° <$ twist $< 4°$)

Fig. 5. The candidate configurations

our requirements. The main geometric parameters of those configurations are listed in Table 2.

The first (initial) configuration for the mini UAV BWB, Fig. 5a, was developed taking into account the requirements mentioned in Sect. 3.1. The surface of the wing and its geometry are determined in order to produce a sufficient lift in cruising flight (at zero angle of attack), to have a sufficient volume for storing all the payload and to make the aerostructure stable. It's noticed on Fig. 6a and b that the aerostructure respects the requirement of mission speed for the level flight and that the maximum lift to drag ratio is at 4°. Meanwhile, Fig. 6c shows

Table 2. Geometric parameters of all configurations

	Section	Chord (m)	y position (m)	x position (m)	x_{np}
Configuration 1	S_0	1 m	0	0	0.5 m
	S_1	1	0.35	0	
	S_2	1	0.45	0	
	S_3	1.1	0.8	0	
	S_4	0.528	1.3	1.072	
	S_5	0.3	1.5	1.5	
Configuration 2	S_0	1	0	0	0.445
	S_1	1	0.35	0	
	S_2	1	0.55	0	
	S_3	1.1	0.9	0	
	S_4	0.3	1.5	1.5	
Configuration 3	S_0	1	0	0	0.568
	S_1	1	0.35	0	
	S_2	1	0.55	0.5	
	S_3	1.1	0.9	0.5	
	S_4	0.3	1.5	1.041	

that the aerostructure verifies the criteria of the longitudinal stability (Sect. 3.2). We note that $x_{cg} = 0.258\,\text{m}$.

In order to have more freedom for maneuvering in flight, the aerostructure twist had to be reduced to the minimum possible while increasing the lift to drag ratio in level flight. For this purpose, the configuration 2, Fig. 5b, has been developed in order to take into account the above modifications. The comparison between the polars of the configuration 2 and the initial one shows that the speed for the cruising flight has increased due to the decrease of the twist, unlike the lift to drag ratio which has been improved, $|C_{m_\alpha}|$ has decreased and $x_{cg} = 0.335\,\text{m}$ has been shifted backwards whereas x_{np} has been pushed forward. So according to these results, we have succeeded to increase the lift to drag ratio for the cruise flight but at the same time speed and stability were affected.

To overcome the disadvantages of the configuration 2, a third one is set up, Fig. 5c, which is based on the achievement of a compromise between the highest lift/drag ratio for the cruise flight and the most aft position that is possible for the neutral point. We can see clearly that the performance is better at the level flight despite the increase in speed and that the stability has been improved with a more static margin ($x_{cg} = 0.430\,\text{m}$ and $x_{np} = 0.568\,\text{m}$).

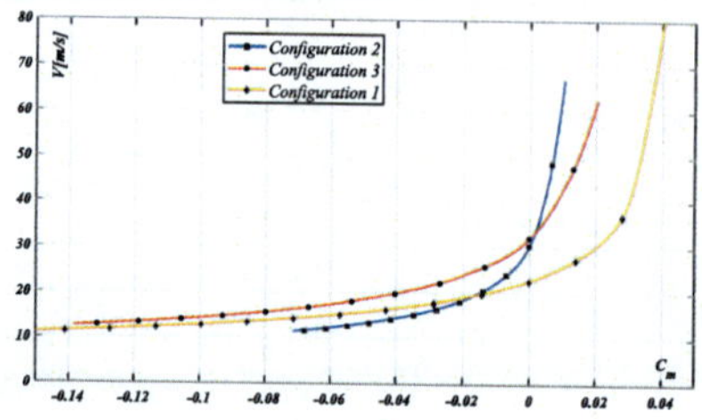

(a) Speed versus pitching moment coefficient

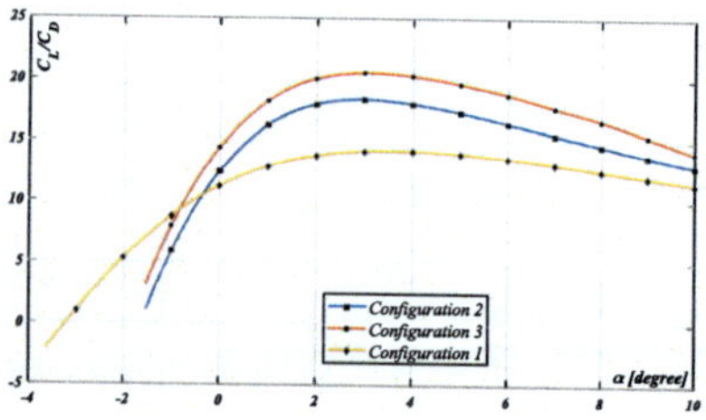

(b) Lift to drag ratio versus angle of attack

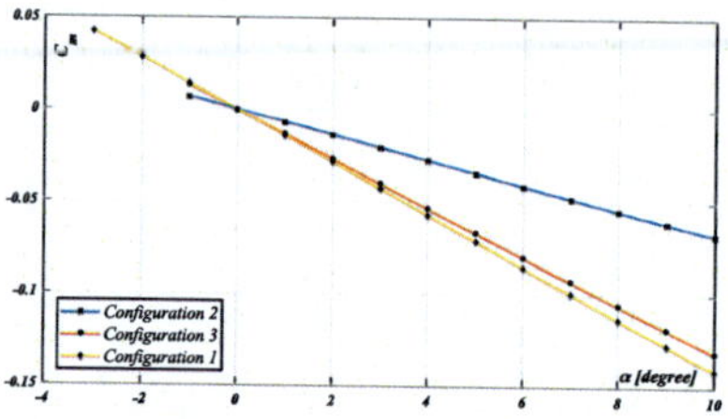

(c) Pitching moment coefficient versus angle of attack

Fig. 6. Polars of the candidate configurations

4 Conclusion

In this work, we were interested in the aerodynamic design study of a mini UAV BWB. An initial design was developed to satisfy the requirements and then improved its aerodynamics and longitudinal stability based on a numerical testing. Graphs and tables are used to show the progression of the configurations that have been developed based on wing sections positions (sweep) and twist variations. The proposed approach has proved to be useful in design tasks, definitions and scheduling, and in performance evaluation efficiency. The resulting configuration is not an optimized design by any means but may constitute a good starting design for more advanced optimization of the external layout using a multidisciplinary optimization approach, and calculation of stability derivatives could be very useful in refining the stability and control characteristics of the UAV.

References

1. https://www.aeroexpo.online/fr/fabricant-aeronautique/drone-voilure-fixe-520.html. Accessed 29 Jan 2022
2. Sacher, P., et al.: Special Course on Engineering Methods in Aerodynamic Analysis and Design of Aircraft. Technical report, Advisory Group for Aerospace Research & Development (AGARD) (1991)
3. Dussart, G., et al.: Flight dynamic modelling and simulation of large flexible aircraft. Flight Physics-Models, Techniques and Technologies, pp. 49–72 (2018)
4. Williams, J.E., Vukelich, S.R.: The USAF stability and control digital DATCOM, vol. II. Implementation of DATCOM Methods. Technical report, Mcdonnell Douglas Astronautics Co St Louis Mo (1979)
5. Hess, J.L., Smith, A.M.O.: Calculation of nonlifting potential flow about arbitrary three-dimensional bodies. J. Ship Res. **8**(04), 22–44 (1964)
6. Katz, J., Plotkin, A.: Low-Speed Aerodynamics. Cambridge University Press (2001). https://doi.org/10.1017/CBO9780511810329
7. Lewis, R.I.: Vortex Element Methods for Fluid Dynamic Analysis of Engineering Systems. Cambridge UP, New York, NY, USA (1991)
8. Liebeck, R.H.: Design of the blended wing body subsonic transport. J. Aircraft **41**(1), 10–25 (2004)
9. Lyu, Z., Martins, J.R.: Aerodynamic design optimization studies of a blended-wing-body aircraft. J. Aircraft **51**(5), 1604–1617 (2014)
10. Wang, K., Zhou, Z.: Aerodynamic design, analysis and validation of a small blended-wing-body unmanned aerial vehicle. Aerospace **9**(1), 36 (2022)
11. Nelson, R.C., et al.: Flight Stability and Automatic Control, 2nd edn. WCB/McGraw Hill, New York (1998)
12. Caughey, D.A.: Introduction to Aircraft Stability and Control Course Notes for M&AE 5070. Sibley School of Mechanical & Aerospace Engineering Cornell University (2011)
13. Lissaman, P.B.S.: Low-Reynolds-number airfoils. Annu. Rev. Fluid Mech. **15**(1), 223–239 (1983)
14. McMasters, J.: An analytic survey of low-speed flying devices-natural and man-made. In: 2nd International Symposium on the Technology and Science of Low Speed and Motorless Flight, p. 1019 (1975)
15. Mcmasters, J., Henderson, M.: Low speed single element airfoil synthesis. Tech. Soar. **6** (1980)
16. Mueller, T.J., Jansen, J.: Aerodynamic measurements at low Reynolds numbers (1982). https://doi.org/10.2514/6.1982-598
17. Hakim, M., Choukri, S.: Détermination de profils aérodynamiques à moyen Reynolds pour une aérostructure à fuselage intégré. 15e Congrès de Mécanique (2022)
18. Deperrois, A.: Guide pour XFLR5. http://www.xflr5.tech/xflr5.htm. Accessed 29 Jan 2022
19. Drela, M.: XFOIL: an analysis and design system for low Reynolds number airfoils. In: Low Reynolds Number Aerodynamics, pp. 1–12 (1989)
20. Günel, O., et al.: CFD vs. XFOIL of airfoil analysis at low Reynolds numbers. In: 2016 IEEE International Conference on Renewable Energy Research and Applications (ICRERA), pp. 628–632 (2016). IEEE
21. Morgado, J., et al.: XFOIL vs CFD performance predictions for high lift low Reynolds number airfoils. Aerosp. Sci. Technol. **52**, 207–214 (2016)

22. Selig, M.S.: Summary of Low-Speed Airfoil Data, vol. 2. SoarTech Publications (1996)
23. Hamada, A., et al.: Design, build and fly a flying wing. Athens J. Technol. Eng. **5**(3), 223–250 (2018)

Integrated Design

Towards a Data-Driven Smart Assembly Design: State-of-the-Art

Amal Allagui[1,2](✉), Imen Belhadj[1], Régis Plateaux[2], Moncef Hammadi[2], Olivia Penas[2], Nizar Aifaoui[1], and Jean-Yves Choley[2]

[1] LGM, ENIM, University of Monastir, 05 rue Ibn Eljazzar, 5019 Monastir, Tunisia
allaguiamal12@gmail.com

[2] Laboratoire Quartz EA7393, ISAE-Supmeca, 3 rue Fernand Hainaut, 93400 Saint-Ouen, France
{regis.plateaux,moncef.hammadi,olivia.penas, jean-yves.choley}@isae-supmeca.fr

Abstract. With the increase of customer requirements and customized products needs, mechanical systems are becoming more sophisticated and equipped with intelligent parts. With this level of intelligence the mechanical systems become more and more complex and also their assembly processes becomes tedious. Therefore, the Industry 4.0 promotes efficient and smart assembly design to support commissioning automation and consequently customer satisfaction. This paper de-scribes the different challenges of assembly design, before presenting the advances on data analysis and Industry 4.0 technologies that can support intelligent assembly design. After analyzing these approaches, an innovative framework is proposed to provide the designer with a tool to address the different challenges of assemblies, based on the one hand on data from smart parts in operation (equipped with RFID sensors) transmitted by the Industrial Internet of Things and Cloud Computing, and on the other hand on the analysis of these massive data and the use of deep learning algorithms.

Keywords: Big data · Deep learning · IIoT · Smart part/assembly design

1 Introduction

The industrial environment is currently undergoing its fourth revolution, named Industry 4.0 (I4.0) (Kagermann et al. 2013). It is used to implement smart services by including new information transmission methodologies through the Industrial Internet of Things (IIoT) for smart factory services (Hoffmeister et al. 2015). These methodologies are notably useful for automatic assembly processes, leading to the Assembly 4.0 (A4.0) concept explained by Cohen et al. (Cohen et al. 2019). By using I4.0 assembly line technologies (Artificial Intelligence (AI), IoT, Augmented Reality (AR), Collaborative Robots (Cobots)), and simulations launched by the Digital Twin (DT), A4.0 allows to better manage quality and maintenance of assembly systems, and thus to increase industrial productivity and profitability (Cohen et al. 2019). Indeed, in order to ensure

L. Azrar et al. (Eds.): CIP 2022, LNME, pp. 343–352, 2023.
https://doi.org/10.1007/978-3-031-23615-0_35

high productivity and flexibility, robots can handle a fair share of the work in A4.0, while human workers keep their supervisory role and the tasks for which robots cannot be "trained" (Khalid et al. 2016). In this context, a preliminary state-of-the-art is required regarding how these different I4.0/A4.0 technologies can fit the challenges of assemblies smart design. As a result, this paper first presents the challenges of CAD part assembly, before focusing on recent advances in data analytics and I4.0 technologies to assess their relevance to the previous challenges. Finally, the innovative framework intended to tackle them is presented and aims to support the design of smart assemblies based on data collected from working smart parts. The last section concludes and presents future work.

2 CAD Assembly Challenges

Over the past decades, a lot of research has been conducted on CAD assemblies with two goals in mind: to facilitate the assembly design steps and to enrich the design platforms with new assembly data.

Considering the first challenge, as the global manufacturing industry shifts toward smart manufacturing, product assembly technology also needs to upgrade from manual/experienced assembly and automated assembly to smart assembly (also called intelligent assembly) (Yi et al. 2021). Figure 1 shows the development of assembly technology from manual to smart assembly, in parallel with the evolution from the 2D paper representations to the 3D digitalization.

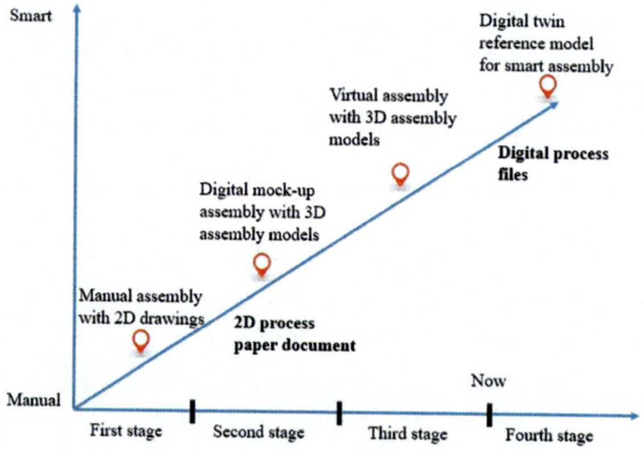

Fig. 1. Product assembly technology development.

In the product life cycle, the disassembly sequence planning (DSP) has three main application phases: Design for Disassembly DfD for the design phase, repair or maintenance procedure for the operational phase and end-of-life EOL disassembly for the disposal phase (Ong et al. 2021). To this respect, Smith and Hung emphasize the importance of DSP in supporting green product design, in addition to being easier and cost-efficient to be disassembled (Smith and Hung 2015).

Among the many existing approaches to improve the assembly process, Zhang et al. proposed an approach based on probabilistic reasoning through a Bayesian Network (BN), which encodes the assembly constraints between parts and then suggests relevant parts. It offers efficient and intelligent assemblies, to show the design efficiency of CAD assemblies, using data-driven method in modeling and assembly treebased method to extract constraints between assembled parts (Zhang et al. 2015). In parallel some authors developed a system that generates assembly sequences based on polychromatic sets (Li et al. 2012; Yu et al. 2013). More recently, using a geometric database, some authors have proposed a new tool to generate disassembly plans (DPs) and quality estimation, based on sub-assemblies plans SUB-DPs and calculation of their quality index (QI), to help users to choose the better SUB-DPs having the highest (QI) (Kheder et al. 2016; Belhadj et al. 2020). Also, Belhadj et al. proposed a tool for parallel DP generation, with the automated provision of the total disassembly time and design efficiency, in order to evaluate the design proposal (Belhadj et al. 2021). Moreover, Li et al. developed a system that generates assembly sequences based on polychromatic sets (Li et al. 2012). Focusing on the second challenge (knowledge enrichment for assembly design), the CAD platforms have been enhanced with geometrical adjustment constraint type that aims to specify the fit type between two parts in order to ensure accuracy in manufacturing process (Allagui et al. 2021). In the same idea, Kloock-Schreiber et al. developed an approach to integrate new information such as part maintenance intervals, part costs, or even index numbers to manage the installation, into CAD model. Authors create an Excel interface by the implementation of a parametric Product Service Systems (PSS) data model based on a CAD application (Kloock-Schreiber et al. 2019). Additionally, based on interference indications and contact points between the assembly components, CAD systems have been enriched with software interfaces to ensure simplicity of CAD-CAE integration (Foucault and Léon 2010). Similarly, STEP standard and its instances have been translated to the Ontology Web Language (OWL) to give a new semantically rich model named Onto-STEP, which can be easily integrated to OWL ontologies. OWL enable to represent knowledge in a computer interpretable form, and then offer in this way to easily combine geometrical and non-geometrical data (Barbau et al. 2012).

With the help of computer tools, work on CAD software has become more and more automated and unsupervised. Yet, the breakthrough of new Industry 4.0 technologies and associated data analytics techniques appear to offer new opportunities that could support smart assembly design, as outlined in the following section.

3 Overview of Data Analytics Advances and I4.0 Technologies to Enhanced Assembly Design

To meet the many challenges of planning assembly/disassembly sequences, many approaches have been developed. However, the quality of the results of these approaches depends essentially on the data extracted from the CAD model. As a result, some authors have proposed a CAD-integrated system to extract geometrical data from a CAD assembly model under exploitable forms (Excel, MATLAB, and XML formats) (Hadj et al. 2018; Gandouzi et al. 2021). Another approach has been pro-posed to allow the retrieval of similar CAD assemblies recovery in a CAD database. Based on an Enriched Assembly Model (EAM), the authors thus helped designers to reuse existing knowledge when creating new products (Lupinetti et al. 2018).

In parallel, with the rise of new information technologies (IT) and of the use of data (Gubbi et al. 2013), smartness in design is becoming a new collaborative challenge between manufacturers and designers to improve design decision-making by using smart techniques from diverse sectors. Due to the profound impacts of the Internet and Internet of Things (IoT), a huge volume of data is being generated from everywhere at an unprecedented rate (Alharthi et al. 2017). Based on this data, design exceeds designer experience to a data-driven design which exploits this data to generate new design or new help framework in design decision making. As a result, much research has focused on artificial intelligence techniques, such as Deep Learning (DL), Machine Learning (ML) and Artificial Neural Network (ANN), to exploit the data available during different phases of the life cycle, whether geometric or otherwise, provided that they can generate some knowledge about the assemblies.

A first interesting research area to address CAD assemblies challenges is the form recognition. To this respect, DL is in the intersection among the research areas of neural networks, artificial intelligence, graphical modeling, optimization, pattern recognition, and signal processing (Deng and Yu 2014). Using DL, Jonathan et al. proposed an approach for the development of decision-making systems based on data processing and object recognition in three heterogeneous formats (3D models, photos, and videos). Authors trained a Convolutional Neural Networks (CNN) to reduce the amount of data to a few indicators or possibilities that an expert can use to produce a higher level decision or operation (Dekhtiar et al. 2018). For the 3D model recognition, many neutral and open formats can be used such as STEP (ISO 10303 2020) and JT (ISO 14306 2017). Data amount is unlimited using Deep Neural Network (DNN); it offers higher performance and fastness in the data treatment process. The goal of learning algorithm is to find the optimal values for the weight vectors to solve a class of problems. Some of the well-known training algorithms are: Gradient Descent, Stochastic Gradient Descent, Momentum, Levenberg Marquardt algorithm, Backpropagation through time. The most used for the 3D models reconstructions are the Adversarial Networks (AN). It uses Tensorflow and Keras as a DL framework and CIFAR10 as a data set; it is one of the most widely used data sets for ML research (Shrestha and Mahmood 2019).

In parallel, a new design framework has been proposed, named Digital Twin Product Design (DTPD) (Tao et al. 2018). This approach is based on real data coming from the operational phase of the considered system: a smart part, defined by Hossain et al. as a component with embedded sensors, for example additive manufacturing parts that contain injected piezoelectric sensor for the detection of vibration and pressure variations (Hossain 2017), or in our case it could be an instrumented assembly. The concept here deals with the collection of real data from the product in use, to monitor the evolution of its performance, its wear, its failures, etc. Users' feedback is used to redesign the product with new performance. But the DTPD framework proposed by Tao et al. is only used for the redesign of existing products (Tao et al. 2018).

To support smart part design, Prasad et al. have proposed a smart part concept that supports the shorter development of new product families from only specifications and reusable smart part templates. It is based on the decomposition of the product into two levels: a system level and series of levels of semi-independent intelligent parts, while ensuring the "intra-parts" communication of the common parameters (constraints and requirements) between the different levels, at the local level for each smart part. Thus, this smart design allows to change the specification values from specifications of a product to different smart parts specifications (Prasad and Rogers 2005).

Finally, a last useful research area addresses the ability to customize products in an intelligent way. Indeed, to increase the turnover of industry and promote smart customization in terms of product designing, manufacturing, pricing, marketing, and service, Zhang et al. used big data analytics to support a decision-making tool for designing custom shoes. The data of the product in use (hours of use, walking habit, and foot landing pressure) are collected by sensors and then analyzed by simulation. Actually, customer feedback and needs are used to support the next generation of footwears (Zhang et al. 2019). Similarly, Zheng and Chen proposed a data-driven design approach, which presents an innovative methodology to develop a smart product "smart respiratory mask". The user is responsible for the choice of the prototype to be manufactured in the cyber world and for the future request for customization of the product in the physical world. This mask contains a readable QR code. The respiratory mask is always connected to the application via Bluetooth 4.0 for monitoring breathing condition. This work involves the user in the product co-design by exploiting real-time information for the improvement of the next products generation (Zheng et al. 2020). The paradigm presented in Fig. 2 is a challenge for data-driven design. Data facilitates the design process and requirements fulfillment, whether for the designer or in an data-driven automated design platform (Cantamessa et al. 2020).

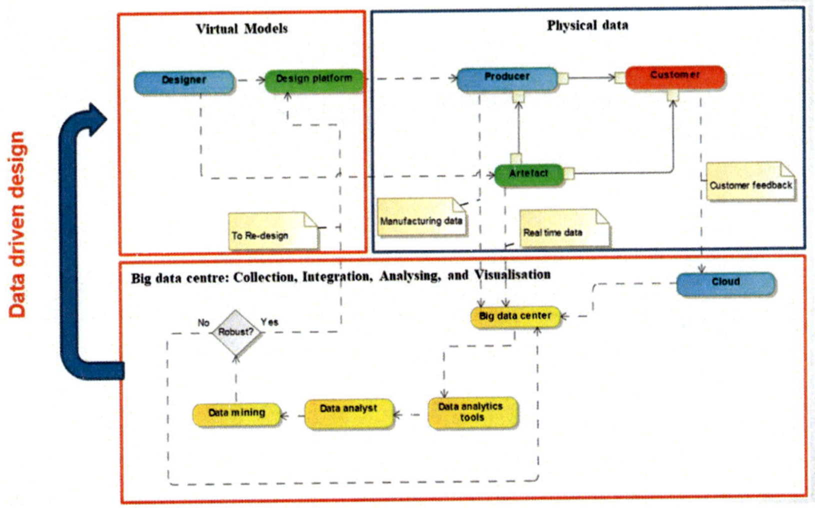

Fig. 2. A new paradigm considering data analysis in the design and manufacturing process.

Similarly, Wang et al. (2013) developed a new paradigm of data-driven smart customization through Digital Twin (DT), for an augmented elevator. The paradigm performs customization in three steps. First, the customization of noise control is driven by the customer DT who can predict the vibration pass-way and the airborne noises and then adjust the design in collaboration with the manufacturer. Second, customer-driven production plans will test machine parameters in the virtual world through a simulation model based on the generated data, in order to reduce the uncertainty caused by machine downtime. Third, IoT and data analytics are used for machine breakdown prediction. The energy efficiency service uses the machine operating context to identify peak-time flow and peak-season flow, and then automatically sets saving energy protocols using ML algorithms (Wang et al. 2021). As a result, designing a smart part, assembly or product needs the collaboration between both virtual and physical worlds, where the latter can include user participation, and notably designer inputs when considering assemblies challenges.

4 Discussion: Assembly Design Between Designer Experience and Data-Driven Based Design

In light of this review, the design process is undergoing a major transformation from designer experience to data-driven based design. Considering the part smartness defined in the literature, as the part ability to collect and monitor all types of data in an assembly. This part can be a sensor with a storage memory in which it will record all its historical and real-time data and exchange them with its design software for product amelioration. Table 1 presents a comparison of selected works in the literature. The comparison is based on tools relying on DL programming, technologies from I4.0 and the degree of design smartness.

Table 1. Comparison of researchers works

References	Programming level	I 4.0	Smart design
(Hoffmeister et al. 2015; Kagermann et al. 2013; Cohen et al. 2019)	+	+	+
(Smith and Hung 2015; Zhang et al. 2015; Ong et al. 2021)	+++	+++	+
(Li et al. 2012; Yu et al. 2013; Kheder et al. 2017; Belhadj et al. 2020; Belhadj et al. 2021)	+	+	+

(*continued*)

Table 1. (*continued*)

References	Programming level	I 4.0	Smart design
(Foucault and Léon 2010; Barbau et al. 2012; Kloock-Schreiber et al. 2019; Allagui et al. 2021)	++	+++	+
(Prasad and Rogers 2005; Gubbi et al. 2013; Deng and Yu 2014; Alharthi et al. 2017; Hossain 2017; Dekhtiar et al. 2018; Hadj et al. 2018; Lupinetti et al. 2018; Tao et al. 2018; Shrestha and Mahmood 2019; Zhang et al. 2019; Cantamessa et al. 2020; Zheng et al. 2020; Wang et al. 2021; Yi et al. 2021; Gandouzi et al. 2021)	+	+++	+++
Scoring definition	+	++	+++
Programming	Use of explicit programming	Use of mixed programming AI + DL	Use of Deep Learning and neural network
I4.0	Work done for traditional industry		Work done for I4.0
Smart design	Designer experience-driven design		Data-driven design

Although most of the work published in the last five years has focused on I4.0 technologies, DL-based intelligent design is attracting researchers. Yet, user customization is limited to trivial choices and so far, designers do not yet have yet a 3D platform that generates appropriate assemblies based on real data collected from working physical systems.

5 Intended Approach

This paper aims to evaluate two types of design approaches: traditional and smart, to propose a decision aid for assemblies designers. Using IT, real-time data, and I4.0

technologies, data can be analyzed to provide optimized design solutions. Similarly, CAD software databases save a lot of time when importing designed parts or assemblies. These databases are useful for proposing a data-driven design ready for implementation. According to the literature, a smart assembly contains smart parts that generate big data and transmit it to the analysis via IoT and Cloud computing technologies, in order to be analyzed later and generate new assembly solutions. This proposed approach will help designers in their decision making for the optimal choice of the assembly.

Figure 3 presents a framework that helps designer to make the best decision. First, RFID sensors monitor real-time data; these data will be transmitted through the IoT and analyzed via data analytics tools such as Tensorflow, Pandas, Numpy… After analyzing data and based on a predefined parts database, a trained neural network will construct some assemblies and display them to designer to give him the opportunity to choose the optimal assembly that satisfies most customer and functional requirements.

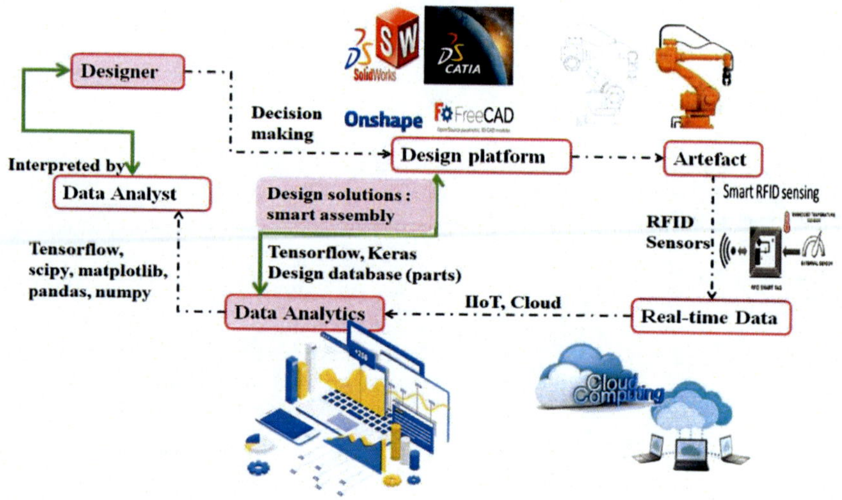

Fig. 3. Data-driven smart assembly design proposed framework

6 Conclusion and Future Work

This paper presents a review of the literature on the challenges of assembly design and the opportunities for finding solutions. It discusses the assembly design development, from traditional to data-driven smart design. Based on works comparison and outcomes, this paper finally presents an innovative data-driven smart assembly design approach to facilitate the designer's decision making. In future work, this approach will be implemented and validated, based on an existing CAD database of assembly parts and the ANN technique.

Acknowledgements. This work is carried out within the Partnership Hubert Curien "PHC Utique" of the French Ministry of Europe and Foreign Affairs, and the Tunisian Ministry of Higher Education and Scientific Research (CMCU 21G1112). The authors gratefully acknowledge their financial support.

References

Alharthi, A., Krotov, V., Bowman, M.: Addressing barriers to big data. Bus. Horiz. **60**(3), 285–292 (2017)

Allagui, A., Belhadj, I., Aifaoui, N., Hammadi, M., Choley, J.-Y.: A CAD-System engineering interoperability by enriching CAD database with functional information. In: 2021 18th International Multi-Conference on Systems, Signals & Devices (SSD). IEEE, pp. 1452–1458 (2021)

Barbau, R., et al.: OntoSTEP: Enriching product model data using ontologies. Comput-Aided Des. **44**(6), 575–590 (2012)

Belhadj, I., Hammadi, M., Aifaoui, N.: Parallel disassembly approach with recycling rate calculation of industrial products. Int. J. Adv. Manufact. Technol. **113**(9–10), 2969–2984 (2021). https://doi.org/10.1007/s00170-021-06830-z

Belhadj, I., Hammadi, M., Trigui, M., Aifaoui, N.: Generation of disassembly plans and quality assessment based on CAD data. Int. J. Comput. Integr. Manuf. **33**(12), 1300–1320 (2020)

Cantamessa, M., Montagna, F., Altavilla, S., Casagrande-Seretti, A.: Data-driven design: the new challenges of digitalization on product design and development. Des Sci **6** (2020)

Cohen, Y., Naseraldin, H., Chaudhuri, A., Pilati, F.: Assembly systems in Industry 4.0 era: a road map to understand assembly 4.0. Int. J. Adv. Manufact. Technol. **105**(9), 4037–4054 (2019). https://doi.org/10.1007/s00170-019-04203-1

Dekhtiar, J., Durupt, A., Bricogne, M., Eynard, B., Rowson, H., Kiritsis, D.: Deep learning for big data applications in CAD and PLM–research review, opportunities and case study. Comput. Ind. **100**, 227–243 (2018)

Deng, L., Yu, D.: Deep learning: methods and applications. Found Trends Signal Process. **7**(3–4), 197–387 (2014)

Foucault, G., Léon, J.-C.: Enriching assembly CAD models with functional and mechanical informations to ease CAE. In: International Design Engineering Technical Conferences and Computers and Information in Engineering Conference, pp. 341–351 (2010)

Gandouzi, G., Belhadj, I., Aifaoui, N., Hammadi, M., Choley, J.-Y.: A CADDEX tool for CAD and CAE integration. In: Conference Proceedings: 2021 IEEE International Systems Conference (SysCon). IEEE, Virtual Conference (2021)

Gubbi, J., Buyya, R., Marusic, S., Palaniswami, M.: Internet of Things (IoT): a vision, architectural elements, and future directions. Future Gener. Comput. Syst. **29**(7), 1645–1660 (2013)

Hadj, R.B., Belhadj, I., Gouta, C., Trigui, M., Aifaoui, N., Hammadi, M.: An interoperability process between CAD system and CAE applications based on CAD data. Int. J. Interact. Des. Manuf. IJIDeM **12**(3), 1039–1058 (2018)

Hoffmeister, M., Festo, A.G., KG, C.: The Industrie 4.0 Component. ZVEIGer Electr Electron Manuf Assoc Frankf Am Main (2015)

Hossain, M.S.: Process development and characterization of smart parts fabricated using powder bed fusion additive manufacturing technologies. The University of Texas at El Paso (2017)

ISO 10303: Industrial automation systems and integration—product data representation and exchange—part 242: application protocol: managed model-based 3D engineering (2020)

ISO 14306: Industrial automation systems and integration—JT file format specification for 3D visualization (2017)

Kagermann, H., Helbig, J., Hellinger, A., Wahlster, W.: Recommendations for implementing the strategic initiative INDUSTRIE 4.0: securing the future of German manufacturing industry; final report of the Industrie 4.0 working group. Forschungsunion (2013)

Khalid, A., Kirisci, P., Ghrairi, Z., Thoben, K.-D., Pannek, J.: A methodology to develop collaborative robotic cyber physical systems for production environments. Logist Res. **9**(1), 1–15 (2016)

Kheder, M., Trigui, M., Aifaoui, N.: Optimization of disassembly sequence planning for preventive maintenance. Int. J. Adv. Manufact. Technol. **90**(5–8), 1337–1349 (2016). https://doi.org/10.1007/s00170-016-9434-2

Kloock-Schreiber, D., Domarkas, L., Gembarski, P.C., Lachmayer, R.: Enrichment of geometric cad models for service configuration. In: CEUR Workshop Proceedings, vol. 2467, pp 22–29. RWTH, Aachen (2019)

Li, Y.L., Gao, J.M., Shi, L., Wang, S.: Assembly modeling for 3d component based on polychromatic sets. Adv. Mater. Res. Trans Tech. Publ. 388–392 (2012)

Lupinetti, K., Giannini, F., Monti, M., Pernot, J.-P.: Multi-criteria retrieval of CAD assembly models. J. Comput. Des. Eng. **5**(1), 41–53 (2018)

Ong, S.K., Chang, M.M.L., Nee, A.Y.C.: Product disassembly sequence planning: state-of-the-art, challenges, opportunities and future directions. Int. J. Prod. Res. **59**(11), 3493–3508 (2021). https://doi.org/10.1080/00207543.2020.1868598

Prasad, B., Rogers, J.: A knowledge-based system engineering process for obtaining engineering design solutions. In: International Design Engineering Technical Conferences and Computers and Information in Engineering Conference, pp. 477–488 (2005)

Shrestha, A., Mahmood, A.: Review of deep learning algorithms and architectures. IEEE Access **7**, 53040–53065 (2019)

Smith, S., Hung, P.-Y.: A novel selective parallel disassembly planning method for green design. J. Eng. Des. **26**(10–12), 283–301 (2015)

Tao, F., Cheng, J., Qi, Q., Zhang, M., Zhang, H., Sui, F.: Digital twin-driven product design, manufacturing and service with big data. Int. J. Adv. Manuf. Technol. **94**(9–12), 3563–3576 (2018). https://doi.org/10.1007/s00170-017-0233-1

Wang, L., Liu, K.: Design and implementation of a data acquisition platform for automated mechanical transmission (AMT). In: Volume 1: 15th International Conference on Advanced Vehicle Technologies; 10th International Conference on Design Education; 7th International Conference on Micro- and Nanosystems. American Society of Mechanical Engineers, Portland, Oregon, USA, p. V001T01A012 (2013)

Wang, X., Wang, Y., Tao, F., Liu, A.: New paradigm of data-driven smart customisation through digital twin. J. Manuf. Syst. **58**, 270–280 (2021)

Yi, Y., Yan, Y., Liu, X., Ni, Z., Feng, J., Liu, J.: Digital twin-based smart assembly process design and application framework for complex products and its case study. J. Manuf. Syst. **58**, 94–107 (2021)

Yu, J., Da Xu, L., Bi, Z., Wang, C.: Extended interference matrices for exploded view of assembly planning. IEEE Trans. Autom. Sci. Eng. **11**(1), 279–286 (2013)

Zhang, C., Chen, D., Tao, F., Liu, A.: Data driven smart customization. Procedia CIRP **81**, 564–569 (2019)

Zhang, K.-K., Hu, K.-M., Yin, L.-C., Yan, D.-M., Wang, B.: CAD parts-based assembly modeling by probabilistic reasoning. In: 2015 14th International Conference on Computer-Aided Design and Computer Graphics (CAD/Graphics), pp. 89–96. IEEE (2015)

Zheng, P., Xu, X., Chen, C.-H.: A data-driven cyber-physical approach for personalised smart, connected product co-development in a cloud-based environment. J. Intell. Manuf. **31**(1), 3–18 (2020)

Numerical Simulation of Weld Thermal Efficiency GTAW Process

Mouad Bensada[1(✉)], Abdellah Laazizi[1], Kaoutar Fri[1], and Jamal Fajoui[2]

[1] SECNDCM-L2MC, ENSAM-Moulay Ismail University, Meknes, Morocco
{mouad.bensada,plek.fri,k.fri}@edu.umi.ac.ma,
a.laazizi@ensam.umi.ac.ma

[2] GeM, IUT-Saint Nazaire 44600, Saint-Nazaire, France
Jamal.Fajoui@univ-nantes.fr

Abstract. Predicting temperature field in a Gas Tungsten Arc Welding process is essential and precondition for any study of thermal or mechanical behavior of the weld. In fact, thermal, material proprieties and physical phenomena have to be taken into account in order to predict welding morphology, residual stresses and the final product quality of the weld. In the case of analytical studies, simulations are done with many simplifications that affect result precision. In this study, a 2D numerical model is developed by taking into account heat transfer and distribution of heat source. Gaussian distribution for heat source is presented and simulated by using Finite Difference Method (FDM) with explicit scheme. This method permits following the evolution of temperature field throughout the studied material during welding process. Also, operating parameters effects on melted zone such as current welding and shielding gas have been discussed.

Keywords: GTAW process · Numerical simulation · Welding · Shielding gas · Explicit scheme

1 Introduction

The use of welding as joining process replacing riveting permits to minimize the weight of structure either in the automotive or the aeronautic industries [1], it contributes also in reducing the production cost. Gas Tungsten Arc Welding (GTAW) is widely used in many industrial applications, it is characterized by high welding quality, compared to other arc welding processes [2]. Therefore, predicting temperature field and controlling defect creation are important for any preventive study of the GTAW process parameters [3]. In addition, temperature gradients are considered one of the main parameters that affects the microstructure of the welded materials then their final mechanical proprieties. As a result, thermal proprieties and physical phenomena have to bc taken into consideration in order to predict welding morphology, residual stresses and the final product quality of the welds [4].

Last decade, it has been challenging to simulate the welding behaviors in whole piece because of software requirements and long computational time [5].

L. Azrar et al. (Eds.): CIP 2022, LNME, pp. 353–362, 2023.
https://doi.org/10.1007/978-3-031-23615-0_36

The aim of the current research is to develop a 2D model of GTAW process by including heat source distribution and the heat transfer. Finite difference method is used to solve numerical model. The effect of welding operating parameters such as welding current and shielding gas are discussed.

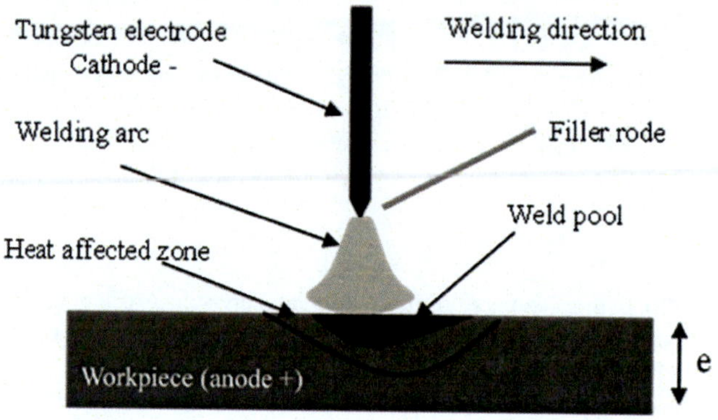

Fig. 1. GTAW process presentation

Nomenclature

C_p	Specific heat (J/Kg.K)	T(t, x, z)	Temperature (°C)
e	Thickness of workpiece(mm)	T_o	Initial temperature, 25 (°C)
h_{conv}	Heat transfer coefficient	t	Time (s)
I	welding current (A)	v	Welding source velocity (m/s)
k	Thermal conductivity (W/m.K)	V	Welding voltage (V)
L	Length of workpiece	x, z	Cartesian coordinats (m)
UTS	Ultimate tensile strength(MPa)	σ	Standard deviation
Q_m	Welding input source (W/m^2)	η	Welding process efficiency
r	radial distance of the Gaussian function (m)	ρ	Density (Kg/m^3)

2 GTAW Process Modeling

As seen in Fig. 1, a non-consumable tungsten electrode is used to create the weld, the tungsten electrode does not melt. An electric arc is produced between the work piece and the tungsten electrode, the heating could reach a temperature up to 19.400 °C [6]. The inert shielding gas is utilized to protect the melted zone and the welding electrode from atmospheric oxidation. GTAW is widely appreciated in fields requiring precision, such as chemical, aircraft and nuclear industry [7].

Rosenthal [8] has been the first to develop an analytical model to calculate temperature distribution during welding by using simplifying assumptions. Heat source was considered as a punctual source with a constant velocity. The calculated temperature field was:

$$T = T_0 + \exp(\frac{-vX}{2K})\psi(x, y, z) \tag{1}$$

With 'T0' is the initial temperature, (x, z) is the coordinate of the punctual heat source, 'v' is the velocity of the heat source and $\boldsymbol{\Psi}(\mathbf{x}, \mathbf{z})$ is a function to be defined.

In the case of the thin plate, one of the most known solutions is [9]:

$$T = T_0 + \frac{\frac{q}{v}}{e(4\pi\lambda\rho ct)^{1/2}}\exp(\frac{-r^2}{4Kt}) \tag{2}$$

where 'e' is the thickness of the plate, 'q' is the welding power and 't' is time. The heat source and melted zone were considered moving with a velocity 'v' and a length 'l' [10].

Nunes [11] has proposed an extended Rosenthal weld model, the temperature field 'T' represented approximately as following:

$$T = T_0 + q\left[F(n = 0) + \sum_{n=1}^{\infty}\left(F(r_n) + F(r'_n)\right)\right] \tag{3}$$

where 'F' is a function to be defined.

Therefore, modelling the heat source as a punctual source does not allow determining the maximal temperature near to the heat source centre, which limits the comparison between the experimental and the analytical study.

3 Heat Source Modelling

In this study, the spatial distribution of the power welding input has been taken into account. The heat source is presented as a Gaussian function distribution as centred normal law with a radial distance 'R', standard deviation '$\sigma = 0.4$'. Figure 2 shows the used heat source distribution [12].

An axisymmetric 2D of a GTAW model is considered. The heat input is depending to the voltage 'V', the current 'I' and the efficiency of the welding process 'η'.

By assuming the energy distribution is Gaussian, the absorbed energy flux on the top surface ($z = 0$) is given as:

$$Q_m = \frac{VI\eta}{2\psi^2}\exp\left(\frac{-R^2}{2\sigma^2}\right) \tag{4}$$

In the case of GTAW the process efficiency "η" takes into account the heat absorption coefficient, since this process is characterized by the presence of an inert gas that protects the molten area. In general, in this process with an inert gas composed of 82% Ar and 18% CO_2, the efficiency of the process is equal to 80% [13].

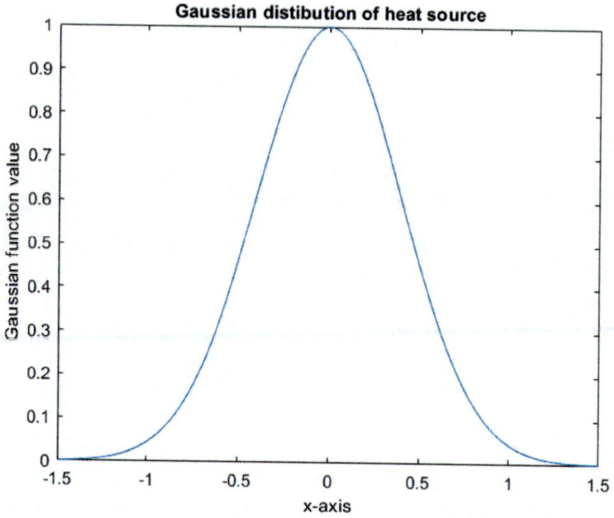

Fig. 2. Heat source presentation

4 Heat Transfer Modeling in GTAW Process

Nowadays, steel is considered as one of the most preferable materials in automotive application due to low cost comparing to others metallic materials. A new generation of high strength steel is now used in the automotive field offering a high capacity of fuel efficiency, enhanced safety a good formability characteristic. Mild 140/270 steel is employed principally on the fuel tank of the car, the structure is assembled by GTAW process, the main characteristic of the used materiel is presented in Table 1.

In this numerical study, a 2D dimensional model is introduced as a plat with a thickness 'e' and width 'L'.

Table 1. Material properties of mild steel

Steel grade	Yield strength (MPa)	UTS (MPa)	Density (Kg/m^3)	K (W/m.K)	Cp (J/K.kg)	hconv (W/m^2.K)
Mild 140/270	140	270	7200	25	300	40

The heat source of welding 'Q_m' has been introduced as a boundary condition (Table 2) in addition to the conviction exchange with the air (Eq. 6).

The governing equations are presented as follow:

$$\rho Cp\frac{\partial T}{\partial t} = k\left(\frac{\partial^2 T}{\partial x^2}+\frac{\partial^2 T}{\partial z^2}\right) \quad (5)$$

The upper surface is in contact with air, it submitted to convective heat transfer as shown below:

$$-k\frac{\partial T}{\partial z} = h_{conv}(T - T_0) + Q_m \quad (6)$$

Cartesian geometry was used; the x–z plane was taken for the surface of the work piece that absorbs the heat generated by electric arc. The length and thickness of the studied piece were $5 \times 5\ mm^2$ respectively.

In order to enhance the accuracy of calculation, the finite difference method was used to solve numerically the mathematical model (5) and (6). The explicit scheme is applied. It builds up an iterative calculation in time. As a result, the temperature is written as:

$$T(t+1, j, k) = T(t, j, k) + \frac{\Delta t}{\rho Cp}\left[\frac{T(t, j-1, k) - 2T(t, j, k) + T(t, j, k)}{\Delta x^2} + \frac{T(t, j, k-1) - 2T(t, j, k) + T(t, j, k+1)}{\Delta z^2}\right] \quad (7)$$

However, the explicit scheme has to respect the following relation (Eq. 8) in order to be stable. For this reason, the choice of Δt is limited:

$$\frac{\Delta t}{\rho Cp}\left(\frac{1}{\Delta x^2} + \frac{1}{\Delta z^2}\right) \leq 1/2 \quad (8)$$

Although the timescale required the use of fine grids and very small-time steps to obtain accurate results, that will lead to longer calculation time. The space grid was taking as $\Delta x = \Delta z = 0.1$ mm.

In this current study, the welding source Q_m was included as Neumann boundary condition. For both lateral surfaces were considered at 25 °C. The model has been developed and simulated in MATALB environment.

Table 2. Boundary conditions of the work piece

Position	Boundary conditions
$x = 0$	$T = T0$
$x = L$	$T = T0$
$z = 0$	$-k\frac{\partial T}{\partial z} = h_{conv}(T - T_0) + Q_m$

5 Results

Figure 3 illustrate the problem-solving flowchart concerning the overall iterative approach for the calculation of temperature field during GTAW process in all points of the workpiece, temperature was calculated at every step of time by using finite difference

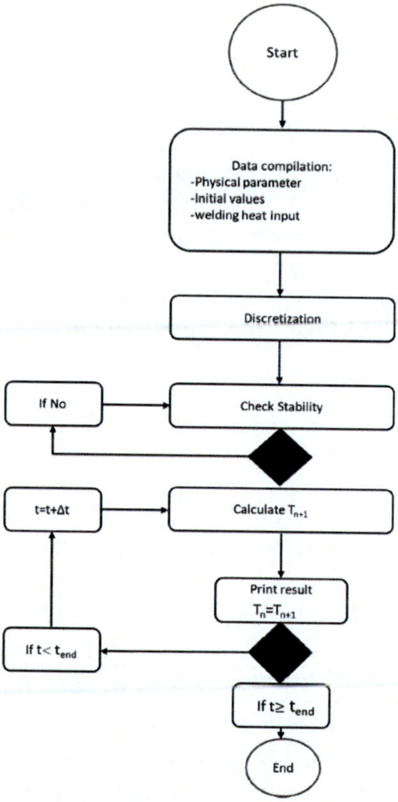

Fig. 3. Problem solving flowchart

method and by checking stability condition at every point of the work piece, then calculate the temperature until finishing the welding time. Figure 4 shows the temperature field after 10 s of GTAW process with current $I = 90$ A and voltage $V = 10$ V.

Using Isotherm diagram (Fig. 5), it was possible to calculate the dimension of the melted zone, for $I = 90$A and $U = 10$V, we have the width = 4.4 mm and depth = 0.7 mm.

With a $V = 15$ V and according to the Fig. 6, varying welding parameters, especially the welding current, affect the dimension of the melted zone. In fact, Increasing the welding current, leads to the growth of the melted zone.

6 Comparison with Experimental Study

In order to ensure of the validity of the developed model, it was necessary to compare with experimental results. KERRICH [14] has studied the effects of various welding parameters on the GTAW process and specially the welding current on the penetration of the melted zone in a structural steel. They showed that increasing the welding current leads to the growth of the melted zone which was in good agreement with the corresponding numerical calculation.

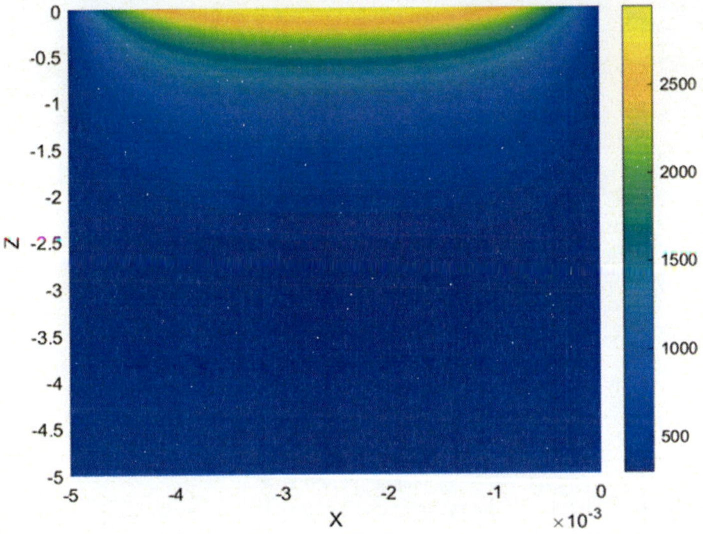

Fig. 4. Temperature field after 10 s of welding process I = 90A, U = 10V

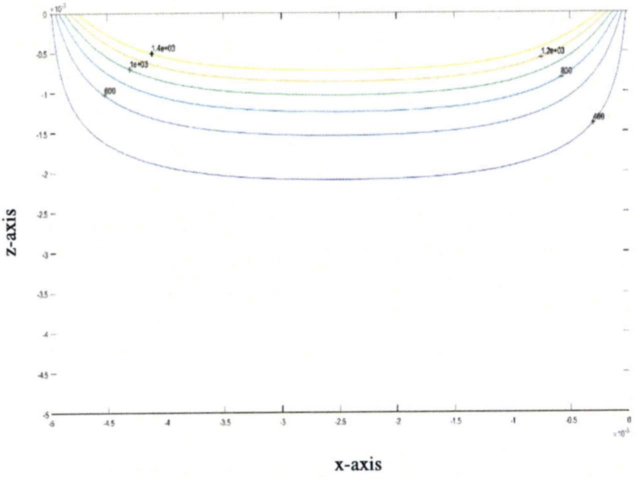

Fig. 5. Isotherm diagram for I = 90 A, V = 10V

According to the Table 3 the comparison with the experimental study reveals a slight difference in terms of the depth of the melted zone, this can be explained by the simplifying assumptions taken in the numerical study including that the weld pool movement were not taking into account.

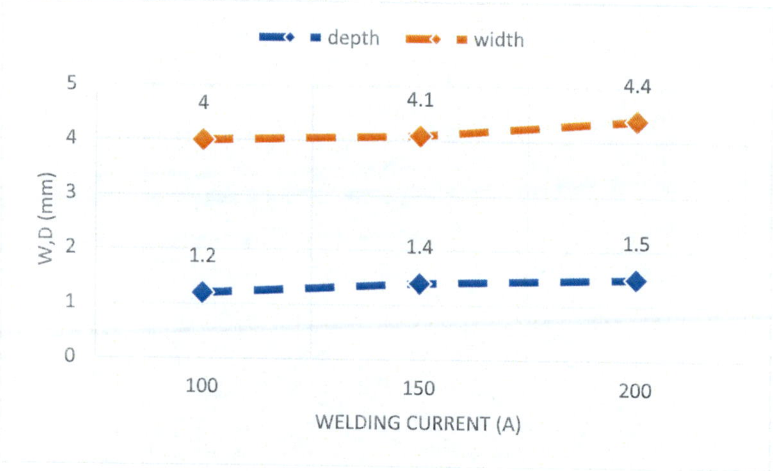

Fig. 6. Evolution of the melted zone dimensions versus welding current.

Table 3. Comparison between experimental and numerical welding penetration

Arc voltage (V)	Welding current (A)	Experimental penetration (mm) [12]	Numerical penetration (mm)
10	180	1.3	1.4
	210	1.8	1.6
	300	2.1	1.8

7 Effect of the Shielding Gas Rate on the Weld Penetration

Due to its inert chemical reactivity with weld pool, Argon (Ar) is used as shielding gas [15]. The effect of gas flow rate of Ar was studied by Anbarasan [16]. He proved that increasing the shielding gas rate leads to the decrease of the process efficiency.

In order to discuss the effect of shielding gas rate on the weld penetration, two values of the process efficiency are implemented in the numerical model (Table 4).

Table 4. Weld penetration in function of the process efficiency

Efficiency	Weld penetration (mm)
0.5	0.5
0.9	0.8

According to the Table 4, increasing the process efficiency leads to the growth of the weld penetration, which mean that increasing the shielding gas rate decrease the weld penetration.

8 Conclusion

A numerical model of GTAW process has been developed to analyze the effect of the welding current and shielding gas rate on the temperature distributions and the shape of weld pool. Finite Difference Method with explicit scheme was used, including a description of a gaussian heat source distribution.

Computational results showed that increasing intensity leads to the growth of the melted zone, which may induce high residual stress and risks of metal failure.

Comparison with experimental results was made in terms of welding penetration. The behavior of measured widths of the fusion zone was in good agreement with the numerical calculations. These analyses can be used to optimize the GTAW power and the efficiency of the process. As an extension of the present research work, we pretend to take into account the weld pool movement in calculating temperature field and simulate the thermal stress.

References

1. Laazizi, A., et al.: Applied multi-pulsed laser in surface treatment and numerical–experimental analysis. J. Opt. Laser Tech. **43**(7), 1257–1263 (2011)
2. Traidia, A.: Multiphysics modelling and numerical simulation of GTA weld pools. PhD thesis Ecole Polytechnique Paris (2011)
3. Zhu, X.K., Chao, Y.J.: Effects of temperature-dependent material properties on welding simulation. J. Comput. Struct. **80**, 967–976 (2002). https://doi.org/10.1016/S0045-7949(02)00040-8
4. Rong, Y., Xu, J., Huang, Y., Zhang, G.: Review on finite element analysis of welding deformation and residual stress. J. Sci. Technol. Welding Joining **23**(3), 198–208 (2018). https://doi.org/10.1080/13621718.2017.1361673
5. Tao, J., Zhang, T.C., Liu, P.T., Li, J.: Numerical computation of a linear friction welding process. J. Mater. Sci. Forum **575–578**, 811–815 (2008). https://doi.org/10.4028/www.scientific.net/msf.575-578.811
6. Kutelu, B.J., Seidu, S.O., Eghabor, G.I., Ibitoye, A.I.: Review of GTAW welding parameters. J. Minerals Mater. Charact. Eng. **6**, 541–554 (2018). https://doi.org/10.4236/jmmce.2018.65039
7. Subodh, K., Shahi, A.S.: Effect of heat input on the microstructure and mechanical properties of gas tungsten arc welded AISI 304 stainless steel joints. Mater. Des. **32**, 3617–3623 (2011). https://doi.org/10.1016/j.matdes.2011.02.017
8. Rosenthal, D.: Mathematical theory of heat distribution during welding and cutting. J. Welding. **20**, 220–234 (1941)
9. Poorhaydari, K., Patchett, B.M., Ivey, D.: Estimation of cooling rate in the welding of plates with intermediate thickness. J. Suppl. Welding. (2005)
10. Song, S., Pei, X., Dong, P.: Analytical interpretation of welding linear heat input for 2d residual stress models. In: ASME 2015 Pressure Vessels and Piping Conference. https://doi.org/10.1115/PVP2015-45753
11. Nunes, Jr. A.C.: An extended Rosenthal weld model. J. Welding J. (1983)
12. Mazuco, F., Teixeira, P.R., Bezerra de Araújo, D.: Two heat source models to simulate welding processes with magnetic deflection. Soldagem & Inspeção (2017)
13. Karadeniz, E., Ozsarac, U., Yildiz, C.: The effect of process parameters on penetration in gas metal arc welding processes. J. Mater. Des. **28**, 649–656 (2007)

14. Kerrich, D.: Contribution a l'etude du parametrage du procede de soudage tig (sans métal d'apport et avec apport de fil chaud): application aux cas des eprouvettes tubulaires. phd thesis universite des sciences & techniques de lille. (1997)
15. Kah, P., Martikainen, J.: Influence of shielding gases in the welding of metals. 1411–1421 (2013). https://doi.org/10.1007/s00170-012-4111-6
16. Anbarasan, N., Jerome, S.: Effect of flow rate and argon-hydrogen shielding gas mixture on weld bead morphology of inconel 718. Mater. Today: Proc. **5**(2018), 26990–26996 (2018). https://doi.org/10.1016/j.matpr.2018.09.002

Robust and Reliable Optimization of a Pair of Gear Wheels

Abderazzak Ziat[1](✉), Hamid Zaghar[2], Abdelmajid Ait Taleb[1], and Sallaou Mohammed[1]

[1] ENSAM, Moulay Ismail University of Meknès, Meknes, Morocco
ab.ziat@edu.umi.ac.ma, {a.aittaleb,m.sallaou}@ensam.umi.ac.ma
[2] FST, Sidi Mohamed Ben Abdellah University, Fes, Morocco
hamid.zaghar@usmba.ac.ma

Abstract. Gear trains are widely used for power transmission in industrial mechanical systems. Taking into account uncertainties in the engineering design process represents a considerable industrial stake in controlling the robustness and reliability. This work consists of minimizing the weight of a pair of gears train by considering the different sources of uncertainty. An optimization technique based on a specific Particle Swarm Optimization algorithm (PSO) is developed to identify the optimal configuration that combines reliability with robustness. The propagation of uncertainty in the proposed method is performed with the Monte Carlo simulation. The design of a pair of gears is a compromise between weight, reliability, and robustness. Therefore, the developed algorithm allows optimization by taking into account the uncertain parameters constituting the pair of gears model while ensuring an acceptable compromise between reliability stability, and performance. In comparison to the concept generally used, the proposed approach offers lower weight and provides superior performance.

Keywords: Robustness · Reliability · Uncertainty · Optimization

1 Introduction

Gears are used in most mechanical systems. Nowadays, there is considerable interest in optimizing the design of such trains, as many power transmission systems require low weight and high reliability and robustness. Several methods have been proposed for optimal gear design. These methods used deterministic design optimization techniques to obtain optimal design parameters [1–4].

Failure to address uncertainties in material properties, service conditions of gears and their manufacturing defects, etc., prompts research in the areas of probabilistic design methodologies applicable to the structural design of machine components.

In this context, some authors have been interested in implementing the concept of robustness and reliability in the design phase [5, 6]. The work presented by Y. M. Zhang et al [5] consists of analyzing the reliability of a pair of gears using the numerical perturbation method and reliability-based design theory.

L. Azrar et al. (Eds.): CIP 2022, LNME, pp. 363–369, 2023.
https://doi.org/10.1007/978-3-031-23615-0_37

In the design of gearbox, uncertainty is explicitly taken into account for a high degree of stability and performance. In addition, mechanical systems are subject to uncertainties in material properties and component shape and size due to processing and manufacturing. Using the theory of probabilistic design optimization, the work of [7] reports on the application of the concept of reliability-based robustness to minimizing the weight of a planetary gearbox.

2 Proposed Approach

In order to optimize the mass of a pair of gears under uncertainties, we propose a stochastic optimization method based on PSO (Particle Swarm Optimization algorithm) and MCS (Monte Carlo Simulation) [8]. This approach consists in coupling an optimization and the propagation of uncertainties. It is an efficient stochastic optimization method for very complex problems (Fig 1).

Table 1 shows the materials, input power and speed, and other parameters of the gearbox considered in the paper.

Table 1. Design Specifications of a pair of gears

Specifications	Symbol	Value
Power [Kw]	P	7.5
Gear ratio	r	4
Input speed [rpm]	N_1	1500
Pressure angle	α	20°
Gear's material	–	20CrMnMo
Shaft material	–	SAE 1045
Reliability	R	99.98%

2.1 Objective Function

The minimization of weight is considered as an objective function (1):

$$F(x) = \frac{\pi}{4}\rho b\left[m^2 Z_1^2\left(r^2+1\right) - d_1^2 - d_2^2\right] \tag{1}$$

With d_1 diameter of pinion shaft 1, d_2 is the diameter of the shaft of wheel 2, m is the modulus of the gears, b the width of the teeth and Z_1 number of teeth of the pinion 1 (Fig. 1).

The robust design optimization (2) is based on the calculation of the mean value μ_F and the standard deviation σ_F of the objective function F_{rob} [9, 10]:

$$F_{rob}(x) = \mu_F + a * \sigma_F \tag{2}$$

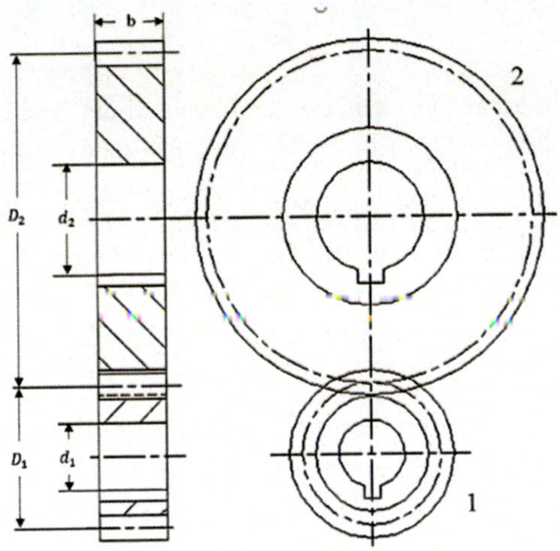

Fig. 1. Pair of gear wheels

2.2 Design Variables

Parameters that affect the mass of a pair of gears are selected as design variables. The vector of design variables X is selected as follows: The 5 design variables associated with the gearbox model are chosen as parameters to be optimized. These variables are listed in Table 2, where their range of variation is specified.

The vector of design variables is identified as follows:

$$X = \{x(1), x(2), x(3), x(4), x(5)\} = \{b, m, , Z_1, d_1, d_2\} \tag{3}$$

2.3 Constrains Functions

A number of constraints must be considered simultaneously. In this section, several design constraints used in the optimization problem considered in this paper are described [1–5]:

$$\sigma_H = Z_H Z_E Z_\varepsilon \sqrt{\frac{2KT(u+1)}{bd^2u}} \leq \sigma_{HP} \tag{4}$$

$$\sigma_F = Y_{Fa} Y_{Sa} Y_\varepsilon \frac{2KT}{bdm} \leq \sigma_{FP} \tag{5}$$

$$d_1^3 \geq \frac{P \times 48.68 \times 10^6}{R_g N_1} \tag{6}$$

$$d_2^3 \geq \frac{P \times 48.68 \times 10^6}{R_g N_2} \tag{7}$$

$$6 < \frac{b}{m} \leq 12 \tag{8}$$

3 Optimization

Based on reliability and robustness criteria, the formulation (9) was developed.

It consists of coupling an optimization and propagation of uncertainties (Fig. 2):

$$\begin{gathered} Minimize\ F_{rob}(\mu f, \sigma f) \\ \text{Under constrainf P } (G(X) > 0) \leq \text{Ptraget} \\ LX \leq X \leq UX \end{gathered} \tag{9}$$

Table 2. Design variables and parameters.

Parameters	Symbol	Value	STD
Tooth width	b	[20 50]	0.5%
Module	m	[2 8]	0.25%
Number of teeth	Z_1	[18 30]	0.5%
Primitive diameter of pingon 1	d_1	[20 50]	0.5%
Primitive diameter of gear 2	d_2	[20 80]	0.5%
Nodal field coefficient	Z_H	2.32	0.0116
Contact ratio coefficient	Z_ε	0.81	0.00405
Elastic coefficient	Z_E	189.8	9.49
Tooth form factor	Y_{Fa}	2.36	0.07788
Stress concentration coefficient	Y_{Sa}	1.75	0.05775
Contact ratio coefficient	Y_ε	0.715	0.003575
Work condition coefficient	K	2.89	0.288
Allowable contact stress	σ_{HP}	1300	156
Allowable bending stress	σ_{FP}	310	62

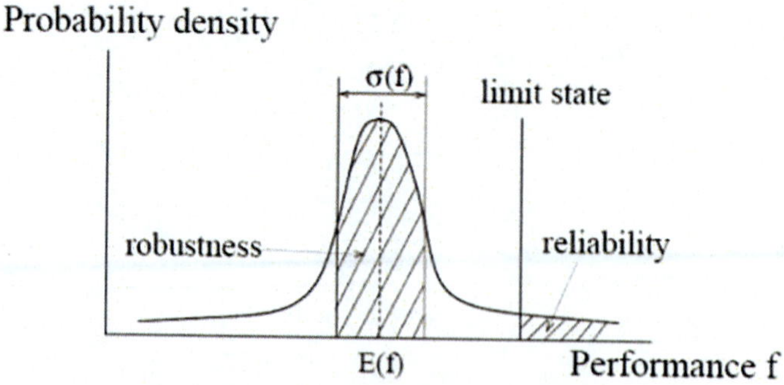

Fig. 2. Robust and reliable design

The developed computer code (9) allows optimization by taking into account the uncertain parameters constituting the pair of gears model while ensuring an acceptable compromise between stability and performance. It also allows a stochastic approach to be taken for a robust and reliable optimization of the gears weight taking into account the uncertainties [9, 10].

The design variables were fitted to a normal distribution with specified (Mean and C.o.V). The statistical data from Zhang et al. [5] and Ziat et al. [7] are shown in the Table 2. The Probability density function type and its probabilistic parameters (mean and standard deviation) are determined based on design tolerances and manufacturing tolerances.

Monte Carlo simulations are used to determine the distribution of output parameters Y when the input parameters X of the model are uncertain. To accomplish this, N values from its probability distribution are randomly selected for each input parameter xi (i = 1,., n)) [11]. Different Moments can also be calculated and reused, including the mean (10) and standard deviation (11).

$$\mu_Y = \frac{1}{N}\sum_{i=1}^{N} G(X_i) \tag{10}$$

$$\sigma_Y = \sqrt{\frac{1}{N}\sum_{i=1}^{N}(G(X_i) - \mu_Y)^2} \tag{11}$$

4 Results

An optimization was performed by PSO is performed with 20 particles and 30 iterations.

The preliminary design of the problem of a pair of gear wheels was made with

$P_f = 1 - R$ (probability of failure) and the same weight (robustness) a = 0.75. In Fig. 3, the evolution of the distribution of fitness values is shown.

The design variables obtained are shown in Table 3.

Several combinations of design variables are found. We will choose the design with the minimum gears center distance.

The design of a pair of gears is a compromise between weight, reliability, and robustness. This paper introduced one model for reliability-based robust optimization design. The model offered a method to help making decisions during the design process.

According to our expectations, this qualitative conclusion is accurate. Although, one can make quantitative judgments based on the complexity of robust reliability contexts, such as the coefficients of variability and the required reliability. Therefore, the probabilistic methods are immediately suggestive of eliminating existing overdesigns or underdesigns.

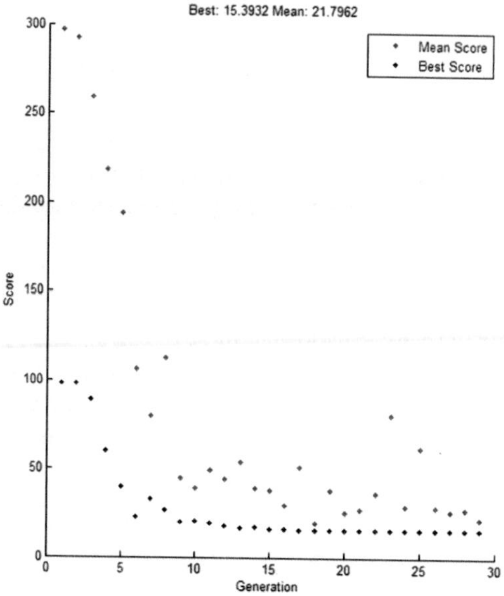

Fig. 3. Evolution of the objective function.

Table 3. Optimization Results

Design Variables	b (mm)	m (mm)	Z_1	d_1 (mm)	d_2 (mm)
Value	37.5	3.5	18	20.2	32.2
Weight				15.39 kg	

5 Conclusion

The theory of probabilistic design optimization is an advanced and advantageous methodology for the design of mechanical systems. This approach encourages the replacement of the safety factor by taking realistic account of the variability of the parameters. It is an optimization approach by simultaneously integrating the factors of reliability and robustness. Each of the variables is considered random with a certain distribution law and the uncertainties are propagated in the mechanical model. Lastly, this work can contribute to the improvement of a continuously variable transmission (CVT), such as in wind-power systems, using a reliable and robust perspective.

References

1. Miler, D., Žeželj, D., Lončar, A., Vučković, K.: Multi-objective spur gear pair optimization focused on volume and efficiency. Mech. Mach. Theory **125**, 185–195 (2018)
2. Golabi, S.I., Fesharaki, J.J., Yazdipoor, M.: Gear train optimization based on minimum volume/weight design. Mech. Mach. Theory **73**, 197–217 (2014)
3. Chong, T.H., Lee, J.S.: A design method of gear trains using a genetic algorithm. Int. J. Korean Soc. Precis. Eng. **1**(1), 62–70 (2000)
4. Patil, M., Ramkumar, P., Shankar, K.: Multi-objective optimization of the two-stage helical gearbox with tribological constraints. Mech. Mach. Theory **138**, 38–57 (2019)
5. Zhang, Y., Liu, Q., Wen, B.: Practical reliability-based design of gear pairs. Mech. Mach. Theory **38**(12), 1363–1370 (2003)
6. Wei, J., Lv, C., Sun, W., Li, X., Wang, Y.: A study on optimum design method of gear transmission system for wind turbine. Int. J. Precis. Eng. Manuf. **14**(5), 767–778 (2013)
7. Ziat, A., Zaghar, H., Ait Taleb, A., Sallaou, M.: Reliable and robust optimization of the planetary gear train using particle swarm optimization and Monte Carlo simulation. SAE Int. J. Mater. Manufact. **15**, 05-15-01-0003 (2021)
8. Kohler, M., Vellasco, M.M., Tanscheit, R.: PSO: a new particle swarm optimization algorithm for constrained problems. Appl. Soft Comput. **85**, 105865 (2019). https://doi.org/10.1016/j.asoc.2019.105865
9. Lelièvre, N., Beaurepaire, P., Mattrand, C., Gayton, N., Otsmane, A.: On the consideration of uncertainty in design: optimization-reliability-robustness. Struct. Multidiscip. Optim. **54**(6), 1423–1437 (2016). https://doi.org/10.1007/s00158-016-1556-5
10. Ziat, A., Sallaou, M., Zaghar, H., Taleb, A.A.: Optimization approach under uncertainty in preliminary design for mechanical engineering. In: 2020 1st International Conference on Innovative Research in Applied Science, Engineering and Technology (IRASET), pp. 1–4. IEEE (2020)
11. Forcina, A., Silvestri, L., Di Bona, G., Silvestri, A.: Reliability allocation methods: a systematic literature review. Qual. Reliab. Eng. Int. **36**(6), 2085–2107 (2020). https://doi.org/10.1002/qre.2675

Check for updates

Toward a Review on Structural Design and Fracture Analysis in Exhaust System

Ouyoussef Nouhaila(✉) and Moustabchir Hassan

Laboratoire d'Ingénierie, Systèmes et Applications, Sidi Mohamed Ben Abdellah University, Fez, Morocco
Nouhaila.ouyoussef@usmba.ac.ma

Abstract. A study of an exhaust system should include multiple aspects as well as vibrations, fracture, fatigue life, thermal analysis, computational fluid dynamic analysis (CFD), and durability. Considering its important rule to guide the burnt and harmful gases whose temperature is extremely high, and discharge the purified exhaust gases at a suitable point of the vehicle away from its occupants, the exhaust systems are mainly developed to reduce emissions and noises, increase durability, minimize corrosion, enhance maintainability, and make it economically viable. So, this present paper gives an overview of the main areas treated in the previous research, this includes the three-dimensional modeling of the exhaust system using CAD Software, mesh optimization, and topology using modal analysis, fracture, vibrations, and fatigue life. This paper also depicts a comparison between different research and identifies the area that is not explored to a greater scope which is the computational fluid dynamics analysis to calculate the propagation of cracks in the exhaust system.

Keywords: Exhaust system · Mesh · Simulation software · Vibration · Fatigue · CAD

1 Introduction

The automotive industry is very competitive, so the increasing need for durability, lighter and cost-effective designs of automotive products has led to more frequent usage of powerful Numerical techniques to solve structural problems. The exhaust system is considered one of the most important components in the vehicle, allowing to route of the gases from the engine to reduce noise and control harmful emissions.

Then this paper aims to present a state of art, of the previous articles published, in which they treat the structural design and fracture analysis in the exhaust system, to show different approaches and domains, and compare them.

This article will be organized into three sections, in which we will expose the previous works conducted in this field, compare them to see the main areas studied, and come out with an overview of the research previously published.

L. Azrar et al. (Eds.): CIP 2022, LNME, pp. 370–376, 2023.
https://doi.org/10.1007/978-3-031-23615-0_38

2 Literature Survey

V. R. Kussam and J. Kumar [1] calculated the stress and displacement of a passenger car exhaust system through the finite element method, considering loads plotted as a function of frequencies and boundary conditions applied at the flange, the bracket, the rear muffler, and the center of the exhaust pipe on the hanger region, as results, they have obtained different natural frequencies on mode shapes and observed Stresses at different location of The Exhaust System came out with the results of maximum stress and displacement figured at the bracket location, due to the excitation produced from the engine.

P. Sasikumar et al. [2] explained how to calculate the fatigue life for mounting brackets using CEA comparing with physical testing, the exhaust system was meshed with 2D elements, Using MSC/NASTRAN to identify the root cause for high stresses in critical placements and predominant mode causing failures.

D. Jai Balaji et al. [3] present a modal analysis of an automotive exhaust system for a passenger car considering only the inline of the exhaust system in this structural analysis, with a mixture of shell and solid elements to get better results in the modal frequency domain. This exhaust system was designed using Catia and discretized using Hypermesh, solved using ABAQUS solver. The results were viewed in HYPERVIEW. Consequently, The mode shapes observed from FEA are mostly in transverse directions, the excitations points were found out at the hanger locations where the displacement is minimum.

Jian Min Xu et al. [4] analyze the vibration characteristic of the exhaust system generated by the engine and the road also; The finite element of the design was set to obtain the different mode shapes for the exhaust system to be able to identify and optimize the mounting of the suspension points employing an average driving DOF displacement. The vibration amplitude of the exhaust system at low-speed operating conditions is relatively bigger than that of the exhaust system in high-speed conditions, as result, optimizing the mounting of the suspension points helps also to prevent from having noises caused by the vibrations in the exhaust system.

Faraz Ahmad et al. [5], the purpose of this paper is the thermo-mechanical analysis of tractor exhaust manifold, the material used is Austenitic Stainless steel-321. This study studied the response of exhaust manifold submitted to higher temperatures. The thermal effect was considered to evaluate the FEA simulation results. They found the fundamental frequencies by executing the free vibration-based modal analysis. The solid model of the exhaust manifold was designed using Pro-E. Finite element analysis was performed using Ansys 14.5.

The natural frequency and vibration mode of the exhaust manifold were gotten in modal analysis and vibration characteristics of the exhaust manifold were studied.

R. N. Akshay et al. [6] a static and dynamic analysis was performed for the exhaust system using Ansys as a tool of CAE, to accurize the maximum stress and displacement of the parts of the exhaust system considering different loads (self-weight and bad road, they noticed that, for the self-weight condition, a hanger has more reaction than others, henceforth the mass is not distributed alike. Otherwise, for the bad road condition, the maximum stress was noticed in the rear muffler shell hanger.

R. Manoj Kumar M.E. et al. [7] studied the Catalytic convertor, intending to reduce the harmful gases, so they have focused on the structure design and tried to change the design of the structure with the spinel from the honeycomb. Consequently, by using Ansys, they found that the spinel structure shows a larger area of contact, higher efficiency, and longer lifetime compared to the honeycomb.

M. H. Shojaeifard et al. [8] have used the method of average driving degrees of freedom displacement (ADDOFD) to accurize and optimize the location of 5 hanger points used in this exhaust system, So they have designed the exhaust system, the finite element method was built, the Free-Free modal and dynamic analysis are used to calculate the transmitted forces that should be less than the standard requirement 10N, as a result, this paper depicts that ADDOFD method was helpful to optimize the points of the location to reduce the maximum possible vibration.

Dr. B. H. Maruthi et al. [9] the objective of this paper is to design and analyze the hangers, the finite element model was used in MSC Nastran to perform the modal analysis considering the range from 1 to 200 Hz, to optimize the best hanger. Then a method called ADDOFD (average driving degree of freedom displacement) was used to identify the displacements, as result, the location of the hanger positions is placed where the displacements are low and where the reaction forces are minimum. This is then validated analytically.

E. Ikpe Aniekan et al. [10] This paper treats the selection of the material used for the automotive exhaust system by using CES, which is based on some specific criteria; high temperature, mass, and corrosion resistance. As result, stainless steel was considered as the best choice of material due to its suitable price and density, acceptable strength at high temperatures, and excellent corrosion resistance it possesses of the protective film of chromium oxide which shapes on the surface of the metal.

Rui F. Martins et al. [11] depict the propagation of cracks in the exhaust system for naval gas turbines, usually occurring on the weld toes of butt and fillet weld joints positioned near the lower support ring of the assembly. So, FEA was made on this structure by using Ansys Software, The Von mises stress and the nominal principal stresses were calculated. As a result, the Maximum stress interval induced in the new geometry is lower than the current. The design and fatigue life are expected to increase to 6x104 cycles in the critical region.

F. Hidayanti et al. [12] explained how to develop a suitable exhaust system for motorsport, using CAD software, and get the fluid velocity data based on computational fluid dynamics (CFD) software, also describe the exhaust fluid flow, to decrease the value of initial counterpressure, a megaphone is fixed at the tip of the exhaust, in order to avoid the impact on exhaust.

Hongjun Liu and Shuya zhi [13] analyze the problem of vibration in the exhaust system and proposed an optimization method in order to reduce stress, so they got the optimization scheme to minimize exhaust system vibration by changing the exhaust system hook lug installation location. Knowing that if the lug position is well adjusted carefully, the optimization result will be improved. Adding to that, another study was developed to reduce the vibration and displacement caused by the exhaust system using the simulation analysis and as a conclusion, the exhaust muffler and the tailpipe have the largest amplitude. So, to optimize the design, they have set lifting the lugs on the position of the biggest displacement in the exhaust system, the way as the further FEM analysis gives satisfying results.

Mohammad Amin Salehnejada et al. [14], analyzed the crack failure based on the finite element method (FEM, a crack was applied to the exhaust manifold model they applied the tension test at a high temperature, Then, the critical fracture force was determined. In the end, Ki was compared with Kic, and propagation of the fracture was discussed.

Guoquan Xia [15], has depicted the three-dimensional transient simulation, in which he determined the boundary conditions between fluid and solid, then a transient simulation for the heat transfer characteristics of this vehicle exhaust system, the discipline between the steady heat transfer characteristics and the periodic pulsating movement of the exhaust gas velocity and temperature were found.

G Varun Ram et al. [16] have tried in this present article to analyze the exhaust system for an ultra-fuel-efficient vehicle, with the aim to minimize at maximum the amount of the backpressure building, those flows are analyzed using the ram induction theory, The CFD has been used to make easier the study of the turbulent flow of the exhaust to get a smooth flow, accordingly, the design of the exhaust has been optimized in terms of materials and geometry.

Yang, Y et al. [17] The objective of this article is to optimize the Hanger location of the exhaust system, using the method of averaged driving DOF displacement (ADDOFD), in which we determine and optimize the exhaust hanger location, dynamic analysis was carried out based on the finite element method, to calculate the force for each hanger to enhance the NVH (Noise, vibration, and harshness) performance.

Marupilla Akhil Teja et al. [18], have presented a model of the exhaust manifold, the finite element method was applied to this structure to find and calculate the values for the input boundary conditions and analyze the flow patterns induced by the flow of the exhaust manifold and determine the velocity and pressure distribution at a maximum flow rate. And finally, they studied the static pressure drops and energy loss in the flow pattern created in the exhaust manifold.

3 Discussion and Results

This table synthesis the different researches conducted in this area based in:

From the analysis of these previous articles, the exhaust system is studied in multiple aspects including vibrations, fracture, and optimizations from the geometric model using modal analysis, so researchers have tried to analyze all possible parameters and optimize their designs (Table 1).

But one area that is not explored to a greater extent is the approach toward designing and optimizing the exhaust system using the computational fluid dynamics analysis to calculate the propagation of cracks in the exhaust system.

As can be seen in Fig. 1, based on a sample review of articles, The major researchers that have been published are the Indian and Chinese searchers.

As can be deduced in Fig. 2, during the years 2011 and 2012, we have fewer articles published, after that, we notice that there is a remarkable increase in the search, especially in the year 2017.

Table 1. Researches comparison in different areas

Articles	3D modeling	Vibration and noises and stress	CFD analysis	Materials engineering	Chemical reactions	Fatigue life and fracture	Optimization
V. R. Kussam and J. Kumar [1]	Δ	Δ					
P. Sasikumar et al. [2]	X	X				X	Δ
D. Jai Balaji et al.[3]	O	Δ		X			
Jian Min Xu et al. [4]	O	Δ					Δ
Faraz Ahmad et al. [5]	O	Δ		O			
R. N. Akshay et al. [6]	X	Δ					
R. Manoj Kumar M.E. et al. [7]		Δ		Δ	Δ		
M.H.Shojaeifard et al. [8]	O	Δ					Δ
B. H. Maruthi et al. [9]	O	Δ					
E. Ikpe Aniekan et al. [10]				Δ		Δ	
Rui F. Martins et al. [11]	O					Δ	
F. Hidayanti et al.[12]	X		Δ				
Hongjun Liu and Shuya zhi [13]	X	X					Δ
Mohammad Amin Salehnejada et al. [14]	Δ	Δ				Δ	
Guoquan Xia [15]	X		Δ				
G Varun Ram et al. [16]	X		Δ				
Yang, Y et al. [17]	X	Δ					O
Marupilla Akhil Teja et al. [18]	X	X	Δ				

O: low, X: Medium, Δ: Strong

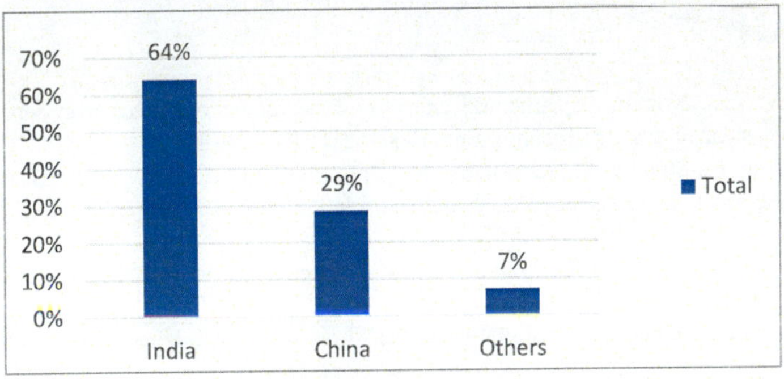

Fig. 1. Paper regions

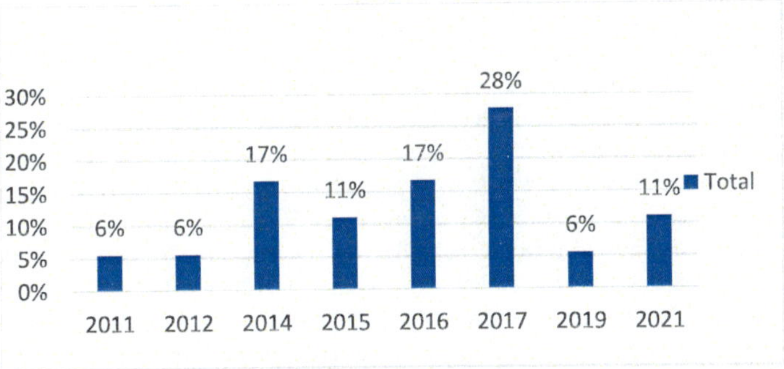

Fig. 2. Publication Paper/Year

4 Conclusion and Future Works

This article depicts a state of art for the previous works presented in this field, this includes the three-dimensional modeling of the exhaust system using CAD Software, mesh optimization using digital simulation software, application of boundary conditions, and launching of calculations and interpretations of the results with the constraints. They have studied multiple features like vibrations stress, fracture, and optimization to attenuate noises and emissions, reduce corrosion in the exhaust system, and make it economically sustainable, yet the computational fluid dynamics analysis was not really deployed in the optimization of the Exhaust system and determining the propagation of cracks.

An optimization in the exhaust system design can be performed using computational fluid dynamics analysis to depict the propagation of cracks in the exhaust system.

References

1. Kussam, V.R., Kumar, J.: Structural analysis of passenger car exhaust system by using hypermesh. Int. J. Emerg. Technol. Eng. Res. (IJETER) **3**(2) (2015)
2. Sasikumar, P., Sujatha, C., Chinnaraj, K.: Transient fatigue analysis of exhaust system mounting brackets for commercial vehicle correlation (No. 2017-01-1333). SAE Tech. Pap. (2017)
3. Balaji, D.J., Srihari, P.V., Sheelvanth, V.B.: Model frequency analysis of automotive exhaust system. IJMERR (2014)
4. Xu, M.,Zhou, S., Chen, S.X.: An analysis of the vibration characteristics of automotive exhaust systems and optimization of suspension points. Open Mech. Eng. J. **8**, 574–580 (2014)
5. Ahmad F., Tomer, V., Kumar, A. Patil, P.P.: FEA Simulation Based Thermo-Mechanical Analysis of Tractor Exhaust manifold
6. Akshay, R.N., Ravindra Babu, G., Gowda, R.D., Madhu, B.P.: Application of CAE in durability assessment of automotive exhaust system under static loads and comparison of natural frequency between coupled and lumped mass. Int. J. Eng. Res. Technol. (IJERT)
7. Manoj Kumar, R., Arulkumar, S., Harish Kumar, M, Sasidharan, B.: A study of exhaust emission techniques and changing the structure in catalytic converter. Int. J. Sci. Res. (IJSR). ISSN (Online): 2319-7064
8. Shojaeifard, M.H., Ebrahimi-Nejad, R., Kamarkhani, S.: Optimization of exhaust system hangers for reduction of vehicle cabin vibrations. Int. J. Autom. Eng. **7**(1) (2017)
9. Maruthi, B.H., Channakeshavalu, K., Mallika, K., Shwethasree, B.: Design and analysis of hangers for exhaust system. Int. J. Eng. Res. Technol. (IJERT). ISSN: 2278-0181
10. Ikpe Aniekan, E., Kelly, O.E., Abdulsamad, G.: Engineering material selection for automotive exhaust systems using CES software. Int. J. Eng. Technol. IJET **3**(2) (2017)
11. Martins, R.F., Viegas, J.C., Cruz, H.J.: Fatigue life assessment of an exhaust system for naval gas turbines. Procedia Eng. **10**, 2548–2553 (2011)
12. Hidayanti, F., Kreshna, A.M., Adi, Wati, E.K.: Developing racing exhaust system performance using computational fluid dynamics software
13. Liu, H., Zhi, S.: Exhaust system finite element analysis and optimizing design. Adv. Mater. Res. 538–541, 591
14. Salehnejada, M.A., Mohammadi, A., Rezaei, M., Ahangari, H.: Cracking failure analysis of an engine exhaust manifold at high temperatures based on critical fracture toughness and FE simulation approach. Elsevier (2019)
15. Xiao, G.: Transient simulation of heat transfers for vehicle exhaust system. Elsevier (2015)
16. Varun Ram, G., Nithesh Sharan, G., Praveen Kumar, T., Tanish, P.: Design and analysis of exhaust system for ultra-fuel efficient vehicles
17. Yang, Y., Yang, Y., Sun, Y., Dong, Z. et al.: Optimization of vehicle exhaust system hanger location. SAE Tech. Pap. 2016-01-0448 (2016). https://doi.org/10.4271/2016-01-0448
18. Teja, M.A., Ayyappa, K., Katam, S., Anush, P.: Analysis of exhaust manifold using computational fluid dynamics

Reverse Engineering for Aeronautical Products: State of the Art and Proposition

Philippe Williatte[1(✉)], Alexandre Durupt[1], Sebastien Remy[2], and Matthieu Bricogne[1]

[1] Université de Technologie de Compiègne, Roberval, Centre de Recherche Royallieu, CS 60 319, 60 203 Compiègne Cedex, France
{philippe.williatte,alexandre.durupt,matthieu.bricogne}@utc.fr

[2] Université de Technologie de Troyes, UR LASMIS, 12 rue Marie Curie, 10010 Troyes, France
sebastien.remy@utt.fr

Abstract. Reverse Engineering (RE) is an activity that consists in digitizing a real part in order to create a numerical or virtual model. Main industrial finalities of RE consist in manufactured parts control, CAD models redesign, and finally product data retrieval in database. RE technologies helps to maintain the integrity of digital models and more generally of enterprise data through the products lifecycles. The aeronautical context presents several challenges for RE activities, such as large database of CAD models, complex shape and structures of the parts, and restrictive constraints on the products definition. Thereby, RE activities for aeronautical product requires a specific framework along with efficient geometry processing and knowledge management tools. This article deals with the study of RE related works presented in the literature in order to identify knowledge limitations and technological gap when considering specific constraints and challenges related to aeronautical systems. A general method based on optimized associated tools is proposed as a Reverse Engineering solution for Aeronautics.

Keywords: Reverse engineering · Shape matching · Computer aided design · Product lifecycle management · Knowledge management

1 Introduction

Reverse engineering (RE) activities usually consists in digitizing a physical part to create a numerical or virtual model of it (Durupt et al. 2009). It applies on components that do not have any Computer Aided Design (CAD) model, or only a semantically poor 3D representation. CAD files are considered as "rich" or "semantically rich" if parametric, i.e. modifications capabilities are available, and modeling features (construction operations with parentage links) are organized in the modelling tree. The designer implements knowledge in such model through the definition of features, rules, and relations between parameters.

In the industrial world, the Product Lifecycle Management (PLM) approach aims at using a coherent set of applications to ensure the collaborative management of products information. CAD models, related heterogeneous data and dependency links are managed in the Information System (IS) by Product Data Management (PDM) solutions,

L. Azrar et al. (Eds.): CIP 2022, LNME, pp. 377–387, 2023.
https://doi.org/10.1007/978-3-031-23615-0_39

enabling Knowledge Management (KM) throughout the life cycle of an evolving Product model. Nowadays, Industrial Engineering is facing the «Industry 4.0» challenges, where the Digital Mock Up (DMU) of products and systems is central, not only for design activities but also as Digital Twins allowing real time data processing and analysis. It is therefore essential for a company to sustain the integrity of enterprise data. However, product data may sometimes not be available to the designer, or no longer corresponds to the real geometry of the manufactured object itself (Buonamici et al. 2018). An approach that combines 3D models retrieval in database and remodeling activities can be a solution to compensate obsolete and/or missing data in IS. Thereby, RE do not only consider CAD (re)modelling from a digitized part, but also tends to create a form of Digital Twin by retrieving and capitalizing in an evolutionary way every DMU, versioned CAD model, and coherent heterogeneous data related to the studied system.

The aeronautical context presents several challenges to remodeling activities and product KM. Geometry processing tools reach their limits on the complex shapes and structures of the products, and aerodynamic properties (intrinsic to the definition of the geometry) require high precision in surfaces remodeling. Moreover, easy and coherent access to data stored in PDM systems is necessary to recover semantically rich CAD file, but still a research question for such industrial database.

This article deals with a review of the scientific literature about RE and the proposition of an optimized solution for the aeronautic sector. First, state of the art methods and tools limitations when considering aeronautical components will be identified. Then, a geometric approach is developed for recognition, analysis and rebuilt of complex surfaces in proprietary components. A KM-based approach is combined to improve the semantic of the new CAD file by making reuse of existing knowledge stored in database more efficient.

2 State of the Art

2.1 Reverse Engineering Process

According to (Buonamici et al. 2018), providing a consistent digital representation of a physical object stay an open question, especially for mechanical engineering needs. Common RE framework in the literature consist in following activities:

Digitizing is the first step to acquire a digital model from a physical component. Resulting data are generally point-clouds in a Cartesian 3D space.

Pre-processing starts with the generation of a 'low level' 3D model, such as a mesh. Then, the geometry is fixed (outliers deletion, holes filling) and simplified (decimation of the vertices, mesh surface smoothing).

Segmentation is the process of (sub-)dividing a 3D mesh into distinct regions, according to geometric criteria. The quality of the segmentation is crucial, since it represents essential information for the modeling steps and influences the overall reconstruction process. Ideally, a segmentation produces regions corresponding to the features of the original CAD model's construction tree (which defines the topology of the model). More precisely, each identified region can be easily associated with a construction operation or a typical surface (Buonamici et al. 2018). In that case, segmentation is called semantic. Many methods exist and are referenced in (Nguyen et al. 2013; Theologou et al. 2015; Kaiser et al. 2019).

Modelling operations are used to generate 3D models compatible with CAD-CAM needs. Freeform approaches are distinguished from Feature-based methods. The former is an explicit modelling approach, which consist in describing a solid by its surfaces (b-rep model). The term features is widely and differently used in the literature about implicit modelling, which seeks to recover a parametric and semantic model by using a procedural approach with sequences of constructions operations with parenting relations.

2.2 Modelling Methods

2.2.1 Freeform Methods

Basis method for 3D modelling consists in fitting freeform surfaces directly to the vertices of the mesh. In several CAD software, direct fitting tools are able to fit NURBS (Non Uniform Rational Basis Splines) curves and B-splines surfaces to generate complex surfaces as close as possible to the data (Ben Makhlouf et al. 2019; Tsai et al. 2009). Freeform surfaces fitting is almost fully automatic (at least for a "light" mesh with a fine pre-processing step) and can represent the real geometry with high precision and details. However, these representations are considered as non-fully semantic because the created surfaces are "freeze", without parameters and sketches, and associations with features is not always possible.

2.2.2 Feature-Based Methods

Feature-based methods try to recover high level design knowledge, such as geometric properties, relations, and design intents. (Wang et al. 2013).

- Independent surfaces fitting:

Primitives fitting consists in primitive shapes (i.e. planes, spheres, cylinders, cones and torus) recognition and adjustment to the segments. Additional knowledge is available such as the nature of the surfaces (label), their dimensions, and control parameters (Bénière et al. 2013; Attene et al. 2010; Li et al. 2011). For (Várady et al. 1997), primitive fitting requires first to know in advance the type of the segment, or to compute intrinsic shape descriptors to determine its class. The Random Sample Consensus (RANSAC) algorithm (Fischler et al. 1981; Schnabel et al. 2007) iteratively try to optimize the parameters of the primitive equations (quadrics in canonical forms), to minimize a distance function between the surface and a set of vertices. However, if primitives shapes represent the large majority of mechanical components (Kaiser et al. 2019), it is insufficient to represent complex geometries of some applications. Works as (Vergeest et al. 2003) introduced the *features* concept to the freeform area as generic shapes (b-rep non-primitive surfaces) to which a designer associate attributes and properties (Song et al. 2005). Freeform features are fitted to the mesh with a distance function minimization that optimizes control parameters and affine transformation. Future challenge for researches in this area is the improvement of surfaces fitting in terms of completeness and consistency, in particular with the introduction of more complex shapes, which should remain generic and not specific to the data (Kaiser et al. 2019).

- 2D sections:

The sectional feature strategy is an important solution for reverse modeling of complicated objects, where primitive fitting is not possible. 2D sketches are the basis of most CAD construction operations (sweeps, multi-sections, revolutions extrusions, etc.…), and can be recovered from plane section in the mesh. The choice of section planes is often left to the user and must remain consistent with the operations that follow (Ke et al. 2006).

- Template-based:

The majority of researches on CAD reconstruction addresses either the retrieval of geometric relations between the generated CAD features, or the study of fast/accurate algorithms of surface fitting (Buonamici et al. 2020). However, in CAD modeling, design functions, sketches, as well as parameters and relationships represent the real high-level knowledge implemented in the model. If available, a *template* makes the remodeling task easier and ensure a semantically consistent model since it contains useful and structured knowledge (Shah et al. 2020). Useful templates could be a generic surface model (Erdős et al. 2014), a more complex one (Song et al. 2005), a Constructive Solid Geometry (CSG) tree (Du et al. 2018; Sharma 2018; Wu et al. 2018) or even a native CAD (Buonamici et al. 2020; Shah et al. 2020). Template-based methods raise the question of the semantic insufficiency of other modeling methods, which ultimately only recover geometric information.

- Knowledge Based Reverse Engineering:

Additionally to design expert knowledge implemented in CAD models, manufacturing process and functional specifications of the product are information that a CAD expert analysis can recover. A purely geometric approach is not sufficient for obtaining such knowledge on the product (Durupt et al. 2008, Fisher et al. 2004). On that basis, Knowledge-Based Reverse Engineering (KBRE) is a set of methods that consist in capitalizing heterogeneous information in an upstream stage for reuse optimization. The in-situ interpretation of this knowledge by the user brings necessary information and greatly improves the semantic level in RE activities. In this consideration, (Durupt et al. 2019) developed a product model implementing functional aspects and production constraints in the creation of CAD model's features. PHENIX project developed a methodology for the capitalization of parametric geometric references and their adjustment to the 3D mesh. METIS project (Bruneau et al. 2016; Ouamer-Ali et al. 2016) integrates the use of heterogeneous data (drawings, images, instructions, templates) to the RE process. One of the main problematic is knowledge management in general (extraction, capitalization, research and reuse), with the question "how to easily access coherent and useful knowledge stored in database?". A solution is brought with the "signature" concept, i.e. the extract of useful information in a chosen formalism (Bruneau et al. 2016), which can link raw data with capitalized ones, therefore making reuse of support data more convenient and bringing semantic knowledge to the RE activity.

Current tendency in RE researches seems to seek more and more integration of already available data in the framework, which is more appropriate to the industrial world where CAD and related heterogeneous database indexing and research has become a main challenge.

2.3 CAD Retrieval

In a global industrial RE logic, the identification of known components in 3D data has become essential to improve knowledge exchange and reuse. Methods and tools for CAD recognition and search in PDM systems have been widely treated and one can find a synthesis in (Lupinetti et al. 2019).

- Shape descriptors:

Surface feature recognition is based on the approximation of intrinsic differential geometric properties extracted from the data points (Ke et al. 2006). 3D shape descriptors (SD) are computed and compared to evaluate the similarity between shapes (Osada et al. 2002, Novotni and Klein 2004). Approaches like (Gal and Cohen-Or 2006; Mousa 2011) use local SD based on surface curvature to recognize specific local surfaces. In the matter of limited computation time, SD often deals with low differential properties (order 1 or 2), which is enough to detect similarities more than precisely recognize complex shapes. Machine Learning and Deep-Learning tools works on global and local recognition and labelling for mesh or point clouds (Xiao et al. 2020; Gezawa et al. 2020). Based on public datasets of simple shapes, unsupervised learning makes the classical analysis of shape descriptors more reliable, even if the technology ability to apply local labeling and semantic segmentation of complex shapes still needs to be proved.

- Topological descriptors:

Geometric descriptors reach their limits when considering part in assembly retrieval. Topological descriptors are concerned with relationships between components, either the kinematic link or the type of assembly constraint (Lupinetti et al. 2019), or by the adjacency relations between surfaces in a CAD model. Topological signatures usually take the form of graphs that are globally or partially compared to match common parts and assembly structures in database. Approaches using such method are usually concerned by the recognition of DMU made up of distinct components (not digitized data), and containing complementary information such as attributes, metadata and constraints (Lupinetti et al. 2018).

2.4 Assessments and Limitations

Well known software packages offer tools for RE of CAD models from point clouds or low-level 3D models. A synthesis can be found in (Buonamici et al. 2018).

Segmentation tools can either cluster groups of vertices based on curvature variation (therefore, no label is added to the segment, and the result highly depends on parameters of sensibility), or compute segments corresponding to primitives surfaces. The semantic segmentation of complex meshes -with the extraction of segments corresponding to local freeform surfaces- remains a technical challenge in geometry processing. As for modeling tools, freeform surfaces fitting usually are automatic and reliable, but most of the so-called feature-based methods require a strong assistance from the user if one wants to recover a CAD model with high semantic level. The study of components with a complex shape typology requires a lot of expertise from the user, or an available template that contains both the geometry and the knowledge of the original designer.

Considering an industrial context where many product data are available in the PDM system, it seems relevant to focus on 3D mesh recognition methods, and CAD retrieval in database. Either for 3D mesh segmentation or part feature recognition, no technologies seems able to identify a complex local shape in 3D mesh or even in low level CAD model. Development of geometry processing tools using complex shapes descriptors is still an open area of research.

3 Research Project

3.1 Industrial Needs

This research project is carried out in the context of a PhD thesis in partnership with industrials from the aeronautic sector. Studied RE activities consists in recreating CAD models from digitized data, which are required in downstream applications. Following specificities about RE for aeronautical components and software limits have been identified: first, 3D mesh shall not be simplified during pre-processing, to keep it as representative of the manufactured part as possible. Consequently, digitized models are composed of several million of points. Geometry of the reversed model shall be close to the digitized component so that any deviation to the functional definition can be identified and further analyzed. Moreover, due to the complex topology and geometry of the parts, each multi-instantiated freeform features has to be treated separately. Applying semantic segmentation is then necessary, and require a performant tool able to identify and isolate segments corresponding to the surfaces of the definition model. Then, modeling consists in b-rep patches fitting and/or surface templates fitting, which can be performed well by commercial solutions. Each definition template are stored in the PDM system, but not necessary in the same CAD model. Efficient RE process therefore requires the ability to search a specific surface (with a labelled mesh segment as query) in a large database.

In the following, a specific method and associated tools for RE of aeronautical components is proposed, based on the KBRE logic and new Shape Descriptor for freeform surfaces. First, a specialized geometry-processing tool is developed for complex local aeronautic surfaces signature. Those signatures are used for semi-automatic shape identification and labelling in raw data, linking the 3D mesh to high-level CAD models and associated knowledge stored in the PDM system.

3.2 Theoretical Approach

Main technical proposition stands in the use of high degree analytic surface models for freeform shape descriptor: every smooth surface S in a Cartesian space can be described by an explicit equation:

$$f : \mathbb{R}^2 \rightarrow \mathbb{R}, S = \left\{P(x, y, z) \in \mathbb{R}^3; f(x, y) = z\right\} \tag{1}$$

As described in (Pouget 2006), the truncated Taylor expansion of the height function (called a n-jet) over a surface yields:

$$f(x, y) = \sum\nolimits_{k=0}^{n} H_{B,k}(x, y) \quad \text{with:} \tag{2}$$

$$H_{B,k}(x, y) = \sum\nolimits_{j=0}^{k} B_{k-j,j}\, x^{k-j} y^{j} \tag{3}$$

On a smooth non-generic CAD surface *S*, a set of points $P \in \mathbb{R}^3$ in a local space can be derived by surface discretization. The surface model is approximated by fitting an n-jet to the discrete data points with a multi-polynomial regression of B_{k-j} parameters. Maximum degree (n) of the polynomial is adjusted to best fit data points. Now, considering an unknown local region of a mesh, a transformation optimization algorithm can be used to fit the already approximated models on it. If the fitting score is under a predefined threshold, the local region can be labelled as surface *S*. Therefore, capitalized surfaces models can be used as shape descriptors for complex aeronautic surfaces recognition in digitized data.

3.3 Reverse Engineering Method

This section presents the proposed method for RE activities on aeronautic components. Associated with previously proposed shape descriptors, this method allows identification of complex shapes in raw scan data and easy access to support data and knowledge stored in a knowledge base (Fig. 1).

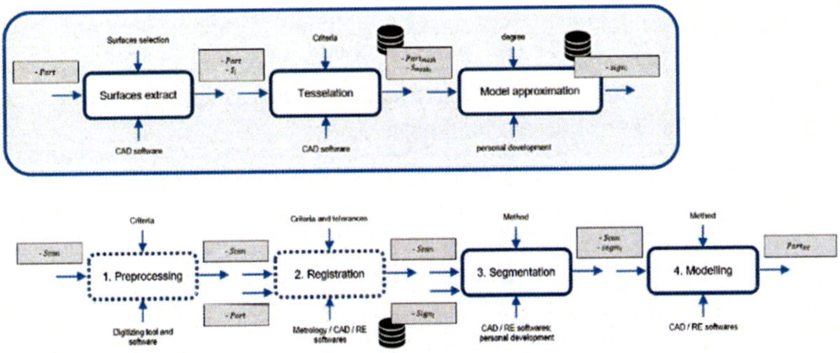

Fig. 1. RE method

In an upstream stage, knowledge is capitalized on CAD models.

(i) Characteristic surfaces are extracted and tessellated.
(ii) Model approximation is computed (cf. 3.2) on the discrete points.
(iii) Local surface signature (association of the surface model with associated metadata) is stored in a knowledge base.

After digitizing a real manufactured component, a 3D mesh is available as input data. RE method is as follow:

(i) The mesh is pre-processed with adjusted parameters.
(ii) Semantic segmentation is applied as follow: surfaces models stored in database are iteratively fitted on random groups of points (a specific surface is identified if points coordinates verifies the model equation). If model fitting is successful, the segment is extend to every neighboring points that belongs to it.
(iii) Semantic segments can be processed separately. Knowledge associated with the labelled segment through the feature signature helps to model it within a CAD software. For example, a freeform b-rep surface interpolation is computed, or local template may be adjusted by global deformation. If the new model has to be parametric, constructions elements (like sections planes, axis) can be prepared and reused. Reconstructed surfaces are assembled as a solid model, or added to a CAD model.

3.4 Discussions

In our research work, we chose the use of high degree mathematical equations as shape descriptors within a "signatures" of complex rigid surfaces. Motivations for this choice are: (i) aeronautical components geometry is composed of smooth freeform surfaces, which cannot be approximated by quadrics equations. Inspired by RANSAC method, we made the assumption that high degree polynomials can be approximated and fitted on data points, to act as shape descriptors. (ii) By computing high degree surfaces models before a potential RE application, the in-situ fitting of candidate models in raw data is lighter than computing high order differential quantities. (iii) First experiments proved that proposed surface model fitting is precise enough to differentiate surfaces with close shapes, and to labell local shapes. Finally, capitalization step makes our approach less generic and specific for only "prepared" components, but the proposition is coherent with the industrial context of this research project.

4 Conclusion

In this article, we introduced our research path to propose a RE solution for aeronautical components. Main objective of this work consist in the development a specific solution for complex geometry processing and shape matching in large industrial database. This theoretical approach is based on a literature review as well as experience feedbacks on RE of aeronautical components with commercial solutions. A global framework conciliate a new tool for local shape recognition of complex surfaces (based on high degree polynomials as analytic surface model) with the Knowledge Based Reverse Engineering area of research to improve useful data and associated knowledge recovering during RE activities. Evaluations will be performed on real industrial applications with aeronautic components, and compared with available tools and latest method based on machine learning capabilities.

References

Durupt, A., Remy, S., Ducellier, G., Guyot, E.: A new reverse engineering process, the combination between the knowledge extraction and the geometrical recognition techniques. In: 2009 International Conference on Computers and Industrial Engineering, CIE 2009, pp. 1367–1372 (2009). https://doi.org/10.1109/iccie.2009.5223773

Buonamici, F., Carfagni, M., Furferi, R., Governi, L., Lapini, A., Volpe, Y.: Reverse engineering modeling methods and tools: a survey. Comput.-Aided Des. Appl. **15**, 443–464 (2018). https://doi.org/10.1080/16864360.2017.1397894

Nguyen, A., Le, B.: 3D point cloud segmentation: a survey. In: 2013 6th IEEE Conference on Robotics, Automation and Mechatronics (RAM), pp. 225–30. IEEE (2013). https://doi.org/10.1109/RAM.2013.6758588

Theologou, P., Pratikakis, I., Theoharis, T.: A comprehensive overview of methodologies and performance evaluation frameworks in 3D mesh segmentation. Comput. Vis. Image Underst. **135**, 49–82 (2015). https://doi.org/10.1016/j.cviu.2014.12.008

Kaiser, A., Ybanez Zepeda, J.A., Boubekeur, T.: A survey of simple geometric primitives detection methods for captured 3D data. Comput. Graph. Forum **38**, 167–196 (2019). https://doi.org/10.1111/cgf.13451

Ben Makhlouf, A., Louhichi, B., Mahjoub, M.A., Deneux, D.: Reconstruction of a CAD model from the deformed mesh using B-spline surfaces. Int. J. Comput. Integr. Manuf. **32**, 669–681 (2019). https://doi.org/10.1080/0951192X.2019.1599442

Tsai, Y.-C., Huang, C.-Y., Lin, K.-Y., Lai, J.-Y., Ueng, W.-D.: Development of automatic surface reconstruction technique in reverse engineering. Int. J. Adv. Manuf. Technol. **42**, 152–167 (2009). https://doi.org/10.1007/s00170-008-1586-2

Wang, J., Gu, D., Gao, Z., Yu, Z., Tan, C., Zhou, L.: Feature-Based solid model reconstruction. J. Comput. Inf. Sci. Eng. **13** (2013).https://doi.org/10.1115/1.4023129

Bénière, R., Subsol, G., Gesquière, G., Le Breton, F., Puech, W.: A comprehensive process of reverse engineering from 3D meshes to CAD models. Comput. Aided Des. **45**, 1382–1393 (2013). https://doi.org/10.1016/j.cad.2013.06.004

Attene, M., Patanè, G.: Hierarchical structure recovery of point-sampled surfaces. Comput. Graph. Forum **29**, 1905–1920 (2010). https://doi.org/10.1111/j.1467-8659.2010.01658.x

Li, Y., Wu, X., Chrysathou, Y., Sharf, A., Cohen-Or, D., Mitra, N.J.: GlobFit: consistently fitting primitives by discovering global relations. In: ACM SIGGRAPH 2011 Papers on—SIGGRAPH ', vol. 11, p. 1. ACM Press, New York, New York, USA (2011). https://doi.org/10.1145/1964921.1964947

Várady, T., Martin, R.R., Cox, J.: Reverse engineering of geometric models—an introduction. Comput. Aided Des. **29**, 255–268 (1997). https://doi.org/10.1016/S0010-4485(96)00054-1

Fischler, M.A., Bolles, R.C.: Random sample consensus: a paradigm for model fitting with applications to image analysis and automated cartography. Commun. ACM **24**, 381–395 (1981). https://doi.org/10.1145/358669.358692

Schnabel, R., Wahl, R., Klein, R.: Efficient RANSAC for point-cloud shape detection. Comput. Graph. Forum **26**, 214–226 (2007). https://doi.org/10.1111/j.1467-8659.2007.01016.x

Vergeest, J.S.M., Spanjaard, S., Song, Y.: Directed mean Hausdorff distance of parameterized freeform shapes in 3D: a case study. Vis. Comput. **19**(7–8), 480–492 (2003). https://doi.org/10.1007/s00371-003-0213-3

Song, Y., Vergeest, J.S.M., Bronsvoort, W.F.: Fitting and manipulating freeform shapes using templates. J. Comput. Inf. Sci. Eng. **5**, 86–94 (2005). https://doi.org/10.1115/1.1875592

Ke, Y., Fan, S., Zhu, W., Li, A., Liu, F., Shi, X.: Feature-based reverse modeling strategies. Comput. Aided Des. **38**, 485–506 (2006). https://doi.org/10.1016/j.cad.2005.12.002

Buonamici, F., Carfagni, M., Furferi, R., Volpe, Y., Governi, L.: Reverse engineering by CAD template fitting: study of a fast and robust template-fitting strategy. Eng. Comput. **37**(4), 2803–2821 (2020). https://doi.org/10.1007/s00366-020-00966-4

Shah, G.A., Polette, A., Pernot, J.-P., Giannini, F., Monti, M.: Simulated annealing-based fitting of CAD models to point clouds of mechanical parts' assemblies. Eng. Comput. **37**(4), 2891–2909 (2020). https://doi.org/10.1007/s00366-020-00970-8

Erdős, G., Nakano, T., Váncza, J.: Adapting CAD models of complex engineering objects to measured point cloud data. CIRP Ann. **63**, 157–160 (2014). https://doi.org/10.1016/j.cirp.2014.03.090

Du, T., Inala, J.P., Pu, Y., Spielberg, A., Schulz, A., Rus, D., et al.: InverseCSG: automatic conversion of 3D models to CSG trees. ACM Trans. Graph. **37**, 1–16 (2018). https://doi.org/10.1145/3272127.3275006

Sharma, G.: CSGNet : Neural shape parser for constructive solid geometry supplementary material. In: Cvpr, pp. 5515–5523 (2018)

Wu, Q., Xu, K., Wang, J.: Constructing 3D CSG models from 3D raw point clouds. Comput. Graph. Forum **37**, 221–232 (2018). https://doi.org/10.1111/cgf.13504

Durupt, A., Remy, S., Ducellier, G., Eynard, B.: From a 3D point cloud to an engineering CAD model: a knowledge-product-based approach for reverse engineering. Virtual Phys. Prototyping **3**, 51–59 (2008). https://doi.org/10.1080/17452750802047917

Fisher, R.B.: Applying knowledge to reverse engineering problems. Comput. Aided Des. **36**, 501–510 (2004). https://doi.org/10.1016/S0010-4485(03)00158-1

Durupt, A., Bricogne, M., Remy, S., Troussier, N., Rowson, H., Belkadi, F.: An extended framework for knowledge modelling and reuse in reverse engineering projects. Proc. Inst. Mech. Eng. Part B: J. Eng. Manuf. **233**, 1377–1389 (2019). https://doi.org/10.1177/0954405418789973

Ouamer-Ali, M.I., Laroche F., Remy S., Bernard A.: Knowledge and Information Structuring in Reverse Engineering of Mechanical Systems. In: Bouras, A., Eynard, B., Foufou S., Thoben, K.D. (eds.) Product Lifecycle Management in the Era of Internet of Things. PLM 2015. IFIP Advances in Information and Communication Technology, vol. 467. Springer, Cham (2016).https://doi.org/10.1007/978-3-319-33111-9_39

Bruneau, M., Durupt, A., Vallet, L., Roucoules, L., Pernot, J.-P.: A three-level signature by graph for reverse engineering (RE) of mechanical assemblies. In: Tools and Methods of Competitive Engineering—Conference, vol. 1, pp. 669–681 (2016)

Lupinetti, K., Pernot, J.-P., Monti, M., Giannini, F.: Content-based CAD assembly model retrieval: survey and future challenges. Comput. Aided Des. **113**, 62–81 (2019). https://doi.org/10.1016/j.cad.2019.03.005

Osada, R., Funkhouser, T., Chazelle, B., Dobkin, D.: Shape distributions. ACM Trans. Graphics **21**, 807–832 (2002). https://doi.org/10.1145/571647.571648

Novotni, M., Klein, R.: Shape retrieval using 3D Zernike descriptors. Comput. Aided Des. **36**, 1047–1062 (2004). https://doi.org/10.1016/j.cad.2004.01.005

Gal, R., Cohen-Or, D.: Salient geometric features for partial shape matching and similarity. ACM Trans. Graph. Assoc. Comput. Mach. **25**, 130–50 (2006).https://doi.org/10.1145/1122501.1122507

Mousa, M.H.: Matching 3D objects using principle curvatures descriptors. In: Proceedings of 2011 IEEE Pacific Rim Conference on Communications, Computers and Signal Processing pp. 447–52. IEEE (2011). https://doi.org/10.1109/PACRIM.2011.6032935

Xiao, Y.-P., Lai, Y.-K., Zhang, F.-L., Li, C., Gao, L.: A survey on deep geometry learning: from a representation perspective. Comput. Vis. Media **6**(2), 113–133 (2020). https://doi.org/10.1007/s41095-020-0174-8

Gezawa, A.S., Zhang, Y., Wang, Q., Yunqi, L.: A review on deep learning approaches for 3d data representations in retrieval and classifications. In: IEEE Access (Vol. 8, pp. 57566–57593). Institute of Electrical and Electronics Engineers Inc. (2020). https://doi.org/10.1109/ACCESS.2020.2982196

Lupinetti, K., Giannini, F., Monti, M., Pernot, J.-P.: Multi-criteria retrieval of CAD assembly models. J. Comput. Des. Eng. **5**, 41–53 (2018). https://doi.org/10.1016/j.jcde.2017.11.003

Pouget, M.: Geometry of surfaces: from the estimation of local differential quantities to the robust extraction of global differential features. https://tel.archives-ouvertes.fr/tel-00102998 (2006)

Numerical Modeling of the Non-linear Elastic Behavior of a Helical-Shaped Bilayer Composite Spring

F. Boussaoui(✉), H. Lahmam, and B. Braikat

Laboratoire d'Ingénierie et Matériaux (LIMAT), Faculté des Sciences Ben M'Sik, Université Hassan II de Casablanca, BP 7955, Sidi Othman, Casablanca, Maroc
fatima.boussaoui1@gmail.com

Abstract. The aim of this work is to propose an analysis of the non-linear elastic behavior of a non-homogeneous spring, made of several successive layers, the goal is to reduce the weight of the spring while keeping the same stiffness of the steel spring. The resolution of the non-linear problem is done is done by high-order algorithm based on the steps of the asymptotic numerical method (ANM). By using a curvilinear finite element, defined along the mean line of the spring. The kinematics adopted in our theoretical formulation takes into account the hypotheses of Timoshenko. The comparative study is based on a parameter "e" which represents the percentage of the outer layer in relation to the inner layer, the results are compared to the homogeneous case. Numerical tests using a simple example showed the possibility of reducing the weight of the homogeneous steel spring, by replacing it with a composite two-layer steel-carbon spring, where the carbon content does not exceed 10%.

Keywords: Composite spring · Timoshenko cinematic · Curvilinear finite element method · Asymptotic Numerical Method (ANM)

1 Introduction

Springs are frequently used in industry for various applications such as scales, watches, brakes, clutches etc. These mechanical elements exist in several forms; cylindrical, conical, spiral. They are generally designed to work in traction, compression or torsion. There are several research works in the literature that have been developed to study the elastic behavior of these curvilinear structures in the static and dynamic cases (Boussaoui and al., 2021) and (Boussaoui and al., 2022). Nowadays, we notice a great evolution in the manufacture of springs to meet the required needs such as, the reduction of weight, performance and resistance to corrosion, the use of composite materials. (Sequeira et al., 2016) Made a comparison of the mechanical characteristics of a composite spring and a homogeneous steel spring in terms of stiffness and weight reduction. (Stephen et al., 2019) Made studied the mechanical behavior of structural steel, epoxy S-Glass and epoxy-glass coil springs. Carbon under axial loading. (Budan and Manjunatha 2010) checked the possibility of changing from the metal coil spring to the composite coil

L. Azrar et al. (Eds.): CIP 2022, LNME, pp. 388–394, 2023.
https://doi.org/10.1007/978-3-031-23615-0_40

spring. (Bakhshesh and Bakhshesh 2012) investigated the exchange of the steel coil spring with three different composite coil springs.

In this paper, a numerical model is proposed to study the elastic response in the static case of a non-homogeneous spring consisting of two layers with the same mean line, taking into account the geometrical non-linearity. The problem is solved using the algorithm developed in (Boussaoui et al., 2021) based on the Numerical Asymmetric Method (Cochelin et al., 2007).

Our analysis is focused on the study of the composite resort external layer thickness influence considered on the different response curves and this in com-parison with the case of the homogeneous resort (Boussaoui et al., 2021).

2 Position of the Problem

We are working on a composite helical spring with section A, made up of several layers of the same mean line as shown in Fig. 1, or $A(A_1 + A_2 + \dots\dots\dots A_N)$ with N_c is the number of layers in the cross section.

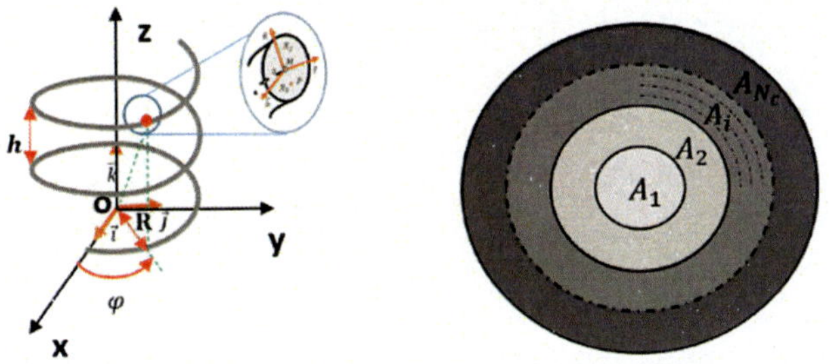

Fig. 1. Mean line and cross section of a composite helical spring1

3 Kinematic

The displacements of the point P located on the section are given by the follow-ing relations (Boussaoui and al., 2021) and (Boussaoui et al., 2022):

$$\begin{cases} U_1(s, X_2, X_3) = u(s) - X_2\theta_3(s) + X_3\theta_2(s) \\ U_2(s, X_2, X_3) = v(s) - X_3\theta_1(s) \\ U_3(s, X_2, X_3) = w(s) + X_2\theta_1(s) \end{cases} \quad (1)$$

s is the curvilinear variable on the mean line, $u, v, w, \theta_1, \theta_2, \theta_2$ are the displacements and rotations at a M.

4 Variational Formulation

The principle of virtual jobs translates into the following relationship (Boussaoui et al., 2021):

$$\delta W_d(U) - \lambda P_{ext}(\delta U) = 0 \tag{2}$$

where $\delta W_d(U)$ represents the variation of the energy of elastic deformations, $P_{ext}(\delta U)$ is the virtual work of external forces and λ is a loading parameter.

Or again:

$$\begin{cases} \int\limits_L \langle \delta\theta \rangle \left([H]^t + [A(\theta)]^t \right) \{S\} ds = \lambda \{P_{ext}\} \\ \{S\} = [D]\left([H] + \frac{1}{2}[A(\theta)] \right)\{\theta\} \end{cases} \tag{3}$$

The matrix [H], $[A(\theta)]$ and the vector $\{\theta\}$ are defined in (Boussaoui et al., 2021). The matrix [D] is the generalized behavior matrix given by.

$$[D] = \sum_{i=1}^{N_c} \begin{bmatrix} E_iA_i & 0 & 0 & 0 & 0 & 0 & 0 & 0 & -\rho_1E_iI_{3i} & E_iI_{3i} & E_iI_{2i} & 0 & \rho_1E_ia_{2i} & \rho_1E_ia_{1i} & 0 \\ 0 & \mu_iA_i & 0 & \rho_1\mu_iI_{3i} & 0 & 0 & 0 & 0 & 0 & 0 & 0 & 0 & 0 & 0 & 0 \\ 0 & 0 & \mu_iA_i & 0 & 0 & \rho_1\mu_iI_{3i} & 0 & 0 & 0 & 0 & 0 & 0 & 0 & 0 & 0 \\ 0 & \rho_1\mu_iI_{3i} & 0 & \mu_iI_{3i} & 0 & 0 & 0 & 0 & 0 & 0 & 0 & 0 & 0 & 0 & 0 \\ 0 & 0 & 0 & 0 & \mu_iI_{2i} & 0 & 0 & 0 & 0 & 0 & 0 & 0 & 0 & 0 & 0 \\ 0 & 0 & \rho_1\mu_iI_{3i} & 0 & 0 & \mu_iI_{3i} & 0 & 0 & 0 & 0 & 0 & 0 & 0 & 0 & 0 \\ 0 & 0 & 0 & 0 & 0 & 0 & \mu_iI_{2i} & 0 & 0 & 0 & 0 & 0 & 0 & 0 & 0 \\ 0 & 0 & 0 & 0 & 0 & 0 & 0 & E_iI_{3i} & 0 & 0 & 0 & \rho_1E_ia_{1i} & 0 & 0 & E_ia_{1i} \\ -\rho_1E_iI_{3i} & 0 & 0 & 0 & 0 & 0 & 0 & 0 & E_iI_{2i} & \rho_1Ea_2 & -\rho_1Ea_1 & 0 & -E_ia_{2i} & -E_ia_{1i} & 0 \\ E_iI_{3i} & 0 & 0 & 0 & 0 & 0 & 0 & 0 & -\rho_1E_ia_{2i} & E_ia_{2i} & E_ia_{1i} & 0 & \rho_1E_ia_{3i} & \rho_1E_ia_{4i} & 0 \\ E_iI_{2i} & 0 & 0 & 0 & 0 & 0 & 0 & 0 & -\rho_1E_ia_{1i} & E_ia_{1i} & E_ia_{2i} & 0 & \rho_1E_ia_{4i} & \rho_1E_ia_{4i} & 0 \\ 0 & 0 & 0 & 0 & 0 & 0 & 0 & \rho_1E_ia_{1i} & 0 & 0 & 0 & E_ia_{1i} & 0 & 0 & \rho_1E_ia_{4i} \\ \rho_1E_ia_{2i} & 0 & 0 & 0 & 0 & 0 & 0 & 0 & -E_ia_{2i} & \rho_1E_ia_{3i} & \rho_1E_ia_{4i} & 0 & E_ia_{3i} & E_ia_{4i} & 0 \\ \rho_1E_ia_{1i} & 0 & 0 & 0 & 0 & 0 & 0 & 0 & -E_ia_{1i} & \rho_1E_ia_{4i} & \rho_1E_ia_{4i} & 0 & E_ia_{4_i} & E_ia_{4i} & 0 \\ 0 & 0 & 0 & 0 & 0 & 0 & 0 & E_ia_{1i} & 0 & 0 & 0 & \rho_1E_ia_{4i} & 0 & 0 & E_ia_{4i} \end{bmatrix} \tag{4}$$

Or μ_i and E_i are respectively the second Lamé coefficient and the Young modulus of the layer i $1 \leq i \leq N_c$ and ρ_1 is the first principal curvature defined in (Boussaoui et al., 2021).

And

$I_{1i} = \int\limits_{Ai} (X_2^2 + X_3^2) dA_i = \frac{\pi r_i^4}{2}$ is the polar quadratic moment. $I_{2i} = \int\limits_{Ai} X_3^2 dA_i = \frac{\pi r_i^4}{4}$ is the quadratic moment with respect to the axis X_2.

$I_{3i} = \int\limits_{Ai} X_2^2 dA_i = \frac{\pi r_i^4}{4}$ is the quadratic moment with respect to the axis X_3.

$a_{1i}, a_{2i}, a_{3i}, a_{4i}$ are forms of additional integrals given by:

$$a_{1i} = \int\limits_{Ai} X_2^2 X_3^2 dA_i = \frac{\pi r_i^6}{24} \quad a_{2i} = \int\limits_{A_i} X_2^4 dA_i = \frac{\pi r_i^6}{8}$$

$$a_{3i} = \int\limits_{Ai} X_2^6 dA_i = \frac{5\pi r_i^8}{64} \quad a_{4i} = \int\limits_{A_i} X_2^2 X_3^4 dA_i = \frac{\pi r_i^8}{64}$$

5 Finite Element Discretisation

The discrete form of the problem in the following form (Bathe 2006):

$$\begin{cases} \sum_{e=1}^{N_e} \int_{L^e} \langle \delta q^e \rangle [B(\theta)]^t \{S\} ds = \lambda P_{ext} \\ \{S\} = [D]\left[[H] + \frac{1}{2}[A(\theta)]\right][G]\{q^e\} \end{cases} \tag{5}$$

with $[B(\theta)] = ([H] + [A(\theta)])[G]$

6 Taylor Series Development

The unknowns of the problem are sought in the form of an integer series truncated at order N with respect to a parameter a defined by:

$$\begin{cases} \{q\} = \{q_0\} + \sum_{p=1}^{N} a^p \{q_p\} \\ \lambda = \lambda_0 + \sum_{p=1}^{N} a^p \{\lambda_p\} \\ \{S\} = \{S_0\} + \sum_{p=1}^{N} a^p \{S_p\} \end{cases} \tag{6}$$

where $\{q_0\}$, $\{S_0\}$ and $\{\lambda_0\}$ are known vectors at the start.

the parameter a is defined by:

$$a = (\langle q \rangle - \langle q_0 \rangle)\{q_1\} + (\lambda - \lambda_0)\lambda_1 \tag{7}$$

After identification according to the powers of "a", we obtain a system of recur-rent equations with the same tangent stiffness matrix:

Order 1

$$\begin{cases} [K_t]\{q_1\} = \lambda_1 \{P_{ext}\} \\ \{S_1\} = [D][B(\theta_0)]\{q_1\} \\ \langle q_1 \rangle \{q_1\} + \lambda_1^2 = 1 \end{cases} \tag{8}$$

Order $2 \le p \le N$

$$\begin{cases} [K_t]\{q_p\} = \lambda_p \{P_{ext}\} + \left\{F_p^{nl}\right\} \\ \{S_p\} = [D][B(\theta_0)]\{q_p\} + \left\{S_p^{nl}\right\} \\ \langle q_1 \rangle \{q_p\} + \lambda_1 \lambda_p = 0 \end{cases} \tag{9}$$

where $[K_t]$ is the tangent stiffness matrix, $\left[F_p^{nl}\right]$ is a quadratic form calculated at every order and $\left[\hat{S}_0\right]$ is the matrix of initial stresses that are given in (Boussaoui and al., 2021).

7 Continuation Technique

This technique allows to connect the different asymptotic braches, solutions of the non-linear problem (5), by updating the tangent matrix at each end of step MAN. Thus the new initial point is written:

$$\{q_0\} \equiv \{q(a_{max})\}, \{S_0\} \equiv \{S(a_{max})\} \text{ et } \lambda_0 \equiv \lambda(a_{max}) \tag{10}$$

Where $a_{\max}$ is defined as follows

$$a_{\max} = \left(\varepsilon \frac{\|U_1\|}{\|U_N\|}\right)^{\frac{1}{N-1}} \tag{11}$$

where ε is the tolerance parameter.

8 Application

In this application, we propose to study the same spring discussed in (Boussaoui and al., 2021) with a two-layer circular cross-section (Fig. 2) having the physical and mechanical properties mentioned in the Table 1.

Table 1. Physical and mechanical properties of the two layers characterizing the composite spring.

Mechanical characteristics	Axial layer in steel	Outer layer carbon
Young's modulus	$E_1 = 2.110^{11} Pa$	$E_2 = 0.710^{11} Pa$
Poisson coefficient	$\nu_1 = 0.3$	$\nu_2 = 0.1$
Volumic mass	$\rho_1 = 7850 Kg/m^3$	$\rho_2 = 1600 Kg/m^3$

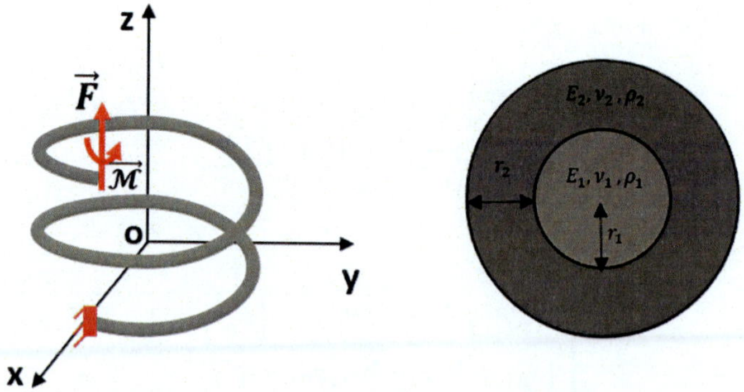

Fig. 2. Two-layer spring and boundary conditions

the mesh used is 59 finite element and the parameters used in the ANM frame-work are: N 10, $\varepsilon = 10^{-10}$ and the ANM steps: $N_{step} = 10$ (Fig. 2).

Using this example, we want to study the influence of the thickness of the outer layer on the nonlinear elastic response of the coil spring under considera-tion. The results obtained will be compared with those of the case of a homoge-neous spring of the same cross-section (Boussaoui and al., 2021).

The parameter representing the percentage of the outer layer in the composite spring is given by

$$e = \frac{r_0 - r_1}{r_1} \times 100 = \frac{r_2}{r_1} \times 100 \tag{12}$$

where is the radius of the section of the homogeneous steel spring (Boussaoui and al., 2021), is the radius of the section of the inner layer and is the radius of the section of the outer layer.

In Figs. 3 and 4 we represent respectively the variation of the displacements U_x, U_y and U_z as well as that of the rotations θ_x, θ_y and θ_z, as a function of the loading perimeter and this for different values of the geometrical perimeter *e*.

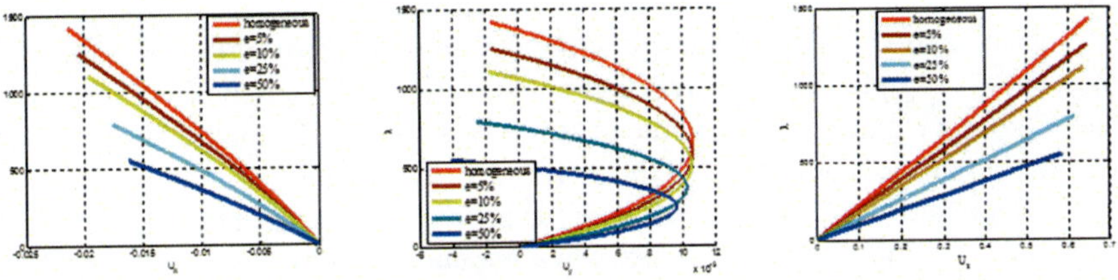

Fig. 3. Load-displacement curves obtained for different values of parameter e

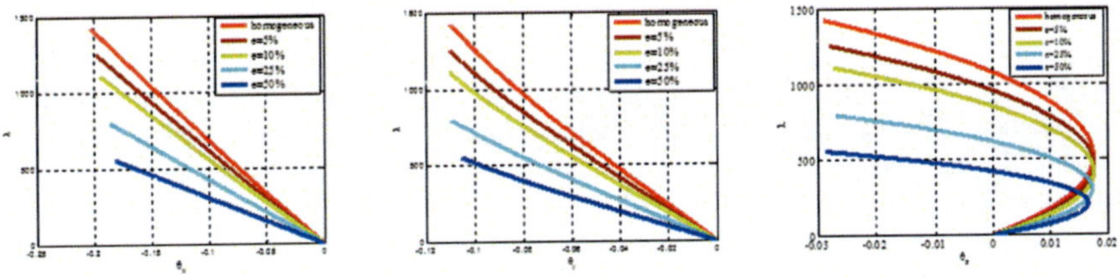

Fig. 4. Load-rotation curves obtained for different values of parameter e

From the previous figures, it can be seen that the use of the external car-bon layer allows to have response curves close to the reference curves associated with the homogeneous case when the value of the geometrical parameter e remains be-low 10%. These results allow us to discuss the reduction of the weight of the studied composite spring as shown in Table 2, while keeping practically the same performance as the homogeneous spring.

With

$$E_m = \frac{m_{\text{homogeneous spring}} - m_{\text{composite spring}}}{m_{\text{homogeneous spring}}} \times 100 \tag{13}$$

Table 2. Variation of the weight of the composite resort as a function of the geometrical parameter e

Geometric parameter e in%	Composite spring volume in m^3	Composite spring mass in Kg	Relative deviation of composite spring mass in %
e = 0	0.1112	873	0
e = 5	0.1112	805	7.76
e = 10	0.1112	741	15.12
e = 25	0.1112	569	34.83
e = 50	0.1112	351	59.71

9 Conclusion

In this work, an analysis of the non-linear response of a helical compo-site spring has been proposed. To show the efficiency of the algorithm performed in this framework, the numerical example of a two-layer composite station was chosen. The central layer of this station is made of steel and the outer layer is made of carbon. Numerical tests carried out on this example showed the possibility of reducing the weight of the steel spring by choosing a suitable percentage of the geometrical parameter, representing the carbon content in the composite spring, while maintaining almost the same non-linear elastic response to the loads imposed on it.

References

Boussaoui, F., Lahmam, H., Braikat, B.: Numerical high-order model for the nonlinear elastic computation of helical structures. Modell. Simul. Eng. (2021)

Boussaoui, F., Lahmam, H., Braikat, B.: An implicit high-order model for dynamic analyses of helical structures. Arch. Appl. Mech. **92**(4), 1381–1395 (2022). https://doi.org/10.1007/s00419-022-02115-3

Sequeira, A., Singh, R.K., Shetti, G.K.: Comparative analysis of helical steel springs with composite springs using finite element method. J. Mech. Eng. Autom. **6**(5A), 63–70 (2016)

Stephen, C., Selvam, R., Suranjan, S.: A comparative study of steel and composite helical springs using finite element analysis. In: 2019 Advances in Science and Engineering Technology International Conferences (ASET), pp. 1–6. IEEE (2019)

Budan, D.A., Manjunatha, T.S.: Investigation on the feasibility of composite coil spring for automotive applications. Int. J. Mech. Mechatron. Eng. **4**(10), 1035–1039 (2010)

Bakhshesh, M., Bakhshesh, M.: Optimization of steel helical spring by composite spring. Int. J. Multi. Sci. Eng. **3**(6), 47–51 (2012)

Cochelin, B., Damil, N., Potier-Ferry, M.: Méthode asymptotique numérique (p. 297). Hermes Lavoissier (2007)

Bathe, K.J.: Finite element procedures. (2006)

Quantifying the Number of Engines and Endurance Effect on the Initial Geometry of an Unmanned Aerial Vehicle Using an Adapted Pre-sizing Method

Amina Kottat(✉) and Mohamed El Amine Ait Ali

Université Mohammed V de Rabat, Ecole Mohammadia d'Ingénieurs, ERG2(ME), Av. Ibn Sina, Rabat, Morocco

amina.kottat@research.emi.ac.ma, aitali@emi.ac.ma

Abstract. A pre-sizing method is necessary to produce an initial geometry of an unmanned aerial vehicle (UAV) to meet the requirements. This geometry is then used in detailed dimensioning of the UAV. Due to the number of formulas used in the pre-sizing method, it is difficult to analyze the effect of the requirement parameters on the initial geometry. In this work we quantify the effect of two important parameters on the results of the pre-sizing method (initial UAV's geometry): number of engines and endurance. We started by describing an adapted pre-sizing method for an UAV and gave all the necessary formulas and how to choose each parameter's value in these formulas. We applied this method in four cases, where we change only one parameter in each case. As a result, the takeoff weight increases when we increase the endurance and decreases when we increase the UAV's number of engines. This effect quantification will help an UAV's designer in better choosing the requirement's parameters for a specific mission.

Keywords: Unmanned aerial vehicle · Pre-sizing of UAV · Aircraft sizing · Conceptual design · Number of UAV's engines effect

1 Introduction

The unmanned aerial vehicle denoted (UAV) are present in many areas, both for military and civilian use. The design of UAVs for a type of application starts with specifications, then the pre-sizing phase to generate an initial geometry of the UAV, based on semi-empirical formulas, and then the detailed dimensioning phase, which consists of the analysis, modification and final validation of the UAV (Niu 1999; Perry and Azar 1982; Megson 2010).

Among the difficulties in the design of a UAV is pre-sizing (Basset et al. 2015). This phase is necessary to define a first geometry that requires, ideally few modifications in the next design phases. To carry out this pre-sizing, it is necessary to have several input data such as: flight mission, payload weight, range, endurance, flight ceiling, rate of climb, cruise and stall speed, takeoff and landing distances. As an output, we evaluate the

L. Azrar et al. (Eds.): CIP 2022, LNME, pp. 395–405, 2023.
https://doi.org/10.1007/978-3-031-23615-0_41

takeoff mass, the needed power and the initial geometry of the UAV which answers all the imposed requirements. Gundlach (2012), Sadraey (2013) and Raymer (2018) proposed a pre-sizing method adapted for planes. The method uses semi empirical formulas and equations to solve (Banal and Ubando 2015), in order to have the results mentioned above. Using these formulas, it is possible to interpret only qualitatively each requirement parameter on the final result. However, for a UAV designer it is more interesting to have a direct quantitative estimation of the effect of each parameter in order to better choose their values. The objective of this work is to adapt this pre-sizing method for UAV first, apply the method for our specific requirement and then quantify the effect of two parameters on the pre-sizing method's results: the number of engines chosen (one or two engines) and the endurance. The choice of these two parameters is based on the need for making the right choice (Palaia and Cipolla 2019) of smaller engines and having a longer endurance (Moelyadi and Zulkarnain 2021). This need makes these two parameters more importance comparing to the other parameters.

In the first part, we present a detailed description of each step of the adapted method of pre-sizing. For each step, we show how to choose each parameter in the semi empirical formulas. Then, we present a case study, with our specific requirements, to illustrate the results of the developed method. In the second part, to quantify the effect of the parameters on the method's results, we perform an analysis of three other cases to compare the effect of two parameters, endurance and number of engines. Theses analysis will help the designer in making a better compromise in choosing the requirements of the UAV for a specific mission.

2 Description of the Adapted Pre-sizing Method and Application to a Case Study

The goal of pre-sizing is to estimate different parameters of the UAV, namely weight to power ratio, wing load, takeoff weight, and to produce a first geometry that should be slightly modified by detailed dimensioning. We use the method proposed by (Gundlach 2012), (Sadraey 2013) and (Raymer 2018) for plane in the case of UAVs. There are six steps: - step 1 defining the requirements – step 2 estimating the weight to power ratio – step 3 estimating the wing loading – step 4 estimating the maximum takeoff weight – step 5 calculating the needed power – step 6 specifying the initial geometry parameters of the UAV. We will describe each step to follow thoroughly. For each step we define each parameter and show how to choose its value.

2.1 Step 1: Specifications' Definition

The first step is to specify the UAV's requirements data including: choice of the number of engines, the weight of the rated payload W_{PL}, the range R, the endurance E, the maximum cruise speed V_{max}, the ceiling H, the stall speed V_S, the takeoff distance d_{TO}, the landing distance d_L and the rate of climb dh/dt. The specification of these parameters depends on the mission and the needs expressed by the final user.

2.2 Step 2: Estimation of Takeoff Weight to Power Ratio

In this step of pre-sizing, an estimate is made according to the category of the aircraft that we want to design. The formula (1) (Taylor 1976) is used to find the value of takeoff weight, W_{TO}, to needed power ratio W_{TO}/P for an UAV with one engine and an UAV with two engines.

$$\frac{P}{W_{TO}} = a \cdot V_{max}^{c} \tag{1}$$

where a = 0.02 and c = 0.22 in the case of an UAV with a single engine and a = 0.03, c = 0.32 in the case of an UAV with two engines (Taylor 1976).

2.3 Step 3: Estimation of Wing Loading

In this step, we estimate wing loading for the UAV (weight of the UAV W divided by the wing area S) in five situations: minimum stall speed, maximum cruising speed, during takeoff, during landing and during climbing (with maximum climbing rate). The result of this step is the minimum wing loading in all these situations. For a minimum stall speed, formula (2) (Sadraey 2013) is used to find the value of wing loading. Where ρ is air density at the specified altitude and C_{Lmax} maximum lift coefficient of the UAV.

$$\frac{W}{S} = \frac{V_s^2 \times \rho \times C_{Lmax}}{2} \tag{2}$$

According to (Gundlach 2012), we should take $C_{Lmax} = 1.6$ in the case of an UAV with one engine and 1.5 in the case of an UAV with two engines.

When the UAV is at its maximum cruising speed we use formula (3) (Loftin 1980) to estimate wing loading. For this calculation we need - the relative density denoted σ (the air density ρ at the takeoff altitude divided by the density at sea level) - V_{max} to find the power factor Ip (Ip is V_{max} divided by 156.25) and the result of step 2 (W_{TO}/P).

$$\frac{W}{S} = I_p^3 \times \sigma \times \left(\frac{W_{TO}}{P}\right) \tag{3}$$

During the takeoff phase we use formula (4) (Raymer 2018). To find this value, we need the ratio found in step 2, the maximum takeoff lift coefficient C_{LTO} (the maximum lift coefficient multiplied by 1.21) and TOP the we calculate using Eq. (5) (Raymer 2018). TOP is a characteristic parameter of takeoff.

$$\frac{W}{S} = TOP \times \sigma \times C_{LTO} \times \left(\frac{P}{W_{TO}}\right) \tag{4}$$

$$d_{TO} = 8.134 \times TOP + 0.01494 \times TOP^2 \tag{5}$$

During the landing of the UAV we use formula (6) (Raymer 2018), to find the needed wing loading to meet a given landing distance requirement.

$$\frac{W}{S} = \frac{d_L \times \rho \times C_{LmaxL}}{2 \times 0.26 \times 1.93} \tag{6}$$

where d_L is given by specifications, and C_{LmaxL} is the maximum landing lift coefficient (we will take the value 1.65 in the case of a one engine and 1.55 in the case of two engines) (Raymer 2018).

During the climbing of the UAV, we characterize it by the rate of climb dh/dt. Which is defined with the height to be climbed as a function of time or as the average vertical speed during the climb. According to Gundlach (2012), Raymer (2018) and Roskam (2003) the rate of climb is maximal when the ratio $(C_L^{\frac{3}{2}}/C_D)$ is maximal. This ratio is given by formula (7).

$$\left(\frac{C_L^{\frac{3}{2}}}{C_D}\right)_{max} = \frac{(3 \times C_{D0} \times \pi \times AR \times e)^{3/4}}{4 \times C_{D0}} \tag{7}$$

To find $(C_L^{\frac{3}{2}}/C_D)$ it is necessary to calculate the Oswald coefficient defined by formula (8), where aspect ratio AR is 7.6 for the UAV with one engine and 7.8 for the UAV with two engines. C_{D0} is the drag coefficient at zero state (is a non-dimensional parameter that relates to drag force of an aircraft to its size, flight altitude and speed) given by formula (9). C_{fe} is the skin friction coefficient (0.0055 for the UAV with one engine and 0.0045 for the UAV with two engines) and S_{wet}/S_{ref} is the ratio of two surfaces: S_{wet} the wetted surface and S_{ref} the reference surface of the wing (equal to 4 for the UAV with one engine and 5 for the UAV with two engines). Formulas (7) and (8) and parameters (AR, S_{wet}/S_{ref}, C_{fe}) are recommended by (Raymer 2018).

$$e = 1.78 \times \left(1 - 0.045 \times AR^{0.68}\right) - 0.64 \tag{8}$$

$$C_{D0} = C_{fe} \times \frac{S_{wet}}{S_{ref}} \tag{9}$$

Then, after determination of $(C_L^{\frac{3}{2}}/C_D)$ we calculate wing loading for the climbing rate using formula (10). Where the propellant efficiency η is 0.8, and W_{TO}/P is found in step 2. And finally, as a result of this step, we identify the minimum value (formula 2, 3, 4, 6 and 10) of wing loading.

$$\left(\frac{W}{S}\right) = \left(\left(\frac{dh}{dt} - \frac{\eta}{\left(\frac{W_{TO}}{P}\right)}\right) \times \left(\frac{C_L^{\frac{3}{2}}}{C_D}\right)\right)^2 \times \frac{\rho}{2} \tag{10}$$

2.4 Step 4: Estimation of Takeoff Weight

To have a first estimation of W_{TO}, we will start by estimating the mass fraction of the fuel denoted by FM_{ca} and the empty mass fraction denoted by FM_{vide}, using formula (11) and (14) given by (Gundlach 2012).

$$FM_{ca} = 1 - \prod_{i=1}^{Nsegs} \left(1 - FM_{ca,i}\right) \tag{11}$$

With N_{seg} is the number of mission segments and $FM_{ca,i}$ the fuel mass fraction for different flight phases. In our applications we chose to have 5 segments in the UAV's mission. $FM_{ca,i} = 0.03$ in the takeoff segment, 0.0015 in the climb segment, 0.005 in the landing segment (Raymer 2018). For the cruising segment we use formula (12) and in the loiter segment we use formula (13).

$$FM_{ca,i} = 1 - e^{\left(\frac{-R.SFC}{\eta\cdot\left(\sqrt{\frac{AR\cdot\pi\cdot e}{4\cdot C_{D0}}}\right)}\right)} \tag{12}$$

where R is the range of action (given by specifications), SFC is the specific fuel consumption (0.5 l/h in the cruising segment, 0.4 l/h in the loiter segment).

$$FM_{ca,i} = 1 - e^{\left(\frac{-E\cdot V_{max}.SFC}{\eta\cdot\left(\sqrt{\frac{AR\cdot\pi\cdot e}{4.C_{D0}}}\right)}\right)} \tag{13}$$

Formula (14) allows us to calculate the mass fraction of the uav's empty mass.

$$FM_{vide} = a + b\cdot W_{TO}^{C1}\cdot AR^{C2}\cdot (P/W_{TO})^{C3}\cdot (W_{TO}/S)^{C4}\cdot V_{max}^{C5} \tag{14}$$

The values of the constants are statistically estimated based on the type of the UAV, and they are given in Table 1 (Roskam 2003).

Table 1. Value of the coefficients for the empty mass.

Number of engine	a	b	c1	c2	c3	c4	c5
One engine	– 0.25	1.14	– 0.2	0.08	0.05	– 0.05	0.27
Two engines	– 0.9	1.32	– 0.1	0.08	0.05	– 0.05	0.2

To calculate the total takeoff weight, we combine formulas (11), (14) and (15), then we solve the obtained non-linear equation of W_{TO}, using the Newton Raphson algorithm (using Matlab).

$$W_{TO} = \frac{W_{PL}}{1-(FM_{ca}+FM_{vide})} \tag{15}$$

2.5 Step 5: Estimating Needed Power

In this step we calculate the needed power P by the trivial formula (16). We used both the takeoff weight W_{TO} and the takeoff weight to needed power ratio W_{TO}/P calculated in the previous steps.

$$P = \frac{W_{TO}}{W_{TO}/P} \tag{16}$$

2.6 Step 6: Definition of the UAV's Initial Geometry

To define the geometric configuration of the different parts of the UAV, we start by calculating the wing area noted Sref (W_{TO} divided by the minimum wing loading). Then we calculate the span, b, by formula (17) that defines the aspect ratio.

$$AR = \frac{b^2}{S_{ref}} \tag{17}$$

To calculate the chord at the root level C_0 we use formula (18), and the chord at the wing tip, C_e, by the expression $C_e = \lambda * C_0$ where $\lambda = 0.5$.

$$C_0 = \frac{2 \cdot S_{ref}}{b \cdot (1 + \lambda)} \tag{18}$$

Then we find the mean aerodynamic chord noted CAM by formula (19) and its position, P_{CAM}, by formula (20).

$$CAM = \left(\frac{2}{3}\right) \cdot C_0 \cdot \left(\frac{1 + \lambda + \lambda^2}{1 + \lambda}\right) \tag{19}$$

$$P_{CAM} = \left(\frac{b}{6}\right) \cdot \left(\frac{1 + 2 \cdot \lambda}{1 + \lambda}\right) \tag{20}$$

After defining the wing's parameters (Raymer 2018), we calculate the fuselage length of the UAV using formula (21) (Roskam 2003).

$$L_f = i \times W_{TO}^j \tag{21}$$

where the typical values used (i, j) is (4.37, 0.23) for the UAV with one engine and (0.86, 0.42) for the UAV with two engines. The next step is to calculate the parameters related to the horizontal and vertical tail (area, span, root chord and tail chord) proposed by (Raymer 2018).

Formula (22) is used to find the surface of the horizontal tail and we calculate the other parameters: S_{ref}, CAM, L_f.

$$S_{HT} = \frac{C_{HT} \times CAM \times S_{ref}}{0.6 \times L_f} \tag{22}$$

where C_{HT} is 0.7 in the case of a one engine and 0.8 in the case of two engines. Then we calculate the span of the horizontal tail, b_{HT}, according to formula (23). The chord at the root level C_{0HT} is calculated by formula (24), and the chord at the end of the horizontal tail noted $C_{eHT} = C_{0HT}/2$.

$$b_{HT} = \sqrt{4 \times S_{HT}} \tag{23}$$

$$C_{0HT} = \frac{2 \times S_{HT}}{b_{HT} \times (1 + \lambda)} \times 0.7 \tag{24}$$

$$S_{VT} = \frac{C_{VT} \times b \times S_{ref}}{0.6 \times L_f} \tag{25}$$

The projected surface of the vertical tail is calculated by formula (25). Where $C_{VT} = 0.04$ in the case of the UAV with a one engine and $C_{VT} = 0.07$ in the case of the UAV with two engines. Then we calculate the span of the vertical tail denoted by b_{VT} using formula (26), the chord at the root level C_{OVT} by the formula (27), and the chord at the end of the vertical tail noted $C_{eVT} = C_{OVT}/2$.

$$b_{VT} = \sqrt{0.64 \times 1.5 \times S_{VT}} \tag{26}$$

$$C_{OVT} = \frac{2 \times S_{VT}}{b_{VT} \times (1 + \lambda)} \times 0.5 \tag{27}$$

2.7 Application of the Pre-sizing Method

To implement the proposed method, we developed a program with an interface in MATLAB to aid in the application of the pre-sizing of UAVs using the described method. It includes all the formulas and the equations mentioned. We applied this method in the case of an UAV with a single engine using the following requirements: endurance 2 h, payload 8.5 kg, $V_{max} = 22.2$ m/s, range 20 km, stall speed 7 m/s, rate of climb 3 m/s, takeoff distance and landing distance 10 m, ceiling 200 m. In step 2, we calculate the value of the ratio $P/W_{TO} = 0.11$ N/W. In step 3, we chose a type of the aircraft (single engine UAV). For each segment we evaluated wing loading. The wing loading that meets all segments requirement $(W/S)_{min} = 18.98$ N/m^2. In step 4 we found the takeoff mass by solving the nonlinear equation. Resulting in $W_{TO} = 29.8$ kg. In step 5 the estimated needed power $P = 1.47$ kW. Then in step 6, we defined the UAV's initial geometry. Table 3 gives the complete parameters necessary to draw the UAV's initial geometry.

3 Study of Number of Engines and Endurance Effect on the Initial Geometry

For an UAV designer quantifying the effect of each requirement parameter on the initial geometry, helps in making a better compromise in choosing these parameters. It also guides him in increasing important parameters like endurance and the payload capacity of the UAV. In this part, we study the effect of the endurance and the number of engines on the initial geometry. We keep all the other requirements' parameters as described in the previous section. Table 4 identifies each case that we will use. Tables 5 and 6 present the results of the 4 cases by the developed application.

3.1 The Effect of Endurance on Initial Geometry of an UAV with One Engine and One with Two Engines

Table 7 synthesizes the variations of the different parameters when the endurance is doubled in both of an UAV with one engine and one with two engines.

Table 3. The results of the total definition of the UAV's geometry

Parameters	S_{ref} (m^2)	b (m)	C_0 (m)	C_e (m)	CAM (m)	PCAM (m)	Lf (m)	SVT (m^2)	bVT (m)	C_{0VT} (m)	C_{eVT} (m)	SHT (m^2)	bHT (m)	C_{0HT} (m)	C_{eHT} (m)
Value	15.69	10.92	1.92	0.96	1.49	2.43	4.2	2.7	1.6	1.12	0.56	6.49	5.09	1.19	0.59

Table 4. Specification of the cases considered

Cases	First case	Second case	Third case	Fourth case
Endurance	2 h	4 h	2 h	4 h
Type of aircraft	Single engine	Single engine	Two engines	Two engines

Table 5. The results of the four case studies: weight, wing parameters, fuselage length

Parameters	WTO (N)	P (kW)	S_{ref} (m^2)	b (m)	C_0 (m)	C_e (m)	CAM (m)	PCAM (m)	Lf (m)
First case	297.80	1.82	15.69	10.92	1.92	0.96	1.49	2.43	4.20
Second case	313.60	1.91	16.52	11.20	1.97	0.99	1.53	2.49	4.25
Third case	225.15	2.87	19.09	12.20	2.09	1.05	1.63	2.71	1.63
Fourth case	230.87	2.94	19.58	12.36	2.11	1.06	1.64	2.75	1.65

Table 6. The results of the four cases: parameters of the horizontal and vertical tail

Parameters	SVT (m^2)	bVT (m)	C_{0VT} (m)	C_{eVT} (m)	SHT (m^2)	bHT (m)	C_{0HT} (m)	C_{eHT} (m)
First case	2.70	1.60	1.12	0.56	6.49	5.09	1.19	0.59
Second case	2.90	1.67	1.16	0.58	6.94	5.27	1.23	0.61
Third case	16.67	4.00	2.78	1.39	25.45	10.09	2.35	1.18
Fourth case	17.11	4.05	2.82	1.41	25.95	10.19	2.38	1.19

Table 7. Effect of endurance on the initial geometry

Parameters	W_{TO} (N)	P (kW)	b (m)	Lf (m)	b_{VT} (m)	b_{HT} (m)
Variation (case 1–case 2) %	+ 5.3	+ 4.9	+ 2.6	+ 1.2	+ 4.4	+ 3.5
Variation (case 3–case 4) %	+ 2.5	+ 2.4	+ 1.3	+ 1.2	+ 1.3	+ 1.0

We notice for the case of an UAV with one engine the takeoff weight increases with 5.3% compared to the first case. The ratios W_{TO}/P and W_{TO}/S do not change when we change the endurance. We also notice that when W_{TO} increases, the needed power increases by 4.9% and the wing area by 5.2%. In the case of an UAV with two engines, we notice an increase in the takeoff weight and the size of the drone. But with a smaller percentage compared to a single engine (W_{TO} increases with 2.5% and the pressure increases with 2.4%).

3.2 The Effect of Number of Engines on the Initial Geometry of the UAV

Table 8 synthesizes the variations of the different parameters when we double the number of engines on the initial geometry. In the case of an UAV with an endurance of E = 2h and one with E = 4h.

Table 8. Effect of the number of engines on the initial geometry

Parameters	W_{TO} (N)	P (kW)	b (m)	Lf (m)	b_{VT} (m)	b_{HT} (m)
Variation (case 1–case 3) %	– 24.4	+ 57.7	+ 11.7	– 61.2	+ 150.0	+ 98.2
Variation (case 2–case 4) %	– 26.4	+ 53.9	+ 10.4	– 61.2	+ 142.5	+ 93.4

We note a reduction of 24.4% for E = 2h and 26.4% for E = 4h of W_{TO} when we increase the number of engines. We also notice an increase in wing's length: an increase of 11.7% for E = 2h and 10.4% for E = 4h. A noticeable increase also of the horizontal tail span: 98.2% for E = 2h and 93.4% for E = 4h. It is the case also for the vertical tail span: an increase of 150.0% for E = 2h and 142.5% for E = 4h. But we notice a reduction of approximately 61.2% of the length of the fuselage when we use two engines instead of one. According to these quantifications, it is best, for these requirement, to use two engines instead of one.

4 Conclusion and Perspectives

In this work, we adapted a pre-sizing method to the case of an UAV. For each of the six steps we presented the formulas used and showed how to choose each parameter in these formulas. We applied this method in a case study with a specific requirement and we gave the initial geometry resulting from this method. To quantify the effect of two important parameters: number of engines in the UAV and its endurance on this initial geometry, we applied this method to three other cases. For each case we changed only one parameter. We quantified the variation of every parameter in the initial geometry. We found, on one hand, that increasing the endurance increases more the takeoff weight in the case of an UAV with one engine (5.3%) comparing to the UAV with two engines (4.9%). On the other hand, we showed that increasing the number of engines significantly reduces (up to 24%) the takeoff weight when we double the endurance. We draw from these results that, for these specifications, it is best to use two engines instead of one. These conclusions will help the UAV's designer to choose better parameters for the requirements necessary for a type of mission. Quantifying the effect of the other parameters will be the logical next step to this work.

References

Banal, L.F., Ubando, A.T.: Fuzzy programming approach to UAV preliminary sizing. In: 8th IEEE International Conference, Humanoid, Philippines (2015)

Basset, P.M., Tremolet, A., Lefebvre, T.: Rotary wing UAV pre-sizing: past and present methodological approaches at Onera. Aerial Robot. (2015). https://doi.org/10.12762/2014.AL08-10

Gundlach, J.: Designing Unmanned Aircraft Systems: A Comprehensive Approach. AIAA, Reston, Virginia (2012)

Loftin, L.K.: Evolution and the Matching of Size to performance. NASA (1980)

Megson, T.H.G.: An introduction to Aircraft Structural Analysis. Elsevier, USA (2010)

Moelyadi, M.A., Zulkarnain, M.F.: HALE UAV ITB perpetual flight. IOP Conf. Ser.: Mater. Sci. Eng. **1109**, 012028 (2021)

Niu, M.C.-Y.: Airframe structural design: practical design information and data on aircraft structures. Adaso Adastra Engineering Center, Hong Kong (1999)

Palaia, G., Cipolla, V.: Preliminary design of a box-wing VTOL UAV. Aircraft Eng. Aerospace Technol. 92(5), 737–743 (2020)

Peery, D.J., Azar, J.J.: Aircraft Structures. McGraw-Hill Book Company, New York (1982)

Roskam, J.: Preliminary Sizing of Airplanes. Design, Analysis and Research Corp. (2003)

Raymer, D.P.: Aircraft Design a Conceptual Approach. AIAA, California (2018)

Sadraey, M.H.: Aircraft Design: A Systems Engineering Approach. Wiley, USA (2013)

Taylor, J.W.R.: Jane's All the World's Aircraft. Jane's London, England, UK (1976)

Green Vehicle Routing Problem (GVRP): State-of-the-Art

Asma Oumachtaq(✉), Latifa Ouzizi, and Mohammed Douimi

Laboratory of Mathematical Modeling and Computer Science, ENSAM, Moulay Ismail University in Meknes, Marjane 2, B.P. 15290, Al-Mansour, Meknes, Morocco
oumachtaqasma@gmail.com

Abstract. Companies today face many demands, not only from customers but also from legislative bodies, associations and other stakeholders. Therefore, sustainable transport management and planning that takes into account economic, societal, legal, and environmental concerns is now a key challenge for public transportation agencies, and an important parameter for company's business continuity. The optimization of the vehicle routes in logistics is a fundamental task for the company in order to reduce the total cost of distribution, fuel consumption and CO_2 emissions. The green vehicle routing problem (GVRP) is an emerging research field that has received recent attention, many researchers have contributed to this emerging research field, studying and analyzing the impact of pollutants—mainly the carbon dioxide equivalent (CO_2) resulting from fleet activities on environment, human health and welfare. This study gives a concise assessment of the literature on GVRP and the major variants that have already been discussed in the literature. Additionally, we provide a mathematical representation of the GCVRP and the method used to calculate the CO_2 emissions matrix that will be used in future projects. We've also highlighted a number of recent 2021 publications on this subject.

Keywords: Green vehicle routing problem (GVRP) · GHG emissions · Literature review · Mathematical model

1 Introduction

The rise in the consumption of goods, the rise in the demand for industrial goods, and the development of technologies in recent years were the main contributors to the evolution of energy consumption and the growth of the quantities of greenhouse gases emitted by the transportation system, which have a significant impact on the environment as they account for between 20 and 25% of global energy consumption and carbon dioxide emissions (World Energy Council 2007). By developing the idea of sustainable transportation and making traditional logistics methods ecologically friendly, this impact can be minimized. Currently, transportation companies and governments are reviewing their processes to address these concerns. Governments and transportation companies are already updating their procedures to solve these issues. They are starting to define their work plans with defined emission reduction goals in mind. The resulting work

L. Azrar et al. (Eds.): CIP 2022, LNME, pp. 406–425, 2023.
https://doi.org/10.1007/978-3-031-23615-0_42

plans must then reduce costs and CO_2 emissions while also striking a balance between the two goals, which are not always positively associated and, in some situations, are completely contradictory.

The VRP vehicle routing problem is the basic model used to serve a set of customers from a depot and find the optimal route that minimizes the total transportation cost. Several studies and researches have devoted themselves to review and classify the VRP problem (Braekers et al. 2006; Eksioglu et al. 2009). The green vehicle routing problem (Green-VRP), a significant and well-studied version of the traditional routing problem, was first described by Sevgı and Mıller-Hooks (2012). Its primary goal is to reduce the overall amount of GHG released from the transportation network.

The purpose of this paper is to present a state of art of green vehicle routing problem considering its more important variants, including the more recent studies published in this field.

The paper is organized as fellows: in Sect. 2 the literature review of capacitated green vehicle routing problem. Next, in the Sect. 3 the GVRP is classified into 7 classes and a brief state of art is presented for each one. Further, recent research published in 2021 are presented in Sect. 4. Finally, Conclusions are drawn in Sect. 5.

2 Classification of GVRP

Due to the significant contribution of freight transportation to GHG emissions and the high level of CO_2 emissions from cars, extensive study has been done to determine how to reduce emissions through effective and environmentally friendly planning of vehicle route.

In the literature GVRP has seven main variants which are Green vehicle routing problem (GVRP)/pollution routing problem (PRP)/The green heterogeneous vehicle routing problem (HVRP)/Energy-minimizing vehicle routing problem (EMVRP)/Time-dependent vehicle routing problem (TDVRP)/Fuel consumption in VRPs (FCVRP)/Electric vehicle routing problem (EVRP), each one cover an aspect of the green Vehicle routing problem. We presented in Table 1 a brief review of each one of those variants (Moghdani et al 2021).

3 Literature Review for Green Capacitated Vehicle Routing Problem (GCVRP)

The Capacitated Vehicle Routing Problem (CVRP) as one of the most important transportation problems in logistics. The standard CVRP is focused on modelling and designing routes that start and end at a single location, identified as a warehouse or distribution center. The objective is to serve a finite set of customers. Considering a fleet of vehicles with homogeneous capacity, while minimizing the total transportation cost without exceeding the capacity of the assigned vehicle (Dantzig and Ramser 1959).

Recently, the routing research community has shown a big interest to the green vehicle routing problem-GVRP as an important variant of VRP (Demir et al. 2014; Lin et al. 2014) with many reviews thinking about its analysis in many real situations. These

Table 1. Literature review for GVRP variants

Variant	Author	Focus of research	Resolution method
GVRP	Schneider et al. (2015)	Investigated the vehicle routing problem with intermediate stops (VRPIS), which necessitates that vehicles make stops at specific locations throughout their route in order to remain functioning	ALNS
	Niu et al. (2018)	Introduced the classic open vehicle routing problem in a sustainable context, with a focus on minimizing costs and emissions	TS
	Messaoud et al. (2018)	Developed a dynamic green vehicle routing problem to minimize the total amount of CO_2 in a dynamic environment	ACO
	Wang and Lu (2019)	Proposed a memetic algorithm with competition for the capacitated green vehicle routing problem, they created a unique decoding approach, and several search operators to solve it	MA with competition (MAC)/KNN
	Dutta et al. (2019)	Introduced a bi-objective GVRP minimizing both distance and CO emissions while considering the demand as a rough variable	NSGA-II

(continued)

Table 1. (*continued*)

Variant	Author	Focus of research	Resolution method
	Ashtineh and Pishvaee (2019)	Introduced new model to evaluate and quantify the effects of alternative fuel cars on an organization's operating costs and the environment by taking into account the amount of toxic compounds that alternative fuels release into the surrounding air	MILP based on WW and TW analysis
	Utama et al. (2021)	Developed an Artificial Bee Colony Algorithm to solve the GVRPTW problem with two objective functions: fuel cost and Delivery costs	ABC
PRP	Demir et al. (2012)	Proposed a heuristic algorithm for the PRP. The algorithm iterates between a VRPTW and a speed optimization problem, they developed an ALNS and a polynomial time technique to solve it	ALNS
	Dabia et al. (2014)	Established a new methodology for PRP in the context of green road freight transportation that incorporates speed and departure time optimization	Branch-and-price algorithm
	Tajik et al. (2014)	Proposed a novel PRP model that incorporates pickup-delivery and time window limitations with uncertain trip times	–
	Kramer et al. (2015)	For the PRP, a fresh speed and departure time optimization technique has been introduced	–

(*continued*)

Table 1. *(continued)*

Variant	Author	Focus of research	Resolution method
	Eshtehadi et al. (2018)	Introduced a mathematical model for the Robust Pollution-Routing Problem using fuel costs and driver pay as the objectives. They used an adaptive large neighborhood search strategy to solve the problem while taking into consideration the degree of data uncertainty	ALNS
	Rauniyar et al. (2019)	Proposed a novel MFEA-based method for resolving the multi-objective PRP. The suggested method simultaneously generates routes by using multi-factorial optimization within NSGA-II	NSGA-II
	Raeesi and Zografos (2019)	Proposed a practical model for PRP in the urban freight distribution network that may balance conventional economic and environmental goals while incorporating a number of fuel-related factors	Path Elimination Procedure (PEP)/MILP
	Yu et al. (2019a)	Developed a mathematical model for heterogeneous fleet PRP (HFPRP) minimizing both the fuel and vehicle variable costs, they used the SA algorithm to solve it	Simulated annealing (SA)

(continued)

Table 1. *(continued)*

Variant	Author	Focus of research	Resolution method
HVRP	Kopfer and Kopfer (2013)	Proposed a new model of the GVRP named as the Emission Minimization Vehicle Routing Problem with Vehicle Categories (EVRP-VC), which takes into account various truck classes with varied CO_2 emission levels	Exact method (CPLEX)
	Koç et al. (2014)	Presented a metaheuristic for the fleet size and mix pollution-routing problem (FSMPRP) was presented out, extending the scope of the pollution-routing problem (PRP) to include the utilization of a heterogeneous vehicle fleet	Hybrid evolutionary metaheuristic
	Zhang et al. (2014)	Proposed a model investigating the problem of fleet size and mixed vehicle routing problem considering CO_2 emissions cost	Numerical simulation/GA
	Moutaoukil et al. (2014)	Fleet size and mix vehicle routing problem (FSMVRP) was studied. The objective was to minimize CO_2 emissions	Exact mathematical formulation/CPLEX
	Yu et al. (2019b)	Introduced a branch-and-price approach using multi-vehicle approximate dynamic programming (MVADP) to precisely solve the heterogeneous HFGVRPTWl. Incorporating various vehicle kinds' price issues into one calculation	TS

(continued)

Table 1. (*continued*)

Variant	Author	Focus of research	Resolution method
	Alinaghian and Zamani (2019)	Incorporating fuel consumption and pollution emissions, a bi-objective model was suggested to solve the inventory routing problem (IRP) during a finite planning period while assuming the fleet composition as heterogeneous	Quantum evolutionary algorithm
	Li et al. (2019)	Proposed a multi-objective linear mathematical model to solve the multi-depot green vehicle routing problem MDGVRP	IACO/L-P metrics approach
EMVRP	Kara et al. (2007)	Provided a model for the Energy Minimizing Vehicle Routing Problem (EMVRP), where cost is determined by multiplying the entire load of the vehicle on the arc by the length of the arc traveled	LP/CPLEX
	Pradenas et al. (2013)	By taking into account the energy needed for each route and evaluating the load and distance between consumers, a model was constructed to evaluate the reduction in greenhouse gas emissions using the vehicle routing problem with backhauls and time windows	Scatter search
	Psychas et al. (2015)	Proposed a hybridization of two metaheuristics to resolve the EMVRP in both pickup and delivery problem	NSGA II/VNS

(*continued*)

Table 1. *(continued)*

Variant	Author	Focus of research	Resolution method
	Cooray and Rupasinghe (2017)	They built a genetic algorithm (GA) to solve the EMVRP after first reviewing the relevant literature on the topic and the usage of metaheuristics as a solution strategy	GA
	Psychas et al. (2018)	Provided a new approach to resolve the Multiobjective Energy Reduction Open Vehicle Routing Problem based on the development of the Influenza Virus in various populations	IVA
	Ghannadpour (2019)	Provided a new model that prioritizes consumers based on time windows for the multi-objective vehicle scheduling and routing problem	NSGA II and CPLEX Solver
TDVRP	Kuo et al. (2009)	A total operating time calculation model was suggested in this study that takes into account time-dependent travel speeds, vehicle loading restrictions, unloading delays, and fuel consumption	TS
	Lewczuk et al. (2013)	Considering changing traffic conditions in urban areas, a mathematical model of the Time-Dependent Vehicle Routing Problem with emission criteria has been presented	Modified version of a GA/a start algorithm

(continued)

Table 1. (*continued*)

Variant	Author	Focus of research	Resolution method
	Norouzi et al. (2017)	Proposed a time-dependent vehicle routing problem (TDVRP) that took traffic congestion into account and lowered carbon emissions along with overall journey time and fuel consumption	PSO algorithm
	Huang et al. (2017)	Presented the time-dependent vehicle routing problem with path flexibility (TDVRP-PF) under traffic congestion given two objectives cost and fuel consumption	–
	Hooshmand and MirHassani (2019)	Presented the vehicle routing problem considering alternative green-fuel powered vehicles. The objective is to minimize CO_2 emissions taking into consideration road congestion and refueling decisions	Hybrid heuristic algorithm
FCVRP	Suzuki (2011)	Creates a novel methodology that reduces fuel use and pollution output. Utilizing the load, speed, and distance as mean parameters	TS
	MirHassani and Mohammadyari (2014)	A fuel consumption model for the capacitated VRP was presented taking into account the vehicle's load as a typical factor affecting CO_2 emissions	Gravitational search algorithm

(*continued*)

Table 1. *(continued)*

Variant	Author	Focus of research	Resolution method
	Zhang et al. (2015)	Presented the problem of optimizing fuel economy while taking into account three-dimensional loading limitations for capacitated vehicles	LS
	Psychas et al. (2016)	Created a Parallel Multi-Start NSGA II algorithm to address two goals of the multiobjective Fuel Consumption Vehicle Routing Problem: minimizing fuel consumption and total travel time	NSGA-II (Non-dominated Sorting GA II)
	Rao et al. (2016)	Proposed a mixed-integer programming model for the FCVRP in which the amount of road gradient is used to evaluate fuel costs	Bi-objective hybrid LS
	Niu et al. (2018)	Outlined the open vehicle routing with time windows problem. The goal is to reduce the overall cost, including fuel emissions costs and driver salaries	Hybrid method based on TS
EVRP	Eguia et al. 2013	With heterogeneous vehicle fleets, a useful variation of the VRP is introduced. In this version, vehicles are characterized by different capacities and prices	Heuristic
	Schneider et al. (2014)	Presented a model for EVRP with time windows considering a restricted battery's capacity and the ability for en route refueling	Hybrid TS/VNS

(continued)

Table 1. (*continued*)

Variant	Author	Focus of research	Resolution method
	Bruglieri et al. (2015)	Proposed a mathematical formulation of the E-VRPTW, optimizing the number of EVs in use as well as the charging, travel, and waiting times	VNS branching
	Yu et al. (2017)	Presented a model for the Hybrid Vehicle Routing Problem (HVRP) minimizing the total cost of travel while considering the utilization of electric and fuel power	Hybrid SA
	Zhao and Lu (2019)	Explores an actual EVRP with time windows considering heterogeneous fleet VRP, with charging stations. The objective is to reduce charging costs and waiting costs	ALNS method
	Basso et al. (2021)	Developed a model for the time-dependent Electric Vehicle Routing Problem with Chance-Constraints (EVRP-CC) and partial recharging also a probabilistic model for energy estimation with Bayesian machine learning was presented	–

MCWSH/DBCA/CIT: the Modified Clarke and Wright Savings heuristic/the Density-Based Clustering Algorithm/customized improvement technique; WW/TW: Well-to-Wheel and Tank-to-Wheel; PSO: particle swarm optimization; IP: integer programming; kNN: k-nearest neighbor; ALNS: adaptive large neighborhood search; LP: linear programming formulations; ILS: iterated local search-based metaheuristic; BOHA: Bi-directional Optimization Heuristic Algorithm; IVA: Influenza Virus Algorithm; LS: the evolutionary local search; IPSO: improved particle swarm optimization algorithm; GA: genetic algorithm; MA: memetic algorithm; ACO: Ant Colony Optimization; SP: Set partitioning procedure studies have been performed studying various types of pollution costs and ecological effects (Kopfer et al. 2014; Küçükoğlu et al. 2013).

In this paper we present a variant of the capacitated vehicle routing problem (CVRP). Toth and Vigo (2002) taking in consideration the environmental issues and GHG emissions, which is known as Green Capacitated Vehicle Routing Problem-GCVRP. The Optimal solution of GVRP is composed of routes minimizing both total traveled distances and fuel consumptions (GHG emission).

It has already been mentioned that the contribution of vehicle routing overviews goes beyond the implicit and unconscious contribution made by reducing travel distance and vehicle numbers. Instead, there are a number of explicit factors related to green transportation issues that can be included in a CVRP model. Apparently, Sbihi and Eglese (2007) and Palmer (2007) was the first to introduce the contribution of CVRP to green transportation. Palmer (2007) proposes an integration of logistical and environmental aspects into one freight demand model to enhance policy analysis.

To extend traditional researches on CVRP with the objective of minimizing fuel consumption and GHG emission (Xiao et al. 2012) consider Fuel Consumption Rate (FCR) as a load dependent function, and add it to the classical CVRP, then a simulated annealing algorithm with a hybrid exchange rule is developed to solve the model.

El Bouzekri, in El Bouzekri and El Hilali (2014), El Bouzekri et al. (2013) and El Bouzekri et al. (2016), presents the Bi-GCVRP problem that aims to find the optimal itinerary for vehicles to serve a set of customers while minimizing the total travelled distance and the total emissions of dioxide of carbon (CO_2). They used the hybrid ant colony, genetic algorithm, evolutionary algorithm and a hybrid metaheuristic for its resolution.

Recently, Shuib and Muhamad (2018) formulated a Mixed Integer Goal Programming (MIGP) model of Green Capacitated VRP (GCVRP) with three objectives which are minimizing the total distance travelled, minimizing the total fuel consumption and minimizing the total Carbon Dioxide emissions. The proposed model is solved by the preemptive GP approach using the MATLAB intlinprog solver.

Because it is impossible to quantify several important variables, such as road slopes, driving behavior, weather conditions, and accidents, the emissions and fuel consumption matrix are frequently computed using approximations (Van Woensel et al. 2009; Palmer 2008). Contemporary research calls for either a fuel-based or distance-based methodology to calculate CO_2 emissions (Palmer 2008; Sheu et al. 2008).

4 Bi-GCVRP Model

The main objective of GVRP is to determine a set of delivery routes with optimal travelled distance and the minimum total emission for all vehicles.

The GCVRP can be modelled mathematically through a complete weighted digraph $G = (V, E)$, where $V = \{0, 1, \ldots, n\}$ is a set of nodes representing the depot or vehicle warehouse (0) and n customers and $E = \{(i, j) \mid i, j \in V\}$ is a set of edges connecting each node, each one with associated a travelled cost Cij. The vehicles are denoted by $K = \{1, 2, \ldots, m\}$.

Further, we adopt the following notation:

qi: Demand of node i.
Q: maximum Capacity of vehicle (all vehicles have the same maximum capacity Q).
Cij: travel cost between i and j (distance between I and j).
eij: amount of CO_2 emitted between i and j.
m: number of vehicles.

The classical linear formulation of the problem is:

x_{ij}^{k} : is a binary variable equal to 1 if customer j is visited immediately after customer i by vehicle k.
$x_{ij}^{k} = 1$: the vehicle k has crossed the road ij.
$x_{ij}^{k} = 0$: otherwise.

The objectives of the problem can be formulated as follows:

$$\min \text{ f1} = \sum_{i=1}^{n}\sum_{j=1}^{n} C_{ij}x_{ij} \tag{1}$$

$$\min \text{ f2} = \sum_{i=1}^{n}\sum_{j=1}^{n} e_{ij}x_{ij} \tag{2}$$

Under the following constraints:

$$\sum_{k \in K}\sum_{i \in V, i \neq j} x_{ij}^{k} = 1 \forall j \in V \setminus \{0\} \tag{3}$$

$$\sum_{j \in V \setminus \{0\}} x_{0j}^{k} = 1 \forall k \in K \tag{4}$$

$$\sum_{i \in V\, i \neq j} x_{ij}^{k} - \sum_{i \in V} x_{ji}^{k} = 0 \forall j \in V, \forall k \in K \tag{5}$$

$$\sum_{i \in V}\sum_{j \in V \setminus \{0\}, i \neq j} q_j x_{ij}^{k} \leq Q \,\forall k \in K \tag{6}$$

$$\sum_{k \in K}\sum_{(i,j) \in S, i \neq j} x_{ij}^{k} \leq |S| - 1 S V \setminus \{0\} \tag{7}$$

$$x_{ij}^{k} \in \{0, 1\} \forall k \in K, \forall (i, j) \in E \tag{8}$$

Our objective function is composed by two different objectives; the objective (1) is to minimize the total distance travelled; objective (2) minimize the CO_2 emissions emitted along the road.

Constraint (3) means "only one visit per vehicle per customer's location" and constraint (4) means "depart from depot", constraint (5) express that the number of vehicles coming in and out of a customer's location is the same, constraint (6) announce that the delivery capacity of each vehicle should not exceed the maximum capacity, constraint

(7) prohibits the creation of sub-tours. Finally, the integrity constraints associated with decision variables are included in (8).

Estimation of CO_2 Emissions

The literature provides a number of studies and methodologies to estimate the CO_2 emission. Among those studies, we discovered the method suggested by Palmer (2008), according to him to minimize CO_2 emissions, it is necessary to establish a CO_2 emission matrix (eij) based on the estimated CO_2 emitted along each arc on the road (Eq. 9)

$$\text{Min} \sum_{i=1}^{n}\sum_{j=1}^{n}\sum_{k=1}^{m} e_{ij}\mathrm{x}_{\mathrm{ij}}^{\mathrm{k}}\,(i \neq j) \tag{9}$$

The emissions and fuel consumption matrix are often calculated using approximations since its estimation requires complex calculations because of the difficulty of quantifying some critical variables as road slopes, driving style, weather conditions, accidents, etc. Contemporary research calls for either a fuel-based or distance-based methodology to calculate CO_2 emissions.

In our case, the total amount of CO_2 emitted by each vehicle depends on the distance traveled and the emission factor (ε) calculated in Table 2.

Table 2. Estimation of the CO_2 emission factor

State of the vehicle	Weight laden (%)	Consumption (l/100 km)	Fuel conversion factor (kg CO_2/l)	Emission factor (kg CO_2/km)
Empty	0	29.6	X2.61	0.773
Lowloaded	25	32.0		0.831
Halfloaded	50	34.4		0.9
High loaded	75	36.7		0.958
Full load	100	39.0		1.018

In the study proposed by Ubeda et al. (2011) (Eq. 10), the authors estimate the matrix (eij) taking account the distance traveled between each pair of nodes (cij) and the emission factor ε(dj) which is computed based on the demand of the arrival node (dj) as follow:

$$e_{ij} = C_{ij} \times \varepsilon\left(d_j\right)\forall i, j \in [1, \ldots, n] \tag{10}$$

5 Recent Studies

The interest in environmental issues and the reduction of CO_2 emissions has grown in recent years. Numerous studies on GVRP and its variations have been published in order to overcome these difficulties, and new methodology and approaches have also been proposed. Among these studies we find that of Dutta et al. (2021), they presented a model for multi-objective green VRP in which he considered two types of vehicles likely company-owned vehicle and third-party logistics in the imprecise environment. The author used a Non-Dominated Sorting Genetic Algorithm II (NSGA-II) to solve the proposed model.

The study in Ferreira et al. (2021) introduced a methodology based on hybrid Meta-heuristic approaches to model and solve the B-objective Green VRP considering two main objectives, minimization of CO_2 emissions and minimization of the difference in demand.

In Di Puglia Pugliese et al. (2021) the author proposed An Adjustable Robust Formulation and developed a row and-column generation solution approach based on a two-stage reformulation for the problem of routing a fleet of electric vehicles in an urban area with Uncertain Waiting Time at Recharge Stations.

Another study of Soysal et al. (2021) was interested in GCVRP by developing a dynamic programming based solution approach involving the ideas of restriction, simulation and online control of parameters to solve the green capacitated routing vehicle problem in case of large-sized instances.

The author in the article Utama et al. (2021) used an Artificial Bee Colony (ABC) algorithm to solve the Green Vehicle Routing Problems with Time Windows (GVRPTW) with an objective of minimizing the total cost of distribution, which involves the total cost of fuel and late delivery costs.

They proposed a spiking neural P system (SN P)-based model with modified rules and learning in conjunction with firefly optimization (FA), which uses the rectified linear unit (reLu) as activation functions to solve the model. The paper Ramachandranpillai and Arock (2021) combined two significant variants of VRP, namely the green vehicle routing problem (GVRP) and the dynamic vehicle routing problem (DVRP).

The study done by Pankaj et al. (2021) presented the green vehicle routing problem with deliveries split into bags and triangular fuzzy travel times, with an objective of minimizing the fuel emissions. They solved the model using a discrete fuzzy-hybridised genetic algorithm.

In the same context (Abdullahi et al. 2021) also studied the vehicle routing problem under environmental aspects, they developed a weighted sum model and an epsilon-constraint model that combine the economic, environmental, and social dimensions, in addition to a biased-randomised iterated greedy algorithm to solve the problem.

Hu Qin presented a state of art for the electric vehicle routing problem, they divided their study into nine classes and reviewed for each one the settings of problem variants and the algorithms used to obtain their solutions, considering the problems in which each vehicle may visit multiple vertices and be recharged during the trip (Qin et al. 2021).

The work of Marrekchi et al. (2021) also contribute in the studies related to green routing problems GRP, they recently proposed a comprehensive review of the current reviews on GRP studies based on a new chronological taxonomy of the literature, gaps for the future studies and the potential impacts for the academic community are presented by the authors.

By contrasting the green vehicle routing problem with simultaneous pickup and delivery G-VRPSPD with the VRPSPD, Olgun et al. (2021) studied the effect of green objective function on total fuel consumption cost. They created a hyper-heuristic (HH-ILS) algorithm based on iterative local search and variable neighborhood descent heuristics to solve the problem.

The green vehicle routing problem was presented by Xiaohui et al. (2021), in which vehicles operate electrically and have a battery with a finite capacity. There are only a certain amount of battery recharging stations available. In order to overcome the issue, this work created a formulation for Mixed Integer Linear Programming and a Large Neighborhood Search.

6 Conclusion

Recent research have looked at environmental concerns in vehicle routing problems in an effort to reduce the amount of GHG emitted, particularly carbon dioxide (CO_2), which has emerged as the major hazard to human health. This article provided a brief literature review of the green vehicle routing problem (GVRP) as one of important variants of VRP, a mathematical model was also presented with a method proposed in the literature to estimate CO_2 emissions. The article focused on research published recently especially from 2016 to 2021.

References

Abdullahi, H., Reyes-Rubiano, L., Ouelhadj, D., Faulin, J., Juan, A.A.: Modelling and multi-criteria analysis of the sustainability dimensions for the green vehicle routing problem. Eur. J. Oper. Res. **292**(1), 143 (2021). https://doi.org/10.1016/j.ejor.2020.10.028

Alinaghian, M., Zamani, M.: A bi-objective fleet size and mix green inventory routing problem, model and solution method. Soft Comput. **23**(4), 1375–1391 (2019). https://doi.org/10.1007/s00500-017-2866-2

Ashtineh, H., Pishvaee, M.S.: Alternative fuel vehicle-routing problem: a life cycle analysis of transportation fuels. J. Clean. Prod. **219**, 166–182 (2019). https://doi.org/10.1016/j.jclepro.2019.01.343

Basso, R., Kulcsár, B., Sanchez-Diaz, I.: Electric vehicle routing problem with machine learning for energy prediction. Transp. Res. Part B: Methodol. 24–55 (2021). https://doi.org/10.1016/j.trb.2020.12.007

Braekers, K., Ramaekers, K., Van Nieuwenhuyse, I.: The vehicle routing problem: state of the art classification and review. Comput. Ind. Eng. **99**, 300–313 (2006). https://doi.org/10.1016/j.cie.2015.12.007

Bruglieri, M., Pezzella, F., Pisacane, O., Suraci, S.: A variable neighborhood search branching for the electric vehicle routing problem with time windows (2015)

Cooray, P., Rupasinghe, T.D.: Machine learning-based parameter tuned genetic algorithm for energy minimizing vehicle routing problem. J. Ind. Eng. **13**, 1–13 (2017). https://doi.org/10.1155/2017/3019523

Dabia, S., Demir, E., Van Woensel, T.: An exact approach for the pollution-routing problem. Relatório Técnico. Beta Res. School Oper. Manag. Logistics (2014)

Dantzig, G.B., Ramser, J.H.: The truck dispatching problem. Manag. Sci. **6**(1), 80–91 (1959). https://doi.org/10.1287/mnsc.6.1.80

Demir, E., Bektas, T., Laporte, G.: An adaptive large neighborhood search heuristic for the Pollution-Routing Problem. Eur. J. Oper. Res. **223**(2), 346–359 (2012). https://doi.org/10.1016/j.ejor.2012.06.044

Di Puglia Pugliese, L., Guerriero, F., Macrina, G.: An adjustable robust formulation and a decomposition approach for the green vehicle routing problem with uncertain waiting time at recharge stations: In: Proceedings of the 10th International Conference on Operations Research and Enterprise Systems - Science and Technology Publication, pp. 72–81 (2021). https://doi.org/10.5220/0010256500720081

Dutta, J., Barma, P.S., Mukherjee, A., Kar, S., De, T., Pamučar, D., et al.: Multi-objective green mixed vehicle routing problem under rough environment. Transport 1–13 (2021). https://doi.org/10.3846/transport.2021.14464

Eguia, I., Racero, J., Molina, J.C., Guerrero, F.: Environmental issues in vehicle routing problems. In: Erechtchoukova, M.G., Khaiter, P.A., Golinska, P. (eds.) Sustainability Appraisal: Quantitative Methods and Mathematical Techniques for Environmental Performance Evaluation, pp. 215–241. Springer, Berlin, Heidclberg (2013). https://doi.org/10.1007/978-3-642-32081-1_10

Eksioglu, B., Volkan Vural, A., Reisman, A.: The vehicle routing problem: a taxonomic review. Comput. Ind. Eng. **57**(4), 1472–1483 (2009). https://doi.org/10.1016/j.cie.2009.05.009

El Bouzekri, A., El Hilali, A.: Evolutionary algorithm for the bi-objective green vehicle routing problem. Int. J. Sci. Eng. Res. **5**(9), 70–77 (2014)

El Bouzekri, A., El Hilali, A.A., Benadada, Y.: A genetic algorithm for optimizing the amount of emissions of greenhouse GAZ for capacitated vehicle routing problem in green transportation. Int. J. Soft Comput. **8**, 406–415 (2013)

El Bouzekri, A., El Hilali, A.A., Benadada, Y.: A hybrid metaheuristic to minimize the carbon dioxide emissions and the total distance for the vehicle routing problem. Int. J. Soft Comput. 409–417 (2016). https://doi.org/10.3923/ijscomp.2016.409.417

Emrah, D., Tolga, B., Gilbert, L.: A review of recent research on green road freight transportation. Eur. J. Oper. Res. 775–793 (2014). https://doi.org/10.1016/j.ejor.2013.12.033

Eshtehadi, R., Fathian, M., Pishvaee, M.S., Demir, E.: A hybrid metaheuristic algorithm for the robust pollution-routing problem. J. Ind. Syst. Eng. **11**(1), 244–257 (2018)

Ferreira, J.C., Teresinha Arns Steiner, M.: A New Approach to the Bi-objective Green Vehicle Routing Problem: Optimization in Newspaper Distribution | Ferreira | Exacta, pp. 1–33 (2021). https://doi.org/10.5585/exactaep.2021.18447

Ghannadpour, S.F.: Evolutionary approach for energy minimizing vehicle routing problem with time windows and customers' priority. Int. J. Transp. Eng. **6**(3), 237–264 (2019)

Hooshmand, F., MirHassani, S.A.: Time dependent green VRP with alternative fuel powered vehicles. Energy Syst. **10**(3), 721–756 (2019). https://doi.org/10.1007/s12667-018-0283-y

Huang, Y., Zhao, L., Woensel, T.V., Gross, J.P.: Time-dependent vehicle routing problem with path flexibility .Transp. Res. Part B: Methodol. 169–195 (2017). https://doi.org/10.1016/j.trb.2016.10.013

Kara, I., Kara, B.Y., Yetis, M.K.: Energy Minimizing Vehicle Routing Problem, Combinatorial Optimization and Applications, pp. 62–71. Springer, Berlin, Heidelberg (2007). https://doi.org/10.1007/978-3-540-73556-4_9

Kopfer, H.W., Kopfer, H.: Emissions Minimization Vehicle Routing Problem in Dependence of Different Vehicle Classes, pp. 49–58. Springer, Berlin, Heidelberg (2013). https://doi.org/10.1007/978-3-642-35966-8_4

Kopfer, H.W., Schönberger, J., Kopfer, H.: Reducing greenhouse gas emissions of a heterogeneous vehicle fleet. Flex. Serv. Manuf. J. **26**(1–2), 221–248 (2014). https://doi.org/10.1007/s10696-013-9180-9

Koç, Ç., Bektas, T., Jabali, O., Laporte, G.: The fleet size and mix pollution-routing problem. Transp. Res. Part B Methodol. **70**, 239–254 (2014). https://doi.org/10.1016/j.trb.2014.09.008

Kramer, R., Maculan, N., Subramanian, A., Vidal, T.: A speed and departure time optimization algorithm for the pollution-routing problem. Eur. J. Oper. Res. **247**(3), 782–787 (2015). https://doi.org/10.1016/j.ejor.2015.06.037

Küçükoğlu, I., Ene, S., Aksoy, A., Öztürk, N.: A green capacitated vehicle routing problem with fuel consumption optimization model. Int. J. Comput. Eng. Res. **3**(7), 16–23 (2013)

Kuo, Y., Wang, C.-C., Chuang, P.-Y.: Optimizing goods assignment and the vehicle routing problem with time-dependent travel speeds. Comput. Ind. Eng. **57**(4), 1385–1392 (2009). https://doi.org/10.1016/j.cie.2009.07.006

Lewczuk, K., Zak, J., Pyza, D., Jacyna-Gołda, I.: Vehicle routing in an urban area: environmental and technological determinants. WIT Trans. Built Environ. **130**, 373–384 (2013)

Li, Y., Soleimani, H., Zohal, M.: An improved ant colony optimization algorithm for the multi-depot green vehicle routing problem with multiple objectives. J. Clean. Prod. **227**, 1161–1172 (2019). https://doi.org/10.1016/j.jclepro.2019.03.185

Lin, C., Choy, K.L., Ho, G.T.S., Chung, S.H., Lam, H.Y.: Survey of green vehicle routing problem: past and future trends. Expert Syst. Appl. 1118–1138 (2014). https://doi.org/10.1016/j.eswa.2013.07.107

Marrekchi, E., Besbes, W., Dhouib, D., Demir, E.: A review of recent advances in the operations research literature on the green routing problem and its variants. Ann. Oper. Res. **304**(1–2), 529–574 (2021). https://doi.org/10.1007/s10479-021-04046-8

Messaoud, E., Idrissi, A.E.B.E., Alaoui, A.E.: The green dynamic vehicle routing problem in sustainable transport. In: 2018 4th International Conference on Logistics Operations Management (GOL), pp. 1–6 (2018). https://doi.org/10.1109/GOL.2018.8378096

MirHassani, S.A., Mohammadyari, S.: Reduction of carbon emissions in VRP by gravitational search algorithm. Manag. Environ. Qual. 766–782 (2014). https://doi.org/10.1108/MEQ-08-2013-0086

Moghdani, R., Salimifard, K., Demir, E., Benyettou, A.: The green vehicle routing problem: a systematic literature review. J. Clean. Prod. 123691 (2021) (2021). https://doi.org/10.1016/j.jclepro.2020.123691

Moutaoukil, A., Neubert, G., Derrouiche, R.: A Comparison of Homogeneous and Heterogeneous Vehicle Fleet Size in Green Vehicle Routing Problem, pp. 450–457. Springer, Berlin, Heidelberg (2014)

Niu, Y., Yang, Z., Chen, P., Xiao, J.: Optimizing the green open vehicle routing problem with time windows by minimizing comprehensive routing cost. J. Clean. Prod. **171**, 962–971 (2018). https://doi.org/10.1016/j.jclepro.2017.10.001

Norouzi, N., Sadegh-Amalnick, M., Tavakkoli-Moghaddam, R.: Modified particle swarm optimization in a time-dependent vehicle routing problem: minimizing fuel consumption. Optim. Lett. **11**(1), 121–134 (2017). https://doi.org/10.1007/s11590-015-0996-y

Olgun, B., Koç, Ç., Altıparmak, F.: A hyper heuristic for the green vehicle routing problem with simultaneous pickup and delivery. Comput. Ind. Eng. **153** (2021). https://doi.org/10.1016/j.cie.2020.107010

Palmer, A.: The Development of an Integrated Routing and Carbon Dioxide Emissions Model for Goods Vehicles, pp. 1–161. School of Management (2007)

Palmer, A.: An integrated routing model to estimate carbon dioxide emissions from freight vehicles. In: Conference Proceedings, University of Hull, pp. 27–32 (2008)

Pankaj, G., Kannan, G., Mukesh Kumar, M., Anisha, K.: Multiobjective capacitated green vehicle routing problem with fuzzy time-distances and demands split into bags. Int. J. Prod. Res. (2021). https://doi.org/10.1080/00207543.2021.1888392

Pradenas, L., Oportus, B., Parada, V.: Mitigation of greenhouse gas emissions in vehicle routing problems with backhauling. Expert Syst. Appl. **40**(8), 2985–2991 (2013). https://doi.org/10.1016/j.eswa.2012.12.014

Psychas, I.-D., Marinaki, M., Marinakis, Y.: A Parallel Multi-Start NSGA II Algorithm for Multiobjective Energy Reduction Vehicle Routing Problem, pp. 336–350. Springer, Cham (2015). https://doi.org/10.1007/978-3-319-15934-8_23

Psychas, I.-D., Marinaki, M., Marinakis, Y., Migdalas, A.: Minimizing the Fuel Consumption of a Multiobjective Vehicle Routing Problem Using the Parallel Multi-Start NSGA II Algorithm, pp. 69–88. Springer, Cham (2016)

Psychas, I.-D., Delimpas, E., Marinaki, M., Marinakis, Y.: Influenza virus algorithm for multiobjective energy reduction open vehicle routing problem. In: Adamatzky, A. (ed.) Shortest Path Solvers. From Software to Wetware, pp. 145–161. Springer, Cham (2018)

Qin, H., Su, X., Ren, T., Luo, Z.: A review on the electric vehicle routing problems: variants and algorithms. Front. Eng. Manag. **8**(3), 370–389 (2021). https://doi.org/10.1007/s42524-021-0157-1

Raeesi, R., Zografos, K.G.: The multi-objective Steiner pollution-routing problem on congested urban road networks. Transp. Res. Part B Methodol. **122**, 457–485 (2019). https://doi.org/10.1016/j.trb.2019.02.008

Ramachandranpillai, R., Arock, M.: A solution to dynamic green vehicle routing problems with time windows using spiking neural P systems with modified rules and learning. J. Supercomput. **77**(9), 9689–9720 (2021). https://doi.org/10.1007/s11227-021-03635-5

Rao, W., Liu, F., Wang, S.: An efficient two-objective hybrid local search algorithm for solving the fuel consumption vehicle routing problem. Appl. Comput. Intell. Soft Comput. **2016**, 7 (2016). https://doi.org/10.1155/2016/3713918

Rauniyar, A., Nath, R., Muhuri, P.K.: Multi-factorial evolutionary algorithm based novel solution approach for multi-objective pollution-routing problem. Comput. Ind. Eng. **130**, 757–771 (2019). https://doi.org/10.1016/j.cie.2019.02.031

Sbihi, A., Eglese, R.W.: The Relationship between Vehicle Routing and Scheduling and Green Logistics - a Literature Survey, pp. 1–25. Department of Management Science, Lancaster University Management School (2007)

Schneider, M., Stenger, A., Goeke, D.: The electric vehicle-routing problem with time windows and recharging stations. Transp. Sci. **48**(4), 500–520 (2014). https://doi.org/10.1287/trsc.2013.0490

Schneider, M., Stenger, A., Hof, J.: An adaptive VNS algorithm for vehicle routing problems with intermediate stops. OR Spectr. **37**(2), 353–387 (2015). https://doi.org/10.1007/s00291-014-0376-5

Sevgi, E., Miller-Hooks, E.: A green vehicle routing problem. Transp. Res. E. 100–114 (2012). https://doi.org/10.1016/j.tre.2011.08.001

Shuib, A., Muhamad, N.A.: Mixed integer multi-objective goal programming model for green capacitated vehicle routing problem. Adv. Transp. Logist. Res. **1**(0), 356–368 (2018). https://doi.org/10.25292/atlr.v1i1.44

Soysal, M., Cimen, M., Cagri, S., Belbag, S.: A heuristic approach for green vehicle routing. RAIRO-Oper. Res. **55**, S2543–S2560 (2021). https://doi.org/10.1051/ro/2020109

Suzuki, Y.: A new truck-routing approach for reducing fuel consumption and pollutants emission. Transp. Res. Transp. Environ. **16**(1), 73–77 (2011). https://doi.org/10.1016/j.trd.2010.08.003

Tajik, N., Tavakkoli-Moghaddam, R., Vahdani, B., Meysam Mousavi, S.: A robust optimization approach for pollution routing problem with pickup and delivery under uncertainty. J. Manuf. Syst. **33**(2), 277–286 (2014). https://doi.org/10.1016/j.jmsy.2013.12.009

Toth, P., Vigo, D.: The Vehicle Routing Problem, pp. 1–17. Society for Industrial and Applied Mathematics, Philadelphia (2002)

Utama, D.M., Fitria, T.A., Garside, A.K.: Artificial bee colony algorithm for solving green vehicle routing problems with time windows. J. Phys.: Conf. Ser. **1933**(1), 1–6 (2021). https://doi.org/10.1088/1742-6596/1933/1/012043

Ubeda, S., Arcelus, F.J., Faulin, J.: Green logistics at Eroski: a case study. Int. J. Prod. Econ. **131**(1), 44–51 (2011)

Wang, L., Lu, J.: A memetic algorithm with competition for the capacitated green vehicle routing problem. IEEE/CAA J. Autom. Sin. **6**(2), 516–526 (2019). https://doi.org/10.1109/JAS.2019.1911405

World Energy Council: Transport Technologies and Policy Scenarios. World Energy Council (2007). Retrieved 26 May

Xiao, Y., Zhao, Q., Kaku, I., Xu, Y.: Development of a fuel consumption optimization model for the capacitated vehicle routing problem. Comput. Oper. Res. **39**(7), 1419–1431 (2012). https://doi.org/10.1016/j.cor.2011.08.013

Xiaohui, L., Peifan, L., Zhao, Y., Yuan, D., Wang, P.: A novel large neighborhood search for solving green vehicle routing (2021). https://doi.org/10.1109/ICAIBD51990.2021.9459066

Yu, V.F., Redi, A.A.N.P., Hidayat, Y.A., Wibowo, O.J.: A simulated annealing heuristic for the hybrid vehicle routing problem. Appl. Soft Comput. **53**, 119–132 (2017). https://doi.org/10.1016/j.asoc.2016.12.027

Yu, V.F., Redi, A.A.N.P., Jewpanya, P., Lathifah, A., Maghfiroh, M.F.N., Masruroh, N.A.: A simulated annealing heuristic for the heterogeneous fleet pollution routing problem. In: Liu, X. (ed.) Environmental Sustainability in Asian Logistics and Supply Chains, pp. 171–204. Springer, Singapore (2019). https://doi.org/10.1007/978-981-13-0451-4_10

Yu, Y., Wang, S., Wang, J., Huang, M.: A branch-and-price algorithm for the heterogeneous fleet green vehicle routing problem with time windows. Transp. Res. Part B Methodol. **122**, 511–527 (2019). https://doi.org/10.1016/j.trb.2019.03.009

Zhang Lu, X., Li, S., Jin, F.: An optimization model on fleet size and mixed vehicle routing problem considering CO_2 emissions cost and its algorithm. In: CICTP 2014: Safe, Smart, and Sustainable Multimodal Transportation Systems, pp. 2715–2725 (2014). https://doi.org/10.1061/9780784413623.260

Zhang, Z., Wei, L., Lim, A.: An evolutionary local search for the capacitated vehicle routing problem minimizing fuel consumption under three-dimensional loading constraints. Transp. Res. Part B Methodol. **82**, 20–35 (2015). https://doi.org/10.1016/j.trb.2015.10.001

Zhao, M., Lu, Y.: A heuristic approach for a real-world electric vehicle routing problem. Algorithms **12**(2), 45 (2019). https://doi.org/10.3390/a12020045

Production Within the Moroccan Cottage Industry: Framing the Problems, Levers and Solutions

Aberkane Mohammed Saad[1](✉) and Youness Farhane[2]

[1] Faculty of Science and Technology, University Sidi Mohammed Ben Abdellah, Fes, Morocco
Mohammedsaad.aberkane@usmba.ac.ma
[2] National School of Applied Sciences, Fes, Morocco
Youness.farhane@usmba.ac.ma

Abstract. The Moroccan artisanal sector covers a large part of the Moroccan population since it guarantees the main source of income; for this, its study and analysis become essential. So, we chose to conduct a survey which helped us to detect the existing failures in the sector as well as the levers which favour the establishment of a technological approach considered as a mean aimed at developing this sector. Thus, the results found showed that the management and communication problems, as well as problems related to the waste of raw materials, remain the most dominant in the sector. Likewise, the respondents indicated an interest in introducing a technological tool (PLM) in their company and finally it was shown through a chi-square test that the introduction of a technological management tool remains independent of the company nature as well as the activity sector.

Keywords: Moroccan craft sector · Management · Communication and organization problems · Problems of waste of raw materials · Chi-square test · Production problems

1 Introduction

It is clear that Morocco is a country which has a diversity of cultures while having a heritage characterized by a mixture of several civilizations which have succeeded one another, namely the Berber, Carthaginian, Roman, Arab-Muslim and Andalusian civilization. This cultural patrimony has allowed the country to have a heritage manifested by historic and monumental cities and which have been classified as a universal and humanitarian heritage (world heritage).

Economically, Morocco is renowned for its various craft activities. Its economy is largely based on the latter. So, apart from agriculture, administration and local authorities, craftsmanship remains an important pillar in terms of production, service provision and job creation.

To tell the truth, the Moroccan handicraft sector remains the second one promoting employment after agriculture. Its trades project the rich heritage of the country namely

L. Azrar et al. (Eds.): CIP 2022, LNME, pp. 426–438, 2023.
https://doi.org/10.1007/978-3-031-23615-0_43

tradition, customs and cultural heritage influenced by all exchanges. For this, more than a hundred trades in double connection with all exchanges as well as with the Moroccan economy have emerged to be taken into consideration: pottery, leather, upholstery, wood, textiles, jewellery, metals, embroidery, zellige, brassware, etc. and which have come together in the form of companies, cooperatives, and mono-craftsmen in order to be able to give a brand image to the country in a lot of events in national and international ones. Among the crafts, we can cite: wood, tapestry, jewellery and metals, pottery, zellige, etc.

In the present work we wanted to reconsider the craft industry as a case-study as well as the management of work within it, namely all the problems faced as well as the proposal of certain technological tools that may be necessary in order to ensure effective management of the latter since the artisanal sector remains an area that requires several measures to ensure progress and modernization because Today, manufacturing management is being re-shaped by a shift from the mass production paradigm to a new on demand [4].

For this, it was considered necessary to carry out a survey within the artisanal sector in Morocco to be able to identify all the needs and problems encountered within this sector where technology can intervene as well as the levers favouring the introduction of the latter to complete our analysis by identifying, determining and framing certain problems that can be solved by the technological tool.

2 Identification of Problems Faced by the Craft Sector

It should be considered that the artisanal industry finds a set of problems limiting the management of all of its products during the first step of identifying needs, several subjects emerged throughout the life cycle of the artisanal product.

- **Environmental problems**: working in the cottage industry generates a multitude of waste that can have a negative effect on the environment whether in the tannery, bone working, metals or brassware. Note that pollution remains the essential factor encountered following all the waste generated in the form of carbon dioxide (atmospheric), liquid discharges or heavy metals or toxic salts.
- **Labour and work management**: to reconsider that artisan work in deplorable conditions which will affect their production because the organization and management of work is ignored in this type of industry.
- **Raw material and energy**: raw material remains an ample problem within the cottage industry with various causes. For example, the insufficiency of the technical means of production as well as the expensive cost prices.
- **Marketing**: It is clear that management strategy ensuring the marketing and dissemination of the craft product in its broad sense is ignored since the sector's participation in fairs and exhibitions remains limited. Also, we can recognize that most artisans remain confined to family structures which is detrimental to the propagation of artisanal art. However, this sector must be more innovative so that it is more attractive and more competitive at the national level or even at the international one. Similarly, foreign products and more specifically the Chinese import harms the marketing of local products since the national market is attacked by the informal sector.

As we have already mentioned, the artisanal industry finds multitude problems (environment, raw materials and energy, marketing, etc.). The artisanal industry is brought back to express its needs in order to remedy all these failures (Fig. 1).

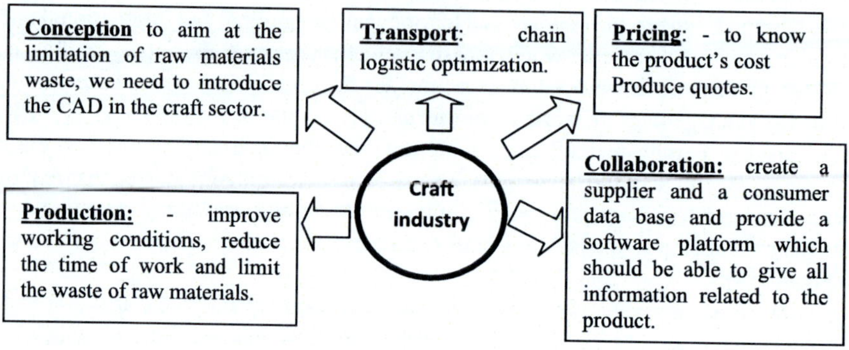

Fig. 1. Needs map for a craft company

We should note that the cottage industry faces a set of problems and failures. Moreover, the management of these found problems remains a challenge that will have to be managed by a set of tools aimed at restructuring the sector in order to breathe new life into the cottage industry while allowing efficient data management through an organization of the sector and an optimal use of the raw material. Furthermore, in order to manage the craft sector, we must take into consideration a real-time response to large orders since craftsmanship has become a cultural heritage of an international character. This will have repercussions on orders for craft products. Likewise, the management of the craft sector remains dependent on the implementation of certain applications (data management, CAD, etc.) which will have benefits on the workforce in order to simplify the tasks of craftsmen; also, this kind of tools will solve a wide range of problems related to product quality, meeting deadlines, optimizing the use of raw materials and the organization of work within the company [5].

Therefore, we were able to identify the problems and levers related to the cottage industry for that we opted for the realization of a survey which will aim to investigate the existence of the problems mentioned in the needs map as well as the levers favouring the implementation of a technological management solution for a certain number of respondents.

3 The Scope of the Study

To contribute to the integration of a technological management solution in the Moroccan artisanal field, it was first necessary to analyse the current situation, namely the problems encountered as well as the levers. That is why, we propose to carry out a survey in order to investigate all the needs and levers pushing craft companies to manage the problems encountered by a technological solution while basing on what was mentioned in the previous paragraph. So, we start by the realization of a certain number of questions which make it possible to identify the problems, the needs and the levers towards technology introduction within the craft companies.

During our study we tried to contact as many companies as possible, but unfortunately, we had answers from 95 companies; also, we tried to differentiate the activity sectors of the companies in order to touch the points studied in a multitude of sectors.

So, after having carried out our survey within the companies mentioned above, we obtained results which will be detailed in the next chapter.

4 Results and Discussion

Before giving a sum up about the results found, we are going to start by a brief description of the companies interviewed.

4.1 Description of the Companies Interviewed

We should note that this survey was conducted among 95 companies in the social economy sector, namely 42 cooperatives, 49 associations and 4 mutuals as shown in Fig. 2.

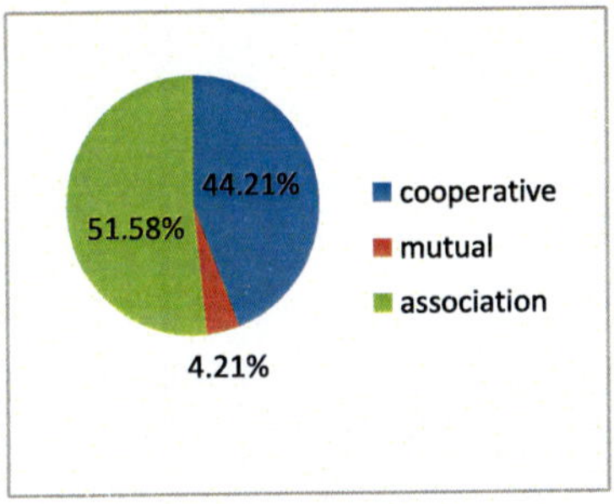

Fig. 2. Nature of the companies interviewed

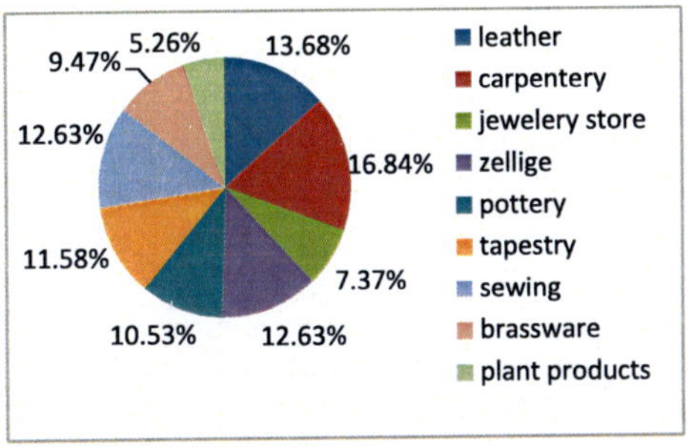

Fig. 3. Breakdown of companies by sector of activity

Likewise, these interviewed companies came from different sectors. So, during our study we tried to multiply the sectors from where they came. To notice that 13.68%

produce in the leather sector, 16.84% in the wood and carpentry sector, 7.37% work in the jewellery sector, 12.63% perform at the zellige sector, 10.53% work in the pottery sector, 11.58% in the tapestry sector, 12.63% in the sewing and embroidery sector, 9.47% in the brass, metal and copper sector and 5.26% are employed in the plant products sector (Fig. 3).

After having carried out a descriptive analysis of all the companies questioned, we move on to the presentation of the problems declared by these companies.

4.2 Presentation of the Problems Encountered by the Interviewed Craft Companies

While analysing the problems encountered by the companies surveyed, we notice the following:

According to Fig. 4 we realize that the companies questioned encounter a variety of problems, namely: 21.05% of companies have management and communication problems, 63.16% have problems relating to the waste of raw materials, 5.26% find problems of transport and logistics chain and 10.53% face costing problems.

After having studied all the problems raised by craft businesses, it is essential to seek a set of solutions through the use of technology, but its use within the cottage industry remains strongly linked to the benefits that technology provides and this point is going to be more discussed in the next paragraph.

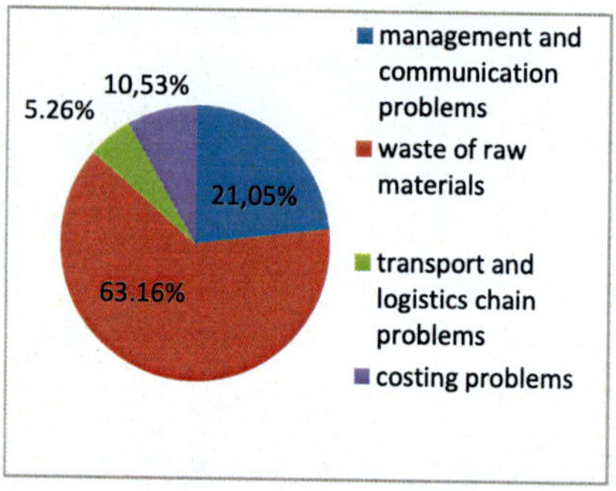

Fig. 4. The problems encountered in a craft company

4.3 Levers and Benefits

The study that we conducted among the companies questioned showed that it is the prospect of the profits to be realized that manifests as the major factor pushing the companies to introduce technology as a mean of management and organization in order to be able to ensure an optimal management of their problems. For that, it was considered essential to see the companies' point of view towards the problems nature that can be unravelled by a technological approach as well as the benefits, the levers and the most influential internal and external factors towards the management of the problems encountered by using a technological approach. Figures 5 and 6 show the results analysis of responses of the companies questioned.

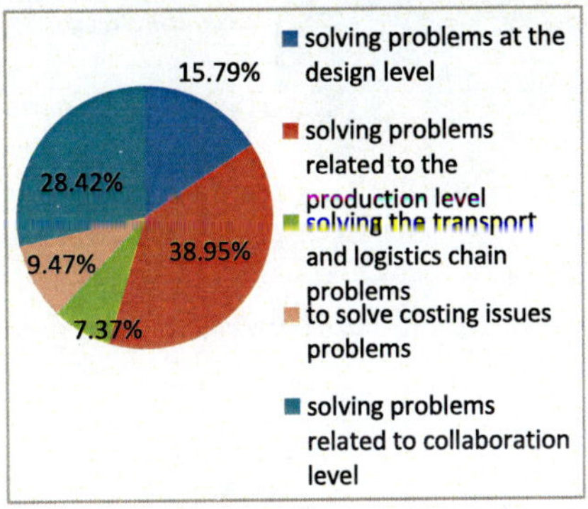

Fig. 5. What can ensure the integration of technological approach to managing the lifecycle of the product into the enterprise

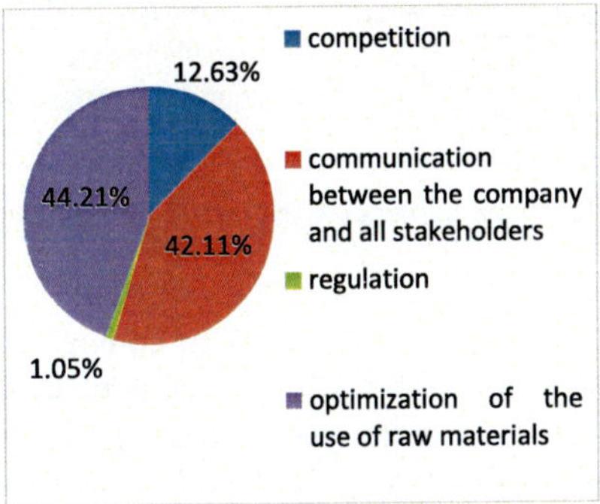

Fig. 6. The external factor pushing enterprises to integrate technology

To consider that 15.79% of the companies surveyed believe that a technological approach to product lifecycle management can solve problems at the design level, 38.95% believe that technology will unravel production problems, 7.37% consider that the technology can opt for the resolution of supplying chain problems, 9.47% state the intervention of technology in the resolution of costing problems and 28.42% in terms of collaboration.

We should note that 12.63% of the companies questioned see that competition is the external factor which can push towards the management of the problems encountered by using technological means, 42.11% prefer the factor relating to the communication between the company and all the stakeholders, 1.05% have opted for regulation and 44.21% have a preference for the factor related to the optimization of the use of raw material (Figs. 7 and 8).

Regarding the internal factor that can push towards management through the integration of technology, we note that 41.05% have a preference for reducing costs and delays,

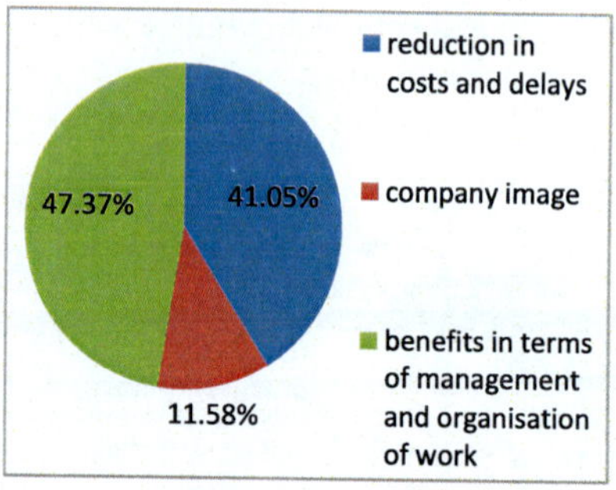

Fig. 7. The internal factor that can push you to integrate the technology

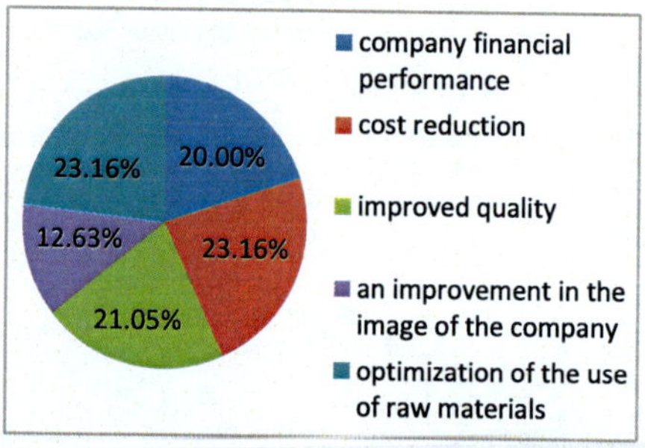

Fig. 8. The levers influencing the integration of technology

11.58% choose the factor relating to the image of the company and 47.37% think about to the benefits in terms of management and organization of work.

Taking into account the levers influencing the integration of technology, we note that 20% of the companies questioned have a preference for financial performance, 23.16% rely on the reduction of costs and 21.05% have opted for the quality improvement, 12.63% select the enhancement of the company image and 23.16% opted for the optimization of the raw material use.

As we have seen previously, the craft sector faces a multitude of problems. But, if we like to solve them we need to test if these problems depend on the companies nature and the activities sector by using the Chi-square test in the next chapter.

4.4 The Chi-Square Test

In order to study and analyse the responses of the companies questioned, it is essential to see if the nature of the company, namely (cooperative, mutual or organization) as well as the sectors of activity will influence the problems found by the company (energy and raw materials, communication, costing, logistics chain). For this, we choose to realize out a chi-square test which relies on the presentation of pivot tables (contingency tables) that cross the two modalities we like to test the relationship between them.

Then the test carried out follows the following steps:

- Ask the hypothesis to be tested: we assume that the two modalities we like to test are independent.
- The creation of a contingency table where we will cross all the modalities taken by the two variables.
- Calculation of the criterion of the observed chi-square test.
- Determination of the significance degree and interpretation of the results. So, for the crossing of the modalities of the two variables (the companies' nature and the problems found) we find Table 1.

Table 1. The contingency table related to the two modalities we like to test (nature of the companies and the problems faced)

	Costing problems	Management and communication problems	Transport and logistics chain problems	Waste of raw materials	Total
Association	5	10	2	32	49
Cooperative	3	8	3	28	42
Mutual	2	2	0	0	4
Total	10	20	5	60	95

After the creation of the pivot table, we move on to the calculation of the criterion of the chi-square test which consists on the application of the following formula allowing the measurement of the difference between the theoretical numbers and the observed ones.

$$\chi^2 = \sum_{j=1}^{L} \sum_{i=1}^{K} \frac{(nij - Tii)^2}{Tij} \tag{1}$$

K: the total of the types of enterprises = 3.

L: the total of the problems faced by the craft industries = 4.

Tij: represent the theoretical numbers that we are going to have if we suppose that the two variables we like to test are independent.

These theoretical numbers are calculated by using the following formula:

Tij = total of row (i) * total of row (j)/total number of observations (95).

nij: represent the observed numbers found in the contingency table (Table 1).

All calculations done; we find Table 2.

After that we can apply the formula 1 to calculate the criterion test, we find a statistic:

$$\chi^2 = \sum_{j=1}^{L} \sum_{i=1}^{K} \frac{(nij - Tii)^2}{Tij} = 11{,}331 \tag{2}$$

while comparing the value of the chi-square criterion found previously (11.331) with the value resulting from the reference table of the chi-square law (we look in the reference

Table 2. The presentation of theoretical numbers Tij

	Costing problems	Management and communication problems	Transport and logistics chain problems	Waste of raw materials	Total
Association	5.157894737	10.31578947	2.578947368	30.94736842	49
Cooperative	4.421052632	8.842105263	2.210526316	26.52631579	42
Mutual	0.421052632	0.842105263	0.210526316	2.526315789	4
Total	10	20	5	60	95

table for the value that corresponds to a probability of 5% with 6 degrees of freedom: 12, 59), we realize that the calculated value is lower than the value taken from the reference table. So, we reject the hypothesis related to independence between the two modalities studied. Hence, we can opt for the integration of a technological tool within the craft industry for solving a specified problem without taking care about the nature of the craft enterprises.

Likewise, we can test if the problems declared by companies depend on the sector of activity. For this, we can use the chi-square test while crossing the two modalities mentioned above, we find the contingency table presented in Table 3.

Table 3. The contingency table related to the two modalities we like to test (the sectors of activities and the problems faced)

	Costing problems	Management and communication problems	Transport and logistics chain problems	Waste of raw materials	Total
Jewellery	0	2	0	5	7
Sewing	2	2	1	7	12
Leather	2	4	0	7	13
Brassware	2	2	0	5	9
Carpentry	2	3	1	10	16
Pottery	0	3	0	7	10
Plant products	1	0	1	3	5
Tapestry	0	3	1	7	11
Zellige	1	1	1	9	12
Total	10	20	5	60	95

After achieving the contingency table that crosses the problems faced by the craft enterprises and the sector of activity, we move on to the calculation of the criterion of the chi square test by using the formula 1 presented above. All calculations done we find the following statistic:

$$\chi^2 = \sum_{j=1}^{L}\sum_{i=1}^{K}\frac{(nij - Tii)^2}{Tij} = 14{,}475 \quad (3)$$

Moreover, we can perform the chi square static by using the SPSS software, we find the following results (Table 4):

Table 4. The Chi square test

	Value	Degree of freedom	Asymptotic significance
Chi square	14,475	24	0.935
Number of observations	95		

While comparing the value of the chi-square criterion found previously (14.475) with the value resulting from the reference table of the chi-square law (we look in the reference table for the value that corresponds to a probability of 5% with 24 degrees of freedom: 36, 42), we realize that the calculated value is lower than the value taken from the reference table. So, we reject the hypothesis related to independence between the two modalities studied. Hence, we can opt for the integration of a technological tool within the craft industry for solving a specified problem without taking care about the sectors of activities related to the craft enterprises.

This means that while implementing a technological tool for solving certain problems within the craft sector, we can choose one sector as a sample (for example the leather one) and the success or the failure of the technological tool isn't going to depend on the sector of activity.

So, it should be noted that our study has shown the existence of a number of problems within the cottage industry and more specifically problems related to the waste of raw material and problems related to collaboration and organization of the sector. For this, the intervention of a new framework for continuous improvement of production process allows improvement in product throughput and product delivery to a customer such as PLM which is a strategic, organizational and technical approach [7].

- Strategic: by characterizing the artisanal product as the only source allowing the artisanal company to generate added value.
- Organizational: by introducing the notion of collaboration and sharing between the company and all the stakeholders.
- Technical: through the introduction of a multitude of IT solutions.

Similarly, it was shown through the chi-square test that the problems raised by the companies questioned remain independent of the company nature as well as the sector of activity, which proves that the implementation of a solution ensuring the resolution of the problems raised, namely the PLM, could be done without taking into consideration the sector of activity or the nature of the company since these characteristics will not have an effect on the problems analyzed.

5 Recommendation of PLM as a Management Approach and an Effective Technological Solution for Problem Solving Within the Cottage Industry

PLM is a multitude of tools, methods and ways that can be effective and useful for the management of the lifecycle of the product [1].

This concept is imposed from day to day in a set of sectors namely the industrial, pharmaceutical, aeronautical, agri-food sector, etc. and which differs according to the needs of the company.

It has been recognized that these systems have regained importance with the advent of Industry 4.0 concepts [13]; this increased recognition stems from the need to integrate technical value chains with commercial chains to help organizations create innovative and personalized products to create markets and competitive advantages [8, 9].

The introduction of the PLM concept in a company is very important. Indeed, this approach makes it possible to highlight the culture of change carried out in a company; also, the PLM does not impose a method to be followed but rather it lights the way while creating a favourable climate for the company to lead in term its goals.

The fact of committing to a PLM approach could play a very important role in the leather sector since the latter will be able to ensure the management of the life cycle of the product from conception to the marketing of the artisanal product as well. It will be able to guarantee collaboration between all the stakeholders, namely the suppliers of raw materials and consumables, the customers, the craftsmen and the master craftsmen as well as all the actors contributing in the production. In addition to that, the PLM will be able to play a role in saving productivity and time since it tries to promote and harmonize creativity and commercial success while trying to optimize all operations throughout the product life cycle as well. The PLM tries to strengthen all the techniques of presentation of the goods (merchandising) as well as their marketing through the use of a software dedicated for the sector which will make it possible to launch on the market an attractive and original product with a planning and a structuring of the work and a sharing of all the tasks in a rational way under a watch on an efficient use of the raw material.

In fact, PLM is a strategy that favours the reduction of time to market and costs while ensuring a preference for values, traditions as well as product quality [11]. Also, its introduction in the sector maintains the circulation of information, its exchange as well as communication with all stakeholders while improving product quality [2, 3]. Note that, the PLM will monitor the costs of all the materials used in the manufacture of the final product (leather valid for use) with a limitation of deadlines and an optimization of transport costs. What is more, the PLM guarantees a complementarity between a product manufactured with a well-qualified professionalism and an innovation in the design while finding a result bringing together authenticity, creativity and modernity and adequate with the trends while eliminating errors and developing models favoured by the customer with efficient management of raw materials and consumables in order to design exceptional products.

Note that the introduction of PLM within a company will have enormous effects which can be presented from different angles as mentioned previously, but it is essential to take into consideration certain barriers before the implementation of such a solution [12] which may look like this:

- Barriers relating to the investigation of technology within the artisanal enterprise. That is to say, barriers relating to the understanding of the stakes of the introduction of the technology as well as its realities which risk according to the companies surveyed to negatively impact the interests of the company [6].
- On the other hand, barriers linked to the integration and acceptance of technology, so we can deduce that this kind of barrier is strongly linked to the lack of know-how and technical knowledge which remain essential for effective integration of technology which leads companies to believe that the integration of technology could negatively impact their product.

6 Conclusion

We should note that, the craft industry remains a cultural heritage which presents the wealth of the country. For this, we have limited the objectives of our work in the specification of the problems, the benefits and the levers within this sector. As a result, we decided to carry out a survey allowing the detection of a set of failures, namely problems relating to management and communication, problems of waste of raw materials. Likewise, we have been able to identify a set of benefits and levers allowing the company to integrate technology within it. To complete our analysis with an interpretation of the results found that confirmed the existence of a predisposition to integrate a technological approach allowing the resolution of the problems encountered which remain independent of the nature of the company and of the sector of activity.

In fact, our study confirmed that 22% of companies face management and communication problems and 63% mentioned the predominance of waste of raw materials. Similarly, the respondents considered that the two external levers pushing the company towards the introduction of technology are communication between the company and all the stakeholders as well as the optimization of the use of the raw material. Regarding internal levers, the structures questioned indicated that it is the prospect of reducing costs and delays as well as the organization of work that can be factors preferring the introduction of technology within the sector.

Throughout all the results and conclusions found, we were able to deduce that the establishment of a technological approach within small and medium-sized enterprises could play an important role through the reduction of the anarchy found during the work as well as the minimization of waste emanating from artisanal work which will present benefits for the society and the environment. Moreover, the establishment of software platforms remains one objective among others targeted by the PLM which will allow efficient data management with the flow of information between all stakeholders in the project and to integrate them in the decision-making [10]. To tell the truth, the PLM remains a tool which favours the coordination and the simplification of the execution of the tasks. Also, it can impact the organization of the company as well as the way of tasks assignment [12].

Integrating the PLM approach, relies primarily on the culture of change. For that, the establishment of a technological approach within the artisanal sector remains a double-edged sword with a total dependence on organizational change while taking into account the resulting risks. Once these recommendations are taken into consideration, a successful PLM approach can be implemented within the craft sector.

References

1. Camba, J.D., et al.: On the integration of model-based feature information in product lifecycle management systems. Int. J. Inf. Manag. **37**(6), 611–621 (2017). https://doi.org/10.1016/j.ijinfomgt.2017.06.002
2. Deuter, A., Rizzo, S.: A critical view on PLM/ALM convergence in practice and research. Procedia Technol. **26**, 405–412 (2016). https://doi.org/10.1016/j.protcy.2016.08.052
3. McKendry, D.A., Whitfield, R.I., Duffy, A.H.B.: Product lifecycle management implementation for high value engineering to order programmes: an informational perspective. J. Ind. Inf. Integr. **26**(2022), 100264 (2022). https://doi.org/10.1016/j.jii.2021.100264
4. Ferreira, F., et al.: Product lifecycle management in knowledge intensive collaborative environments: an application to automotive industry. Int. J. Inf. Manag. **37**(1), 1474–1487 (2017). https://doi.org/10.1016/j.ijinfomgt.2016.05.006
5. Rodríguez, J.C., Gómez, M., Ramírez, K.N.: Competitive advantage in knowledge-based firms of emerging economies: evidence from Mexico. Int. J. Glob. Small Bus. **7**(1) (2015)
6. Chaochotechuang, P., Mariano, S.: Alignment of new product development and product innovation strategies: a case study of Thai food and beverage SMEs. Int. J. Glob. Small Bus. **8**(2) (2016)
7. Sahno, J., Shevtshenko, E., Karaulova, T.: Framework for continuous improvement of production processes. Inzinerineekonomika Eng. Econ. **26**(2), 169–180. ISSN: 1392-2785
8. Singh, S., Misra, S.C.: Identification of barriers to PLM institutionalization in large manufacturing organizations: a case study. Bus. Process Manag. J. **25**(6), 1335–1356 (2019). https://doi.org/10.1108/BPMJ-12-2017-0367
9. Singh, S., Misra, S.C., Kumar, S.: Critical barriers to PLM institutionalization in manufacturing organizations. IEEE Trans. Eng. Manag. (2019). https://doi.org/10.1109/TEM.2019.2911574
10. Soto-Acosta, P., Placer-Maruri, E., Perez-Gonzalez, D.: A case analysis of a product lifecycle information management framework for SMEs. Int. J. Inf. Manag. **36**(2), 240–244 (2015). https://doi.org/10.1016/j.ijinfomgt.2015.12.001
11. Stark, J.: Product Lifecycle Management. 21st Century Paradigm for Product Realisation, vol. 1. Springer, Basel, Switzerland (2015)
12. Tanja, S., Damijan, M.: The optimal selection of internal communication tools during change in organizations. Int. J. Glob. Small Bus. **7**(1) (2015)
13. Wang, D.: Building value in a world of technological change: data analytics and industry 4.0. IEEE Eng. Manag. Rev. **46**(1), 32–33 (2018)

Health, Bioengineering

Investigation of Two Newly Designed Ventricular Assist Device Models

Mohamed Bounouib[1](✉), Hind Benakrach[1], Mourad Taha-Janan[1], Mohamed Es-Sadek Zeriab[2], and Wajih Maazouzi[2]

[1] Laboratory of Applied Mechanics and Technologies, Industrial and Health Science and Technology Research Center (STIS), ENSAM, Mohammed V University in Rabat, Rabat-Institutes, Rabat, Morocco
mohamed.bounouib@um5s.net.ma

[2] Industrial and Health Science and Technology Research Center (STIS), ENSAM, Mohammed V University in Rabat, Rabat-Institutes, Rabat, Morocco

Abstract. Over the years, ventricular assist devices (VADs) have helped save and prolong the lives of thousands of patients with end-stage heart failure. However, many people have had severe complications after transplantation, and some have even lost their lives. This article will focus on one of the most prevalent complications associated with VADs: hemolysis. In VADs, numerous studies have proven that the occurrence of hemolysis is significantly related to both the shear stress and the exposure time. This paper will analyze the performance of two new VADs in terms of hydraulic properties, induced shear stress, exposure time and hemolysis. The simulations results revealed Model B's dominance in terms of hydraulic properties and Model A's dominance in terms of hemolysis. In terms of exposure time, the performance of the two models varied depending on the rotational speed used. Model A showed a shorter exposure time at rotational speeds below 10,000 rpm, but at rotational speeds above 10,000 rpm, Model B's exposure time became the shortest. As for the induced shear stress, there was no apparent difference between the two models. Comparing what was obtained with what was published, the obtained results were very promising and showed the great potential of both models.

Keywords: Ventricular assist device · Hemolysis · Hemodynamics · Shear stress

1 Introduction

1.1 A Subsection Sample

Congestive heart failure is a chronic heart disease in which the heart cannot pump enough blood to the rest of the body. The lack of blood supply affects all primary body functions, which can manifest as excessive fatigue, shortness of breath, loss of appetite, irregular heartbeat, heart palpitations, or sudden weight gain.

As the number of people suffering from it has increased from 33.4 million in 1990 to 64.3 million in 2017 [1], heart failure has become an emerging epidemic. In developed

L. Azrar et al. (Eds.): CIP 2022, LNME, pp. 441–448, 2023.
https://doi.org/10.1007/978-3-031-23615-0_44

countries or countries with very high incidence, unique associations and organizations have been created to assess and monitor the spread of this disease.

The most famous heart failure associations are the American College of Cardiology Foundation (ACCF), the American Heart Association (AHA), and the New York Heart Association (NYHA). Each year, these associations publish statistics on heart disease and stroke and guidelines for managing heart failure. Management guidelines vary depending on the development and progression of the disease or the severity of symptoms.

The ACCF and AHF use the development and progression of heart failure as a criterion and divide heart failure into four stages (A, B, C, and D) [2]. On the other hand, NYHA uses a functional classification based on symptom severity as a criterion and divides heart failure into four categories (I, II, III, and IV) [3].

Because heart failure is an incurable disease, management guidelines generally aim to slow or prevent the disease's progression. The treatments mentioned in the guidelines can be divided into four types:

- Healthy lifestyle changes: Eat a healthy diet and exercise.
- Medications: angiotensin II receptor blockers, beta-blockers, hydralazine.
- Implanted cardiac rhythm control devices: defibrillators, cardiac resynchronization therapy devices, implantable defibrillators.
- Surgery: valve surgery, heart transplantation, ventricular assists device implantation, artificial heart implantation.

In the early stages (A, B, and C) or (I, II, and III), treatments such as a healthy lifestyle, medications, and Implanted cardiac rhythm control devices can help reduce symptoms and improve patients' lives. However, in the end-stage (D) or (IV), these treatments have little to no effect on the patient's condition, so surgery is last.

As previously mentioned, surgical procedures include heart transplantation, ventricular assist device implantation, and artificial heart implantation. Of these surgical procedures, heart transplantation is the best option for patients with end-stage heart failure because the patient receives a new, healthy heart. However, when it comes to organ transplants, the issue of organ donation shortage is always raised. Moreover, the heart is no exception. An alternative had to be found to overcome the organ donation shortage, which is why ventricular assist devices were invented.

Since their invention, ventricular assist devices have undergone a remarkable evolution, with the appearance of many models that differ in terms of functions, capabilities, and purposes of use. This improvement in ventricular assist devices has saved thousands of patients with congestive heart failure. However, despite the improved performance of VADs in recent years, these devices are still not free of defects and can, in some cases, lead to severe complications that can result in patient death. Hemolysis is considered the most common among the complications that follow VAD implantation.

Hemolysis is one of the most studied aspects of mechanically induced blood damage. In general, hemolysis is defined as the rupture of red blood cells and the release of the cytoplasm they contain into the surrounding plasma. Hemolysis occurs when red blood cells are exposed to high-level shear stress, and hemolysis can also occur when red blood cells are exposed to low-level shear stress for an extended period.

This paper will study and analyze two of our newly developed ventricular assist device models. The study will focus on hydrodynamic properties, induced shear stress, exposure time, and hemolysis.

2 Method

Devices: In this study, we will use two newly designed ventricular assist devices. Both devices provide continuous flow and are in the form of two multipart axial pumps [4].

Fig. 1. The final design of Model A

Fig. 2. The final design of Model B

The first model (Model A) consists of three parts: the inductor, rotor, and diffuser, while the second model (Model B) consists of four parts: the inductor, rotor, diffuser, and straightener. The complete design of both models is shown in Figs. 1 and 2.

The initial designs were created using 3D modelling software. Once the designs were completed, the two models were rebuilt using BladeGen. The components of both models were built separately according to the initial designs.

Mesh: TurboGrid was used to create the meshes for all parts of both models (inducer, rotor, diffuser, and straightener). Since TurboGrid was explicitly designed for turbomachinery, it uses an O-grid in the areas around the blades to ensure high element concentration while using an H-grid in the spaces between the blades. We set a clearance gap of 0.15 between the blade tips of the rotor and the inner wall of the housing. This configuration ensures that high-pressure gradients are captured. Figure 3 shows a sample of the obtained mesh.

Computational Fluid Dynamics: Blood flow was calculated using the commercial software ANSYS CFX. All simulations were performed in a steady-state flow condition, with blood modelled as an incompressible Newtonian fluid with a dynamic viscosity of 0.0035 Pa s and a density of 1050 kg m^{-3}.

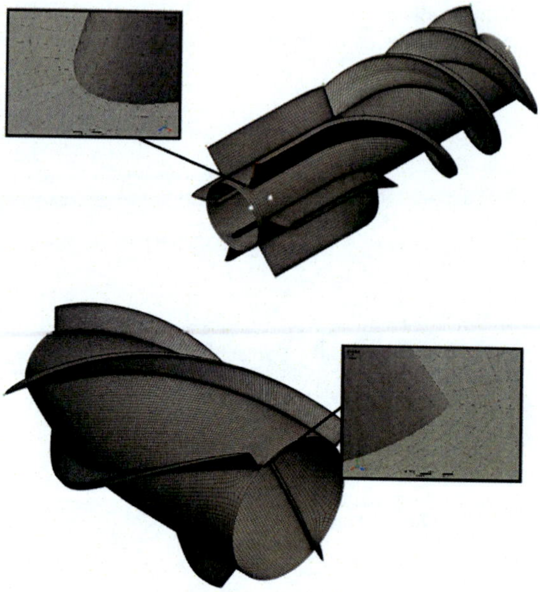

Fig. 3. Meshes used for the VAD calculations (Model A rotor (top), Model B rotor (bottom))

In order to obtain accurate results, the boundary conditions were defined to simulate a critical case of heart failure. Thus, an 80 mmHg static pressure and a 5 L/min flow rate were used at the inlet and outlet. A no-slip condition was used on the walls. The rotor domains were designated as rotational domains with speeds ranging from 7000 to 15,000 rpm. Except for the rotor domains, all the other domains were designated as stationary regions.

Shear Stress: The shear stress was calculated using the three-dimensional numerical model proposed by Bludszuweit [5].

$$\tau = \left[\frac{1}{3}\left(\tau_{ii}^2 + \tau_{jj}^2 + \tau_{kk}^2\right) - \frac{1}{3}\left(\tau_{ii}\tau_{jj} + \tau_{jj}\tau_{kk} + \tau_{kk}\tau_{ii}\right) + \left(\tau_{ij}^2 + \tau_{jk}^2 + \tau_{ki}^2\right)\right]^{1/2} \quad (1)$$

Hemolysis: To predict hemolysis, we used the mathematical model proposed by Giersiepen [6]. This model combines both shear stress and exposure time to estimate hemolysis induced by the ventricular assist device.

$$HI(\%) = 3.62 \times 10^{-5} \times t^{0.785} \times \tau^{2.416} \quad (2)$$

3 Results

To evaluate the hydraulic properties, induced shear stress, exposure time, and hemolysis. Both models were simulated using different rotation speeds, and using different design points allowed us to get a clear view of the potential of each model.

Figure 4 shows the pressure raise curves for both designs. At the same design point, Model B provides a higher pressure rise than Model A. The difference in pressure rise increases as the rotational speed increases.

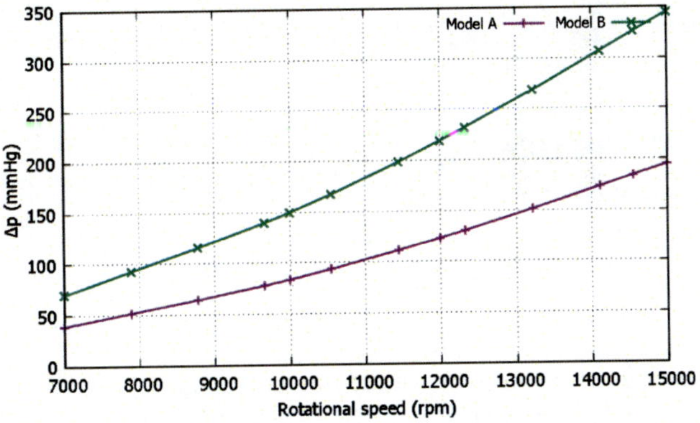

Fig. 4. Pressure rise curves for both models at different rotational speeds

We monitored the shear stress applied to each red blood cell at each time point by following the streamlines. The shear stress was analyzed using mass distribution histograms. Figures 5 and 6 show the mass distribution of shear stress in both models at a rotation speed of 10,000 rpm. From the two figures, we can see that at a rotational speed of 10,000 rpm, both models produce shear stress magnitudes below 300 Pa. As for the low shear stress magnitude, in Model A, 99.97% of the points are subjected to shear stress magnitudes below 150 Pa versus 99.98% in Model B.

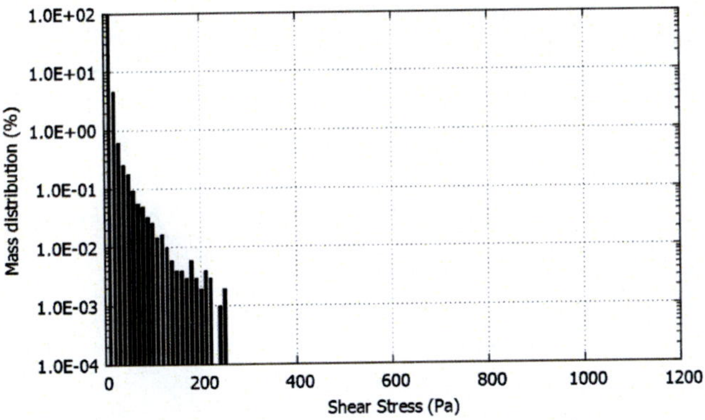

Fig. 5. Mass distribution of shear stress in Model A

Due to the large number of red blood cells used in this analysis, the trajectories followed by each particle differ from each other, resulting in a difference in exposure time. To address this issue, we calculated the average exposure time for both models. Figure 7

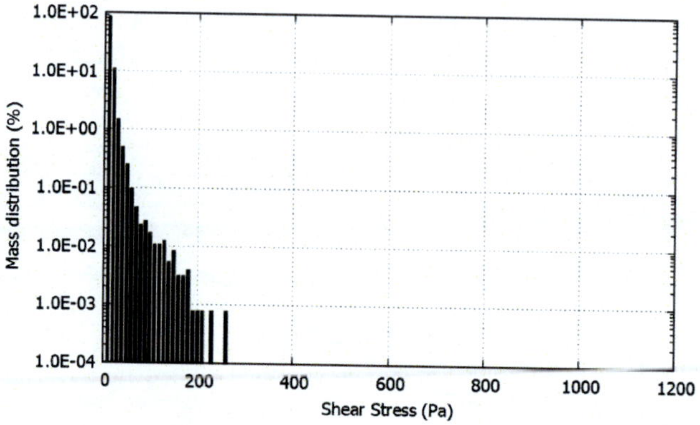

Fig. 6. Mass distribution of shear stress in Model B

compares the average exposure time for the two models at different rotational speeds. It can be seen that at rotational speeds below 10,000 rpm, the exposure time of Model B exceeds that of Model A. However, the difference decreases as the rotational speed increases. At rotational speeds above 11,000, the exposure time of Model B becomes less than that of Model A.

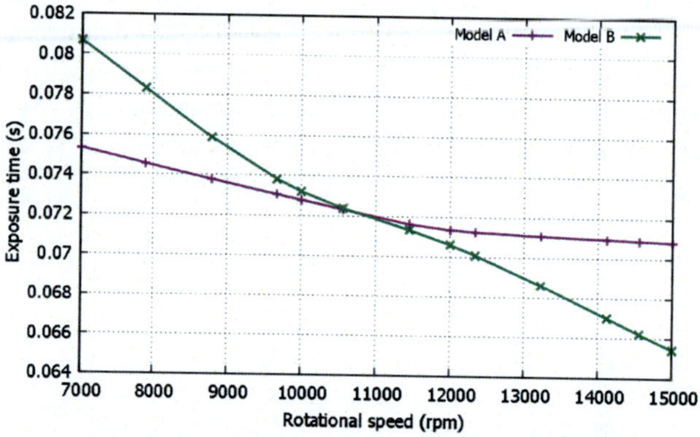

Fig. 7. Exposure time curves for Model A and Model B

For induced hemolysis, it can be seen in Fig. 8 that Model B produces more hemolysis than Model A. The amount of induced hemolysis increases with the rotation speed.

4 Conclusion

This paper investigates the induced shear stress, exposure time, and hemolysis of the two newly designed ventricular assist device models. After performing a series of simulations and analyzing the final results, we found that the performance of the two designs differed depending on the aspect investigated.

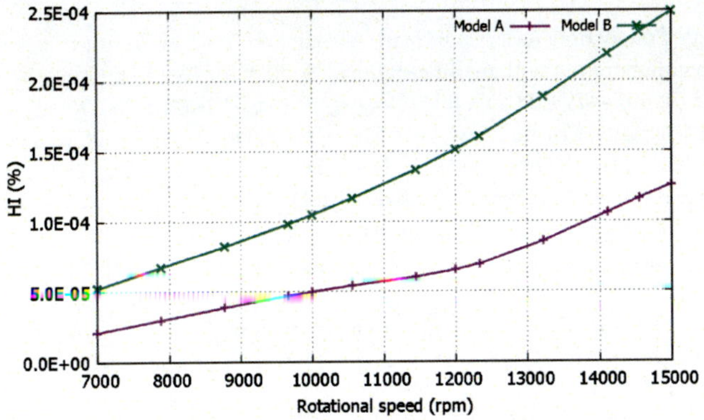

Fig. 8. Hemolysis curves for Model A and Model B

In terms of hydraulic properties, Model B provided a higher pressure raise than Model A when comparing the two models at the same design point, which means that by using Model B, we can achieve the same pressure raise as in Model A by using a lower rotational speed.

In terms of shear stress, the shear stress' mass distribution histograms showed almost the same results in both models at a rotational speed of 10,000 rpm. In ventricular assist devices, shear stress magnitudes above 150 Pa are usually associated with hemolysis [7]. However, in our models, less than 1% of the investigated points are subjected to shear stress magnitudes $<$ 150 Pa, which means that the performance of both models was good.

For exposure time, at rotational speeds below 100,00 rpm, the average exposure time of Model A was shorter than that of Model B. However, when rotational speeds exceed 11,000 rpm, the exposure time of model B becomes shorter, which means that the particles passing through Model B are exposed to the device-induced shear stress for a shorter time than particles passing through Model A.

In terms of hemolysis, although the hemolysis index increased when the rotational speed was increased, the performance of Model A was better across all design points.

If we compare the performance of the two models with that obtained when analyzing VADs of the same type [8]. It can be said that the performance of the two models was exceptional in all aspects. At the same operating conditions, Our models generated high pressure compared to the 3 axial flow VAD investigated by Fraser et al. This is also valid for shear stress, exposure time, and hemolysis index.

In summary, Model B was chosen to be the subject of a more profound investigation.

References

1. Bragazzi, N.L., Zhong, W., Shu, J., Abu Much, A., Lotan, D., Grupper, A., Younis, A., Dai, H.: Burden of heart failure and underlying causes in 195 countries and territories from 1990 to 2017. Eur. J. Prev. Cardiol. (2021). https://doi.org/10.1093/eurjpc/zwaa147
2. Hunt, S.A., et al.: 2009 focused update incorporated Into the ACC/AHA 2005 Guidelines for the Diagnosis and Management of Heart Failure in Adults: a report of the American College

of Cardiology Foundation/American Heart Association Task Force on Practice Guidelines: developed in collaboration with the International Society for Heart and Lung Transplantation. Circulation **119**, 391–479 (2009). https://doi.org/10.1161/CIRCULATIONAHA.109.192065

3. The Criteria Committee of the New York Heart Association: Nomenclature and criteria for diagnosis of diseases of the heart and blood vessels. J. Am. Med. Assoc. **153**, 891–891 (1953). https://doi.org/10.1001/jama.1953.02940260115033
4. Bounouib, M., Benakrach, H., Es-Sadek Zeriab, M., Taha-Janan, M., Maazouzi, W.: Numerical study of a new ventricular assist device. Artif. Organs. **44**, 604–610 (2020). https://doi.org/10.1111/aor.13635
5. Bludszuweit, C.: Three-dimensional numerical prediction of stress loading of blood particles in a centrifugal pump. Artif. Organs. **19**, 590–596 (1995). https://doi.org/10.1111/j.1525-1594.1995.tb02386.x
6. Giersiepen, M., Wurzinger, L.J., Opitz, R., Reul, H.: Estimation of shear stress-related blood damage in heart valve prostheses - in vitro comparison of 25 aortic valves. Int. J. Artif. Organs. **13**, 300–306 (1990). https://doi.org/10.1177/039139889001300507
7. Alemu, Y., Bluestein, D.: Flow-induced platelet activation and damage accumulation in a mechanical heart valve: numerical studies. Artif. Organs. **31**, 677–688 (2007). https://doi.org/10.1111/j.1525-1594.2007.00446.x
8. Fraser, K.H., Zhang, T., Taskin, M.E., Griffith, B.P., Wu, Z.J.: A quantitative comparison of mechanical blood damage parameters in rotary ventricular assist devices: shear stress, exposure time and hemolysis index. J. Biomech. Eng. **134**, 081002 (2012). https://doi.org/10.1115/1.4007092

Toward a Dielectric Modeling of Ovarian Tumors Using the Mathematical Models of the Blood-Based Biomarker CA125 and the Blood-Borne Tumor-Shed Biomarker SEAP

Oumaima El Hassani(✉) and Adil Saadi

Control, Piloting and Supervision of Systems, Ecole Nationale Supérieure d'Arts et Métiers, Moulay Ismail University of Meknès, Meknes, Morocco
oumaima.elhassani@edu.umi.ac.ma, a.saadi@ensam-umi.ac.ma

Abstract. Ovarian cancer is still the deadliest gynecologic malignancy, with a five-year survival rate of about 47% in all stages. Early detection remains the key to reaching survival gains. Through the present study, we aim to exploit the latest developments in mathematical studies dedicated to the detection of epithelial ovarian tumors by using two different biomarkers: CA125, which is a blood-based biomarker actually used in clinical analysis, and SEAP, which is exclusively tumor-shed and currently under study. CA125 is secreted in plasma (one compartment) and studied for four different scenarios. SEAP is secreted in plasma and the periphery (two compartments) and allows the study of the aggressiveness of the ovarian tumor in the two media. Using the one- and two-compartment models, we can specify the detection times of ovarian tumors for each scenario with respect to current clinical imaging modalities and their corresponding tumor sizes and concentrations of biomarkers. Those results are exploited in the dielectric study of the biomarkers to define the dielectric strength or increment for each model and scenario of the development of the ovarian tumor in order to be used in further dielectric modeling of ovarian tumor biomarkers.

Keywords: Ovarian cancer · Biomarker · CA125 · SEAP · One-compartment model · Two-compartment model · Sensitivity · Specificity · Dielectric strength · Clinical imaging · Plasma · Periphery · Detection time

1 Introduction

Time is a matter of life and death for cancer patients (Hazelton and Luebeck, 2011). The detection of tumors before they migrate to distant areas is thus an important priority in cancer research (Cohen et al., 2018). Effective early invasive cancer screening is, without an uncertainty, the best hope for lowering cancer mortality and morbidity (Kalinich and Haber, 2018). Ovarian cancer is a heterogeneous group of malignant tumors that affect various parts of the peritoneal cavity (Mehra et al., 2011). In 2018, 295,000 new cases

L. Azrar et al. (Eds.): CIP 2022, LNME, pp. 449–461, 2023.
https://doi.org/10.1007/978-3-031-23615-0_45

of ovarian cancer were detected worldwide, resulting in the deaths of almost 180,000 women (Ferlay et al., 2019). Early-stage (I, II) patients with ovarian cancer have a 5-year survival rate of over 90%, but late-stage (III, IV) Only 20–40% of ovarian cancer patients survive the disease for five years (Zalewski et al., 2015). Despite advancements in treatment, the 5-year survival rate for someone with ovarian cancer is about 45%, making it the second most deadly gynecological malignancy (Charkhchi et al., 2020). Only about 75% of invasive epithelial ovarian cancer patients have symptoms in the early stages, whereas early stage patients only exhibit a few minor, non-specific symptoms (Hennessy et al., 2009). Expense and non-invasiveness have naturally given rise to biomarkers in blood, urine, and other body fluids (Hazelton and Luebeck, 2011). Several biomarkers for ovarian cancer have been the subject of substantial research. Cancer Antigen 125 (CA125), also known as Carbohydrate Antigen 125, has been the most important biomarker in ovarian cancer screening, detection, and management over the past four decades. CA125, also called mucin 16 or MUC16, is a large membrane glycoprotein that belongs to the mucin family and is encoded by the MUC16 gene (Scatena, 2015). On the surface of ovarian cancer cells, it's detected. The antigen is then shed and measured in ovarian cancer patients' serum samples (Charkhchi et al., 2020). Several methods have been used for early detection of ovarian cancer: Imaging (transvaginal ultra-sonography, MRI), Circulating biomarkers (proteins, auto-antibodies, circulating tumor DNA, Methylated DNA, mi RNA), Tumor proximate fluids (Tampons, cervical swabs, endometrial lavage), Multimodal assessments [ROCA, Cancer SEEK (Cohen et al., 2018; Kalinich and Haber, 2018; Elias et al., 2018)]. Furthermore, recent research has proven its efficiency in tumor detection via circulation cancer blood biomarkers (Lokshin et al., 2021), such as Wong et al. (2019), Cohen et al. (2018), Eftimie and Hassanein (2018), Root (2019), Phan et al. (2020), and others have gone beyond tumor recurrence after therapeutic analysis, such as Hori (2021).

Among the different methods that have been used for the detection of biomolecules such as tumor biomarkers, we can find biosensing methods suggested in the work of (Mehrotra et al., 2019). Microwave biosensors have gained popularity due to a number of benefits, the most notable of which are that they are non-invasive and label-free, as well as high sensitivity, selectivity, and real-time monitoring of biological processes (Mehrotra et al., 2019). The nature of electromagnetic waves' interactions with biomatter opens up endless possibilities in biosensing. Since biomatter is a dielectric, all changes are measured in terms of its dielectric characteristics (Mehrotra et al., 2019).

In the same context, the present study is a first step toward the detection of epithelial ovarian cancer in its early stages by considering biosensing techniques. This could be done through a good understanding of the dielectric behavior of the tumor biomarkers and their relation to the development of the tumor. To achieve this target, an analysis of the mathematical models of tumor development in relation to the secretion of biomarkers CA125 and SEAP (one and two-compartment models) as early suggested by Hori et al. for different scenarios was of great interest. This modeling is then used through the dielectric strength/increment model of glycoprotein to characterize the dielectric behavior of CA125 and SEAP for biosensing purposes.

2 Theory

2.1 One-Compartment Model (Plasma of Blood)

S. Hori and S. Gambhir established a mathematical model to explain the relationship between dynamic plasma biomarker kinetics and tumor growth, starting with a single cancer cell (Hori and Gambhir, 2011; Konforte and Diamandis, 2013). The model was fed data on CA125 shedding and ovarian tumor growth, for which published literature is easily accessible for tumor growth characteristics and shedding rates (Hori and Gambhir, 2011), to illustrate a situation where both malignant and non-cancerous cells shed the biomarker CA125. In this investigation, there were two growth models employed. The first was modeled by a mono-exponential growth equation, which made the assumption that the tumor's growth strictly increased from its genesis to time zero. A Gompertzian equation was used to represent the second model, which assumed Mono-exponential growth and tumor decay. The authors investigated two scenarios for each growth model. A biomarker was 100% tumor specific in one scenario and was secreted by both tumor and normal cells in the other (Konforte and Diamandis, 2013). Since it is assumed that biomarkers in whole-body plasma are well mixed and homogeneous, one-compartment model kinetics can be used to characterize it (Hori and Gambhir, 2011).

The following differential equation represent the biomarker shedding model:

$$\frac{dq_{pl}(t)}{dt} = U_T(t) + U_H(t) - K_{el} * q_{pl}(t) \tag{1}$$

U_H and U_T are the influx of biomarker in plasma shed respectively by healthy and tumor cells.

Where

$$U_T(t) = f_{Pl,T} * R_T * N_T(t) \tag{2}$$

$$U_H(t) = f_{Pl,H} * R_H * N_H(t) \tag{3}$$

The number of tumor cells is assumed by the two following two models:

- **Mono-exponential function**

$$N_T(t) = N_{T,0} e^{K_{GR*} t} \tag{4}$$

- **Gompertzian function**

$$N_T(t) = N_{T,0} e^{\frac{K_{GR}}{K_{decay}} *(1-e^{-K_{decay} * t})} \tag{5}$$

Mathematical resolutions of the differential equation for different scenarios are as follow:

Biomarker shed by tumor cells only:

- **Gompertzian function**

$$q_{pl}(t) = \frac{f_{Pl,T} * R_T * N_{T,0}}{K_{EL}} * e^{\frac{K_{GR}}{K_{decay}} * \left(1 - e^{-K_{decay} * t}\right)} * (1 - e^{-K_{el} * t}) \tag{6}$$

- **Mono exponential function**

$$q_{pl}(t) = \frac{f_{Pl,T} * R_T * N_{T,0}}{K_{el}} * e^{K_{GR} * t} * (1 - e^{-K_{el} * t}) \tag{7}$$

Biomarker shed by tumor and healthy cells:

- **Gompertzian function**

$$q_{pl}(t) = \left[\frac{f_{Pl,T,} * R_T * N_{T,0} * e^{\frac{K_{GR}}{K_{decay}} * \left(1 - e^{-K_{decay} * t}\right)} + f_{Pl,H} * R_H * N_{H,0}}{K_{el}}\right] * (1 - e^{-K_{el} * t}) \tag{8}$$

- **Mono exponential function**

$$q_{pl}(t) = \left[\frac{f_{Pl,T,} * R_T * N_{T,0} * e^{K_{GR} * t} + f_{Pl,H} * R_H * N_{H,0}}{K_{el}}\right] * (1 - e^{-K_{el} * t}) \tag{9}$$

Table 1 presents the parameters of the one-compartment model, their physical meanings and units as well as the baseline values.

2.2 Two-Compartment Model (Plasma + Periphery)

In a further study, Hori et al. established a mathematical model of the dynamics of cancer biomarkers in various compartments that took into consideration the shedding of biomarkers from tumor and healthy cells. Using only blood biomarker sample data, this model can assist in deciding whether to intervene on a possibly aggressive cancer,

Table 1. Parameters of the one-compartment model (Hori and Gambhir, 2011)

Parameter	Description (units)	Baseline value
$q_{Pl}(t)$	Mass of biomarker in whole-body plasma (U)	–
R_T	Biomarker-shedding rate per tumor cell (U/day/cell)	4.5×10^{-5}
$N_{T,0}$	Initial number of biomarker-shedding tumor cells	1
K_{GR}	Growth rate of tumor cell population (day^{-1})	5.78×10^{-3}
K_{el}	Elimination rate of biomarker from plasma (day^{-1})	0.11
K_{decay}	Rate at which tumor growth rate decreases (day^{-1})	10^{-4}
$F_{Pl,H} * R_H * N_{H,0}$	Healthy cell shedding influx (U/day)	4.56×10^3
d	Detection limit of assay (U/ml)	1.5
c	Plasma biomarker cut-off for healthy and disease states (U/ml)	34.11
$f_{Pl,T}$	Fraction of biomarker entering tumor vasculature	0.1
V_{pl}	Mean plasma volume in a 70 kg female patient (ml)	3150

when to continue with active monitoring of a non-aggressive cancer, and how to continue blood sampling in a patient whose disease state is uncertain (Hori et al., 2017). Human embryonic kidney cells (293t) and ovarian cancer cell line A2780 (epithelial ovarian carcinoma) were cultivated and utilised in their research. Due to the fact that a single vector expressing SEAP-FLAG and FL-eGFP was used to transfect them, they are referred to as A2780-SEAP-FL-eGFP and 293T-SEAP-FL-rGFP, respectively (Hori et al., 2017). At basic pH levels, the Alkaline Phosphatase is a membrane-bound glycoprotein that catalyzes the hydrolysis of phosphate7 monoesters (Sharma et al., 2012). A2780 and 293t cell lines produce SEAP, a secreted alkaline phosphatase biomarker. It was developed utilizing a three-part compartmental modeling approach based on weight balance fundamentals (Hori et al., 2017).

The peripheral compartment may represent a non-physical mechanism of biomarker sequestration or biomarker uptake or storage into a physical site (such as interstitial or extravascular space) (e.g. binding or interaction with other proteins in plasma or extravascular space) (Hori et al., 2017).

The following system represent the biomarker shedding model in the case of two-compartment model:

$$\frac{dq_{pl}(t)}{dt} = U_T(t) + U_H(t) - \left(K_{el} + K_{periph,pl}\right) * q_{pl}(t) + K_{pl,periph} * q_{periph}(t) \quad (10)$$

and

$$\frac{dq_{periph}(t)}{dt} = K_{periph,pl} * q_{pl}(t) - K_{pl,periph} * q_{periph}(t) \quad (11)$$

where:

$$U_T(t) = R_T N_T(t) = R_T N_{T,0} e^{K_{GR*t}} \quad (12)$$

$$U_H(t) = R_H N_H(t) = U_{H,ss} = K_{el} q_{pl,ss};\ K_{GR} = \frac{\ln(2)}{t_{dt}} \tag{13}$$

Specific examples of aggressive, moderately aggressive, and non-aggressive tumor growth have been simulated using doubling periods t_{dt} of 2, 6, and 18 months, respectively, for the purposes of this study (Hori et al., 2017).

There are several methods for the resolution of this different system, namely algebraic and numerical methods such as Euler and Rung Kutta. We opted for the algebraic method based on the eigen values of the matrix constituted by the coefficients of the different system to derive the equations for $q_{pl}(t)$ and $q_{periph}(t)$ based on the parameters shown in Table 2 for different scenarios.

The results found are shown in Sect. 3 in terms of concentrations vs. time.

Table 2 presents the parameters of the two-compartment model, their physical meanings and units as well as baseline values.

Table 2. Parameters of the two-compartment model (Hori et al., 2017)

Parameter	Description (units)	Baseline value
$q_{Pl}(t)$	Mass of biomarker in whole-body plasma (U)	–
$q_{Periph}(t)$	Mass of biomarker in peripheral compartment (ng)	–
$k_{Pl,Periph}$	Rate of biomarker movement from general periphery to plasma (day^{-1})	0.31
$k_{Periph,Pl}$	Rate of biomarker movement from plasma to general periphery (day^{-1})	0.43
K_{GR}	Growth rate of tumor cell population (day^{-1})	Calculated
$N_{T,0}$	Initial number of biomarker-shedding tumor cells	1
K_{el}	Elimination rate of biomarker from plasma (day^{-1})	0.36
V_{Pl}	Plasma volume of distribution (ml)	7446
R_T	Tumor cell biomarker shedding rate (ng/day/cell)	10^{-4}
R_H	Healthy cell shedding rate (ng/day/cell)	–
$U_{H,ss}$	Influx of plasma biomarker from healthy cells (ng/day)	3.71×10^3
$q_{pl,ss}$	Amount of mass of biomarker in SteadyState conditions (ng)	1.03×10^4

3 Dielectric Modeling of a Biomarker Protein

The dielectric characteristics of biological systems are noteworthy and have long been explored in biophysics (Feldman et al., 2003). For more than 75 years, dielectric spectroscopy has been performed to investigate the dielectric characteristics of cells and organelles (Trainito et al., 2019). Oncley explored the dielectric characteristics of globular proteins in aqueous solutions, building on the pioneering work of Debye, Onsager, and Kirkwood on polar molecules (Feldman et al., 2003).

It has recently been demonstrated that using time domain dielectric spectroscopy, also known as time domain TDS spectroscopy (Feldman et al., 2003), was successful in obtaining fresh data on the structural and kinetic characteristics of protein globules during conformation transition processes.

The electrical polarization's decline in a TDS experiment can be represented as follows:

$$G(t) = \Delta\varepsilon * \Gamma(t) \tag{14}$$

where $\Delta\varepsilon$ is the dielectric strength and $\Gamma(t)$ is a macroscopic dipole correlation function (DFC). It has been demonstrated that the protein dipoles are responsible for the amplitude of the dielectric polarization, which is defined by the Onsager-Oncley expression (Feldman et al., 2003):

$$\Delta\varepsilon = \frac{\mu^2 * NA * C}{2 * K * T * \varepsilon_0 * M} \tag{15}$$

where M in Dalton (Da) is the molecular weight of the protein, N (6.02×10^{23} mol^{-1}) is the Avogadro number, C is the concentration of protein in solution (g/l), ε_0 (8.85×10^{-12} F m^{-1}) is the permittivity of free space, μ (C m) is the dipole moment of molecule, K (1.38×10^{-23} J K^{-1}) is Boltzmann's constant, T (°K) is the absolute temperature (Feldman et al., 2003). This model could be used in the study of dielectric strength of glycoproteins like CA125 and SEAP.

4 Results

Considering the one-compartment model, the simulation results were carried out in MATLAB for different scenarios of the development of ovarian tumor cells in relation to the concentration of the biomarker CA125 vs time and are shown in Fig. 1. The parameters of the study are taken from the literature (see Table 1) (Hori and Gambhir, 2011). The concentration is defined as $C_{pl} = q_{pl}(t)/V_{Pl}$.

Considering the secretion of biomarker CA125 by tumor cells and no back secretion by healthy cells, from Fig. 1 (a), the detection time is $t_1 = 3816$ days (10.6 years) for Gompertzian model, and $t_2 = 3168$ days (8.8 years) for Mono-exponential model. Assuming the clinical CA125 enzyme linked immunosorbent assay's current detection limit (ELISA assays) d = 1.5 U/ml, we found the concentration of $C_{pl1} = 1.19$ U/ml and the number of tumor cells $N_{T1} = 9.18 \times 10^7$ cells using the Gompertzian model and $C_{pl2} = 1.16$ U/ml and $N_{T2} = 6.5 \times 10^6$ cells using the Mono-exponential model.

Considering the secretion by tumor and healthy cells, from Fig. 1 (b), by using the Gompertzian model, the detection time is approximately $t_3 = 4536$ days (12.6 years), while by using the Mono-exponential model, the detection time is $t_4 = 3636$ days (10.1 years). Assuming the cut off limit c = 34.11 U/ml, we found the concentration of $C_{pl3} = 13.16$ U/ml and $N_{T3} = 1.42 \times 10^9$ cells for Gompertzian model, and $C_{pl4} = 30.56$ U/ml and $N_{T4} = 4.48 \times 10^7$ cells for Mono-exponential model. This study shows an improvement in tumor detected diameter (or cells number) of 4.5-fold or 90-fold improvement over the median tumor volume detected using ultrasound at clinical diagnostic (Hori and Gambhir, 2011).

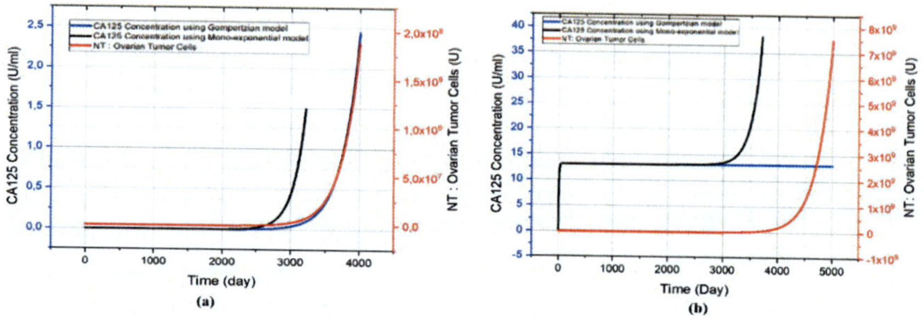

Fig. 1. Development of ovarian tumor cells in relation to the concentration of the biomarker CA125 vs time (a) Considering the secretion by tumor cells only using Gompertzian and Mono-exponential models (b) Considering the secretion by tumor and healthy cells using Gompertzian and Mono-exponential models

Hori et al. suggested a new blood-born tumor-shed biomarker dedicated to the study of ovarian cancer in two-compartments of blood (plasma and periphery). As for the two-compartment model, the simulation results of the mathematical differential system in MATLAB for different scenarios of the development of ovarian tumor cells in relation to the concentration of the biomarker SEAP versus time for three cases of tumor aggressiveness that are shown in Fig. 2. The parameters of the study are taken from the literature (see Table 2) (Hori et al., 2017). The concentrations are defined as $C_{plasma} = q_{pl}(t)/V_{Pl}$ and $C_{periph} = q_{periph}(t)/V_{Pl}$.

Figure 2 (a) shows the aggressive case of ovarian tumors and biomarker reaching the plasma and periphery, Fig. 2 (b) shows the moderately aggressive case of ovarian tumors and Fig. 2 (c) shows the non-aggressive case.

From Fig. 2 (a) in the case of an aggressive ovarian tumor, the detection time assumed until the tumor became imaging detectable is $t_1 = 51$ months (1530 days), corresponding to a tumor size $N_{T1} = 9.4 \times 10^7$ cells and a detection limit of biomarker concentration in plasma, $C_{plasma,1} = 0.26$ ng/ml and in the periphery, $C_{periph,1} = 0.36$ ng/ml. From Fig. 2 (b), in the case of moderately aggressive ovarian tumors, the detection time is $t_2 =$ 12.5 years (4500 days) until the tumor becomes imaging detectable, corresponding to a tumor size of $N_{T2} = 3.34 \times 10^7$ cells and a detection limit of biomarker concentration, in plasma $C_{plasma,2} = 14.66$ ng/ml and in the periphery $C_{periph,2} = 20.83$ ng/ml. According to Fig. 2 (c), the detection time until the tumor becomes imaging detectable in the case of non-aggressive ovarian tumor is $t_3 = 37$ years (13,320 days), corresponding to a tumor size of $N_{T3} = 2.54 \times 10^7$ cells and a detection limit of biomarker concentration, in plasma $C_{plasma,3} = 131.1$ ng/ml and in the periphery $C_{periph,3} = 185.4$ ng/ml.

Regarding the dielectric study, and as mentioned earlier in the theory section, we have used the model of globular protein for dielectric strength suggested by Onsager-Oncley to investigate the behavior of the two biomarkers by considering the two models studied for the detection of epithelial ovarian carcinoma.

Figure 3 illustrates the behavior of CA125 in terms of dielectric increment by considering the one-compartment model for the different scenarios of the development of an ovarian tumor. (Gompertzian Fig. 3 (a) and Mono-exponential Fig. 3 (b)). In fact, an increasing dielectric increment is concluded for both cases, and the medium is more

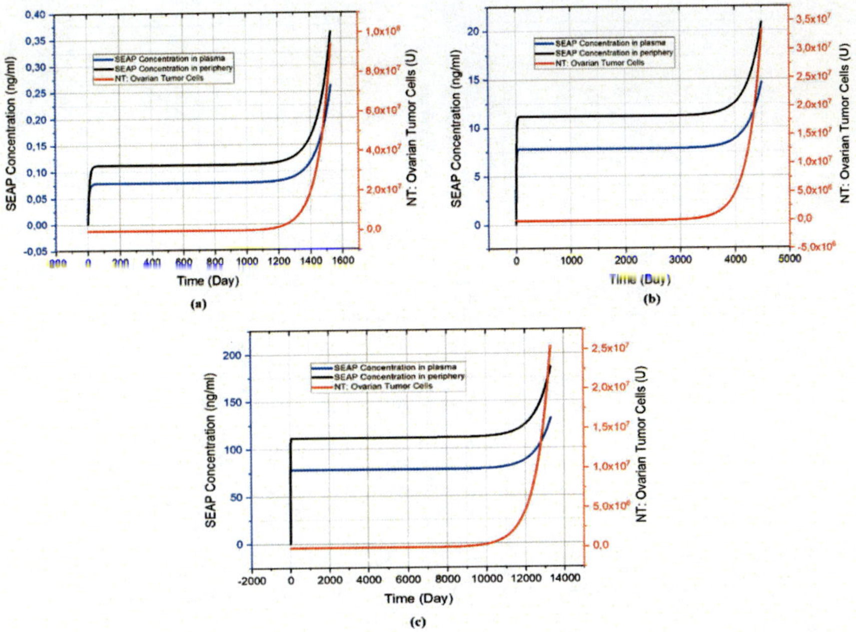

Fig. 2. Development of ovarian tumor cells in relation to the concentration of the biomarker SEAP vs time, (a) case of aggressive tumor, (b) case of moderately aggressive tumor and (c) case of non-aggressive tumor

dielectric by considering the realistic case of secretion of biomarkers by tumor and healthy cells using the Gompertzian model. The mono-exponential model shows an increasing dielectric increment too. Analogically, the medium is more dielectric using a mono-exponential model in the realistic case of secretion of biomarkers by tumor and healthy cells.

Figure 3 (c) illustrates the variation of dielectric increment versus temperature for different concentrations of CA125. The dielectric increment decreases with increasing temperature, and we can conclude that by increasing the concentration of CA125, the medium becomes more dielectric. Figure 3 (d) illustrates the dielectric increment versus concentration of CA125 for different temperatures. By increasing the concentration of CA125, the dielectric increment also increases. We can conclude that the temperature doesn't influence the dielectric behavior of CA125 much.

Figure 4 (a), (b), and (c) illustrates the dielectric increment of SEAP versus the number of tumor cells developed considering the two-compartment model. We can conclude an increase in dielectric increment when the tumor becomes bigger over time. The amplitude is higher in the periphery than in the plasma. The medium becomes more dielectric in the case of an aggressive tumor than in a non-aggressive case. Figure 4 (d) shows the dielectric increment curves versus temperature for different concentrations of SEAP. As for the CA125, while increasing the concentration, the medium shows higher values of dielectric increment, but this one decreases with the increase of temperature. Figure 4 (e) illustrates the dielectric increment versus concentration of SEAP for different

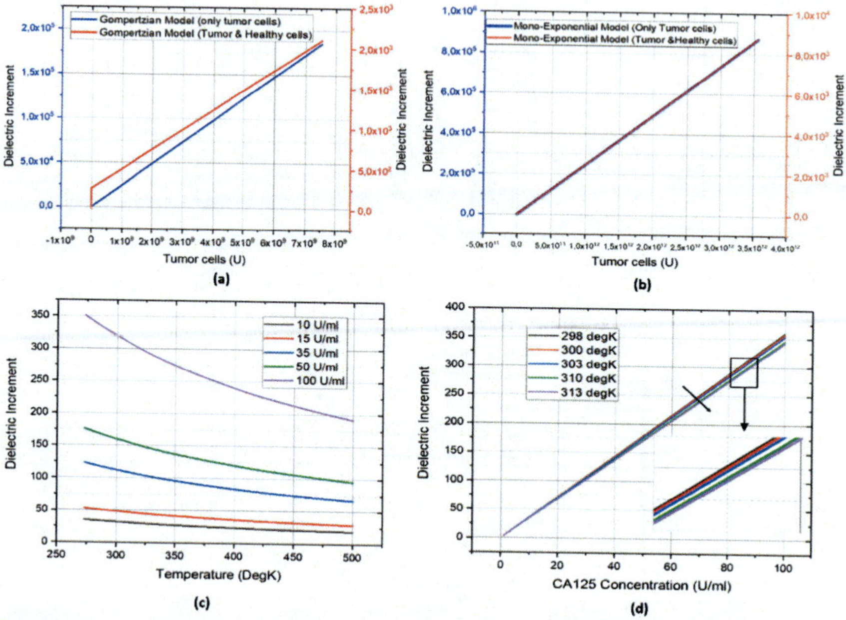

Fig. 3. Dielectric analysis of CA125 (a) Dielectric increment vs. Number of developed tumor cells using Gompertzian model. (b) Using the Mono-exponential model. (c) Dielectric increment vs temperature for different concentrations of CA125. (d) Dielectric increment vs CA125 concentration for different temperatures of the medium

temperatures; while increasing the concentration, the variation of the dielectric increment isn't much influenced by the variation of the temperature.

As a result, the study of the dielectric increment of the two biomarkers could be the key to further studies and modeling of their dielectric behavior in blood, thus, getting a better understanding of dielectric characterization for biosensing purposes.

5 Discussion

Lutz et al. conducted an investigation of the model's sensitivity (ability to detect cancer) (Hazelton and Luebeck, 2011) and specificity (ability to prevent false positives) with respect to various probabilities of false positive test findings (Lutz et al., 2008). It was hypothesized that the mean percentages of secreted tumor biomarkers that entered the intravascular space ranged from 0.1% to 20%. By combining a multi-marker approach that has shown promising improvements in sensitivity with longitudinal biomarker algorithms like the Risk of Ovarian CAncer algorithm (ROCA), which has shown improvements in lead time and specificity, Simmons et al. developed, in their recent work, a longitudinal biomarker panel for the treatment of ovarian cancer. They identified 83.2% sensitivity and 98% specificity for ovarian cancer early diagnosis (Simmons et al., 2019). In a more general study, Cohen et al. tried to implement cancer gene ctDNA sequencing with tumor-associated serum protein quantification for earlier tumor identification

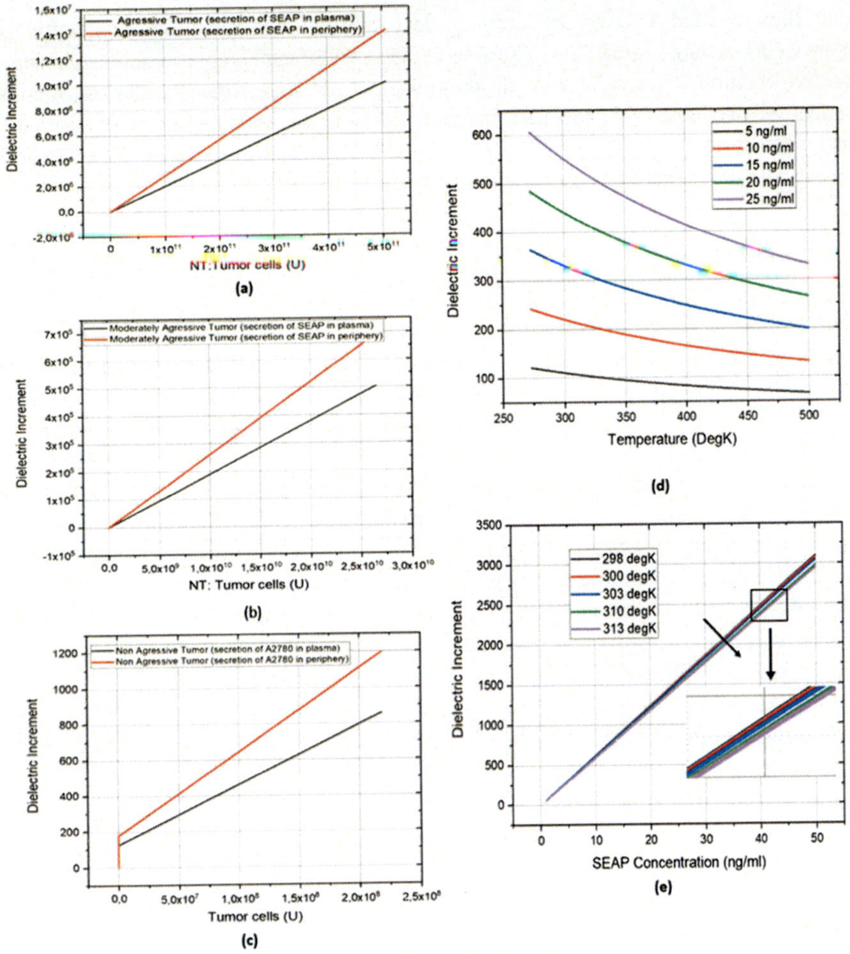

Fig. 4. Dielectric increment of SEAP vs number of developed tumor cells using two-compartment model (a) case of aggressive tumor, (b) case of moderately aggressive tumor and (c) case of non-aggressive tumor, (d) dielectric increment vs temperature for different concentrations of SEAP, (e) dielectric increment vs SEAP concentration for different temperatures of the medium

and localisation. The Omega score is calculated using plasma-based sequencing of 16 cancer genes and eight cancer-associated serum proteins to determine the probability of getting one of eight different types of cancer. Using a machine learning system, the data is then integrated with 31 additional serum proteins, the gender of the patient, and the tissue of origin. For breast, lung, colon, esophagus, pancreatic, stomach, liver, and ovarian cancers, test sensitivity ranged from 33 to 98% among 1005 patients and 812 healthy controls. The accuracy of tissue of origin predictions ranged from 39 to 84% (Cohen et al., 2018). Wong et al. investigated at several supervised learning algorithms for identifying various cancer kinds from results of multianalyte blood tests. Eight circulating protein markers and one circulating DNA mutation score are present. As a result, they selected A1DE, deep learning, decision trees, and naive Bayes as multiclass supervised learning algorithms. There have been significant improvements. The most effective method (CancerA1DE) can diagnose cancer better than the current method (Cancer SEEK) with 99% specificity and can double the existing sensitivity for early cancer detection from 38% to 77% (i.e., Stage I). It can even reach 90% for Stage II cancers across a variety of cancer types (Wong et al., 2019).

6 Conclusion

In the present paper, an analysis of the mathematical models recently developed for the screening of epithelial ovarian carcinoma is carried out for different scenarios of the development of tumors and the secretion of corresponding biomarkers. In fact, the one-compartment model is analyzed by considering the biomarker CA125 for Gompertzian and Mono-exponential models as well as for the secretion by tumor cells on the one hand and by tumor and healthy cells on the other hand. Furthermore, the analysis of the two-compartment model for biomarker SEAP shed by A2780/293t cell lines was suggested to take into account the aggressiveness of the ovarian tumor on one hand and analyzing the shed of biomarker in plasma and periphery on the other hand. Finally, we considered the analysis of dielectric strength or increment of the two biomarkers for different scenarios, as well as the analysis considering the variation of the concentrations and temperature of the medium. This one is more dielectric for a realistic case of shedding of biomarker CA125 by tumor and healthy cells, also in the case of aggressive tumors by considering the biomarker SEAP. The increase in concentrations of biomarkers increases the dielectric strength, but the temperature hasn't a great influence on the dielectric of the biomarkers' mediums. As a result, dielectric characterization of blood biomarkers could be of great utility for further biosensing studies.

References

Charkhchi, P., Cybulski, C., Gronwald, J., Wong, F.O., Narod, S.A., Akbari, M.R.: CA125 and ovarian cancer: a comprehensive review. Cancers **12**, 3730 (2020). https://doi.org/10.3390/cancers12123730

Cohen, J.D., et al.: Detection and localization of surgically resectable cancers with a multi-analyte blood test. Science **359**, 926–930 (2018). https://doi.org/10.1126/science.aar3247

Eftimie, R., Hassanein, E.: Improving cancer detection through combinations of cancer and immune biomarkers: a modelling approach. J. Transl. Med. **17** (2018)

Elias, K.M., Guo, J., Bast, R.C.: Early detection of ovarian cancer. Hematol. Oncol. Clin. North Am. **32**, 903–914 (2018). https://doi.org/10.1016/j.hoc.2018.07.003

Feldman, Y., Ermolina, I., Hayashi, Y.: Time domain dielectric spectroscopy study of biological systems. IEEE Trans. Dielect. Electr. Insul. **10**, 728–753 (2003). https://doi.org/10.1109/TDEI.2003.1237324

Ferlay, J., Colombet, M., Soerjomataram, I., Mathers, C., Parkin, D.M., Piñeros, M., Znaor, A., Bray, F.: Estimating the global cancer incidence and mortality in 2018: GLOBOCAN sources and methods. Int. J. Cancer **144**, 1941–1953 (2019). https://doi.org/10.1002/ijc.31937

Hazelton, W.D., Luebeck, E.G.: Biomarker-based early cancer detection: is it achievable? Sci. Transl. Med. **3** (2011). https://doi.org/10.1126/scitranslmed.3003272

Hennessy, B.T., Coleman, R.L., Markman, M.: Ovarian cancer. Lancet **374**, 1371–1382 (2009). https://doi.org/10.1016/S0140-6736(09)61338-6
Hori, S.S.: A mathematical model of tumor regression and recurrence after therapeutic oncogene inactivation. Sci. Rep. **14** (2021)
Hori, S.S., Gambhir, S.S.: Mathematical model identifies blood biomarker-based early cancer detection strategies and limitations. Sci. Transl. Med. **3** (2011). https://doi.org/10.1126/scitranslmed.3003110
Hori, S.S., Lutz, A.M., Paulmurugan, R., Gambhir, S.S.: A model-based personalized cancer screening strategy for detecting early-stage tumors using blood-borne biomarkers. Cancer Res. **77**, 2570–2584 (2017). https://doi.org/10.1158/0008-5472.CAN-16-2904
Kalinich, M., Haber, D.A.: Cancer detection: seeking signals in blood. Science **359**, 866–867 (2018). https://doi.org/10.1126/science.aas9102
Konforte, D., Diamandis, E.P.: Is early detection of cancer with circulating biomarkers feasible? Clin. Chem. **59**, 35–37 (2013). https://doi.org/10.1373/clinchem.2012.184903
Lokshin, A., Bast, R.C., Rodland, K.: Circulating cancer biomarkers. Cancers **13**, 802 (2021). https://doi.org/10.3390/cancers13040802
Lutz, A.M., Willmann, J.K., Cochran, F.V., Ray, P., Gambhir, S.S.: Cancer screening: a mathematical model relating secreted blood biomarker levels to tumor sizes. PLoS Med. **5**, e170 (2008). https://doi.org/10.1371/journal.pmed.0050170
Mehra, K., Ning, G., Drapkin, R., McKeon, F., Xian, W., Crum, C.: STICS, SCOUTs and p53 signatures; a new language for pelvic serous carcinogenesis. Front. Biosci. (Elite Ed.) **3**, 625–634 (2011)
Mehrotra, P., Chatterjee, B., Sen, S.: EM-wave biosensors: a review of RF, microwave, mm-wave and optical sensing. Sensors **19**, 1013 (2019). https://doi.org/10.3390/s19051013
Phan, T., Crook, S.M., Bryce, A.H., Maley, C.C., Kostelich, E.J., Kuang, Y.: Review: Mathematical modeling of prostate cancer and clinical application. Appl. Sci. **29** (2020)
Root, A.: Mathematical modeling of the challenge to detect pancreatic adenocarcinoma early with biomarkers. Challenges **15** (2019)
Scatena, R. (ed.): Advances in Cancer Biomarkers. Advances in Experimental Medicine and Biology. Springer, Netherlands, Dordrecht (2015). https://doi.org/10.1007/978-94-017-7215-0
Sharma, A., Abtin, F., Shepard, J.-A.O.: Image-guided ablative therapies for lung cancer. Radiol. Clin. North Am. **50**, 975–999 (2012). https://doi.org/10.1016/j.rcl.2012.06.004
Simmons, A.R., et al.: Complementary longitudinal serum biomarkers to CA125 for early detection of ovarian cancer. Cancer Prev. Res. **12**, 391–400 (2019). https://doi.org/10.1158/1940-6207.CAPR-18-0377
Trainito, C.I., et al.: Characterization of sequentially-staged cancer cells using electrorotation. PLoS ONE **14**, e0222289 (2019). https://doi.org/10.1371/journal.pone.0222289
Wong, K.-C., et al.: Early cancer detection from multianalyte blood test results. iScience **15**, 332–341 (2019). https://doi.org/10.1016/j.isci.2019.04.035
Zalewski, et al.: The new FIGO staging system for ovarian, fallopian tube and primary peritoneal cancer—2014 update. Oncol. Clin. Pract. **11**, 129–134 (2015)

Additive Manufacturing

Correlation Between Tribological Behaviour and Hardness of Mollusc Shell-UHMWPE Composite

Besma Sidia(✉) and Walid Bensalah

Ecole Nationale d'Ingénieurs de Monastir, Laboratoire de Génie Mécanique (LGM), Université de Monastir, Rue Ibn Eljazzar 5019, Monastir, Tunisie
Sidia.besma@gmail.com

Abstract. This study addresses the effect of hardness in the tribological behaviour of Mollusc Shell (MS)-UHMWPE composite. FTIR analysis was performed for the chemical characterization of the MS particles and the UHMWPE powder. Hot compression molding was used to prepare the composites with different weight contents of MS, namely: 0, 5, 10, 15, 20, and 25 wt%. Wears tests were conducted using a linear reciprocating tribometer. The results showed that the addition of MS particles to UHMWPE has a positive effect on its tribological and mechanical properties. Indeed, a considerable reduction in the friction coefficient was obtained by reinforcing the polymer with the MS particles. In addition, a considerable increase in Rockwell hardness was measured with the use of MS particles. However, it has been shown that the hardness of the composite is not the main factor affecting its wear resistance but rather its friction coefficient.

Keywords: MS-HDPE biocomposite · Wear and tribology · Spectroscopy · Hardness · Correlation

1 Introduction

Ultra-high molecular weight polyethylene (UHMWPE) is an important industrial thermoplastic widely used for various mechanical and medical applications (Chen et al., 2021; Sakoda et al., 2020; Terrani, 2016; Wang et al., 2021). This is because it has very high wear resistance and a low coefficient of friction (Vadivel et al., 2021). Despite these advantages, UHMWPE remains an artificial material that damages the human body in such a medical application. In the case of the hip joint prosthesis, as a specific field, UHMWPE was commonly used as the acetabular cup material of the prosthesis (Saikko, 2019). Thus, UHMWPE debris ejected into the human body leads to implant degradation (Xu et al., 2019). Therefore, reinforcing the polymer with fibers or particles was an appropriate approach to increase the life of the implant. In fact, composite materials are characterized by a better mass-rigidity ratio, a low sensitivity to fatigue and good mechanical and tribological properties (Efe et al., 2020; Hussein et al., 2015). These advantages are given due to the combination of more than two materials.

L. Azrar et al. (Eds.): CIP 2022, LNME, pp. 465–473, 2023.
https://doi.org/10.1007/978-3-031-23615-0_47

Several authors have adopted different biomaterials as reinforcement for polymers, such as natural coral (Ge et al., 2009), hydroxyapatite (Fang et al., 2006), zeolite (Chang et al., 2013), eggshell (Gbadeyan et al., 2020), etc. Mahshuri and Amalina (2016) reinforced the polyester with clamshell particles. In their study, the developed biocomposite exhibited strong mechanical properties in terms of hardness and a compressive modulus. Hussein et al. (2011). Used eggshell as reinforcement to high-density polyethylene. The results revealed that eggshells improved the mechanical and tribological properties of the polymer. In addition, 10% eggshell by weight provided the best tribological properties. Therefore, the addition of more than 10% eggshells has an inverse effect on the tribological properties of the polymer. Wang et al. (2018) reinforced UHMWPE with hydroxyapatite particles. The results showed that Young's modulus of the composite was increased with the increase of filler content.

Investigating the bio-waste as sea shells has a point of interest given its good tribological properties (Sidia and Bensalah, 2021), biocompatibility (Fleischli et al., 2008), and high resistance to fracture (Jagatheeshwaran et al., 2016), etc.

This work is therefore aimed at investigating the tribological properties of mollusc shell (MS) particles reinforced UHMWPE composites. The different weight percent of MS particles was employed. FTIR analysis was performed for the chemical characterization of the MS particles and the UHMWPE powder. Then, to evaluate the developed composite, a correlation between friction, wear, and hardness of the composites was performed.

2 Experimental Details

2.1 Composite Preparation Process

For the tribological tests, M30NW stainless steel was used as the pin material (Table 1). The pin has a hemispherical shape with a 10 mm diameter (Sidia and Bensalah, 2020). Biocomposite discs were elaborated by the hot compressing molding process as described in our previous work (Sidia and Bensalah, 2021). 0, 5, 10, 15, 20, and 25 are the different weight percent of MS particles added to the UHMWPE. All the experimental details of the MS particles preparation are presented in our previous work (Sidia and Bensalah, 2021).

Table 1. Chemical composition of M30NW stainless steel.

M30NW						
Element	C	Cr	Ni	Mn	Mo	N
Composition (wt. %)	≤ 0.06	21	9.5	4	2.2	0.4

2.2 Tribological Tests

Tribological tests were carried out using the linear reciprocating pin on the disc tribometer presented by Sidia et al. (2020). All the friction tests were repeated three times and achieved under dry condition at room temperature. The parameters of the tribological tests are resumed in Table 2.

Table 2. Tribological parameters.

Sliding speed	30 mm/s
Frequency	1 Hz
Normal load	20 N
Number of cycles	15,000 cycles

2.3 2D Profilometers

Taylor Hobson Surtronic 2D profilometer was used to measure the cross-sectionam area of the composite wear track generated after tribological tests. The wear volume is then determined by multiplying the obtained cross-sectional area by the distance of the wear track.

2.4 Rockwell Hardness

Rockwell hardness of the MS-UHMWPE composites was conducted using AFFRI universal hardness tester. According to the ASTM D785-03 standard (ASTM, 2008), the parameters of the test are fixed. For the repeatability of the results, five measurements were conducted for each sample.

2.5 Fourier Transform InfraRed, FTIR, Spectroscopy

FTIR analysis of MS particles were performed using a Perkin Elmer Spectrum TwoT-MIR spectrometer. During this study, FTIR spectra were recorded in the range of 400–4000 cm^{-1} of wave number. Thus, the results were taken in transmittance mode.

3 Results

3.1 FTIR Analysis

FTIR was used for the chemical characterization of the MS particles. Figure 1 shows the FTIR spectra of MS particles in terms of transmittance mode.

In view of this figure, it is clear that the MS are mainly composed of the carbonate anion. Indeed, four major modes of carbonate calcium vibration between the range of 600–1600 cm^{-1} of wave number are detected (Table 3). The result shows a good agreement between the spectrum of MS particles and that of human bone and aragonite explored by Bayarı et al. (2020).

Table 3. Infrared band positions and symbols with their vibration assignments for the mollusc shell particles.[a]

Wave number (cm^{-1})	Symbol	Vibration assignment
698–711	ν_4	Planar bending vibration mode of $CaCO_3$ molecule
852	ν_2	Out of plane bending vibration mode of $CaCO_3$ molecule
1080	ν_1	Symmetrical stretching vibration mode of $CaCO_3^{2-}$ ion
1474	ν_3	Antisymmetric stretching vibration mode of $CaCO_3$ molecule

[a] *Sources* Linga Raju et al. (2002) and Sidia and Bensalah (2020)

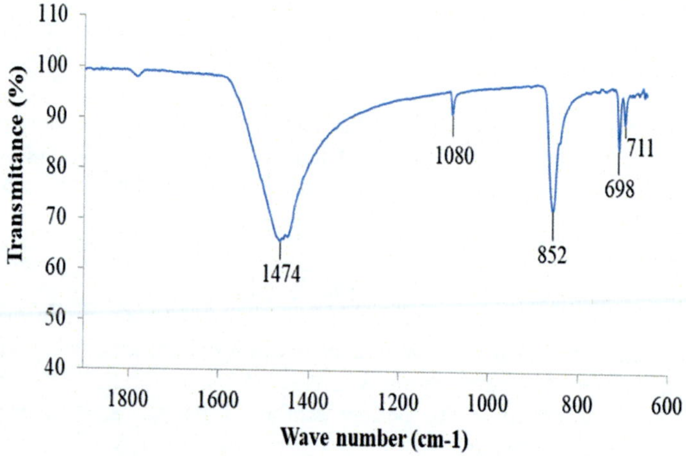

Fig. 1. FTIR spectra of mollusc shell particles.

The FTIR spectrum of UHMWPE is presented in Fig. 2. This figure highlights the vibrational energy of the different bonds found in the different functional groups studied. Four fundamental bands for UHMWPE could be found. As shown in Fig. 2 the two doubled peaks at 2919 cm^{-1} and 2851 cm^{-1} with very high intensity, a band with high intensity at 1462 cm^{-1} and the peak at 719 cm^{-1} with high intensity are detected in the FTIR spectra of UHMWPE (Table 4). These peaks are consistent with those of HDPE or UHMWPE polymers (Bayarı et al., 2020; Hamzah et al., 2018).

3.2 Correlation Between Rockwell Hardness and Tribological Behaviour of the Composites

Figure 3 shows the correlation between the coefficient of friction and the wear volume generated on the different biocomposite discs, after friction tests against M30NW pins. From this figure, it is highlighted that the addition of the MS particles to the UHMWPE has decreased the coefficient of friction. In fact, the lowest coefficient of friction was obtained with 15 and 25 wt.% of MS-UHMWPE. From Fig. 3 it is investigated that the

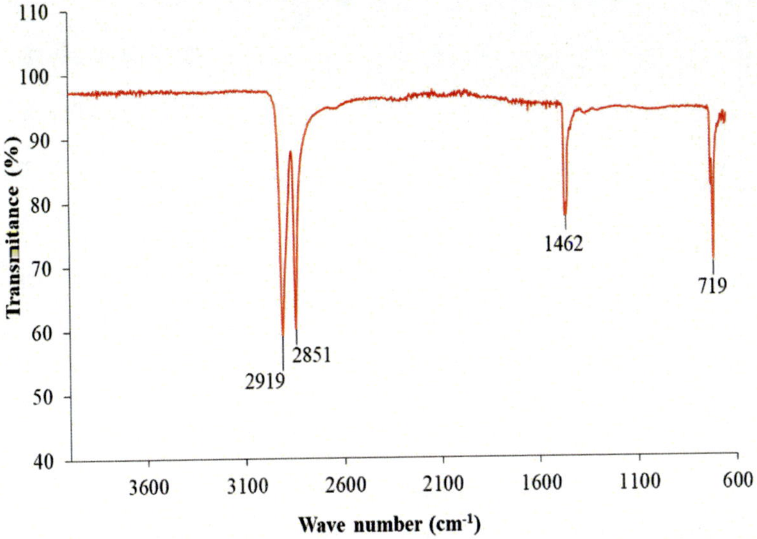

Fig. 2. FTIR spectra of UHMWPE powder.

Table 4. Infrared band positions and symbols with their vibration assignments for UHMWPE powder.[a]

Wave number (cm^{-1})	Symbol	Vibration assignment[a]
719	ν_1	Deformation vibration by tilting the CH2 group of amorphous bonds
1462	ν_3	Bending deformation vibration of the CH2 group of methylene in amorphous phase
2851	ν_2	Symmetrical elongation vibration of C-H bonds
2919	ν_4	Antisymmetric elongation vibration of C-H bonds

[a] *Sources* Bayarı et al. (2020) and Hamzah et al. (2018)

lowest coefficient of friction is not necessarily related to the lowest wear volume. In fact, the lowest wear volume was for 5 wt.% of MS-UHMWPE and the highest wear volume corresponds to 20 and 25 wt.% of MS-UHMWPE. Therefore, MS debris are more important in the contact to reduce the friction between the composite and the metal.

The representative Fig. 4 shows the relationship between the Rockwell hardness of the different biocomposites and their generated wear volumes after friction tests. From this figure, it is evident that the addition of MS particles has a significant effect on both Rockwell hardness and wear volume of the pure UHMWPE. On the one hand, the MS particles have increased the Rockwell hardness of the UHMWPE, where it evolved from 60 to 85 HRR for 0 and 25 wt.% MS-UHMWPE, respectively. In fact, the MS is a hard material so the hardness of the hole composites was increased (Mundel and Sankar,

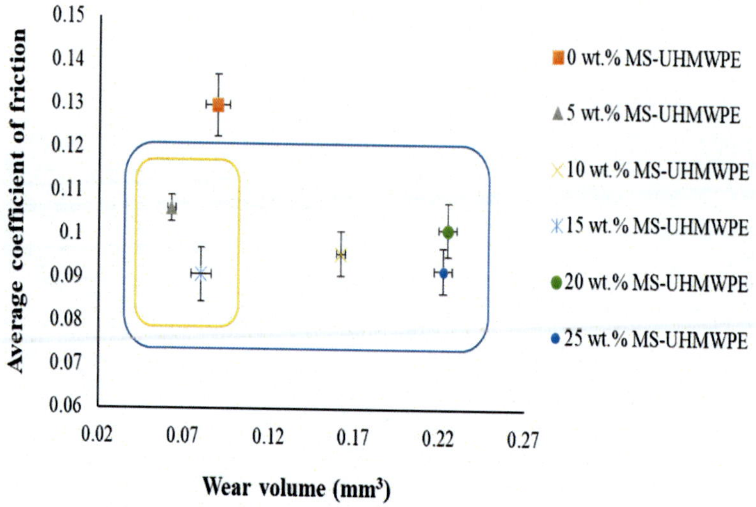

Fig. 3. Evolution of average coefficient of friction as a function of biocomposites wears volumes.

2014). On the other hand, a different trend was shown for the wear volume. Indeed, only 5 and 15 wt.% MS-UHMWPE have a positive effect on the wear volume compared to the 0 wt.% MS-UHMWPE. The point is that the hardness of the composite is not the main factor affecting its wear resistance. Similar observations are made in the literature (Mahshuri and Amalina, 2016; Qi et al., 2014).

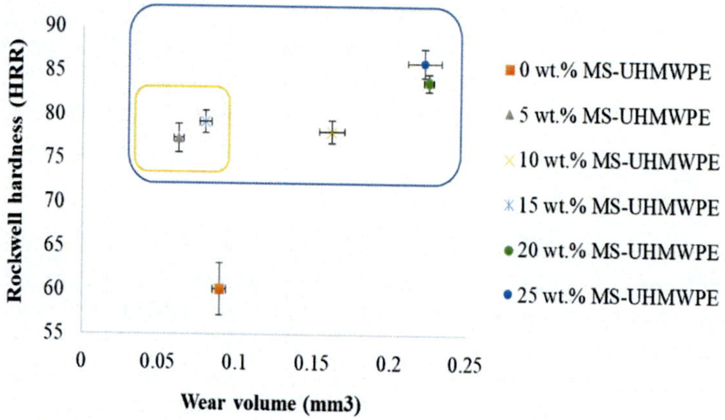

Fig. 4. Evolution of Rockwell hardness as a function of wear volume of the biocomposites.

Figure 5 presents the relationship between the Rockwell hardness of each biocomposite and its coefficient of friction obtained after friction tests against M30NW pins. From this figure, it is properly highlighted that the coefficients of friction of the samples are significantly related to their hardness. In fact, it can be found that the lowest coefficient of friction is obtained with the addition of the MS particles to the UHMWPE. Consequently, increasing the hardness of a composite may reduce its coefficient of friction when sliding against Stainless steel pin. These results are in concordance with those from the literature (Xu et al., 2015) where the adhesion between composite and metal decreases.

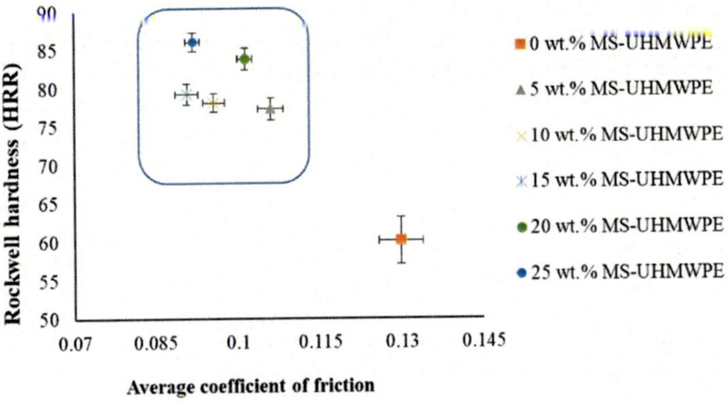

Fig. 5. Evolution of Rockwell hardness as a function of the average friction coefficient of the composites.

4 Conclusion

In this study, the tribological and mechanical properties of composites obtained by hot pressing were examined. The results revealed that the addition of MS particles to UHMWPE has a beneficial effect on its tribological and mechanical properties. Indeed, a significant decrease in the friction coefficient was observed when the polymer was reinforced with the MS particles. In addition, a significant increase in Rockwell hardness was measured with the reinforcement by MS particles. Therefore, a consistent relationship between material hardness and friction coefficient was found. However, it is concluded that the hardness of the composite is not the main factor affecting its wear resistance.

References

ASTM: D785-03-Standard Test Method for Rockwell Hardness of Plastics and Electrical Insulating. Annual Book of ASTM Standards, pp. 1–6. ASTM International, United States (2008). https://doi.org/10.1520/D0785-08.2

Bayarı, S.H., Özdemir, K., Sen, E.H., Araujo-Andrade, C., Erdal, Y.S.: Application of ATR-FTIR spectroscopy and chemometrics for the discrimination of human bone remains from different archaeological sites in Turkey. Spectrochim. Acta - Part A: Mol. Biomol. Spectrosc. **237** (2020). https://doi.org/10.1016/j.saa.2020.118311

Chang, B.P., Akil, H.M., Nasir, R.M.: Mechanical and tribological properties of zeolite-reinforced UHMWPE composite for implant application. Procedia Eng. **68**, 88–94 (2013). https://doi.org/10.1016/j.proeng.2013.12.152

Chen, X., Zhang, S., Zhang, L., Zhu, P., Zhang, G.: Design and characterization of the surface porous UHMWPE composite reinforced by graphene oxide. Polymers **13**, 1–14 (2021). https://doi.org/10.3390/polym13040482

Efe, G., Altınsoy, İ., Türk, S., Bindal, C.: An investigation on UHMWPE-HAp composites manufactured by solution-gelation method. Sakarya Univ. J. Sci. **24**, 1–9 (2020). https://doi.org/10.16984/saufenbilder.396984

Fang, L., Leng, Y., Gao, P.: Processing and mechanical properties of HA/UHMWPE nanocomposites. Biomaterials **27**, 3701–3707 (2006). https://doi.org/10.1016/j.biomaterials.2006.02.023

Fleischli, F.D., Dietiker, M., Borgia, C., Spolenak, R.: The influence of internal length scales on mechanical properties in natural nanocomposites: a comparative study on inner layers of seashells. Acta Biomater. **4**, 1694–1706 (2008). https://doi.org/10.1016/j.actbio.2008.05.029

Gbadeyan, O.J., Adali, S., Bright, G., Sithole, B., Omojoola, A.: Studies on the mechanical and absorption properties of achatina fulica snail and eggshells reinforced composite materials. Compos. Struct. **239**, 112043 (2020). https://doi.org/10.1016/j.compstruct.2020.112043

Ge, S., Wang, S., Huang, X.: Increasing the wear resistance of UHMWPE acetabular cups by adding natural biocompatible particles. Wear **267**, 770–776 (2009). https://doi.org/10.1016/j.wear.2009.01.057

Hamzah, M., Khenfouch, M., Rjeb, A., Sayouri, S., Houssaini, D.S., Darhouri, M., et al.: Surface chemistry changes and microstructure evaluation of low density nanocluster polyethylene under natural weathering: a spectroscopic investigation. J. Phys: Conf. Ser. **984**, 1–14 (2018). https://doi.org/10.1088/1742-6596/984/1/012010

Hussein, A.A., Salim, R.D., Sultan, A.A.: Water absorption and mechanical properties of high-density polyethylene/egg shell composite (2015)

Hussein, A.A., Salim, R.D., Sultan, A.A.: Water absorption and mechanical properties of high-density polyethylene/egg shell composite water absorption and mechanical properties of high-density polyethylene / egg shell composite. J. Basrah Res. **37**, 36–42 (2011)

Jagatheeshwaran, M.S., Elayaperumal, A., Arulvel, S.: The role of calcinated sea shell particles on friction-wear behavior of electroless NiP coating: fabrication and characterization. Surf. Coat. Technol. **304**, 492–501 (2016). https://doi.org/10.1016/j.surfcoat.2016.07.053

Linga Raju, C., Narasimhulu, K.V., Gopal, N.O., Rao, J.L., Reddy, B.C.V.: Electron paramagnetic resonance, optical and infrared spectral studies on the marine mussel Arca burnesi shells. J. Mol. Struct. **608**, 201–211 (2002). https://doi.org/10.1016/S0022-2860(01)00952-8

Mahshuri, Y., Amalina, M.A.: Hardness and compressive properties of calcium carbonate derived from clam shell filled unsaturated polyester composites Hardness and compressive properties of calcium carbonate derived from clam shell filled unsaturated polyester composites. Mater. Res. Innov. **18**, 291–294 (2016). https://doi.org/10.1179/1432891714Z.000000000970

Mundel, G., Sankar, M.R.: Preparation and tribological characterization of linear low density poly-ethylene sea shell (LLDPE/Sea Shell) bio composite. In: 5th International & 26th All India Manufacturing Technology, Design and Research Conference (2014)

Qi, S., Fu, Z., Yun, R., Jiang, S., Zheng, X., Lu, Y., et al.: Effects of walnut shells on friction and wear performance of eco-friendly brake friction composites. Proc. Inst. Mech. Eng. Part J: J. Eng. Tribol. **228**, 511–520 (2014). https://doi.org/10.1177/1350650113517112

Saikko, V.: Wear and friction of thin, large-diameter acetabular liners made from highly cross-linked, vitamin-E-stabilized UHMWPE against CoCr femoral heads. Wear **432–433**, 202948 (2019). https://doi.org/10.1016/j.wear.2019.202948

Sakoda, H., Okamoto, Y., Haishima, Y.: In vitro estimation of reduction in strength and wear resistance of UHMWPE for joint prostheses due to lipid-induced degradation. J. Biomed. Mater. Res. - Part B Appl. Biomater. 1–7 (2020). https://doi.org/10.1002/jbm.b.34641

Sidia, B., Bensalah, W.: Tribological and mechanical behaviors of mollusk shell-ultrahigh molecular weight polyethylene bio-composite. Polym. Compos. **42**, 1 (2021). https://doi.org/10.1002/pc.26051

Sidia, B., Bensalah, W.: Tribological properties of high density polyethylene based composite: the effect of mollusc shell particles under dry condition. J. Compos. Mater. **55** (2020). https://doi.org/10.1177/0021998320965655

Terrani, K.A.: UC Berkeley Electronic Theses and Dissertations. DNA Mediated Assembly of Protein Heterodimers on Membrane Surfaces, p. 67 (2016)

Vadivel, H.S., Bek, M., Šebenik, U., Perše, L.S., Kádár, R., Emami, N., et al.: Do the particle size, molecular weight, and processing of UHMWPE affect its thermomechanical and tribological performance? J. Market. Res. **12**, 1728–1737 (2021). https://doi.org/10.1016/j.jmrt.2021.03.087

Wang, C., Bai, X., Guo, Z., Dong, C., Yuan, C.: Friction and wear behaviours of polyacrylamide hydrogel microsphere/UHMWPE composite under water lubrication. Wear **477**, 203841 (2021). https://doi.org/10.1016/j.wear.2021.203841

Wang, J., Gao, H., Gao, L., Cui, Y., Song, Z.: Ratcheting behavior of UHMWPE reinforced by carbon nanofibers (CNF) and hydroxyapatite (HA): experiment and simulation. J. Mech. Behav. Biomed. Mater. **88**, 176–184 (2018). https://doi.org/10.1016/j.jmbbm.2018.08.022

Xu, J.Z., Muratoglu, O.K., Oral, E.: Improved oxidation and wear resistance of ultrahigh molecular weight polyethylene using cross-linked powder reinforcement. J. Biomed. Mater. Res. Part B Appl. Biomater. **107**, 716–723 (2019). https://doi.org/10.1002/jbm.b.34165

Xu, S., Akchurin, A., Liu, T., Wood, W., Tangpong, X.W., Akhatov, I.S., et al.: Mechanical properties, tribological behavior, and biocompatibility of high-density polyethylene/carbon nanofibers nanocomposites. J. Compos. Mater. **49**, 1503–1512 (2015). https://doi.org/10.1177/0021998314535959

Prediction of Progressive Damage of Graphite/Epoxy Laminates Under Three-Point Bending

Hamza El idrissi[1(✉)], Youssef Benbouras[2], Mouad Bellahkim[2], Abbass Seddouki[1], and Jamal Echaabi[2]

[1] Mechanical Engineering Laboratory, Faculty of Sciences and Techniques, Sidi Mohammed Ben Abdellah University, B.P. 2202, Route d'Imouzzer, Fez, Morocco
{hamza.elidrissi4,abbass.seddouki}@usmba.ac.ma
[2] Engineering Mechanics and Innovation Laboratory, Applied Research Team on Composites, Management and Innovation, ENSEM, Hassan II University of Casablanca, BP 8118, Oasis, Casablanca, Morocco
mouad.bellahkim@ensem.ac.ma, j.echaabi@ensem-uh2c.ac.ma

Abstract. To maintain structural integration in such composite material applications, it is necessary to fully understand the behaviour of the material under various mechanical loads and model its responses. The purpose of this study is to use the progressive damage approach to assess the failure of a composite laminate subjected to a bending load. To simulate the initial and successive damage, Puck's failure criterion and the gradual degradation model were utilized. Through the finite element software Abaqus, a user-defined material (UMAT) subroutine has been developed and implemented. Furthermore, the impact of geometrical parameters has been investigated. As a result of this study, the ratio (l/h) and width play a determining role in the progressive damage after the onset of macroscopic failure where the ultimate load of the composite laminate, which defines the strength of the laminate, is reduced by decreasing the width for the specimens. However, the estimated results of the numerical modelling have a significant correlation with the experimental test.

Keywords: Three-point bending test · Puck's failure criteria · Element weakening method · Progressive damage model · User-defined material model (UMAT) · Damage mode

1 Introduction

Composites have emerged as excellent candidate materials for structural applications such as aircraft and vehicles, due to their high specific stiffness, high specific strength and low specific weight. However, the Engineering methods for composite laminates are still conservative. To assure structural safety, several costly and time-consuming tests have to be performed. The need for more robust failure theories and damage propagation

L. Azrar et al. (Eds.): CIP 2022, LNME, pp. 474–482, 2023.
https://doi.org/10.1007/978-3-031-23615-0_48

methodologies is imperative to successfully reduce time and cost and maximise the potential gains from composite structures.

Recent studies have made considerable progress in the development of effective and reliable prediction tools for composite materials [1–4]. Su et al. [5] have used a computational model for progressive damage modelling which is based on the model suggested by Ridha et al. [6] to investigate the damage progression and size effects on composite laminates subjected to compressive loading. Their model takes into account the spread crack and cohesive zone models to predict intralaminar and interlaminar damage, respectively.

In addition, the three-point bending test provides a simple experiment that can be used to validate the numerical modelling adopted for laminated composites. Liu et al. [7] developed a FE model that includes both an intralaminar damage model and a cohesive surface model. Benbouras et al. [8] conducted several three-point bending tests on a typical stacking sequence used in composite structures. It was found that successive failures are dependent on the stacking sequence, with orthogonal sequences having a considerable impact on ply delamination.

However, the availability of these progressive damage models does not make the task easy, as the failure concept of laminates and their response to structural behaviour is not easy to predict. In this research, a progressive damage model for laminated composites subjected to three-point bending is developed to predict the bending behaviour and progressive damage. This is accomplished by building a user-defined subroutine (UMAT) in Abaqus, which makes failure analysis simple and accessible without incurring high computational costs. The competence of the FE model constructed to predict the mechanical response and progressive failure of the composite was assessed by comparing experimental and numerical results, such as failure modes, load versus deflection curves, and the influence of different sizes.

1.1 Finite Element Model

The finite element model created in Abaqus is shown in Fig. 1. The UMAT subroutine developed was embedded to simulate the three-point bending events performed on the composite samples with lay-up $\left[[+45°/-45°/90°/0°]_3\right]_s$. The virtual specimen was discretized using 3 nodes triangular (S3) elements with a size of 1 mm × 1 mm. In addition, the laminate was subjected to displacement-controlled loading. A linear constraint equation was used to simulate the loading surface with displacement-controlled loading applied to a reference point RP and named at the underside of the loading nose to ensure that displacements would be uniform across all nodes on this loaded surface and the reference point RP. The precision of the simulation was set to double precision to reduce the accumulation error during the compilation. A consistent time increment of less than 1e−02 s was obtained, giving an approximate run time of less than 1 h on a 12 CPU workstation.

1.2 Properties for Model Input

We utilized several composite laminate specimens in this investigation to ensure the applicability and efficacy of the approach. The mechanical properties of graphite/epoxy,

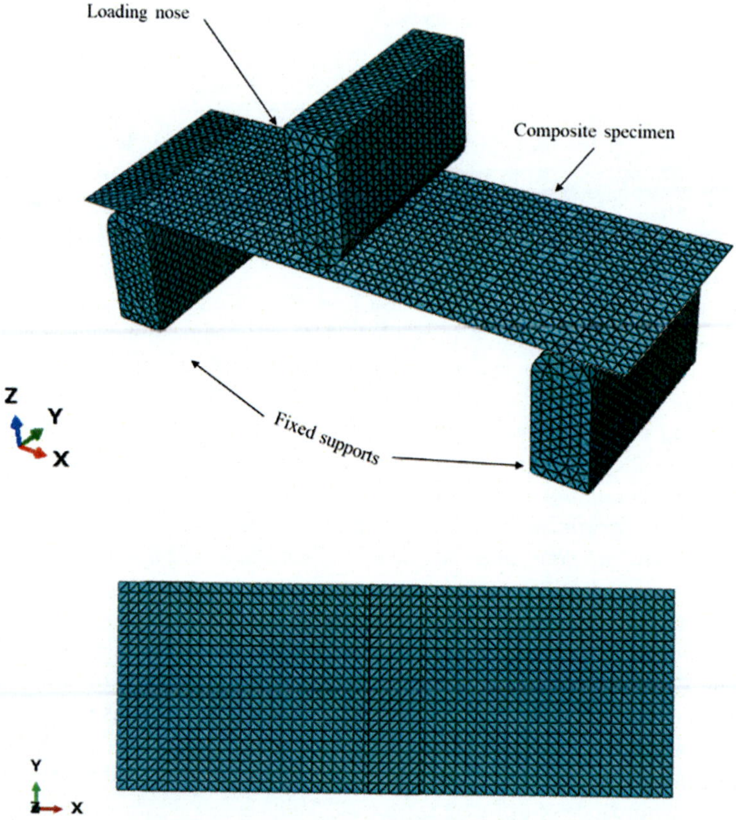

Fig. 1. The FE models

and the dimensions of the specimen were obtained from the work presented by Irhirane et al. [9], which is presented in Tables 1 and 2.

1.3 Model Implementation

A thorough description of the computational analysis procedure using the FEM, including the methodology applied in the present work, has been provided. To predict the initial failure of composite laminates subjected to three-point bending, the Puck failure criteria inputted stresses, strains and material properties for each element. Following detection of the failure, the material stiffness of the initially failed element was degraded using the progressive failure method mentioned above Puck-EWM. Moreover, we repeatedly calculated the stresses and strains until the affected areas were completely failed (catastrophic failure).

In the finite element software, it should be noted that convergence problems may appear when calculating the failed elements. Nevertheless, the elements must be degraded to a value close to zero and the values of the damage variables are updated according to the failure mode.

Table 1. Material properties of graphite/epoxy specimens

Properties	Graphite/epoxy
E_{11} (GPa)	116
E_{22} (GPa)	6.9
G_{12} (GPa)	5.6
υ_{12}	0.3
υ_{12}	0.3
E_f (MPa)	230,000
υ_{f12}	0.2
X_T (MPa)	1860
X_C (MPa)	1590
Y_T (MPa)	59
Y_C (MPa)	207
S_{12} (MPa)	110

Table 2. Dimensions of the specimens used in this study in (mm)

Material	Label	Length	Width	Thickness	Distance l between supports	Ratio l/h
Graphite/epoxy	A	75	25	3.6	57.5	16
	D	75	10	3.6	57.5	16

In this study we have implemented the theory of the Puck failure criterion and the material stiffness degradation model, including the EWM, by using a user-defined subroutine UMAT in the commercial FEA code (Abaqus) [10, 11]. Besides, the solution dependence of state variables (STATEV) presents the failure index and damage variables for the failure modes in the mentioned UMAT. The concept of the symbols, within the UMAT subroutine, has been displayed in Table 3. Also, the algorithm of computation is presented in Fig. 2.

2 Results and Discussion

The modelling allows us to deduce the successive failures, the failure modes, the damage initiation and evolution and the macroscopic curves. Figures 3 and 4, to analyse the load-deflection behaviour, illustrate the prediction of the three-point bending results of all labels (A, D) using the applied method and the experimental test.

During the first stage of a linear problem, there is no initial degradation of stiffness or material properties. Once the failure criteria and degradation patterns are reached, due to a certain amount of damage in one layer, the damage can lead to further redistribution

Table 3. Marking

Symbol	Concept
SDV1	Puck's fiber failure in tension
SDV2	Puck's fiber failure in compression
SDV3	Puck's matrix failure in tension
SDV4	Puck's matrix failure in compression
SDV6	Damage variable for fiber tensile
SDV7	Damage variable for fiber compressive
SDV8	Damage variable for matrix tensile
SDV9	Damage variable for matrix compressive

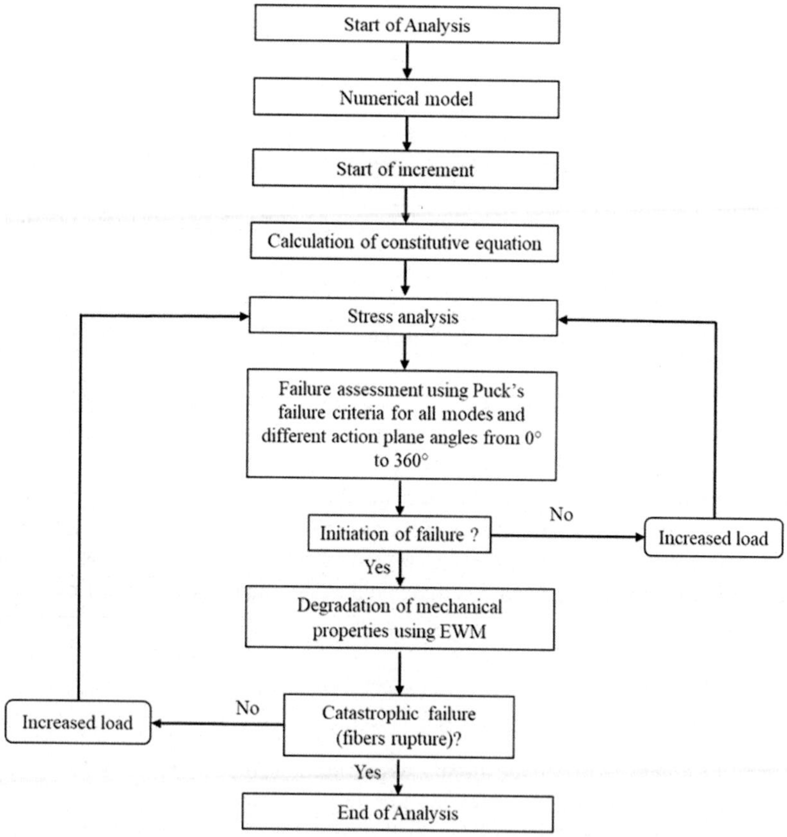

Fig. 2. The computational algorithm

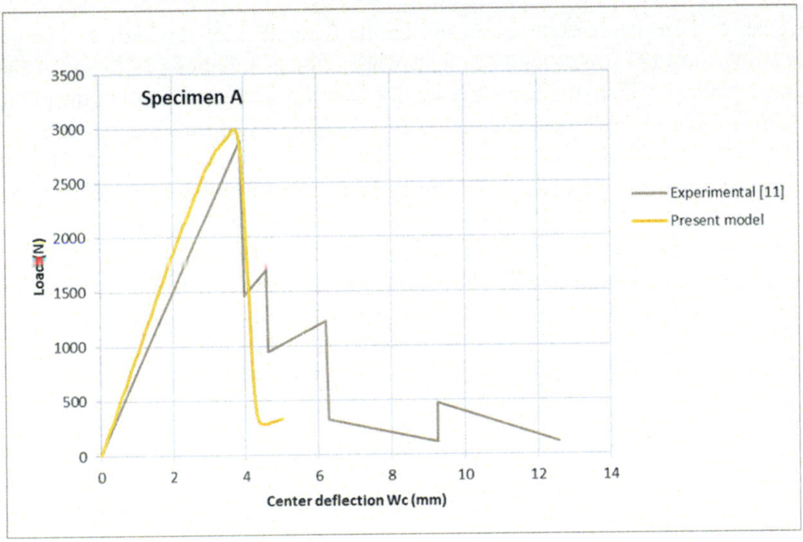

Fig. 3. Load-deflection curves of specimen A

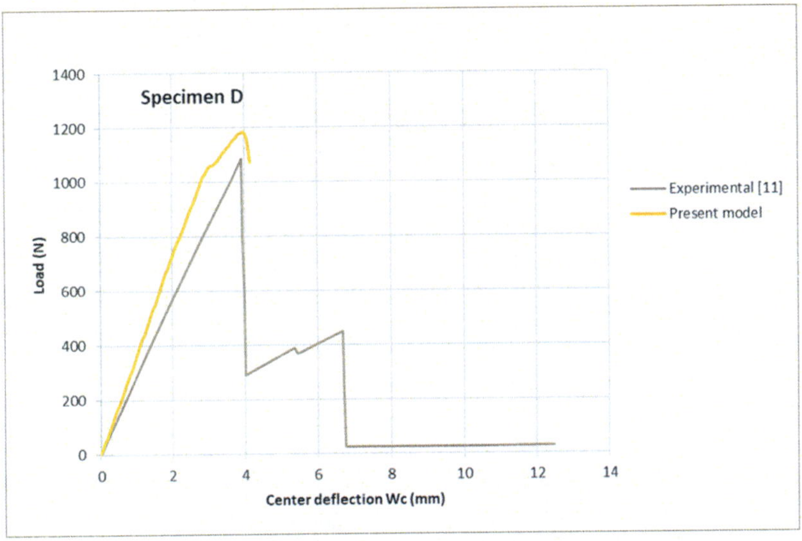

Fig. 4. Load-deflection curves of specimen D

of stresses in the laminate. Another layer will fail, although the load may increase and thus the stiffness properties decrease again. Furthermore, it is found that even after the first layer has failed, the laminate still has a lot of strength. The reduction in stiffness can be observed for both composite materials, where the curves shown in the graph are an indicator of the occurrence of damage and therefore the material behaviour becomes non-linear.

Arguably, there is a difference in the predictive ability of the load vs. deflection curves using the Puck-Ewm with the test result, which is noticed in the final failure deflection, which is due to the further reduction in stiffness and strength of the elements leads to the elimination of these elements, thus, a faster divergence can be seen.

Table 4 shows the successive failures that occurred in the specimens (A, D) to better understand the damage process. The failure starts at the bottom and develops upwards for both specimens (A, D) where the matrix cracking appears first in the 90° plies and grows to the 45° plies. Moreover, the fiber failure is noticed in the 0° plies which the first failure appears in the two upper 0° plies and is preceded by matrix cracking. The main difference in the result between the specimen (A, D) is the succession of damage after the occurrence of macroscopic failure and is due to the different widths of the specimens.

In peak load, the curves of load vs. deflection exhibit a vertical jump that corresponds to macroscopic failures in plies orientated at 0 degrees and this value vary with the width. Consequently, catastrophic failure occurs across the layers due to the decrease in all material properties, resulting in the complete failure of the composite laminate. Those predictions are in good correlation with the experimental result reported by Echaabi et al. [12].

Table 4. Results obtained by finite element method for test specimens A, D

Specimen	Failure sequence	F (N)	Layer number	Angle (°)	Type of failure
A	1	1583.63	3	90	SDV3
	2	1630.08	1	45	SDV3
	3	1766.41	2	– 45	SDV3
	4	2264.25	5	45	SDV3
	5	2473.57	7	90	SDV3
	6	2556.94	21	0	SDV2
	7	2596.71	24	45	SDV4
	8	2596.71	6	– 45	SDV3
	9	2630.04	23	– 45	SDV4
	10	2886.22	20	45	SDV4
	11	2974.87	17	0	SDV2
	12	2991.36	All remaining layers	–	SDV2, SDV4
D	1	655.432	3	90	SDV3
	2	655.432	1	45	SDV3
	3	706.3	2	-45	SDV3
	4	934.573	5	45	SDV3

(continued)

Table 4. (*continued*)

Specimen	Failure sequence	F (N)	Layer number	Angle (°)	Type of failure
	5	981.677	24	45	SDV4
	6	1027.84	7	90	SDV3
	7	1027.84	21	0	SDV2
	8	1042.16	6	– 45	SDV3
	10	1042.16	23	– 45	SDV4
	11	1147.56	17	0	SDV2
	12	1182.64	All remaining layers	–	SDV2, SDV4

3 Conclusion

In the present work, different composites laminated specimens have been examined under three-point bending to predict the process of damage starting from the onset of damage until the final failure. On the one hand, we proposed a finite element model, including puck's failure criteria and element weakening method, which accurately describes the behaviour of the specimens used.

Furthermore, there is no need to calibrate any of the parameters of the Puck failure criteria. However, it can reliably predict the initial and final failure load of composite laminates when coupled with a degradation model. Structural failure assessments of hugely complicated composite constructions may be possible. When constructing fiber-reinforced composites, it's critical to choose the right matrix and fiber material qualities to maximize the reinforcing effect and breaking strength of the fibers.

Acknowledgements. The authors also gratefully acknowledge the helpful comments and suggestions of the reviewers, which have improved the presentation.

References

1. Kim, J.S., Lee, S.Y., Roh, J.H.: Failure analysis of laminated composites subjected to flexural loadings. Adv. Compos. Mater. **29**(3), 301–316 (2020). https://doi.org/10.1080/09243046.2020.1724467
2. Han, K.N., Zhou, W., Qin, R., Yang, S., Ma, L.H.: Progressive damage analysis of carbon fabric-reinforced polymer composites under three-point bending. Fibers Polym. **22**(2), 469–479 (2021)
3. Benbouras, Y., Bellahkim, M., Maziri, A., Mallil, E., Echaabi, J.: An analytical and experimental study of the nonlinear behaviour of a carbon/epoxy under a three-point bending test. Plast. Rubber Compos. 1–9 (2021).https://doi.org/10.1080/14658011.2021.1950457
4. El Idrissi, H., Seddouki, A.: Damage modeling of composites laminate containing stress concentrations. In: 2022 2nd International Conference on Innovative Research in Applied Science, Engineering and Technology (IRASET), 2022, pp. 1–5. https://doi.org/10.1109/IRASET52964.2022.9737799

5. Su, Z.C., Tay, T.E., Ridha, M., Chen, B.Y.: Progressive damage modeling of open-hole composite laminates under compression. Compos. Struct. **122**, 507–517 (2015). https://doi.org/10.1016/j.compstruct.2014.12.022
6. Ridha, M., Wang, C.H., Chen, B.Y., Tay, T.E.: Modelling complex progressive failure in notched composite laminates with varying sizes and stacking sequences. Compos. Part A **58**, 16–23 (2014). https://doi.org/10.1016/j.compositesa.2013.11.012
7. Liu, H., et al.: Modelling damage in fibre-reinforced thermoplastic composite laminates subjected to three-point bend loading. Compos. Struct. **236**, 111889 (2020). https://doi.org/10.1016/j.compstruct.2020.111889
8. Benbouras, Y., Bellahkim, M., Maziri, A., Mallil, E., Echaabi, J.: Nonlinear modeling of the failure of a graphite epoxy under a three-point bending test. Polym. Polym. Compos. **28**(2), 119–139 (2020). https://doi.org/10.1177/0967391119866597
9. Irhirane, E. H., Echaabi, J., Aboussaleh, M.,Hattabi,M., Trochu, F.:Matrix and Fibre Stiffness Degradation of a Quasi-isotrope Graphite Epoxy Laminate Under Flexural Bending Test. J. Reinf. Plast. Compos. 28(2), 201–223 (2009). https://doi.org/10.1177/0731684407084213
10. Puck, A., Schürmann, H.: Failure analysis of FRP laminates by means of physically based phenomenological models. Compos. Sci. Technol. **62**(12–13, Special Issue), 1633–1662 (2002). https://doi.org/10.1016/S0266-3538(01)00208-1
11. Lee, C.S., Yoo, B.M., Kim, M.H., Lee, J.M.: Viscoplastic damage model for austenitic stainless steel and its application to the crack propagation problem at cryogenic temperatures. Int. J. Damage Mech. **22**(1), 95–115 (2013). https://doi.org/10.1177/1056789511434816
12. Echaabi, J., Trochu, F., Pham, X.T., Ouellet, M.: Theoretical and experimental investigation of failure and damage progression of graphite-epoxy composites in flexural bending test. J. Reinf. Plast. Compos. **15**(7), 740–755 (1996). https://doi.org/10.1177/073168449601500707

Microscopic and Macroscopic Damage of CFRP [0]12 Composite Laminates Under Three-Point Bending Test

M. Bellahkim(✉), Y. Benbouras, K. Kimakh, A. Maziri, El. Mallil, and J. Echaabi

National Superior School of Electricity and Mechanics, Laboratory of Mechanics, Engineering and Innovation (LM2I), Hassan II University of Casablanca, Casablanca, Morocco
mouad.bellahkim@ensem.ac.ma

Abstract. The interlaminar failure, also called delamination, is considered among the catastrophic failures that occur in laminated composites because of high interlaminar stresses. This work aims to present an experimental study of the effect of the span-thickness (l/h) ratio on the interlaminar damage in carbon/epoxy laminated composites under the three-point bending test. Two specimens are studied with the thickness are respectively equal to 2 mm and 4 mm. Other specimens were also tested with different values of span-thickness (6, 8, 10, 12, 14, and 16). An MTS tensile machine was equipped with a bending fixture to assure the three-point bending test. A digital microscope is used to follow the microscopic failure mechanisms during loading. Microscopic observations show that delamination accompanied by fiber buckling is the main mode of damage for relatively low values of l/h. This study showed that there is a relationship between the thickness of the specimen and the number of delaminations observed. The same remark was revealed for the macroscopic behavior.

Keywords: Delamination · Laminated composites · Interlaminar damage · Microscopic damage

1 Introduction

Laminated carbon/epoxy composite material is widely used in applications where mass reduction is critical. The wider the use of a material, the greater the likelihood of eventual failure. The ability to characterize fractures, for example in terms of identifying failure modes, characteristic parameters, or critical failure values, is essential to ensure the integrity of parts in service and for the design of future products (Bathias 2002). Failure of laminated composites can occur in several very complex ways. The failure modes depend on the stacking sequence, direction of loading relative to fiber orientation and on the geometrical parameters such as the span-thickness ratio (l/h). (Echaabi et al. 1996) studied the effect of the l/h ratio on the bending behavior of a quasi-isotropic unidirectional laminate. (Benbouras et al. 2021) examined, experimentally and analytically, the bending behavior of a fabric laminate ($[45/0]_{2s}$ and $[0/90]_6$). Out-of-plane

L. Azrar et al. (Eds.): CIP 2022, LNME, pp. 483–489, 2023.
https://doi.org/10.1007/978-3-031-23615-0_49

stresses lead to the initiation of delamination in the interface due to the difference in elastic properties of adjacent plies (Herakovich 1981).

The stacking sequence is one of the main parameters that influence the occurrence of delamination near the free edges. (Czarnek et al. 1983) and (Herakovich et al. 1985) used the moiré technique at the free edge side of a symmetrical laminate designated [(θn/−θn]s where n = 2 and θ = 10°, 25°, 30°, 45°; they highlight important displacement gradients at the + θ/− θ interface. (Herakovich 1989a, Herakovich 1989b) compared the tensile behavior of the clustered [θn/− θn]s and the alternating [(± θ)n]s laminates with θ = 10°, 30°, 45°; They show that the stress required to initiate the delamination of the alternating laminates is greater than that in which the clustered laminates. This is what is commonly called the thickness effect, two plies of the same orientation stacked successively behaving like a single ply of twice the thickness. (Diaz and Caron 2006) observed by an optical microscope, for θ = 10°, 20° and n = 1, 2, 3, 4, that microcracks or matrix plasticity due to shearing may appear at the + θ/− θ interfaces depending upon the fiber density at the interface. This sliding is observed before delamination onset.

In addition, the thickness of the composite structures influences the mode of damage. (Yuan et al. 2017) studied experimentally and numerically by the finite element method the effect of ply thickness on the variation of interlaminar stresses, where they tested specimens with different fiber areal weight (20, 60, 120 g/m^2). The results found show that an increase in the thickness of the plies leads to an increase in the interlaminar shear stress σ_{13}. Unlike tensile loading, the w/h ratio exhibits the effect of thickness in the case of bending. Caminero et al. 2016 observed a delamination failure under bending loading for $[0]_{12}$ stacking sequence with l/h = 9.25.

In this paper, the microscopic and macroscopic damage of CFRP composite laminates under three-point bending test are investigated for many l/h values (8, 10, 12, 14 and 16) and two thickness (2 and 4). The experimental protocol for the bending test is adopted based on the ASTM D70 recommendations. A digital microscope is used to observe the microscopic failure mechanisms during loading.

2 Experimental Protocol

2.1 Specimens Preparation

The specimens are made from a unidirectional Carbon T300–3000/Epoxy prepreg, the stacking sequences and dimensions are shown in Table 1. The polymerization is carried out in an autoclave, where the parameters provided by the supplier are listed in Table 2.

2.2 Cutting of Specimens

The specimens are cut form the plates of dimensions (300 mm * 400 mm * 2 mm). In order to avoid any type of defect, a specific CNC milling machine for composite materials is used to cut the plates and the cutting parameters are presented in Table 3.

Table 1. Geometric characteristics of the specimens.

Specimen	Orientations	Length (mm)	Width (mm)	Thickness h (mm)	Span l (mm)	l/h
A1	[0]24	36.8	10	4	24	6
A2	[0]24	44.8	10	4	32	8
A3	[0]12	32.8	10	2	20	10
A4	[0]12	36.8	10	2	24	12
A5	[0]12	40.8	10	2	28	14
A6	[0]12	44.8	10	2	32	16

Table 2. Polymerization parameters.

Temperature (°C)	Pressure (MPa)	Vacuum (MPa)
125	1.2	0.1

Table 3. Cutting parameters.

Cutter diameter	2mm
Feed rate	5–6 mm/s
Depth of cut	2 mm
Spindle speed	18000 rpm

2.3 Polishing of Specimens

In order to obtain a good side view to observe by microscopic, All specimens were polished using silicon carbide abrasive paper (with a grit P300, P400, P600 and P1200 respectively) and alumina powder of size 0.05 μm (see Fig. 1).

2.4 Microscopic Observation

A digital microscope, the characteristics of which are indicated in Table 4, was used to observe the microscopic damage during the loading of the specimens (Fig. 2). The microscope is fixed directly using a magnetic base on the surface of the test piece to follow the damage as the loading progresses.

3 Results and Discussion

The objective of this work is to study the effect of the l/h ratio on the interlaminar damage of laminated Carbon/Epoxy composites under the three-point bending test. For this, a

Fig. 1. Polishing machine with alumina powder

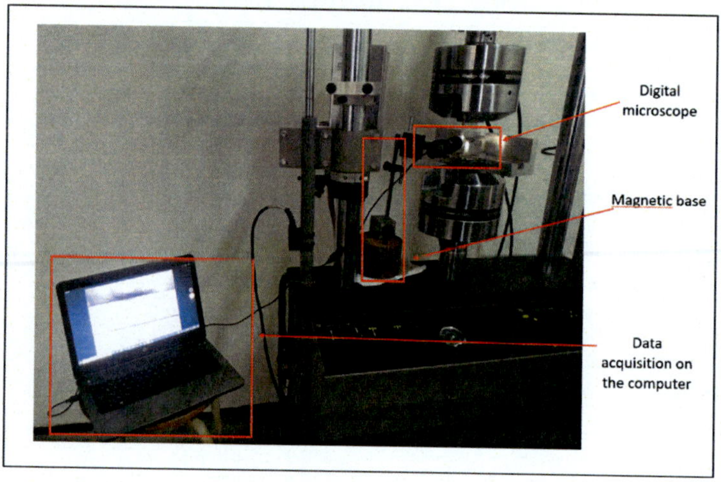

Fig. 2. Digital microscopic

Table 4. Characteristics of digital microscopic.

Characteristics	Value
LED	8
Resolution	12 MP
ZOOM	× 800

fixture is adopted to take into consideration the recommendations of the three-point bending standard ASTM D790 (see Fig. 3). A digital microscope is used to follow the microscopic failure mechanisms during loading. The MTS machine with a 100 KN load cell and a loading speed of 0.2 mm/min is operated to carry out the bending tests.

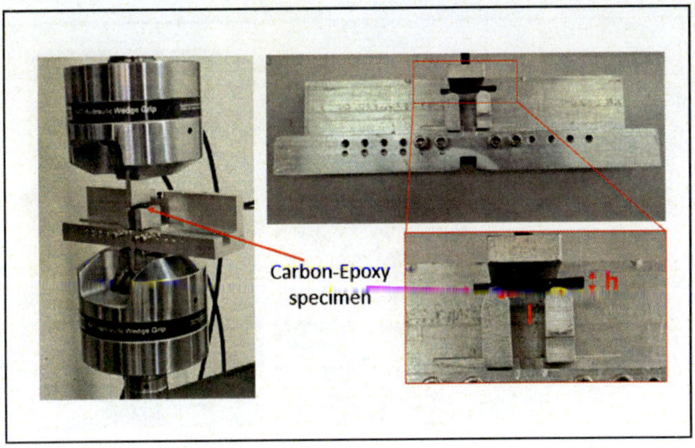

Fig. 3. Three-point bending test

Figure 4 shows the load-displacement curves of different values of l/h ratio for the stacking sequence $[0]_{12}[0]_{24}$. For the specimens A1 and A2 with h = 4 mm, the macroscopic curve shows several drops in stiffness. However, the macroscopic curves for the specimens A3 to A6 with h = 2 present fewer drops of stiffness.

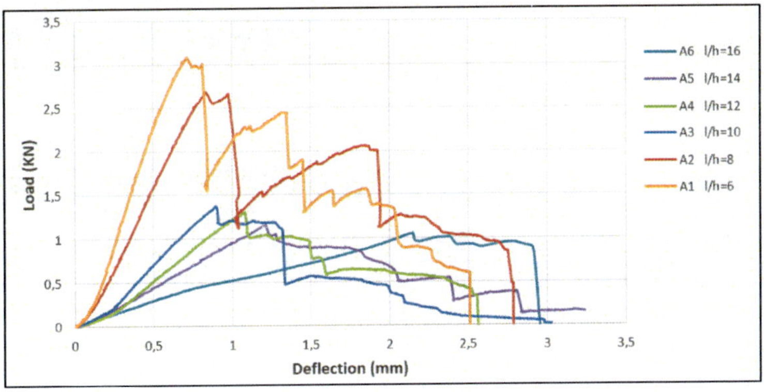

Fig. 4. Load-deflection curves

Figures 5 and 6 present respectively the succession and the mechanisms of microscopic failure for the specimens A1 and A3 with the thickness equal respectively 2 and 4 mm. These microscopic observations show that the specimen breaks by jumps of approximately 6 plies. Interlaminar shear failure by means of delaminations and fiber breakages are observed in the micrographs of side view. These delaminations cause load drops on the load-displacement curves, see Fig. 5. The microscopic damage mode is the same for the others specimens (A2, A4, A5 and A6).

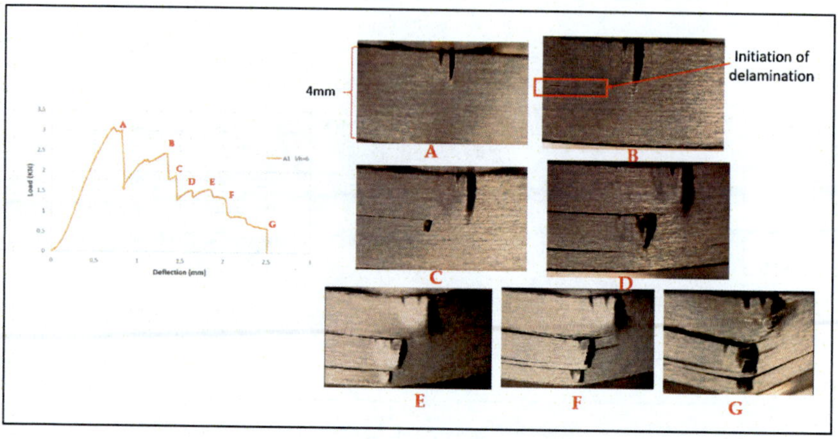

Fig. 5. The microscopic successive failure for A1 specimen

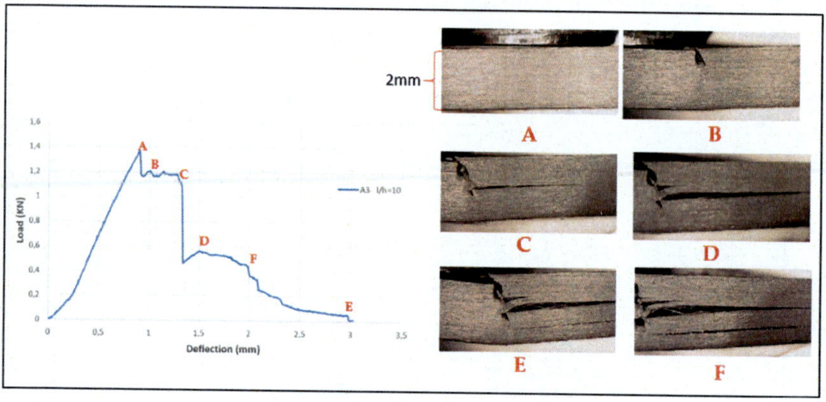

Fig. 6. The microscopic successive failure for A3 specimen

4 Conclusion

An experimental study was made to study the effect of the small values of l/h ratio on the interlaminar damage of laminated composites under the three-point bending test. The macroscopic curves show, on the one hand, that the behavior is linear before the first failure drop, and on the other hand, that the specimens whose thickness equals 4 mm present several drops of rigidity. These drops indicate the fiber breakages and delaminations observed by the microscope. The mode of damage and the succession failure are linked mainly with the orientation of the plies and geometric parameters such as the l/h ratio. In perspective, it would therefore be wise to duplicate this experimental approach for other orientations.

Acknowledgements. We want to send our sincere thanks to Prof Abdelkerim Chouaf and Mrs khadija Kimakh for all the technical and experimental help.

References

Bathias, C.: Matériaux composites, 2eme Editi. Dunod (2002)

Echaabi, J., Trochu, F., Pham, X.T., Ouellet, M.: Theoretical and experimental investigation of failure and damage progression of graphite-epoxy composites in flexural bending test. J. Reinf. Plast. Compos. **15**(7), 740–755 (1996)

Benbouras, Y., Bellahkim, M., Maziri, A., Mallil, E., Echaabi, J.: An analytical and experimental study of the nonlinear behaviour of a carbon/epoxy under a three-point bending test. Plast. Rubber Compos., pp. 1–9 (2021)

Benbouras, Y., Bellahkim, M., Maziri, A., Mallil, E., Echaabi, J.: Nonlinear modeling of the failure of a graphite epoxy under a three-point bending test. Polym. Polym. Compos. **28**(2), 119–139 (2019)

Herakovich, C.T.: On the relationship between engineering properties and delamination of composite materials. J. Compos. Mater. **15**(4), 336–348 (1981)

Czarnek, R., Post, D., Herakovich, C.: Edge effects in composites by Moiré interferometry. Exp. Tech. **7**(1), 18–21 (1983)

Herakovich, C.T., Post, D., Buczek, M.B., Czarnek, R.: Free edge strain concentrations in real composite laminates: experimental-theoretical correlation. J. Appl. Mech. **52**(4), 787–793 (1985)

Herakovich, C.T.: Edge effects and delamination failures. J. Strain Anal. Eng. Des. **24**(4), 245–252 (1989)

Herakovich, C.T.: Failure modes and damage accumulation in laminated composites with free edges. Compos. Sci. Technol. **36**(2), 105–119 (1989)

Diaz, A.D., Caron, J.F.: Interface plasticity and delamination onset prediction. Mech. Mater. **38**(7), 648–663 (2006)

Yuan, Y., Yao, X., Liu, B., Yang, H., Imtiaz, H.: Failure modes and strength prediction of thin ply CFRP angle-ply laminates. Compos. Struct. **176**, 729–735 (2017)

Caminero, M.A., Rodríguez, G.P., Muñoz, V.: Effect of stacking sequence on Charpy impact and flexural damage behavior of composite laminates. Compos. Struct. **136**, 345–357 (2016)

ASTM D790.: Standard test methods for flexural properties of unreinforced and reinforced plastics and electrical insulating materials. Am. Soc. Test. Mater. (1997). https://doi.org/10.1520/D0790-10

Application of a Design for Additive Manufacturing Methodology to Optimize the Mechanical Performance of a PLA Test Sample

Salem Houcine[1](✉), Abouchadi Hamid[2], and El Bikri Khalid[1]

[1] M2SM, Research Center STIS, Department of Mechanical Engineering, ENSAM, Mohammed V University, Rabat, Morocco
{h.salem,k.elbikri}@um5r.ac.ma

[2] PCMT, Research Center STIS, Department of Mechanical Engineering, ENSAM, Mohammed V University, Rabat, Morocco

Abstract. Additive manufacturing processes are widely used in fabrication laboratories in order to prove a concept, they are also used in maintenance to re-engineer defective parts in homes or in factories. Additive manufacturing brings undeniable advantages over other manufacturing processes, such as the complexity of the parts that can be designed, and also the different possibilities that it allows like the ability to build irremovable assemblies or multi-materials and multicolor products. That said, to take full advantage of the added value of these processes, specific manufacturing parameters must be anticipated, in terms of the necessity to add supports to assure the feasibility of the design, or in term of the anticipation of the optimum orientation during the fabrication to assure the mechanical performance of the designed part. Thus, it is necessary to rethink the design of the parts and to adapt it to additive manufacturing processes. In this context, a methodology was created to take into account the specificities of these processes in order to optimize the manufacturing of the parts. This work consists of using this methodology to optimize the mechanical performance of a 3D printed test specimen. The objective is to rule on the importance of the methodology to optimize the design and production of 3D printed parts.

Keywords: Additive manufacturing · 3d printing · Design · Methodology · Traction · Test sample

1 Introduction

Additive manufacturing brings undeniable benefit compared to other manufacturing processes. In particular, it allows the production of non-removable assemblies. Thanks to the layer-by-layer horizontal design, it allows the imagination of designers to run free, as it allows the design of any possible geometric shape, in addition to the advantages of cost and speed of realization (Shah et al. 2019).

L. Azrar et al. (Eds.): CIP 2022, LNME, pp. 490–499, 2023.
https://doi.org/10.1007/978-3-031-23615-0_50

More and more fields are using additive manufacturing, with the aim of making proof of concept prototypes in design offices, or packaging (Arrieta et al. 2017), or to quickly replace a faulty part in factory maintenance, or in the biomedical field with prostheses (maxillofacial, leg, teeth, etc.), or simply in the aeronautical field where the freedom of design and the materials used brings great added value.

That said, this type of manufacturing brings its share of limitations. For example, for layer-by-layer fabrication of hollow shapes in fused deposition modeling, it is necessary to include fabrication supports. The designer must therefore think about the filling percentage of the material to define the ideal weight and cost (Ceresana 2017), he must also think about how to remove the supports after design. He must also anticipate the staircase effect that can influence the quality of the exterior surface of the part, and many other parameters that are not taken into account during manufacture with standard processes.

For this purpose, a methodology has been created to optimize the design of parts in additive manufacturing. This methodology can be followed systematically to take into consideration the limitations of additive manufacturing while enjoying the advantages of the processes (Wiberg et al. 2019).

To prove the interest of this methodology, it was applied to a standardized tensile specimen. The interest in this case is to optimize the mechanical performance of this specimen (Lubombo and Huneault 2018). The choice fell on this type of part because it can easily assess the difference in mechanical performance. Therefore, the part has been redesigned while respecting the standard and following the methodology cited step by step (Salem et al. 2022). Afterwards, the part was tested to evaluate the effect of the methodology on the mechanical performance (Lunt 1998).

2 Methods

2.1 Design Methodology

Most additive manufacturing processes have the same design chain. This chain begins with the design of a part on 3D software, this part meets well-defined specifications. Then, the model is saved as a '.stl' file; this file corresponds to the discretization of the 3D part in the form of several small triangles. Then, software adapted to the process generates a 'gcode' file that corresponds to the trajectory of the print head or the extruder. This file is then inserted into the manufacturing machine that translates the data into printing operations layer by layer until the final part is obtained. Depending on the surface condition of the part obtained, and the expectations specified in the specifications, a finishing operation can be applied to improve the surface condition and remove any manufacturing supports. The last operation is the test operation, to check the response of the part to the mechanical or assembly constraints requested in the specifications (Fig. 1).

The studied additive manufacturing methodology begins with the consideration of customer specifications, which corresponds to an in-depth study of the client recommendations (Salem et al. 2020).

The second part is the functional domain, which corresponds to the translation of the specifications into a 3D design containing functional surfaces and functional volumes. The next step corresponds to the domain of the process, this domain corresponds to

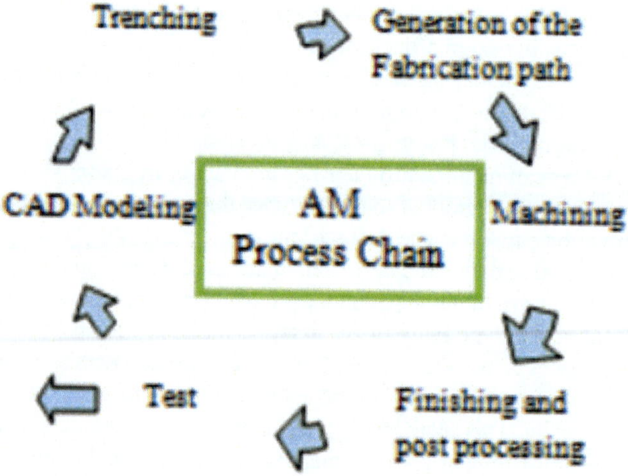

Fig. 1. Classical process of additive manufacturing

the consideration of the specific data of the process, such as the printing volume of the machine or the temperature of the print head or any other information related to the machine. After taking into account the functional domain and the process, the part obtained therefore meets the specifications and is feasible on the chosen process, the next step is then the optimization of the part to save time and cost manufacturing. This optimization is often topological to save material. Then, a simulation is necessary to check the compatibility of the optimized part with the specifications. If the simulation is validated, the last manufacturing step can be started. If not, the DFAM is restarted (Fig. 2).

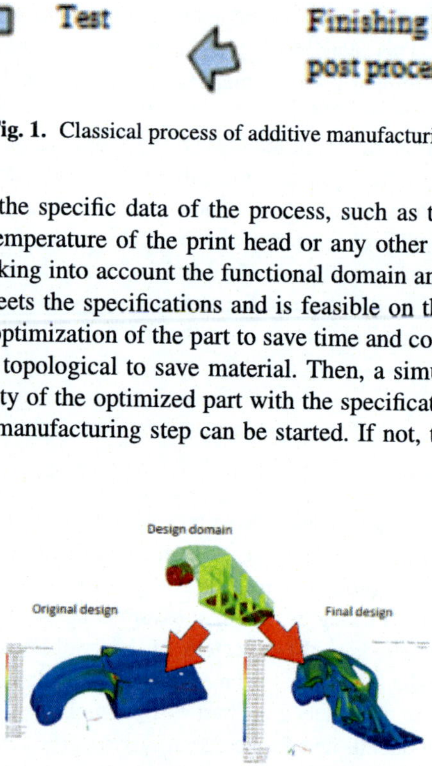

Fig. 2. Illustration of the design optimization

Here is the summary of the methodology in question: (Fig. 3).

The methodology will be followed gradually to build the part: (Fig. 4).

The first step consists of transforming the customer's specifications into mechanical constraints. These specifications correspond to the mechanical properties of the material, the physical properties, the loadings and the mechanical stresses (Aguero et al. 2019). The customer, in the event of assembly, product life cycle management or end-user choice, may request other constraints.

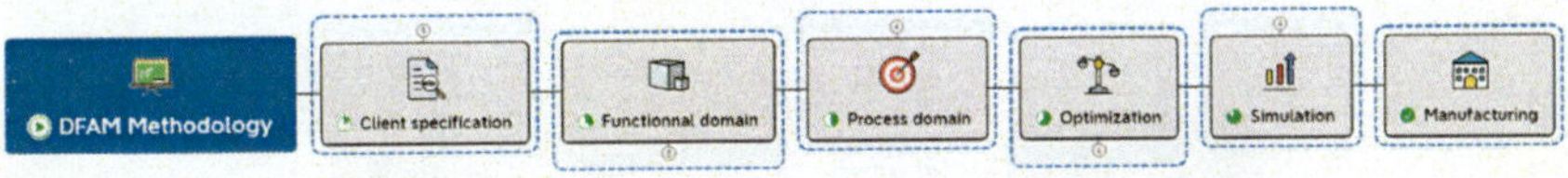

Fig. 3. Design for Additive Manufacturing Methodology

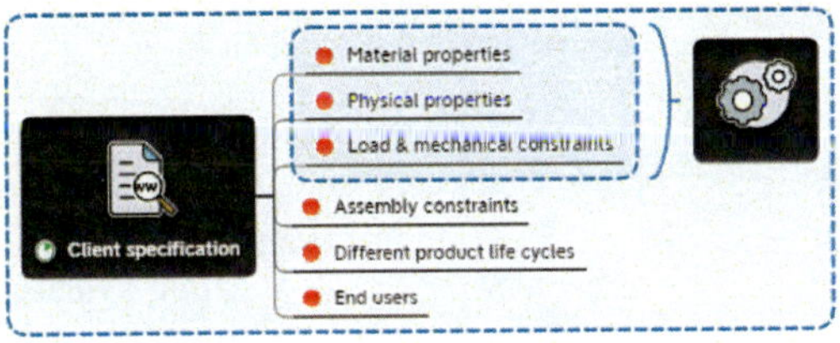

Fig. 4. Client specification

In the case of this part, the customer wants a standardized part to test the mechanical properties of the PLA material. The source material properties are listed below. It is a simple part without assembly and without product life cycle management, because PLA is a recyclable material. (Mikula et al. 2021) (Fig. 5).

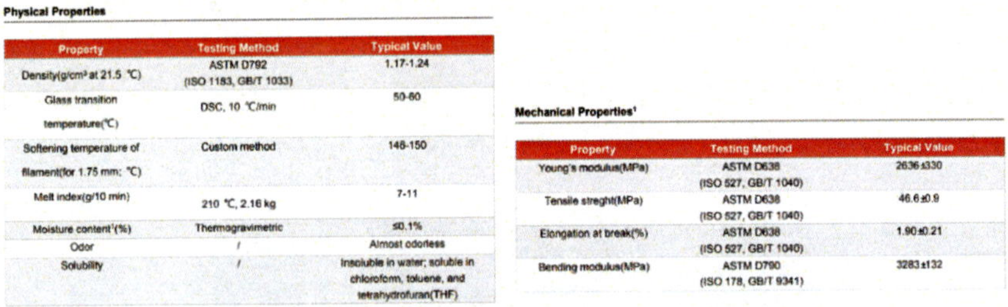

Physical Properties

Property	Testing Method	Typical Value
Density(g/cm³ at 21.5 ℃)	ASTM D792 (ISO 1183, GB/T 1033)	1.17-1.24
Glass transition temperature(℃)	DSC, 10 ℃/min	50-60
Softening temperature of filament(for 1.75 mm; ℃)	Custom method	148-150
Melt index(g/10 min)	210 ℃, 2.16 kg	7-11
Moisture content[1](%)	Thermogravimetric	≤0.1%
Odor	/	Almost odorless
Solubility	/	Insoluble in water; soluble in chlorofom, toluene, and tetrahydrofuran(THF)

Mechanical Properties[1]

Property	Testing Method	Typical Value
Young's modulus(MPa)	ASTM D638 (ISO 527, GB/T 1040)	2636±330
Tensile streght(MPa)	ASTM D638 (ISO 527, GB/T 1040)	46.6±0.9
Elongation at break(%)	ASTM D638 (ISO 527, GB/T 1040)	1.90±0.21
Bending modulus(MPa)	ASTM D790 (ISO 178, GB/T 9341)	3283±132

Fig. 5. Material properties

The most important property in this case is the filling parameter that must be 100% to be compared with other processes.

The mechanical loading of the part will be done in a standardized way with a traction machine. This loading is defined with the standard ASTM D638.

No assembly or life cycle constraints are specified.

On the other hand, in order to compare the results with tests carried out on other specimens of the same size and same material, a constraint is specified on the orientation of the fibers (Fig. 6).

The second phase concerns the functional area of the part, i.e. the surfaces and volumes that will be in contact with the outside. These surfaces must meet the dimensions and tolerances of the ASTM D638 standard. The most important functional surfaces are those in contact with the jaws of the traction machine.

For the functional areas, the methodology specifies the dimensions and tolerances of the part (Table 1 and Figs. 7 and 8).

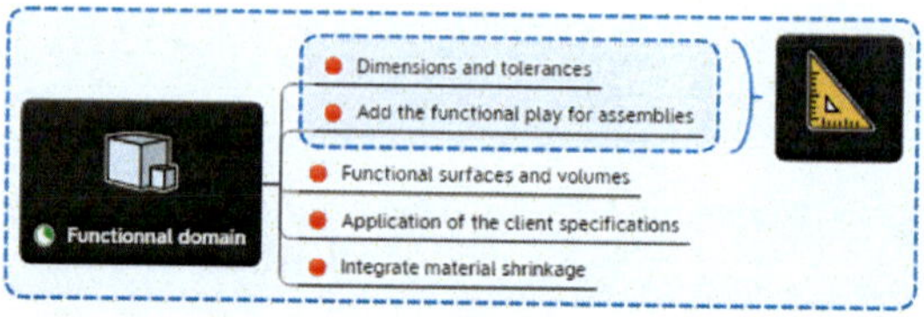

Fig. 6. Functional domain

Table 1. Test sample dimensions from the ASTM standard

Dimension	Value mm	Symbol
Thickness	3	T
Width	10	W
Parameter3	60	L

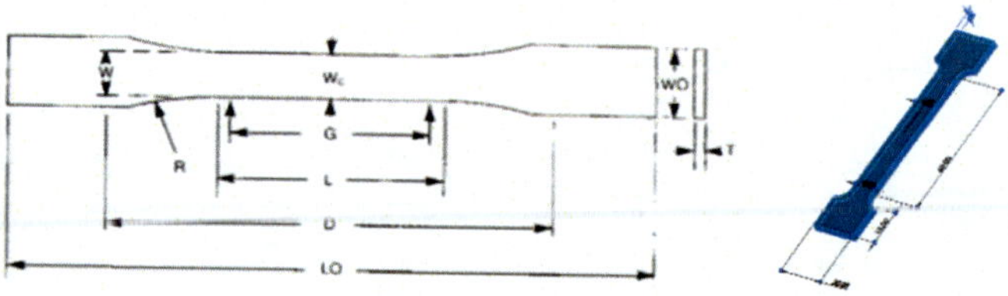

Fig. 7. Test specimen

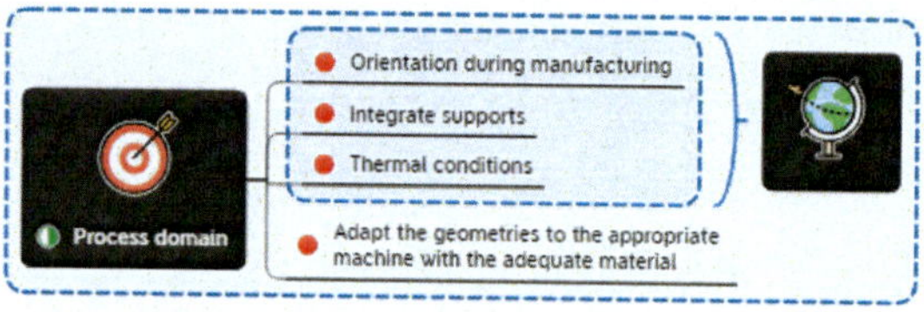

Fig. 8. Process domain

This phase concerns the domain of the process. This involves taking into consideration the characteristics of the manufacturing process to meet customer specifications.

In this case, the orientation of the fibers during manufacturing is a key element. Automatic orientation on 100% infill is 0 and 90°. With a layer of longitudinal fibers followed by a perpendicular layer until the final piece is obtained.

Here the choice that will be made is related to an orientation other than the one defined by the manufacturing software. Indeed, studies have proven that cracks propagate faster on a 0–90 orientation than on a $+ 45 - 45$ orientation. During the simulation step we will choose the most interesting solution to have better mechanical properties (Fig. 9).

The design domain stage consists of the inclusion of the two previous domains to define the final design of the part (Fig. 10).

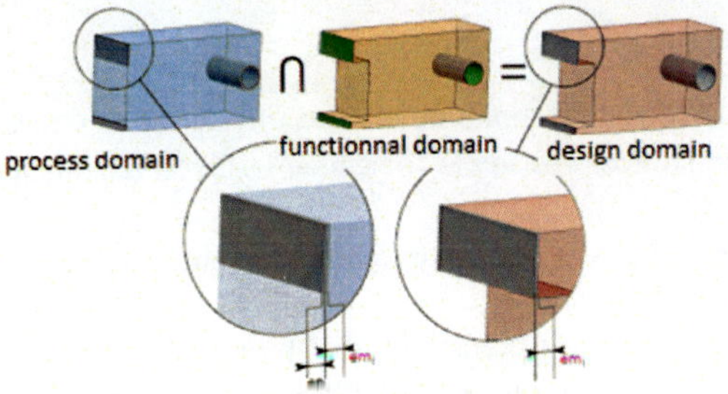

Fig. 9. Design domain

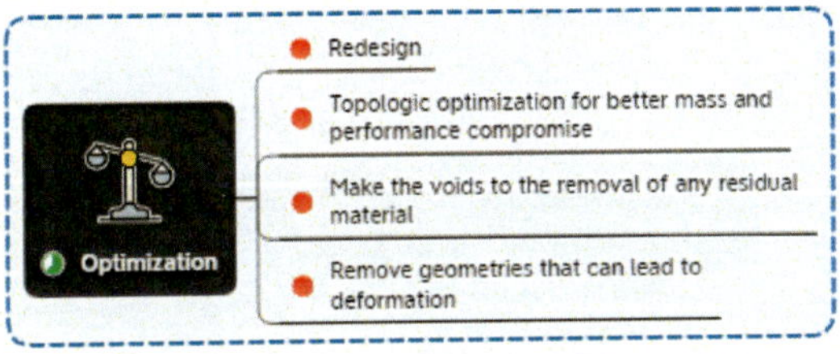

Fig. 10. Optimization phase

During the optimization step, the part can be topologically optimized to save weight or manufacturing cost. But in this case the volume and the dimensions are fixed by the standard. The only optimization can be done on the process domain with the modification of the orientation of the fibers during manufacturing (Fig. 11).

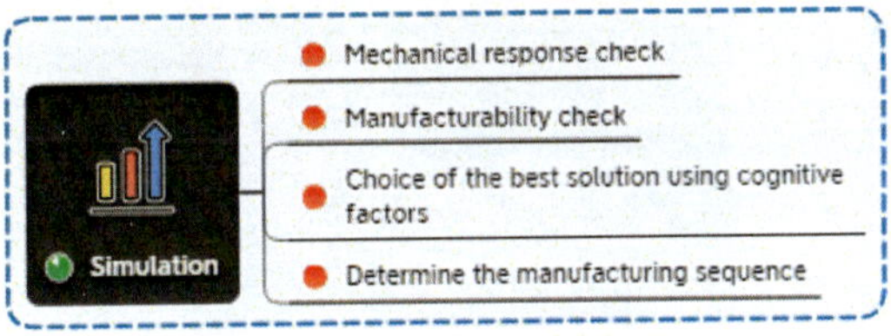

Fig. 11. Simulation phase

The simulation step makes it possible to verify the predictions established during the optimization phase. This analysis made it possible to verify that the results obtained with a manufacturing orientation of + 45 − 45 are better than the results of the orientation of 0 90.

The simulation is carried out with CURA 3D software, which takes into account the parameters of the manufacturing machine to check the feasibility of the part.

2.2 Test Specimen

To prove the interest of the methodology, it was necessary to choose a part that can be easily tested. The choice was made for a standardized tensile specimen for testing thermoplastic materials.

The part meets the ASTM D638 standard; this standard defines the design, production and test criteria for the specimen.

In an article currently being published, this type of test piece was made to check the performance of PLA Polylactic Acid before and after going through the printing machine. In this study, the criteria of the standard will always be respected, but by following the manufacturing methodology step by step.

Here are the dimensions of the part and the characteristics defined by the standard: (Fig. 12).

Fig. 12. Test specimen

The design was made with Solidworks 2016 software.

The realization was carried out with the following machine according to the conditions defined by the standard: (Fig. 13).

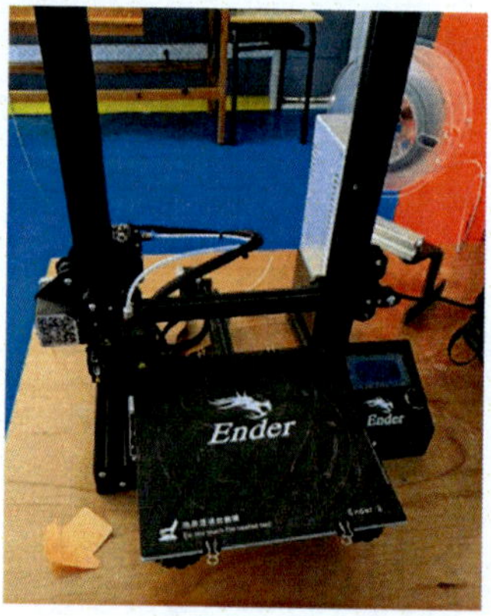

Fig. 13. Printing machine

3 Results

Here are the results obtained for the part without DFAM: (Table 2).

Here are the results after DFAM: (Table 3).

Table 2. Test result without DFAM

Measured properties	Symbol	Unit	Test 1	Test 2	Test 3	Test 4	Test 5	Av	SV	CV
Thickness	T	mm	2.86	2.9	2.92	2.88	2.88	2.888	0.023	0.008
Length	L	mm	60.45	59.7	59.7	60.45	59.8	60.02	0.395	0.007
Width	W	mm	10.095	10.1	10.02	10.05	10.07	10.066	0.032	0.003
Surface	S	mm^2	28.872	29.28	29.26	28.94	29	29.070	0.185	0.006
Load at yield point	Py	N	1190	1110	1100	1210	1370	1196	108.5	0.09
Load at elastic limit	Pe	N	1180	1010	1070	1170	1290	1144	108.1	0.094
Displacement to Py	Dy	Mm	2.61	2.42	1/98	2.24	2.1	2.27	0.251	0.111
Displacement to Pe	De	mm	2.32	1.78	1.76	2	1.88	1.95	0.229	0.117
Elastic modulus	E	MPa	1147.3	1277	1395	1344	1455	**1323.74**	118.3	0.089
Resistance to yield point	sy	MPa	41.21	37.9	37.6	41.8	47.2	41.14	3.88	0.094
Elastic limit	se	MPa	40.87	34.5	36.6	40.4	44.5	39.37	3.91	0.099
Strain at yield point	ey	%	4.33	4.054	3.317	3.706	3.512	3.78	0.41	0.108
Strain at elastic limit	ee	%	3.83	2.982	2.948	3.309	3.144	3.24	0.36	0.111

Firstly, the real dimension of the specimens are lower than expected, this behavior is common in the fused deposition modeling process, because there is a small void between each layer that cannot be filled with material, indeed cylindrical fibers deposited on each other cannot fill a volume 100%.

Secondly, the load at yield point and elastic limit is higher in the + 45 – 45 print orientation; on the other hand, the displacements are lower. This is explained by the correlation of the load and the displacement following the Hook law.

The data shows that simply changing the orientation during manufacturing has greatly impacted the performance of the part. The average elastic modulus raised by 7%. It went from 1323.74 to 1422.4 MPa.

By following the design methodology, the mechanical performances of the part have increased noticeably. Therefore, the methodology has been validated for the fused deposition modeling process.

Table 3. Test result with DFAM

Measured properties	Symbol	Unit	Test 1	Test 2	Test 3	Test 4	Test 5	Av	SV	CV
Thickness	T	mm	2.89	2.93	2.4	2.85	2.86	2.7828	0.022	0.08
Length	L	mm	60.3	60	60.3	60.8	60.2	60.308	0.32	0.01
Width	W	mm	10.1	10.1	10.2	10.1	10.1	10.127	0.07	0.01
Surface	S	mm^2	29.2	29.5	24.6	28.7	28.9	28.169	2.02	0.07
Load at yield point	Py	N	1300	1230	1200	1230	1240	1240	36.7	0.03
Load at elastic limit	Pe	N	1260	1220	1130	1180	1110	1180	62	0.06
Displacement to Py	Dy	mm	2.39	2.12	1.92	2.02	2.31	2.152	0.2	0.09
Displacement to Pe	De	Mm	2.09	1.91	1.63	1.76	1.78	1.834	0.17	0.09
Elastic modulus	E	MPa	1357	1323	1643	1417	1372	**1422.4**	128	0.1
Resistance to yield point	sy	MPa	42.2	39.3	45.8	40.3	40.4	41.6	2.53	0.06
Elastic limit	se	MPa	40.8	39	42.9	38.6	35.9	39.446	2.6	0.07
Strain at yield point	ey	%	4.03	3.6	3.25	3.38	3.9	3.6346	0.33	0.009
Strain at elastic limit	ee	%	3.53	3.25	2.77	2.96	3.02	3.1073	0.29	0.09

4 Conclusion

In this article, a design methodology is used for the production of parts in additive manufacturing. The interest is to show that this methodology makes it possible to increase the mechanical performance of the parts.

Standardized tensile specimens have been redesigned following the methodology. It turned out that the limitations of the fused deposition modeling process were not all anticipated by the ASTM standard, especially the orientation of the fiber deposit during manufacture. Wire deposition does not allow 100% filling because of the diameter and shape of the deposited filament. So the orientation of the deposit becomes an important parameter in terms of mechanical performance.

After respecting the methodology and changing the orientation of the part, an improvement in the mechanical performance of the specimens has been noticed.

Thanks to this study, the interest of the additive manufacturing methodology to design parts with simple shapes has been demonstrated. The next step is to use the methodology on more complex parts, with the aim of optimizing their weights, surface qualities and other parameters that represent the benefit of additive manufacturing. Then, the interest of the methodology on other additive manufacturing processes must be shown.

References

Agüero, A., et al.: Study of the influence of the reprocessing cycles on the final properties of polylactide pieces obtained by injection molding. Polymers **2019**, 11 (1908)

Arrieta, P.M., Samper, D.M., Aldas, M., López, J.: On the use of PLA-PHB blends for sustainable food packaging applications. Materials **10**, 1008 (2017)

Ceresana. "Bioplastics—Study: Market, Analysis, Trends—Ceresana". www.ceresana.com. Archived from the original on 4 November 2017. Retrieved 9 May 2018

Mikula, K., et al.: 3D printing filament as a second life of waste plastics—a review. Environ. Sci. Pollut. Res. **28**, 12321–12333 (2021)

Lubombo, C., Huneault, M.: Effect of infill patterns on the mechanical performance of lightweight 3D-printed cellular PLA parts. 2018 Department of Chemical and Biotechnological Engineering, Université de Sherbrooke, Sherbrooke, Canada (2018)

Lunt, J.: Large-scale production, properties and commercial applications of polylactic acid polymers. Polym. Degrad. Stab. **59**(1–3), 145–152. https://doi.org/10.1016/S0141-3910(97)00148-1. ISSN 0141-3910 (1998)

Salem, H., Abouchadi, H., Elbikri, K.: Design for additive manufacturing, J. Theoretical Appl. Inf. Technol., ISSN 1992-8645 (2020)

Salem, H., Abouchadi, H., Elbikri, K.: PLA mechanical performance before and after 3D printing. Int. J. Adv. Comput. Sci. Appl. West Yorkshire 13(3), (2022). https://doi.org/10.14569/IJACSA.2022.0130340

Shah, J., Snider, B., Clarke, T., Kozutsky, S., Lacki, M., Hosseini, A.: Large-scale 3d printers for additive manufacturing: design considerations and challenges. Int. J. Adv. Manuf. Technol. **104**, 3679–3693 (2019)

A. Wiberg, J. Persson, J. Ölvander.: Design for additive manufacturing—a review of available design methods and software. Rapid Prototyping J., ISSN: 1355-2546 (2019)

Metallurgical Study of a Material Produced by Selective Laser Melting

Kaoutar Fri[1(✉)], Abdellah Laazizi[1], Iatimad Akhrif[2], Mostapha El Jai[1,2], Abdelmalek Ouannou[1], and Mouad Bensada[1]

[1] ENSAM-Meknes, SECNDCM-L2MC, University of Moulay Ismail, Meknes, Morocco
kaoutar.fri@gmail.com, a.laazizi@ensam.umi.ac.ma

[2] Euromed Polytechnic School, UEMF Euromed Center of Research, Euro-Mediterranean University of Fez, Fez, Morocco

Abstract. Additive manufacturing has become an emerging strategy for a broad range of new applications. Indeed, selective laser melting (SLM) is an extremely promising additive manufacturing process for the manufacturing of 3D and net-shaped metal components with a fairly high quality of finish surface and mass density, the construction done by laser focused computer controlled scanning over a metal powder bed. The use of additive manufacturing is most relevant in aerospace applications because the gain in complexity offered to the geometries of the objects allows to obtain light components (reducing energy consumption) and to consolidate the assemblies by reducing the number of their components (improving reliability). The objective at hand in this investigation is to develop and characterize a new stainless steel material by SLM from a powder by acting on the operating parameters, namely speed and power. For this, metallographic observations and hardness measurements were performed on several samples.

Keywords: 316 Stainless Steel · Additive Manufacturing · 3D Printing · Microstructure · Hardness · Metallurgy

1 Introduction

Currently, additive manufacturing has many advantages and will be part of the manufacturing process solutions in a sustainable way. Therefore, it will have a strong impact on the mechanical industry in the near future. Many experts have seen it as an industrial revolution replacing usual processes and accelerating gains and productivity [1]. The industrial reality is very different, especially the investment and manufacturing costs which are still an obstacle to the expansion of these technologies. In order to make selective choices in additive manufacturing, it is necessary to have a good understanding of existing processes, their constraints and specificities [2, 3]. Three inputs are required for additive manufacturing: materials, CAD model and energy source.

Manufacturing processes can be classified into three categories: Formative, subtractive or additive manufacturing. Additive manufacturing is a very flexible processing technique that can be applied to plastics, metals, ceramics, concrete as well as building

L. Azrar et al. (Eds.): CIP 2022, LNME, pp. 500–508, 2023.
https://doi.org/10.1007/978-3-031-23615-0_51

materials [4]. This technology covers a variety of processes opposite to the philosophy of subtractive manufacturing where material is removed. In additive manufacturing, material is joined or solidified under computer control (CAD model) to create a three-dimensional object that meets the desired shape.

According to the parameters of the process utilized to create components by selective laser melting, the physical properties may change substantially compared to the general properties of the identical alloy processed in a conventional manner [1–4].

According to pull tests made by A.Chniouel [5] the parameters of the reference SLM process (P = 150W, V = 675mm/s) lead to a Elasticity limit of 555 MPa, an Maximum mechanical resistance of 575 MPa and an elongation at break of 84%. These results show that the parts manu-factured by SLM are well above the values required by the RCC-MRx standard Rules for the Design and Construction of Mechanical Equipment for High-Temperature, Experimental and Fusion Nuclear Facilities(Elasticity limit of 190 MPa, Maximum mechanical resistance of 490 MPa, and elongation at break of 45%) and comparable to those of a forged 316 steel (Elasticity limit of 270 MPa, Maximum mechanical resistance of 611 MPa and elongation at break of 74%).

Generally, through the microstructure that metals reveal their different mechanical properties. Therefore, metallographic and hardness tests are essential for any material development study. Thus, in this study, many 316SS samples were developed and analyzed by SLM additive manufacturing for different operating parameters.

2 Selective Laser Melting (SLM)

The strategy chosen has a significant influence on the developing of process and the device design, the maturity of these processes has been significantly improved due to the search for new materials, which lead to strong and stable processes. There are seven categories of additive manufacturing processes identified in ASTM standards.

SLM is among the most standard additives in metal powder manufacturing that produces almost final form parts based on the melting of power materials in a layer-by-layer process pattern. Laser fusion has been used to create samples in this study. For this purpose, a 3D model of the part to be manufactured is adapted to the layer-by-layer cutting. When the powder is deposited in several layers, it is swept by the laser, which melts the powder and fuses it as it cools to obtain a solid metal. The whole operation is repeated to create the motif layer by layer.

Involves selective laser melting as an additive manufacturing process to generate parts of plastic or solid metal based on a powder. Solid metal parts are created by melting and fusing metal powder with a laser in a layer-by-layer manner [6]. In the powder bath SLM process, a light coating of metal powder, Typically 20–40%, is placed on the top side of the machine platform and a laser is scanned over a portion of this coating, fusing the powder which then materializes [7, 8].

The objective is to establish an implementation for various states of the product developing process, and for distinct patterns of products. Like all manufacturing processes, the determination of which fabrication process and materials to use is highly dependent upon the part or system requirements. It involves a trade-off of how to balance the process abilities with the requirements of the designed part or system.

3 Experimental Study and Characterization

In this experimental study, the samples were made by using SLM additive manufacturing and tested on both sides: the F_S side face and the F_T top face. They were produced using different combinations of operating parameters, namely laser speed and power [9]. The choice of parameters was made according to the characteristics of the machine and the shade of the material, in order to increase the rate of absorption of laser energy by the metal powder, an Nd-YAG laser with a length of the 1064 nm wave is used. Two sweep speeds $V_1 = 200$ mm / s and $V_2 = 300$ mm / s and three powers $P_1 = 100$ W, $P_2 = 200$ W and $P_3 = 300$ W have been selected (Table 1).

Table 1. Classification of samples according to the different operating parameters

Speed (mm / s) / Power (W)	V_1	V_2
P_1	P_1V_1	P_1V_2
P_2	P_2V_1	P_2V_2
P_3	P_3V_1	P_3V_2

3.1 Material 316 SS

316 stainless steel is a chromium-nickel based steel, it is known for its resistance to corrosion of many substances thanks to the addition of molybdenum in its composition. Molybdenum makes 316 stainless steel more corrosion resistant, with specific resistance to chlorine pitting. 316SS is also more applicable in high temperature areas than other stainless steels. It has a higher heat resistance than 304 [10, 11].

(a) Chemical Composition by X-RF
Additive manufacturing gives the chemical composition which is very close to that obtained with conventional manufacturing processes. Below is the composition of 316SS produced by SLM obtained by X-RF analysis (Table 2):

Table 2. XRF analysis of 316SS.

Element	Si	P	Ti	Cr	Mn	Mo	Ni	Fe
Wt.%	1	0.01	0.02	17.4	1.2	2.68	11.82	Bal

(b) Metallographic Study
To determine the microstructures of the samples made, they went through several crucial steps, after polishing and reacting the sample using etching the microstructure is finally revealed with an OLMPUS BX6 optical microscope. Figure 1 shows the results of some samples.

Interpretation of Results:

- Compared to conventional stainless steels, we have observed a change in microstructure within the same sample depending on the orientation of the fibers during the processing (lateral surface and the TOP).
- The workpieces produced by SLM have a grain structure that is columnar along the construction direction, exhibiting a microstructure that is substantially different than those usually observed in conventional alloys.
- Also the microstructures differ from one sample to another, this change of structure is due to the fact that the power used as well as the speed of deposition of the layers during the development were different.
- According to SEM analysis (Fig. 2), the non-melting of the powder is approved.

3.2 Hardness Test

H. Zhang et al. [12] established that the decrease in micro-hardness values with the inclusion of the Ni element might be attributed to the transition of phase from ferrite to austenite. As the Ni content increases, the wearing resistance of the deposition samples gradually decreases, which may be ascribed to the reason that the wearing resistance is proportionate to the micro-hardness according to Archard's law. Moreover, the corrosion resistance was significantly improved with increasing Ni content, as well as the improved electrochemical behavior might be ascribed to the reason that the Ni element, possessing outstanding in corrosion resistance, removed the dissolving of the steel in the anode region and moved the corrosion potential to the noble direction (Table 2).

To investigate the hardness of 316SS steel, the study opted for Vickers hardness for its accuracy. The indenter is then dropped on the surface to be analyzed with a force equal to $f = 4.9N$. It is left to rest on the sample for 15 s, then raised. A black square is then observed on the screen using the Toupview software, which corresponds to the print. Thus the lengths of the diagonals of this square is read.

3.3 Study of the Operating Parameters

Many studies have been conducted to determine the most influential parameters on the microstructure of SLM consolidated parts; power, scanning speed, distance between two successive laser strokes and powder bed thickness are those that most affect the microstructure of SLM processed parts [13]. The microstructural properties of 316 SS steel parts are significantly affected by the SLM process parameters. Scanning speed and laser power are often compounded are amongst the crucial parameters of this metal additive manufacturing process [14].

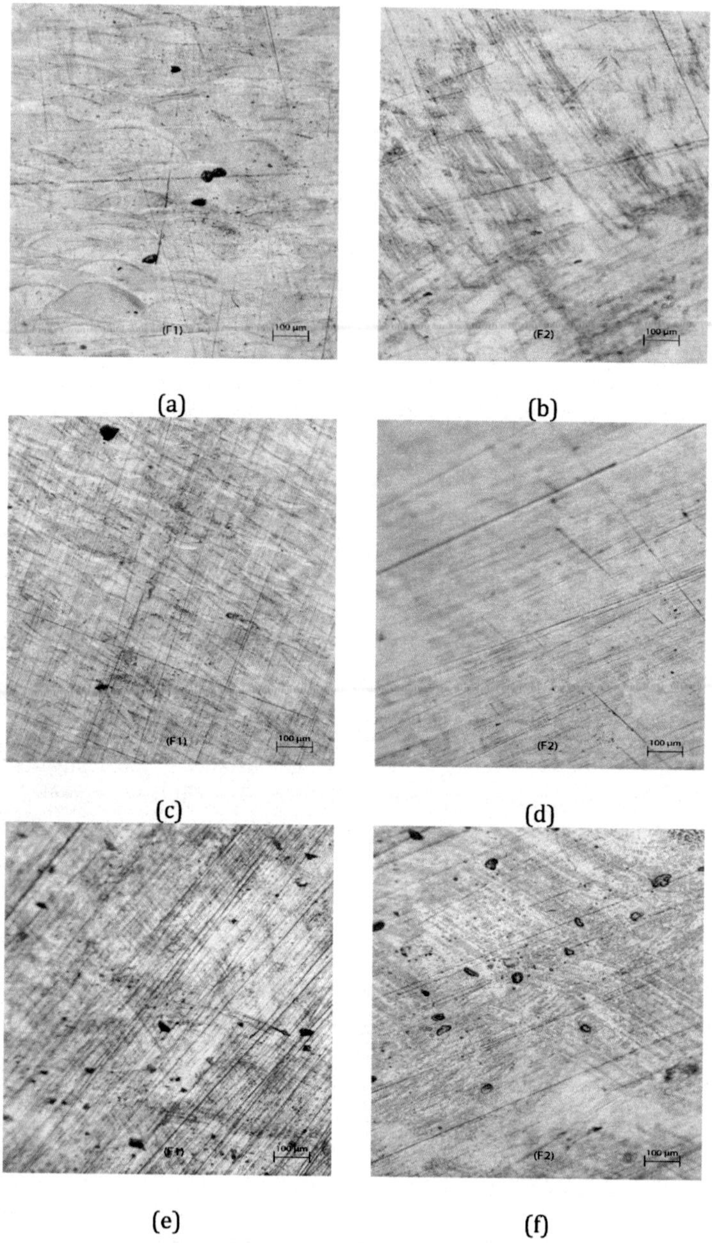

Fig. 1. Microstructures obtained from both sides after chemical etching (**a**, **b**) P_2V_1 sample (**c**, **d**) P_2V_1 sample and (**e**, **f**) P_2V_1 sample.

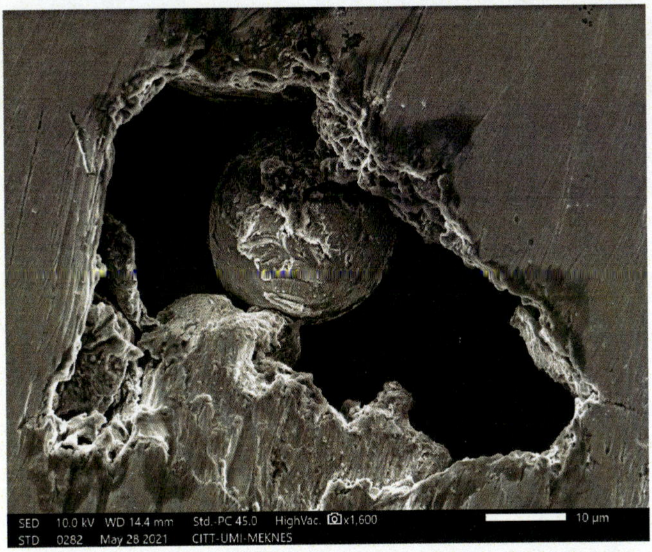

Fig. 2. SEM analysis of the sample P1V1 developed by a power of 100W and a speed of 200mm/s.

i. Effect of power on hardness
The effect of laser power was analyzed. Figure 3 show the hardness as a function of power. The highest value was observed at P_2 with the speed of 300 mm/s.

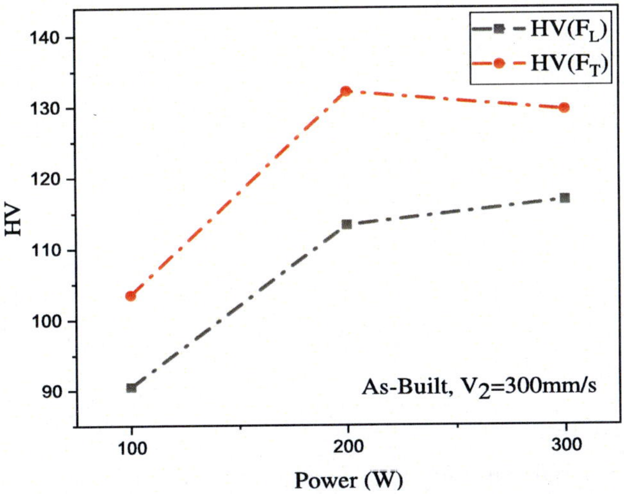

Fig. 3. Hardness values of both sides on (V = 300mm/s).

It is obtained that the hardness of the top face is greater than the side face at V = 300mm/s and the hardness increases with the power of the laser. This result is adequate for both faces of the sample. Thus, it is noted that the hardness of the side face is greater than the hardness on the top face, which summarizes that the sample is not homogeneous.

ii. Effect of scanning speed on hardness.
Several authors note the impact of power on the microstructure. Montero-Sistiaga et al. [15] vary the power from 400 W to 1 kW. When the power reaches 1 kW, the grains are more elongated in the build direction. In contrast, when the power used is low, the microstructure of the 316L steel made by SLM is smaller. Choo et al. [16] found the same result. Decreasing the power from 380 W to 200 W resulted in a refinement of the cell size from 1.5 μm to 0.75 μm. For Kong et al. [17], by increasing the laser power from 120 W to 220 W, the grain width of the steel increases from 32 to 56 μm. Li et al. [18] suggest using low scanning speed and high power to avoid the "balling" effect, i.e. the formation of liquid droplets splashing on the surface during fabrication [19]. These flaws induce porosities that lead to a reduction in the density of the consolidated material [20].

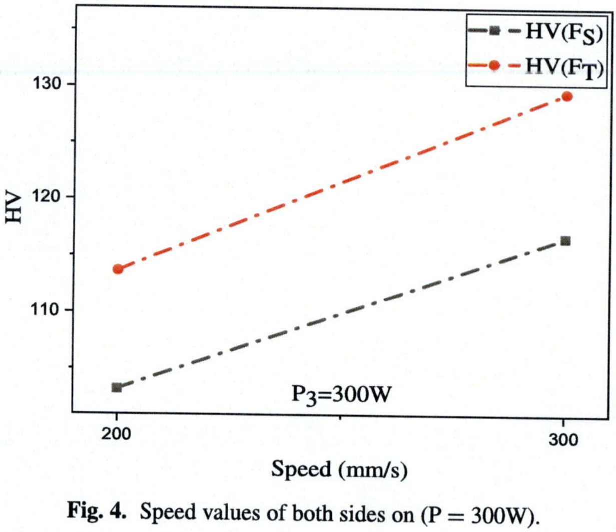

Fig. 4. Speed values of both sides on (P = 300W).

Interpretation of Results:

The hardness of the top face is important compared to the lateral face whatever the power or the speed of elaboration which means the non-homogeneity of our material and it is due to the cooling rate which is different between the two faces because of the strategy of manufacture called "Chess" which is not with the same way on the two faces.

In Fig. 3, it can be seen that in the top face the hardness increases due to several physical phenomena that are involved in this process precisely the hardening of the material, whereas in the side face there is a small decrease of the hardness on the samples elaborated with a power of 200W.

It is clear in Fig. 4 that the hardness increases with the speed at 300W power, and this is due to the high power used as well as the melting time for each powder bed and it implies that the dimensions of the melting bath depend on the process parameters and that the geometry of the melting baths is important because it affects the local cooling speed, which partly controls the shape and size of the grains which explains the results obtained.

4 Conclusions

Selective laser melting additive manufacturing is a promising technology to produce high quality shape materials. It is a very suitable process to produce high strength steel.

The effect of scanning speed and power of SLM were analyzed. The results showed that the power has more effect on the microstructure than the scanning speed.

This work confirmed the strength of the 316SS stainless steel developed by SLM. Metallurgical studies were done and the different structures were observed depending on the manufacturing orientation of samples.

References

1. Santos, V.M.R., et al.: Design and characterisation of an additive manufacturing benchmarking artefact following a design-for-metrology approach. J. Add. Manuf. **32**, 100964 (2020)
2. He, Y., Montgomery, C., Beuth, J., Webler, B.: Melt pool geometry and microstructure of Ti6Al4V with B additions processed by selective laser melting additive manufacturing. J. Mater. Des. **183**, 108126 (2019)
3. C. Montgomery, J. Beuth, L. Sheridan and N. Klingbeil, Process mapping of inconel 625 in laser powder bed additive manufacturing, Proceeddings 2015 Solid Freeform Fabrication Symposium, Austin, TX, 2015
4. Wohlers T, editor. Wohlers Report 2013 - Additive Manufacturing and 3D Printing State of the Industry - Annual Worldwide Progress Report. 18th ed. Fort Collins, CO: Wohlers Associates; 2013
5. Aziz Chniouel - Etude de l'élaboration de l'acier inoxydable 316L par fusion laser sélective sur lit de poudre : influence des paramètres du procédé, des caractéristiques de la poudre, et des traitements thermiques sur la microstructure et les propriétés mécaniques. Submitted on 20 Dec 2019
6. Yap, C.Y., et al.: Review of selective laser melting: materials and applications. J. Appl. Phys. Rev. **2**, 041101 (2015)
7. Frazier, W.E.: Metal additive manufacturing: a review. J. Mater. Eng. Perform. **23**, 1917–1928 (2014)
8. S. Sun, M. Brandt and M. Easton, Powder bed fusion processes, J. Laser Additive Manufacturing (2017) 55–77
9. Laazizi, A., et al.: Applied multi-pulsed laser in surface treatment and numerical–experimental analysis. J. Opt & Laser Tech **43**, 1257–1263 (2011)
10. Sun, Z., Tan, X., Tor, S.B., Yeong, W.Y.: Selective laser melting of stainless steel 316L with low porosity and high build rates. J. Mater. Des. **104**, 197–204 (2016)

11. Kurzynowski, T., Gruber, K., Stopyra, W., Kuźnicka, B., Chlebus, E.: Correlation between process parameters, microstructure and properties of 316 L stainless steel processed by selective laser melting. J. Mater. Sci. Eng. A **718**, 64–73 (2018)
12. H.Zhang et al, Effect of Ni content on stainless steel fabricated by laser melting deposition. J. Optics and Laser Technology. 101 (2018) 363–371
13. A. Leicht, U. Klement, E. Hryha, Effect of build geometry on the microstructural development of 316L parts produced by additive manufacturing, Mater. Charact. (2018)
14. Miranda, G., et al.: Predictive models for physical and mechanical properties of 316L stainless steel produced by selective laser melting. Mater. Sci. Eng. A. **657**, 43–56 (2016)
15. Montero-Sistiaga, M.L., Godino-Martinez, M., Boschmans, K., Kruth, J.-P., Van Humbeeck, J., Vanmeensel, K.: Microstructure evolution of 316L produced by HP-SLM (high power selective laser melting). Addit. Manuf. **23**, 402–410 (2018)
16. H. Choo, K.-L. Sham, J. Bohling, A. Ngo, X. Xiao, Y. Ren, P.J. Depond, M.J. Matthews, E. Garlea, Effect of laser power on defect, texture, and microstructure of a laser powder bed fusion processed 316L stainless steel, Mater. Des. (2018)
17. Kong, D., et al.: Bio-functional and anticorrosive 3D printing 316L stainless steel fabricated by selective laser melting. Mater. Des. **152**, 88–101 (2018)
18. Li, R., Shi, Y., Wang, Z., Wang, L., Liu, J., Jiang, W.: Densification behavior of gas and water atomized 316L stainless steel powder during selective laser melting. Appl. Surf. Sci. **256**, 4350–4356 (2010)
19. X. Zhou, X. Liu, D. Zhang, Z. Shen, W. Liu, Balling phenomena in selective laser melted tungsten, J. Mater. Process. Technol. 222 (2015)
20. Ahmadi, A., Mirzaeifar, R., Moghaddam, N.S., Turabi, A.S., Karaca, H.E., Elahinia, M.: Effect of manufacturing parameters on mechanical properties of 316L stainless steel parts fabricated by selective laser melting: A computational framework. Mater. Des. **112**, 328–338 (2016)

Embedded Systems, Control and Modeling

Nonlinear Rigid-Flexible Manipulator Adaptive Model Predictive Control

Aycha Hannane(✉), Mohammed Bakhti, and Badr Bououlid Idrissi

Ecole Nationale Supérieure d'Arts Et Métiers, Moulay Ismail University, BP 4024, Marjane II, Beni Hamed, 50000 Meknes, Morocco
hannane.aycha96@gmail.com, {m.bakhti,b.bououlid}@umi.ac.ma

Abstract. This paper deals with an active vibration damping while tracking the desired position of a nonlinear rigid-flexible manipulator (RFM) based on an adaptive model predictive control (AMPC) scheme. The Hamilton's principle is used to derive the nonlinear equations of movement, and the assumed modes method is used to approximate the elastic deflection of the manipulator free tip. The system Jacobeans, required at each time step, are derived along the system trajectory to optimize the mechanical torque supplied to the manipulator. The adaptive model predictive controller is then designed using the MPC toolbox that is adjusted by varying its weights and constraints. The effectiveness of the controller is demonstrated via numerical simulation with various tuning parameters assessment carried out using the MATLAB/Simulink environment, while the performance is quantified by computing the root mean square error (RMSE). The results show that the AMPC controller is performant and effective when damping the tip vibration while tracking the angular position of a rigid-flexible manipulator.

Keywords: Active vibration control · Adaptive model predictive control · Rigid-flexible manipulator

1 Introduction

Generally, the two link manipulators are far privileged to the conventional ones, due to their greater payload to manipulator weight ratio, higher operation speed, larger workspace and lower energy consumption. Nevertheless, their control strategies are much more sophisticated since they must also consider the structural flexibility and vibration problem while tracking desired positions (Dwivedy and Eberhard 2006).Their modeling approaches and their control/observation strategies must consider both the rigid body and the flexible degrees of freedom (Sayahkarajy et al. 2016). For the derivation of their nonlinear equations of motion, the flexible manipulators are often modeled based on an energetic approach. The Hamilton's principle is generally used during the modeling process, and it's usually associated to the finite element method or the assumed modes method to approximate the elastic deflection.

Commonly, only first few most significant vibration modes are considered while formulating the dynamic equations of motion (Arkouli et al. 2021).

L. Azrar et al. (Eds.): CIP 2022, LNME, pp. 511–520, 2023.
https://doi.org/10.1007/978-3-031-23615-0_52

On the one hand, most of the active vibration control strategies require the state feedback, and many nonlinear observer formulations have been addressed for the flexible manipulators. On the other hand, many control strategies have been addressed for the simultaneous position-vibration control. A computationally efficient model predictive controller has been compared to a classical one in (Zhang et al. 2016). A multivariable adaptive controller has been proposed for the simultaneous position and force control of the manipulator (Shi et al. 1999), and various robust controllers has been assessed in (Park et al. 2002) and (Theodore and Ghosl 2001). More recently, soft computing approaches such as fuzzy logic and neural network techniques (Tian and Collins 2004) and (Subudhi and Morris 2009) have also been explored.

In the last Three decades, considerable research papers have been devoted to the adaptive controllers for complex systems facing internal and external disturbances and multiple interactions. It is first used by Richalet (1978), he developed his predictive functional control technique for industrial applications in the late of seventies, from that time on, the control schemes continues to develop helping different industries reduce cost and improve energy consumption. Adaptive model predictive control algorithm was presented by Slotine and Li (1988), that was later used by Sponge to represent the adaptive control for flexible manipulators under the assumption of weak joint elasticity (Spong 1989) and for rigid ones in (Ortega and Spong 1989).

In this work, an adaptive model predictive control is formulated with appropriate cost function to be minimized. The cost function allows for a rigid body position tracking as well as a tip vibration minimization. Once, the equations describing the dynamic of the system are derived in Sect. 2, the AMPC controller is detailed in Sect. 3. Based on the iterative linearization of the system, the input torque is optimized over a prediction horizon. In Sect. 4, the simulation results are illustrated to address some relative aspects discussion, and conclusion is outlined in Sect. 5.

2 Rigid-Flexible Manipulator Model

The rigid flexible manipulator has been modeled as an Euler-Bernoulli beam, where the rigid link of length L_1 is rigidly attached to the servo motor, described by the shoulder joint angular position θ_1. The flexible link of length L_2 driven by servo motor, is described by the elbow joint angular position θ_2 and the elastic displacement $w(x, t)$. Note that x is the non deformed point location on the flexible link. And r stands for the radius of the rigid hub.

The geometry and coordinates of the rigid-flexible manipulator are illustrated in Fig. 1:

Since the manipulator acts on a horizontal plane, the gravity is not considered. Moreover, the shear deformation is neglected according to the Euler-Bernoulli assumption, therefore the kinematics of the system can be described by the position vector ϑ in the inertial system (X, Y, Z):

$$\vartheta = \begin{bmatrix} \vartheta_x \\ \vartheta_y \end{bmatrix} = \begin{bmatrix} L_1 \cos(\theta_1) + (x + r)\cos(\theta_1 + \theta_2) - w\sin(\theta_1 + \theta_2) \\ L_1 \sin(\theta_1) + (x + r)\sin(\theta_1 + \theta_2) + w\cos(\theta_1 + \theta_2) \end{bmatrix} \tag{1}$$

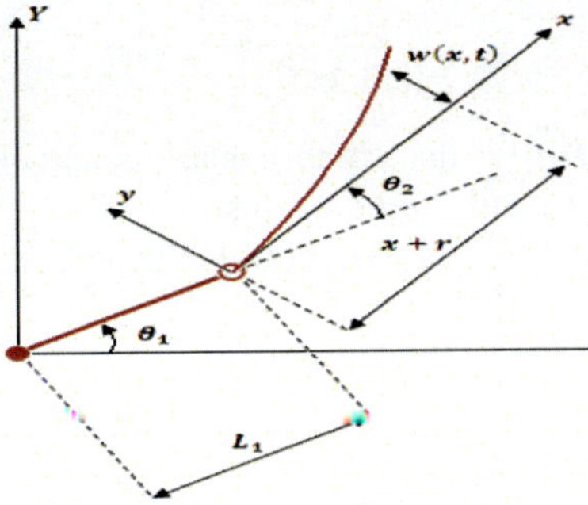

Fig. 1. The rigid-flexible manipulator geometry and coordinates

The kinetic energy of the system can be deducted using the Hamilton's principle:

$$T = \frac{1}{2} I_1 \dot{\theta}_1^2 + \frac{1}{2} I_h \left(\dot{\theta}_1^2 + \dot{\theta}_2^2 \right) + \frac{1}{2} m_h L_1^2 \dot{\theta}_1^2 + \frac{1}{2} \int_0^{L_2} \rho A \dot{\vartheta}^2 dx \quad (2)$$

where ρ is the mass density of the flexible link manipulator, A is its cross-section area and m_h is the elbow hub mass.

In accordance with the Euler Bernoulli assumption, the potential energy of the system is given by (Tokhi and Azad 2008):

$$P = \frac{1}{2} \int_0^{L_2} EI_2 \left(\frac{\partial^2 w}{\partial x^2} \right)^2 dx + \frac{1}{2} \int_0^{L_2} F(x, t) \left(\frac{\partial \vartheta}{\partial x} \right)^2 dx \quad (3)$$

where E is the flexible link Young's modulus and I its moment of inertia.

For a uniform beam, we define $F(x, t)$ as follows (Yigit et al. 1988):

$$F_c(x, t) = \frac{1}{2} \rho \dot{\theta}_2^2 \left(L_2^2 - x^2 \right) + \rho \dot{\theta}_2^2 r (L_2 - x) \quad (4)$$

Based on the extended Hamilton's principle, the system equations can be written as (Dym and Shames 2013):

$$\int_{t_0}^{t_f} (\delta T - \delta P + \delta W) dt = 0 \quad (5)$$

where δW is the work done by the joint torques τ_1 and τ_2 at the shoulder and the elbow joints respectively.

Substituting all energy equations into the Hamilton's equations, which govern the motion of the rigid flexible two link manipulator (Bakhti et al. 2017), and rearranging them into a system behavior model as follows:

$$M(q)\ddot{q} + h(q, \dot{q}) + K(q) = u(t) \quad (6)$$

Note that q is the vector of generalized coordinates representing the rigid body and the elastic modal coordinates and u is the vector of the external forces.

$$q = \left[\theta_1 \ \theta_2 \ q_1 \right]^T \quad (7)$$

$$u(t) = \begin{bmatrix} \tau_1 & \tau_2 & 0 \end{bmatrix}^T \tag{8}$$

where $M(q)$ is a mass matrix, $K(q)$ is the stiffness matrices and $h(q, \dot{q})$ the vector that regroups the nonlinear centrifugal and Coriolis terms.

$w(x, t)$ is approximated using the assumed modes method, and it is given by:

$$w(x, t) = q_1(t)\varphi_1(x) \tag{9}$$

where

$$\varphi_1(x) = \sin(px) - \sigma\cos(px) - \sinh(px) + \cosh(px) \tag{10}$$

$$p = \frac{\sqrt{3.5160}}{L_2} \tag{11}$$

And:

$$\sigma = \frac{\sin(pL_2) + \sinh(pL_2)}{\cos(pL_2) + \cosh(pL_2)} \tag{12}$$

Explicit matrices Formulations are given in the simulation section and the appendix.

We define the modal damping matrix H_d that contains the shoulder servomotor viscous friction coefficient α_m and the flexible structural damping as follows (Hassan et al. 2007):

$$H_d = \begin{bmatrix} \alpha_m & 0 \\ 0 & 2\xi_1 m_{22}\omega_1 \end{bmatrix} \tag{13}$$

where ω_1, ξ_1 and m_{22} are respectively the first elastic mode natural frequency, the modal damping coefficient and the corresponding element of the mass matrix $M(q)$.

3 The Adaptive Model Predictive Controller Algorithm

In this section, a controller is designed for two link manipulator in order to solve an optimization problem at each time step. It is based on minimizing a cost function called J which is linked to the error between the predicted output and the set point and the control signal changes over a prediction horizon of N_p samples, using the N_c samples of control signal.

We define for a discrete time invariant system, the following model:

$$\begin{cases} x(k+1) = Ax(k) + Bu(k) \\ y(k) = Cx(k) + Du(k) \end{cases} \tag{14}$$

The input $u(k)$ is assumed not to affect the output $y(k)$ because of the receding horizon control, thus $D = 0$.

We define a new state vector connecting $\Delta u(k)$ to the output $y(k)$:

$$x(k) = \begin{pmatrix} \Delta x(k) \\ y(k) \end{pmatrix} = \begin{pmatrix} x(k+1) - x(k) \\ y(k) \end{pmatrix} \tag{15}$$

After rearranging the equations (Eqs. 14 and 15), we obtain:

$$x(k+1) = Ax(k) + B\Delta u(k) \tag{16}$$

Matrices A, B and C are defined to optimize the control signal that is designated as a vector of future N_c control samples trajectory:

$$\Delta U = [\Delta u(k), \Delta u(k+1), \ldots \Delta u(k+N_c-1)]^{\mathrm{T}} \tag{17}$$

The cost function is given by:

$$J = (S_p - Y_p)^T Q (S_p - Y_p) + \Delta U^T R \Delta U \tag{18}$$

where Q and R are symmetric semi definite weighting matrices.

With:

$$S_p = \begin{bmatrix} 1\ 1\ \ldots\ 1 \end{bmatrix}_{1\times N_p} r(k) = \overline{S_p} r(k) \tag{19}$$

The control signal is calculated assuming that:

$$\frac{\partial J}{\partial \Delta U} = 0 = -2\Phi^{\mathrm{T}} Q (S_p - Fx(k_i)) + 2\left(\Phi^{\mathrm{T}}\Phi + R\right)\Delta U \tag{20}$$

where:

$$F = \begin{bmatrix} C_a A_a\ C_a A_a^2\ C_a A_a^3 \ldots C_a A_a^{N_p} \end{bmatrix}^T \tag{21}$$

And:

$$\Phi = \begin{bmatrix} C_a B_a & 0 & 0 & \ldots & 0 \\ C_a A_a B_a & C_a B_a & 0 & \ldots & 0 \\ C_a A_a^2 B_a & C_a A_a B_a & C_a B_a & \ldots & 0 \\ \ldots & \ldots & \ldots & \ldots & \ldots \\ \ldots & \ldots & \ldots & \ldots & \ldots \\ C_a A_a^{N_p-1} B_a & C_a A_a^{N_p-2} B_a & C_a A_a^{N_p-3} B_a & \ldots & C_a A_a^{N_p-N_c} B_a \end{bmatrix} \tag{22}$$

The control signal increment within the framework of predictive control is given by: (n denotes the inputs variables number)

$$\begin{aligned} \Delta u(k) &= \left[\left[I_{n\times n}[0]_{n\times n(N_{c-1})}\right]\left(\Phi^{\mathrm{T}}\Phi + R\right)^{-1}\Phi^{\mathrm{T}} Q \overline{S_p}\right] r(k) \\ &- \left[\left[I_{n\times n}[0]_{n\times n(N_{c-1})}\right]\left(\Phi^{\mathrm{T}}\Phi + R\right)^{-1}\Phi^{\mathrm{T}} Q F\right] x(k_i) \\ &= K_r r(k) - K_x x(k_i) \end{aligned}$$

4 Simulation Results

In this section, a simulation is carried out using Matlab/Simulink to anticipate the following state variables; the shoulder angle θ_1, the elbow angle θ_2 and the first modal coordinate q_1 and their derivatives.

The step time has been set to $0.001s$, to overcome the nonlinearities of the plant model. The set points are set to $\theta_{1\_ref} = pi/2$, $\theta_{2\_ref} = pi/4$ and $q_{1\_ref} = 0$. The tuning parameters are defined in Table 2.

Table 1 shows the numerical parameters used for simulation (Table 2).

Table 1. Numerical parameters of the system

Rigid link	Mass	$m_1 = 1$ Kg
	Length	$L_1 = 0.5$m
	Inertia	$I_1 = 0.0834$Kg.m^2
Flexible link	Length	$L_2 = 0.5$m
	Mass density per unit length	$\rho A = 0.15$Kg.m^{-1}
	Flexural rigidity	$EI_2 = 1$ N.m^2
	Quadratic moment	$I = 1.4510^{-9}$m^4
	First mode damping coefficient	$\xi_1 = 0.01$ m
	First elastic mode nature frequency	$\omega_1 = 36.3131$rad/s
Elbow hub	Radius	$r = 0.04$ m
	Mass	$m_h = 0.5$Kg.m^2
Shoulder servomotor and hub	Inertia	$I_h = 0.002$Kg.m^2
Elbow servomotor	Viscous friction coefficient	$\alpha_m = 0.95$ Nm.rd^{-1}.s^{-1}

Table 2. AMPC tuning parameters

Prediction horizon (Number of samples)		Error weighting matrix	Input increment weighting matrix	Control horizon
Case 1	$\mathbf{N_p} = 200$	$\mathbf{Q} = \begin{bmatrix} 10 & 0 & 0 \\ 0 & 10 & 0 \\ 0 & 0 & 5000 \end{bmatrix}$	$\mathbf{R} = \begin{bmatrix} 10 & 0 \\ 0 & 10 \end{bmatrix}$	$\mathbf{N_c} = 2$
Case 2	$\mathbf{N_p} = 500$			
Case 3	$\mathbf{N_p} = 700$			

To assess the performance of the proposed controller and its effectiveness, a simulation was run on different prediction horizons. The Fig. 2 shows the shoulder joint angle response over three different prediction horizons, in case when $N_p = 200$, the response has an overshoot with a very large settling time. However, when the prediction horizon becomes larger, the shoulder joint angle response is much better. In Fig. 3, is illustrated

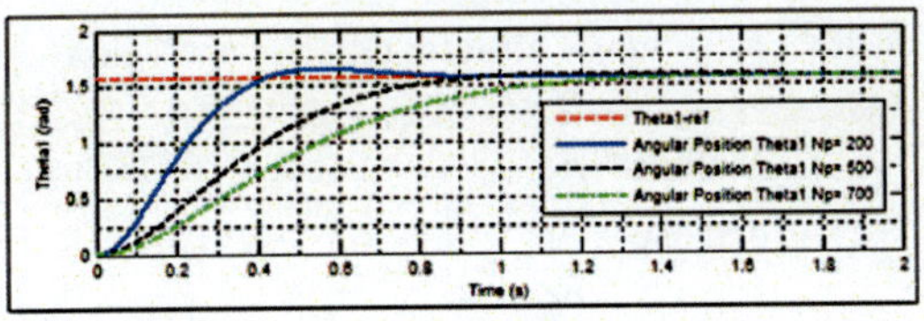

Fig. 2. Shoulder angle for different prediction horizons.

the elbow joint angle motion, where the response is good for the Three prediction horizons, although when $N_p = 200$, a visible vibration is noticed. In Fig. 4, is illustrated the first modal coordinate for different prediction horizons, as shown the response has multiple overshoots and rebound during 0.8second.

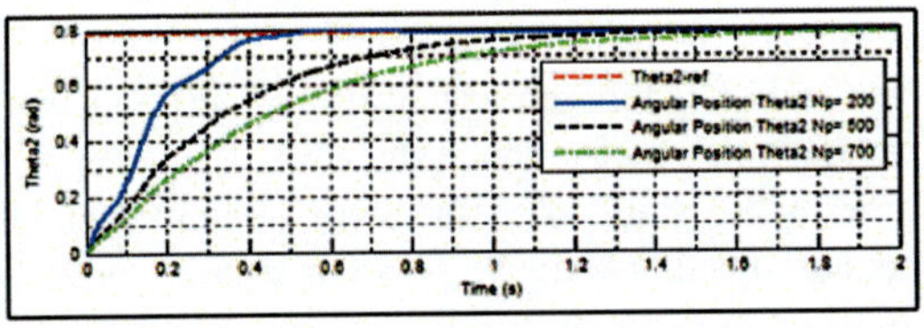

Fig. 3. Elbow angle for different prediction horizons.

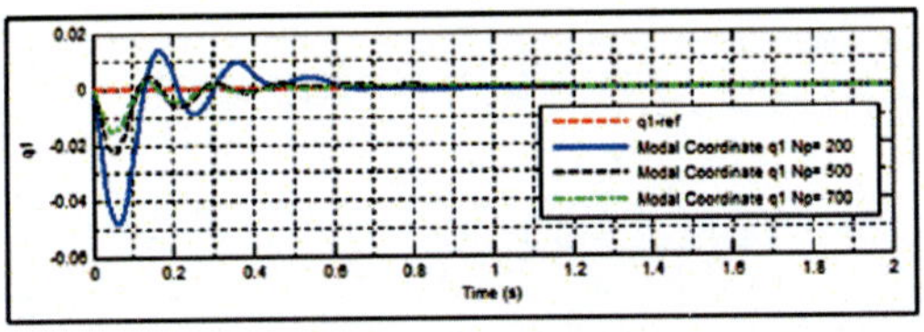

Fig. 4. First modal coordinate for different prediction horizons.

The interpretation of the prediction horizon is more complex than it might be expected, here it refers to how far we wish to predict the future system's outputs which are the shoulder angle, the elbow angle and the first modal coordinate. However, it is crucial to know the current measurement taking into consideration the constraints thus, the optimization is performed with the prediction horizon and the latest available information. It is important to realize that the prediction horizon increase ($N_p = 700$) damp the vibration of the manipulator free tip by 0.4 second, while track the shoulder angle by around 1.7 second and 1.8 second for the elbow angle. The receding control horizon and the weights are required to provide an accurate prediction of these outputs.

The root mean square error *RMSE* is calculated to quantify the AMPC performance (where N_s is the number of samples):

$$RMSE(q(t)) = \sqrt{\frac{1}{N_s}\sum_{k=0}^{N_s-1}\left(q(kT_s) - q_{ref}\right)^2} \tag{24}$$

Table 3. Root mean square error for different prediction horizon

	$\mathbf{Max(\|q(t)\|)}$	$\mathbf{RMSE(q(t))}$
$\mathbf{N_p} = 200$	$\mathbf{8.9568e} - 07$	0.0089
$\mathbf{N_p} = 500$	$\mathbf{1.4733e} - 06$	0.0039
$\mathbf{N_p} = 700$	$\mathbf{2.5904e} - 05$	0.0026

Table 3 shows the performance of the system in term of *RMSE* and the maximum absolute value of the first modal coordinate. According to the results, the prediction horizon has a significant impact on the modal coordinate; a smaller prediction horizon leads to a lower modal coordinate's amplitude ($8.9568e - 07$), but a high $RMSE = 0.0089$.However, the mean root square error is lowered $\boldsymbol{RMSE} = 0.0026$ when the prediction horizon is high $N_p = 700$.

Comparing our results with (Mohammed and Eltayeb 2018)'s work, we can say that the AMPC controller has better performance than PID, however it is less performant than the SMC as it shows in the results obtained by (Mohammed and Eltayeb 2018) by a difference of 1 second, and this due to the high torque used to move the two links of the manipulator, which is 100× higher than the torque used in our work. In brief, our main objective in this work is to damp the tip vibration while tracking the desired angular position of the joints (Figs. 5 and 6).

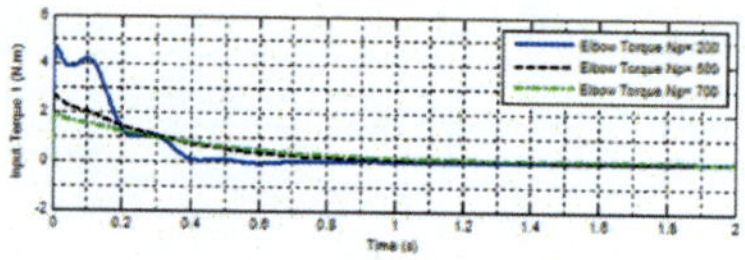

Fig. 5. The elbow torque

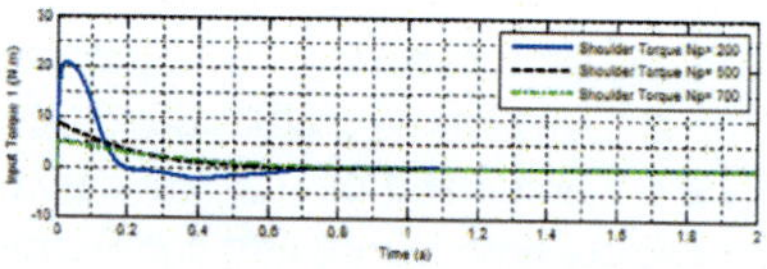

Fig. 6. The shoulder torque

5 Conclusion

The AMPC controller has demonstrated its effectiveness when tracking the desired two angular positions while minimizing and damping the residual vibration of the manipulator free tip. And it has proven that a prediction horizon of samples led to a better performance particularly when $N_p = 700$.

Appendix

Numerical values of model matrices used for simulation.

The elements of mass matrix:

$$M = \begin{matrix} 0.2370 + 0.0218\cos(\theta_2) - 0.08q_1\sin(\theta_2) & 0.0099 + 0.0109\cos(\theta_2) - 0.04q_1\sin(\theta_2) & 0.0323 + 0.04\cos(\theta_2) \\ 0.0099 + 0.0109\cos(\theta_2) - 0.04q_1\sin(\theta_2) & 0.0099 & 0.0323 \\ 0.0323 + 0.04\cos(\theta_2) & 0.0323 & 0.1392 \end{matrix}$$

The stiffness matrix: $K = \begin{bmatrix} 0 & 0 & 0 \\ 0 & 0 & 0 \\ 0 & 0 & 183.52 \end{bmatrix}$.

The elements of vector of nonlinear centrifugal and Coriolis terms:

$$h_1 = \dot{\theta}_2^2(-0.04q_1\cos(\theta_2) - 0.0109\sin(\theta_2))$$
$$-\dot{\theta}_1\dot{\theta}_2(q_1\cos(\theta_2) + 0.0218\sin(\theta_2)) - 0.08\left(\dot{\theta}_1 + \dot{\theta}_2\right)\sin(\theta_2)$$

$$h_2 = \dot{\theta}_1^2(0.04q_1\cos(\theta_2) + 0.0109\sin(\theta_2))$$

$$h_3 = 0.0444q_1\left(\dot{\theta}_1 + \dot{\theta}_2\right)^2 + 0.04\dot{\theta}_1^2\sin(\theta_2)$$

The damping matrix: $H_d = \begin{bmatrix} 0.95 & 0 & 0 \\ 0 & 0.95 & 0 \\ 0 & 0 & 0.1010 \end{bmatrix}$.

References

Arkouli, Z., Aivaliotis, P., Makris, S.: Towards accurate robot modelling of flexible robotic manipulators. Procedia CIRP **97**, 497–501 (2021)

Dwivedy, S.K., Eberhard, P.: Dynamic analysis of flexible manipulators, a literature review. Mech. Mach. Theory, 41 (7), pp. 749–777

Park, N.C., Yang, H.S., Park, H.W., Park, Y.P.: Position/vibration control of two-degree-of-freedom arms having one flexible link with artificial pneumatic muscle actuators. J. Robot. Auton. Syst. **40**, 239–253 (2002)

Sayahkarajy, M., Mohamed, Z., MohdFaudzi, A.A.: Review of modelling and control of flexible-link manipulators. Proc. Inst. Mech. Eng. Part I J. Syst. Control Eng., 230 (8), pp. 861–873

Shi, Z.X, Fung, H.K., Li, Y.C.: Dynamic modelling of rigid-flexible manipulator for constrained motion task control. J. Appl. Math. Model. 23, 509–525

Subudhi, B., Morris, A.S.: Soft computing methods applied to the control of a flexible robot manipulator. J. Appl. Soft Comput. **9**, 149–158 (2009)

Theodore, R.J., Ghosl, A.: Robust control of multilink flexible manipulators. J. Mech. Machine Theory **38**, 367–377 (2003)

Tian, L., Collins, C.: A dynamic recurrent neural network-based controller for a rigid-flexible manipulator system. J. Mechateronics **14**, 471–490 (2004)

Zhang, J., Yin, X., Liu, J.: Economic mpc of deep cone thickeners in coal beneficiation. Can. J. Chem. Eng. **94**(3), 498–505 (2016)

Tokhi, M.O., Azad A.K.M.: "Flexible Robot Manipulators Modelling, simulation and control". The Institution of Engineering and Technology (2008)

Yigit, A.S., Scott, R.A., Ulsoy, A.G.: Flexural motion of a radially rotating beam attached to a rigid body. J. Sound Vib. **121**(2), 201–210 (1988)

Dym, C.L., Shames, I.H.: Solid mechanics, a variational approach. Springer, New York (2013)

Bakhti, M., Tarla, L.B., Idrissi, B.B.: Flexible manipulator state and input estimation using higher order sliding mode differentiator. 2017 14th International Multi-Conference on Systems, Signals & Devices (SSD). IEEE, p. 302–306 (2017)

Hassan, M., Dubay, R., Li, C., Wang, R.: Active vibration control of a flexible one-link manipulator using a multivariable predictive controller. Mechatronics **17**(6), 311–323 (2007)

Richalet, J., Rault, A., Testud, J.L., Papon, J.: Model predictive heuristic control: Applications to industrial processes. **14**(5), 413–428 (1978)

Slotine, J.-J., Weiping, L.: Adaptive manipulator control: a case study. IEEE Trans. Autom. Control **33**(11), 995–1003 (1988)

Spong, M.W.: Adaptive control of flexible joint manipulators. Syst. Control Lett. **13**(1), 15–21 (1989)

Ortega, R., Spong, M.W.: Adaptive motion control of rigid robots: a tutorial. Automatica **25**(6), 877–888 (1989)

Mohammed, A.A., Eltayeb, A.: Dynamics and control of a two-link manipulator using PID and sliding mode control. International Conference on Computer, Control, Electrical, and Electronics Engineering (ICCCEEE) (2018)

Comparative Analysis of Adaptive Model Predictive and Sliding Mode Controllers for Longitudinal Tire Slip Ratio Tracking

Hamza Ben Moussa(✉) and Mohammed Bakhti

Moulay Ismail University, Ecole Nationale Supérieure d'Arts Et Métiers, BP 4024, Marjane II, Beni Hamed, 50000 Meknès, Morocco
benmoussa.hamza95@gmail.com, h.benmoussa@edu.umi.ac.ma, m.bakhti@umi.ac.ma

Abstract. This study aims to evaluate the performance of adaptive model predictive (AMPC) and first order sliding mode controllers (FOSMC) when used to track the longitudinal tire slip ratio. A detailed model for the vehicle longitudinal dynamic will be derived, and an anti-lock braking system (ABS) is set up by continuously adjusting the braking torque. There are various methods to predict the braking torque for the anti-lock braking system (ABS), the models are based on the number of wheels to be treated, this study is based on a model of a quarter vehicle where all four wheels are processed in an identical way. The tire longitudinal slip ratio control during braking ensures a perfect stability for the vehicle. The system model is linearized around the actual trajectory for the AMPC to predict the output over a suitable prediction horizon, and the discontinuous sign function is substituted by a continuous one for the FOSMC to prevent undesirable chattering. The efficiency of the controllers, in terms of output tracking and feasible control signals, is appraised by numerical simulation and detailed comparison.

Keywords: Anti-lock braking system · Adaptive Model Predictive Control · First Order Sliding Mode Control · Optimal braking torque · Longitudinal slip ratio

1 Introduction

The Antilock Braking System (ABS) main objective is to control the longitudinal tire slip ratio, while braking, ensuring adequate vehicle stability (Savaresi and Tanelli 2010). It is the most used technique for braking control, and it has been subject to numerous research papers where many algorithms have been tested to overcome the high nonlinearities arising from the modelling process or the uncertainties related to the tire longitudinal forces models. A gain planning controller has been derived, for the 4 front and rear wheels with braking at each wheel (Vietinghoff and Kiencke 2007). An approach based on a nonlinear backstepping dynamic surface control has been proposed to track the reference slip ratio avoiding the unstable region (Qiu et al. 2015), and an extremum

L. Azrar et al. (Eds.): CIP 2022, LNME, pp. 521–531, 2023.
https://doi.org/10.1007/978-3-031-23615-0_53

seeking algorithm has been proposed (Dinçmen et al. 2010) for ABS control managing to maximize the road-tire friction coefficient by seeking the optimal slip ratio.

Various methods have been reported in the literature for prediction the braking torque for the anti-lock braking system, Wei et al. (2019) has developed a braking system for electric vehicles with one motor for each wheel using AMPC. Harifi et al. (2008) are proposed an anti-lock braking system model based to first order SMC, the system proposed ensures a finite time convergence of the slip ratio to the desired value, in addition to its ability of reducing the system dynamic, sliding mode control (SMC) methods are widely used, due to their flexibility in handling non-linear systems and their robustness against uncertainties (Castro et al. 2013, Hsu 2015). Unfortunately, the sign function, used for the braking torque calculation, causes undesirable chattering and makes the control signal unfeasible. To overcome this problem, a smooth arctangent function is used.

In this study we will predict the braking torque for an ABS system using the AMPC corrector used by Wei et al. (2019), the results will be compared with an anti-lock braking system with first order SMC used by Harifi et al. (2008), both controllers are applied to for a quarter vehicle model (Mirzaeinejad and Mirzaei 2010, Park et al. 2006).

This paper aims to compare an Adaptive Model Predictive Controller (AMPC) to a first order Sliding Mode (SMC) in order to track the desired slip ratio to maximize the longitudinal tire forces based of The Magic formula of the tire model. The comparison of the proposed controllers outlines their abilities to track the slip ratio set point trying to minimize the stopping distance/time given an initial vehicle velocity.

This document will be presented as follows. In Sect. 2, the quarter vehicle model is used to describe the vehicle dynamic given that only the longitudinal displacement is considered (Mirzaeinejad and Mirzaei 2010, Park et al. 2006). For the tire longitudinal force, the Magic formula (Bakker et al. 1987) is proposed. Finally, the dynamic equations of the system are derived, and a state space formulation is explicitly set. In the next section, the AMPC algorithm is briefly introduced, and the process of iterative linearization is summarized. The SMC algorithm is deduced from a Lyapunov stability discussion in Sect. 4. Simulation results along with the comparison details are illustrated in Sect. 5, and Sect. 6 gives the concluding remarks.

2 System Modeling

The external longitudinal forces acting on the vehicle include aerodynamic drag force F_{aero}, gravitational forces, longitudinal tire forces F_x and rolling resistance forces R_x (Rajesh Rajamani 2012). In this study, we will suppose that $F_{aero} = 0$, $R_x = 0$.

The governing equations of the wheel's motion model are as follows by (Mehdi Mirzaei and Hossein Mirzaeinejad 2012):

$$m\dot{V} = -F_x - \text{mg}\sin\theta \tag{1}$$

$$J_w\dot{\omega} = r_{eff}F_x - T \tag{2}$$

where V is the longitudinal velocity of the vehicle, F_x is the longitudinal tire force, ω is the angular velocity of the wheel, J_w is the total moment of inertia of the wheel, r_{eff}

is the average radius of the wheel, T is the braking torque, finally, m is the total mass of the quarter vehicle given by (Mehdi Mirzaei and Hossein Mirzaeinejad 2012):

$$m = \frac{1}{4}m_v + m_w \tag{3}$$

In the previous equation, m_v is the sprung mass of the vehicle and m_w is the wheel mass.

The longitudinal tire force F_x is the friction force from the ground that acts on the tires. Experimental results have established that the longitudinal tire force generated by each tire depends on (Rajesh Rajamani 2012):

- The normal load on the tire F_z.
- The longitudinal slip ratio σ.
- The friction coefficient of the tire-road interface μ.

The normal load acting on the tire depends to two components, a static component due to the distribution of the vehicle mass and a dynamic component due to the load transfer during braking. Therefore, the tire normal load in the quarter model is by (Mehdi Mirzaei and Hossein Mirzaeinejad 2012):

$$F_z = mg - \frac{m_v h}{2l}\dot{V} \tag{4}$$

where h is the height of the sprung mass (gravity center) and l is the vehicle wheelbase.

The difference between the actual longitudinal velocity at the axle of the wheel V and the equivalent rotational velocity $r_{eff}\omega$ of the tire is the longitudinal slip σ. During the braking $V > r_{eff}\omega$, the wheel longitudinal slip is calculated as (Rajesh Rajamani 2012):

$$\sigma = \frac{V - r_{eff}\omega}{\max(V, r_{eff}\omega)} = \frac{V - r_{eff}\omega}{V} \tag{5}$$

Differentiating Eq. (5) respecting the time gives the derivative form:

$$\dot{\sigma} = -r_{eff}\frac{\dot{\omega}V - \dot{\omega}\dot{V}}{V^2} \tag{6}$$

The longitudinal force generated F_x is described as a function of the longitudinal slip. The relationship between the longitudinal force and slip is linear if the longitudinal is small. But the relationship changes from linear to non-linear for high values of the longitudinal slip.

The Magic formula of the tire model allows us to calculate the longitudinal tire forces for a wide large longitudinal slip. The Magic formula model is (Rajesh Rajamani 2012):

$$Magic(\sigma) = D\sin(Ca\tan(B\sigma - E(B\sigma - a\tan(B\sigma)))) \tag{7}$$

where D is the friction coefficient peak value, C is the common value of the shape factor, B is the stiffness factor, and E is the curvature factor.

The parameters B, C, D and E are defined as follows (Rajesh Rajamani 2012):

$$D = \mu \tag{8}$$

$$C = \frac{2}{\pi}\sin^{-1}\left(\frac{F_{xs}}{DF_z}\right) \tag{9}$$

where F_{xs} the asymptotic value of longitudinal force (F_{xs} is the longitudinal force for longitudinal slip values around 1).

$$E = \frac{B\sigma_m - \tan\left(\frac{\pi}{2C}\right)}{B\sigma_m - \tan^{-1}(B\sigma_m)} \tag{10}$$

where σ_m is the value of the longitudinal slip where the longitudinal force is maximal ($F_x = DF_z$).

$$B = \frac{DF_z}{\sigma_m CD} = \frac{F_z}{\sigma_m C} \tag{11}$$

Finally, the longitudinal force generated F_x is:

$$F_x = F_z Magic(\sigma) \tag{12}$$

By using the previous equations and differentiating the longitudinal velocity of the vehicle V respecting the time we got the derivative form $\dot{V}$:

$$\dot{V} = \frac{-mg(Magic(\sigma) + \sin(\theta))}{m - Magic(\sigma)\frac{m_v h_c}{2l}} \tag{13}$$

On the other hand, differentiating the longitudinal slip σ respecting the time we got the derivative form $\dot{\sigma}$:

$$\dot{\sigma} = \frac{-r_{eff}}{V^2}\left[\left(\frac{r_{eff}\, Magic(\sigma)\left(mg - \frac{m_v h}{2l}\dot{V}\right) - T}{J_w}V\right) - \left(\frac{V(1-\sigma)}{r_{eff}}\frac{-mg(Magic(\sigma) + \sin(\theta))}{m - Magic(\sigma)\frac{m_v h_c}{2l}}\right)\right] \tag{14}$$

3 The Adaptive Model Predictive Controller Principle MPC

Based on Eqs. (13) and (14), the nonlinear system is:

$$\begin{cases} \dot{V} = f(\sigma) \\ \dot{\sigma} = g(\sigma, V, T) \end{cases} \tag{15}$$

Considering the state $x = [x_1, x_2]^T = [\sigma, V]^T$, the input $u = T$ and the output $y = x_1 = \sigma$, we first linearize the system before using the AMPC algorithm. The system linearization based to jacobine matrix is:

$$\dot{x} = A_b x + B_b u \tag{16}$$

where: $A_b = \begin{pmatrix} \frac{df}{dx_1}\Big|_{x_{1_0}} & \frac{df}{dx_2}\Big|_{x_{2_0}} \\ \frac{dg}{dx_1}\Big|_{x_{1_0}} & \frac{dg}{dx_2}\Big|_{x_{2_0}} \end{pmatrix}$; $B_b = \begin{pmatrix} \frac{df}{du}\Big|_{u_0} \\ \frac{dg}{du}\Big|_{u_0} \end{pmatrix}$

Then, we have:

$$\begin{cases} x(k+1) = (t_s A_b + I_2)x(k) + t_s B_b u(k) = Ax(k) + Bu(k) \\ y(k) = [10]x(k) = Cx(k) \end{cases} \tag{17}$$

where t_s is sample time and I_2 is the identity matrix.

We will note the difference of control signal and the state variable by (Liuping Wang 2008):

$$\begin{cases} \Delta u(k) = u(k) - u(k+1) \\ \Delta x(k) = x(k) - x(k+1) \end{cases} \tag{18}$$

An augmented state vector is defined:

$$x_a(k) = \begin{pmatrix} \Delta x(k) \\ y(k) \end{pmatrix} = \begin{pmatrix} x(k+1) - x(k) \\ y(k) \end{pmatrix} \tag{19}$$

Then:

$$x_a(k+1) = \begin{bmatrix} \Delta x(k+1) \\ y(k+1) \end{bmatrix}_3 = \begin{bmatrix} A & 0_{2\times 1} \\ CA & 1_{1\times 1} \end{bmatrix}_{(3)\times(3)} \begin{bmatrix} \Delta x(k) \\ y(k) \end{bmatrix}_3 + \begin{bmatrix} B \\ CB \end{bmatrix}_3 \Delta u(k) \tag{20}$$

$$y(k) = \left[0_{1\times 2}1\right]_3 \begin{bmatrix} \Delta x(k+1) \\ y(k) \end{bmatrix}_3 \tag{21}$$

Finally:

$$\begin{cases} x_a(k+1) = A_a x_a(k) + B_a \Delta u(k) \\ y(k) = C_a x_a(k) \end{cases} \tag{22}$$

Matrices A_a, B_a and C_a define the augmented state model, N_p is the Prediction horizon samples, we will note the future and N_c is the Control horizon samples (Liuping Wang 2008):

$$\Delta U = [\Delta u(k), \Delta u(k+1), \ldots \Delta u(k+N_c-1)]^T \tag{23}$$

$$\Delta X = \left[x(k+1), x(k+2), \ldots x\left(k+N_p\right)\right]^T \tag{24}$$

$$Y_p = \left[y(k+1), y(k+2), \ldots y\left(k+N_p\right)\right]^T \tag{25}$$

Y_p can be given by:

$$Y_p = Fx(k) + \Phi \Delta U \tag{26}$$

where:

$$F = \left[C_a A_a \; C_a A_a^2 \; C_a A_a^3 \ldots C_a A_a^{N_p} \right]^T \tag{27}$$

$$\Phi = \begin{bmatrix} C_a B_a & 0 & 0 & \ldots & 0 \\ C_a A_a B_a & C_a B_a & 0 & \ldots & 0 \\ C_a A_a^2 B_a & C_a A_a B_a & C_a B_a & \ldots & 0 \\ \ldots & \ldots & \ldots & \ldots & \ldots \\ \ldots & \ldots & \ldots & \ldots & \ldots \\ C_a A_a^{N_p-1} B_a & C_a A_a^{N_p-2} B_a & C_a A_a^{N_p-3} B_a & \ldots & C_a A_a^{N_p-N_c} B_a \end{bmatrix} \tag{28}$$

For a given set-point signal $r(k)$ within a prediction horizon N_p, the data vector S_p within in prediction horizon N_p is (Liuping Wang 2008):

$$S_p = \left[1 \; 1 \; \ldots \; 1 \right]_{1 \times N_p} r(k) = \overline{S_p} r(k) \tag{29}$$

We define the cost function J that reflects the control objective as (Liuping Wang 2008):

$$J = \left(S_p - Y_p\right)^T Q \left(S_p - Y_p\right) + \Delta U^T R \Delta U \tag{30}$$

where Q and R are symmetric semidefinite weighting matrices.

The optimal control signal that will minimize the cost function J is (Liuping Wang 2008):

$$\frac{\partial J}{\partial \Delta U} = 0 = -2\Phi^T Q\left(S_p - Fx(k_i)\right) + 2\left(\Phi^T \Phi + R\right)\Delta U \tag{31}$$

Finally, the predictive control at time sample k is (Liuping Wang 2008):

$$\begin{aligned} \Delta u(k) = & \left[[1\,0\,0\ldots0]_{1\times N_c} \left(\Phi^T \Phi + R\right)^{-1} \Phi^T Q \overline{S_p} \right] r(k) \\ & - \left[[1\,0\,0\ldots0]_{1\times N_c} \left(\Phi^T \Phi + R\right)^{-1} \Phi^T Q F \right] x(k) \end{aligned} \tag{32}$$

$$\Delta u(k) = K_r r(k) - K_x x(k) \tag{33}$$

The AMPC algorithm is a potential remedy for dealing with a nonlinear model. Every time step, the process model is linearized around the actual state and control signal. Although the issue is still a convex optimization, it should be modified iteratively because the process model is time-variant.

4 Sliding Mode Controller Design

Based on the nominal plant derived in Sect. 2, a first order sliding mode controller formulation is detailed in this section to control the longitudinal tire slip ratio.

The nonlinear equation of the system, given by Eq. (15), are rearranged into the following general form:

$$\dot{x} = \begin{pmatrix} f_1(x) \\ f_2(x) \end{pmatrix} + \begin{pmatrix} g_1(x) \\ g_2(x) \end{pmatrix} u \tag{34}$$

where $x = \begin{bmatrix} \sigma & V \end{bmatrix}^T$ and u is the braking torque. The longitudinal tire slip ratio is σ_{ref} and the sliding surface is $s = \sigma - \sigma_{ref}$.

T slip setpoint is supposed constant, and its value is obtained from the magic formula so as to maximize the longitudinal tire force.

The control signal is given by:

$$u = u_{eq} + u_{sw} \tag{35}$$

The equivalent control u_{eq} is obtained as follows:

$$\dot{s} = \dot{\sigma} = f_1(x) + g_1(x)u_{eq} = 0 \tag{36}$$

Then:

$$u_{eq} = -\frac{f_1(x)}{g_1(x)} \tag{37}$$

And the switching control u_{sw} is given by condition $s\dot{s} < \mathrm{f}(s) < 0$ and based on the Lyapunov candidate function $V = \frac{1}{2}s^2$.

Using the time derivative, we have:

$$\begin{aligned}\frac{dV}{dt} &= s\dot{s} = (f_1(x) + g_1(x)u)s \\ &= \left(f_1(x) + g_1(x)\left(u_{eq} + u_{sw}\right)\right)s \\ &= \left(f_1(x) + g_1(x)\left(-\frac{f_1(x)}{g_1(x)} + u_{sw}\right)\right)s \\ &= g_1(x)u_{sw}s \end{aligned} \tag{38}$$

So, let:

$$u_{sw} = -\frac{ksign(s)}{g_1(x)} \tag{39}$$

Then:

$$\frac{dV}{dt} = -g_1(x)\frac{ksign(s)s}{g_1(x)} = -k|s| \tag{40}$$

the use of the sign function ($sign(s)$) causes an undesirable chattering in the control signal. Many techniques may be used to lower or eliminate the chattering; however, this is beyond the scope of this paper.

5 Simulation Results

In this section, the simulation results are illustrated. The performance of both controllers, AMPC and SMC, are alyzed regarding the tracking of the longitudinal tire slip ratio tracking.

For the AMPC algorithm, the linearized nominal model, over the system trajectory, is used for the optimization of the braking torque on the prediction horizon.

Then, the results are compared to the first order SMC. The controllers' algorithms were implemented in Matlab/Simulink environment, and the model simplifying as well as the Jacobians derivation was carried out using the Matlab Symbolic Math Toolbox. Small time steps for numerical integration have been required due to the nonlinearities of the nominal plant. The step time has been set to 10^{-3} S, for the linearization of the nonlinear equations, and so for the AMPC algorithm, while a smaller time step of 10^{-5} S has been appropriate for the SMC.

Simulations have been carried out for the quarter car model based on the Magic formula tire model, and Table 1 shows the numerical parameters that have been adopted. The vehicle is supposed travelling along a straight path on a flat dry road, and the longitudinal tire force is supposed to be at its maximum for $\sigma = 0.1$, which is set to be the desired slip ratio.

Table 1. Numerical parameters of the quarter vehicle model and the tire model.

The quarter vehicle model		The tire model	
Average radius of the wheel	$r_{eff} =$ 0.326 m	$Magic(\sigma) = D\sin(Ca\tan(B\sigma - E(B\sigma - a\tan(B\sigma))))$	
Wheelbase	$l = 2.5$ m		
Center of gravity height	$h = 0.5$ m	$B = 12.12$	$C =$ 1.65
Wheel mass	$m_w = 40$ Kg	$D = 0.7$	$E =$ −0.58
vehicle sprung mass,	$\frac{1}{4}m_v =$ 415 kg	Angle of inclination of the road on which the car is moving	
Total moment of inertia of wheel	$J_w = 1.7$ N.m^2	$\theta = 0$	

Initially, the vehicle has a longitudinal velocity equal to 30 m.s^{-1}, and no braking torque is applied. The initial slip ratio is the zero.

Table 2 summarizes the tuning parameters for both controllers, and Fig. 1 illustrates the results for the ABS system. Both controllers ensure the vehicle stopping without skidding, and that the setpoint slip ratio is well tracked. However, the settling time for the SMC is much better than for the AMPC controller. The required time, as well as the distance, for the braking are almost equal between the two controllers.

Table 2. Numerical parameters of the controllers.

AMPC controller		SMC controller	
Control increment weight	$u_w = 0$	Controller gain	$k = 1$
Slip ratio error weight	$e\_w = 1$		
Prediction horizon (samples)	$Np = 120$		
Control horizon (samples)	$Nc = 1$		

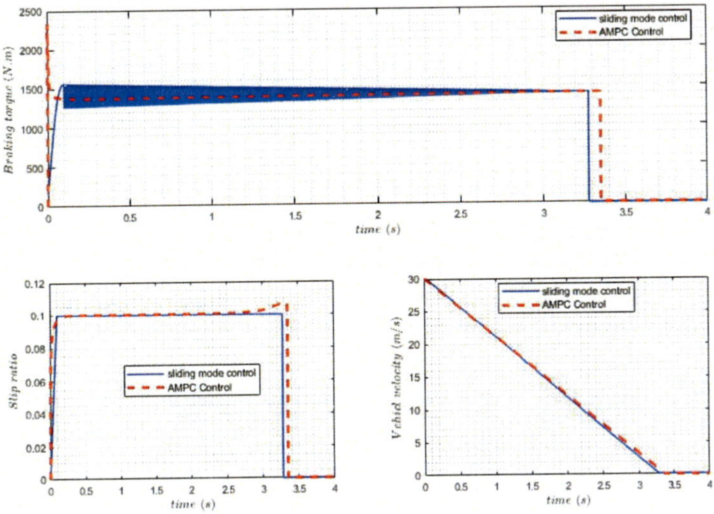

Fig. 1. Sign function-based SM control compared to AMPC control

Nevertheless, the control signal given by the AMPC controller is far better than that of the SMC. The sign function has generated unfeasible braking torque illustrated by torque values characterized with very high switching.

Table 3. Comparative simulation results.

	AMPC controller	SMC controller	
		Using sign function	Using arctangent function
Maximum braking torque	$u_{\max} = 2.310^3$ N.m	$u_{\max} = 1.5610^3$ N.m	$u_{\max} = 1.5410^3$ N.m
Required time for braking	$t_{stop} = 3.293$ s	$t_{stop} = 3.279$ s	$t_{stop} = 3.269$ s
Stopping distance	$d_{stop} = 50.46$ m	$d_{stop} = 49.86$ m	$d_{stop} = 49.56$ m

When the arctangent function is used instead of the sign function, the SMC also gives a smooth and feasible braking torque profile as shown in Fig. 2. Comparative results are summarized in Table 3, when the stopping distance is evaluated by integrating the vehicle velocity over the period of braking time.

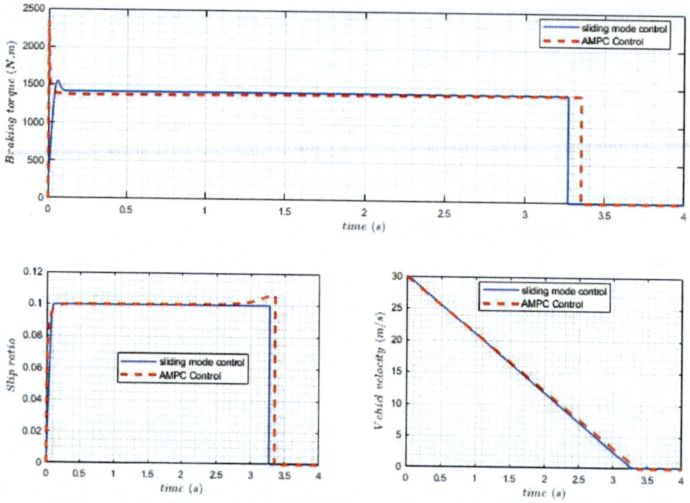

Fig. 2. Arctangent function-based SM control compared to AMPC control

6 Conclusion

An Adaptive Model Predictive Controller and a first order Sliding Mode Controller have been proposed for an ABS control strategy. A simplified quarter vehicle model has been used for the dynamic modelling and the nonlinear equations deriving, while the magic formula has been used to model the tire.

The longitudinal tire slip ratio set point has been set to the value that maximize the longitudinal tire force, and both controllers showed satisfying tracking abilities. In spite of a higher computation load, the AMPC algorithm gave the most feasible control signal even if its maximum value is relatively high.

The SMC ensured a perfect tracking for the slip ratio set point, giving a much lower stopping time/distance. Unfortunately, the required control signal is unfeasible when the sign function is used, yet the arctangent function result is far better.

References

Bakker, E., Nyborg, L., Pacejka, H.B.: Tyre modelling for use in vehicle dynamics studies. SAE Tech. Paper 870421 (1987)

Castro, R., Araujo, R.E., Freitas, D.: Wheel slip control of EVs based on sliding mode technique with conditional integrators. IEEE Trans. Ind. Electron. **60**(8), 3256–3271 (2013)

Dinçmen, E., Acarman, T., Aksun, B.G.: ABS control algorithm via extremum seeking method with enhanced lateral stability. IFAC Proc. Vol. **43**(7), 19–24 (2010)

Harifi, A., Aghagolzadeh, A., Alizadeh, G., Sadeghi, M.: Designing a sliding mode controller for slip control of antilock brake systems. Trans. Res. Part C **6**(6) (2008)

Hsu, C.-F.: Intelligent exponential sliding-mode control with uncertainty estimator for antilock braking systems. Neural Comput. Appl. **27**(6), 1463–1475 (2015). https://doi.org/10.1007/s00521-015-1946-4

Wang, L.: Model Predictive Control System Design and Implementation Using MATLAB, pp. 30–42. Springer (2008)

Mirzaei, M., Mirzaeinejad, H.: Optimal design of a non-linear controller for anti-lock braking system, p. 3 (2012)

Mirzaeinejad, H., Mirzaei, M.: A novel method for non-linear control of wheel slip in anti-lock braking systems. Control Eng. Pract. **18**(8), 918–926 (2010)

Qiu, Y., Liang, X., Dai, Z.: Backstepping dynamic surface control for an anti-skid braking system. Control Eng. Pract. **42**, 140–152 (2015)

Park, E.J., Stoikov, D., Luz, L.F.D., Suleman, A.: A performance evaluation of an automotive magnetorheological brake design with a sliding mode controller. Mechatronics **16**(7), 405–416 (2006)

Rajamani, R.: Vehicle Dynamics and Control, Second Ed. pp. 87, 115, 380, 386–387, 406–409. Springer (2012)

Savaresi, S., Tanelli, M.: Active Braking Control Systems Design for Vehicles. Springer-Verlag, New York (2010)

Vietinghoff, A., Kiencke, U.: Gain scheduling for combined four-wheel steering and individual wheel braking. IFAC Proc. Vol. **40**(10), 327–333 (2007)

Xu, W., Chen, H., Zhao, H., Ren, B.: Torque optimization control for electric vehicles with four in-wheel motors equipped with regenerative braking system (2019)

A New Denoising Method for Motor Fault Diagnosis

Dinh-Khoa Tran[1(✉)], Ho-Si-Hung Nguyen[1], Hai-Canh Vu[2], Nassim Boudaoud[2(✉)], The-Dung Vo[1], and Duc-Hanh Dinh[1]

[1] University of Science and Technology - The University of Danang, Da Nang, Viet Nam
tdkhoa@dut.udn.vn
[2] Roberval laboratory, University of Technology of Compiègne, Compiègne, France
nassim.boudaoud@utc.fr

Abstract. In recent years, Deep Learning (DL) has recently become the key to success in the manufacturing industry. Motor faults diagnosis based on the vibration data, one of the deep learning applications in the modern manufacturing model, is recently a tendency in the scientific community. Given that the vibration data is highly sensitive to a number of noises. The unnecessary motions of background can have a negative impact on the information input for the acceleration sensor. That's reason why cleaning vibration signal might be consider as a first phase for for bearing machines fault diagnostics. In this paper, a new denoising method based on Fast Fourier Transform (FFT) and K-means clustering is firstly proposed to improve the performance of the motor fault diagnosis. Convolutional Neural Network (CNN) is then applied to classify the motor faults. To validate the performance of the proposed approach, the open-source Case Western Reserve University (CWRU) data-set is considered. The experimental results confirm the advantages of the proposed denoising method when compared to the other existing methods.

Keywords: Signal denoising · Vibration signal · Deep learning · Convolutional neural network · Motor fault diagnosis

1 Introduction

In many manufacturing sectors, equipment failure can result in significant production losses or even endanger the lives of employees. Therefore, a mechanical health evaluation is crucial to making the best maintenance choices and preventing machine failures. Machine learning has been widely used and developed into a potent tool for machine diagnosis, particularly for motor fault classification based on vibration data, as a result of the development of monitoring techniques and data. Indeed, a number of machine learning techniques such as Support Vector Machine (SVM), Decision Trees (DT), K-nearest neighbors algorithm (KNN), Principal Component Analysis (PCA), Support Vector Machine

L. Azrar et al. (Eds.): CIP 2022, LNME, pp. 532–540, 2023.
https://doi.org/10.1007/978-3-031-23615-0_54

(SVM) [1], Decision Trees (DT) [2], K-nearest neighbors algorithm (KNN) [3], Principal Component Analysis (PCA) [4] ,...have been successfully applied to the different diagnosis problems. However, they are usually applied in cases of pure data-set without considering the existing noises. The performance of these methods is then quite limited. To overcome the problem of motor diagnosis with noisy data, we first propose a new denoising method which is a combination of the fast Fourier transform and the K-means clustering, to denoise the data. The Convolutional Neural Network (CNN) [5] is then used to classify the motor faults.

In the next sections of the paper, we will discuss in more detail the existing denoising methods (Sect. 2), our denoising method (Sect. 3), the CNN model (Sect. 4), and finally, the application and the validation of the proposed model is done on the open-source Case Western Reserve University (CWRU) dataset.

2 Existing Denoising Methods

In this section, we will present in more detail the existing denoising methods.

Wavelet Transform

Wavelet transform [6] is a technique for time-frequency domain analysis. For the different types of denoising methods based on the Wavelet transform, a soft threshold was defined for specific applications to get rid of the noises existing in the data. There exist a number of denoising methods based on the Wavelet transform, for example fixed threshold, level-dependent threshold, soft and hard threshold [7]. In the paper, the hard threshold algorithm will be compared to our proposed denoising method. According to this method, the hard threshold is fixed as follows.

$$\alpha_m(x) = \begin{cases} 1, & \text{if } |x| \geqslant \text{T} \\ 0 & \text{if } |x| < \text{T} \end{cases} \tag{1}$$

From the above equation, a new frequency band can be obtained

$$\tilde{x} = \sum_{m=0}^{N-1} \rho_T(X_B[m]) g_m \tag{2}$$

with

$$\rho_T = a_m(x) * x = \begin{cases} x, & \text{if } |x| \geqslant \text{T} \\ 0 & \text{if } |x| < \text{T} \end{cases} \tag{3}$$

Singular Value Decomposition

Singular Value Decomposition (SVD) [8] is An artificial method to establish low-dimensional data to high-dimensional data approximation in terms of the prevailing patterns. The SVD builds a new architectural visualization of the data in terms of the original domain. Its dimensional reduction produced several amazing applications. In this study, we examine the SVD's denoising potential using a threshold for singular values. A wonderful way to get rid of contaminants as much as possible was discussed in the paper [8]. The hard threshold method, which releases the lower rank in the component of singular value decomposition of the data matrix Y, selects the best estimate of rank r. Two instances are included in the hard threshold method, depending on the length of each dimension of the matrix data for Y. For instance, the secret formula for the n-by-n matrix and the m-by-n matrix may be used to identify an appropriate rank r for truncating the singular value. From this reasonable rank r, the input signal of low volume represented for noise will be moved. The character signal is still present, on the other hand.

Savitzky-Golay Filter
High-frequency noise can be effectively smoothed using the [9], but the noise retains its original form. The fundamental principle of this approach is to fit each point of a least-square to a collection of high-order polynomials across an odd-sized window centered at the point. A moving average is first used to stabilize erratic data. In this process, a certain number of points must be aligned on their ordinates in order to get an average ordinate. Second, the idea of discrete convolution operation is put into practice.

The main drawback of the current approaches is the need to manually configure a threshold for each individual application. As a result, in this work we suggest a novel denoising approach based on the FFT and K-means, which will be described in the following part, to automatically pick the threshold and to enhance the performance of the denoising methods.

3 Proposed Denoising Method

Our proposed denoising method is based on the FFT and K-means.

Fast Fourier Transform (FFT)
The idea of FFT [10] is to optimize the coefficient setting and therefore decrease the computation time when compared to the Discrete Fourier Transform (DFT). Indeed, the FFT algorithm is based on the following equations:

$$Forward\ X_F(k) = \sum_{n=0}^{N-1} x(n) w_N^{kn}, k = 0, 1, ..., N-1 \tag{4}$$

$$Inverse\ x(n) = \frac{1}{N} \sum_{n=0}^{N-1} X_F(k) w_N^{-kn}, k = 0, 1, ..., N-1 \tag{5}$$

Where N is the number of data samples; $x(n)$ is a uniformly-spaced sample sequence; X_F is a set of complex numbers in the same length sequence of equally-spaced sample; w_N^{kn} is the Nth root of the kth DFT coefficients. In the other words, w_N^{kn} is a set of kn^{th} unity w_N

$$w_N = exp(\frac{-j2\pi}{N}) \tag{6}$$

Denoising Method Based on FFT and K-Means (DFK)

The denoise process contains four main steps. First, each sample of input data is inverted forward by FFT. The converted signal now is a set of a real number. Secondly, the power spectrum is easily computed from the previous set of a real number. Next, from the obtained power spectrum, noise and pure signal should be distinguished based on their amplitudes. A threshold is normally defined for this purpose. In this paper, K-means clustering has been chosen to play the role of this threshold to separate the power spectrum into two classes (K=2). One class represents the noise and the other represents the pure signal. At the last step, the pure signal is obtained by reconverting the pure power spectrum part.

For more details, let take an example where the data $f(x)$ is generated as a combination of pure signal $g(x)$ and a Gaussian white noise N

$$f(x) = g(x) + N \tag{7}$$

$$= \sin 2\pi f_1 t + \sin 2\pi f_2 t + N \tag{8}$$

where, $f_1 = 60$ (Hz) and $f_2 = 100$ (Hz).

First, f(x) is converted to F(x) by FFT according to the Eq. 4. By the definition [11], power spectrum $P(x)$ is effortlessly computed based on F(x):

$$P(x) = \langle | F(x) |^2 \rangle \tag{9}$$

It is obvious that The amplitude spectrum of the pure signal and noise are considerably different in magnitude. Hence, pure signals and noise may be automatically separated by K-means with K is set to 2 classes (see Fig. 1). One class (in orange) represents the power spectrum of the pure signal and the other class (in blue) represents the power spectrum of noise. Finally, the F_i of the first class with high amplitude is reconverted back to the original signals by Inverse Fast Fourier Transform (IFFT) Eq. 5. It should be noted that the other clustering methods can be utilized in stead of K-means. In addition, this example is relatively simple, the advantage of K-means will be demonstrated more clearly in the next section with a real data set.

4 Motor Diagnosis Using DFK and CNN

In this section, our proposed method DFK is applied to reduce the noise in the CWRU dataset. The Convolutional Neural Network (CNN), a powerful ANN model for classification problem, is used to classify the different motor faults based on the denoising data.

Data-set

The CWRU dataset is a readily available dataset and open-source, claims the publication [12]. On the CWRU website, which offers access to the bearing data for both healthy and unhealthy bearings, the created dataset is saved and made available. Data were gathered for normal bearings, single point drive-end (DE), and fan-end (FE) faults and entered into this database. A 2 hp Reliance electric

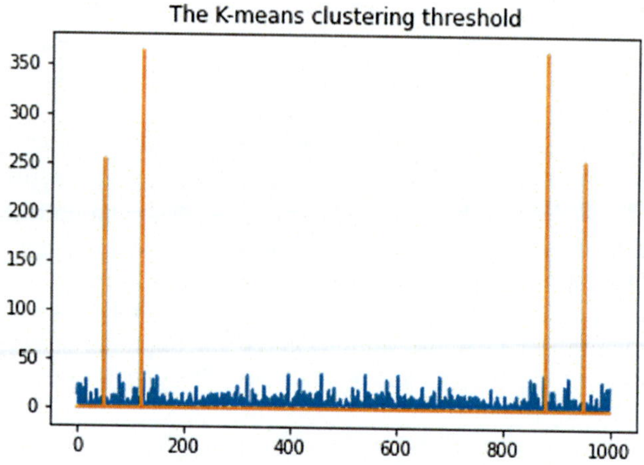

Fig. 1. K-means clustering for the separation of pure signal and noise

induction motor, a torque transducer, a dynamometer, and control electronics, which are not shown in the figure-make up the bearing test rig setup that was used to collect the CWRU dataset. The motor shaft is supported by the test bearings. Through a dynamo-meter and electronic control system, torque is supplied to the shaft. Each defective bearing was restored on the test rig after the faults were seeded on the REBs, the IR, and the OR. The test bearings were subjected to electro-discharge machining to introduce single point defects with fault sizes of 7, 14, 21, 28, and 40 mils. 0.001 inches is equivalent to one mil. For the 7, 14, and 21 mil faults, SKF bearings were utilized, whereas NTN comparable bearings were used for the 28 mil and 40 mil faults. Except for the inner-race faulty bearing with a diameter of 0.028 inches, the outer-race faulty bearing with a diameter of 0.040 inches, and the ball bearing fault with a diameter of 0.028 inches, all the bearings had a fault depth of 0.011 inches. Both the 0.028-inch inner-race defective bearing and the 0.040-inch outer-race faulty bearing have a fault depth of 0.050 inches. Additionally, according to [13], the depth of the 0.028-inch-diameter ball-bearing fault is 0.150 inches.

To validate our model, 7 and 21 mils diameter faults data will be used in this paper. The experiment data-set was split into two sets for the work of training and validating. The percentage of train data is 75 and the percentage of validation data is 25. Each sample was cutoff into small samples equally, the length of each small sample is 200.

CNN Parameters

The CNN model proposed in [14] is adopted. The training process would be more stable and low down the risk of under-fitting the data. Because Mean Squared Error (MSE) [15] played the role of loss function for our model. Furthermore, Adam optimizer was used to improve the accuracy score and speed up training time. Each turn was trained 100 epochs, It seems like a race against time to get the best suitable combination between denoising methods and this CNN model for the CWRW dataset. The denoising methods used for this experience were torch-on in the Sect. 3.

To evaluate the performance of the result in each iteration and final iteration. Accuracy, Precision, Recall, and F1 Score metrics were used to know what is the better model. According to the paper [16], Accuracy, the formula of Precision, Recall, and F1 score is implemented in order to evaluate the performance of each method. For make sure the denoising ability of each method, the performance of each method was evaluated based on S/N values. The main idea of S/N values was got from the paper [17]. The lower value of S/N is, the fewer concentration of noise is.

Table 1. The performance of denoising methods

Denoising methods	S/N
Original	20.62
SVD	20.62
Wavelet	20.6
Savitzky-Golay	20.51
DFK	**17.2**

The Table 1 shown that, The ability of the DFK was far better than others. Especially, this is more clear when the DFK was tested on the CWRU dataset. The sound of this dada was extremely affected by noise from other engine parts.

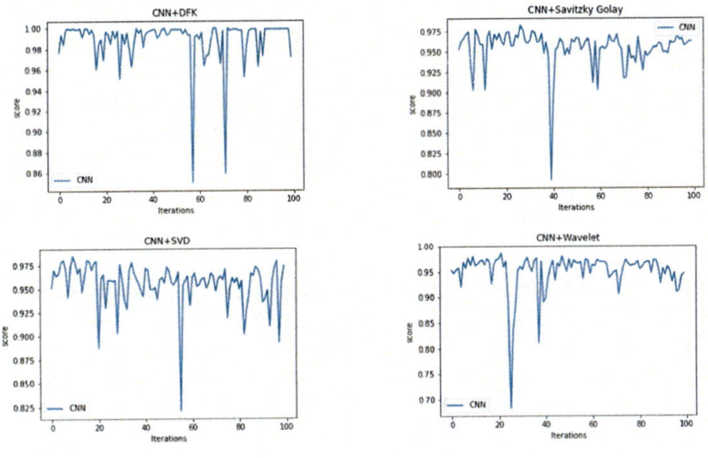

Fig. 2. Validating methods in the data-set of normal baseline and 21 mils diameter faults data

To figure out the best methods for processing the CWRU dataset, each denoising method was combined with the CNN model. Hence, the best amalgamation would have been figured out on the statistic table by validating method. To get the overall view, the performance of each combination was evaluated by the accuracy score on the validation dataset.

The data of 21 mils diameter faults are firstly used and the obtained results are shown in the Fig. 2. According to this figure, the model accuracy score throughout iterations was not stabilized in all cases. But the performance of the DFK method had been greatest at all with the accuracy score swinging from the percentage of 86% to 100%. On the other hand, the model accuracy score dropped to 70% when compared to the Wavelet denoising method.

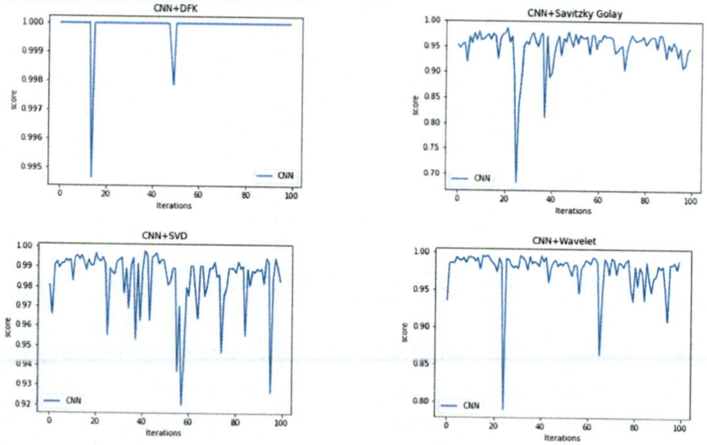

Fig. 3. Validating methods in the data-set of normal baseline and 7 mils diameter faults data

We also test the performance of the different methods by using the 07 mils diameter faults dataset. The obtained results are reported in Fig. 3. According to the figure, all of the denoising methods had a soaring accuracy score, which is from 80% to 100%. Furthermore, the upper left graph caused great surprise through its performance. The accuracy score had reached approximately 100%. The accuracy score on the validation dataset reached up to 100 percent in the first iteration. The pure signal and noise have been well separated thanks to the performance of DFK. From that, the CNN model had done a good job in training and predicting data. It is true that the DFK method forces the CNN model to be more robust.

It's obvious that the CNN model has taken advantage of our proposal to get 100% test accuracy score in the final iteration. Comparing our achievement to the paper [18], the DFK contributed to improving the test accuracy score from 93.88% to 100% with the same default setting parameter.

5 Conclusions

In this paper, we propose a new denoising method DFK. The DFK and CNN are then utilized to diagnose motor faults. The proposed model is tested on the CWRU dataset. The experiments shown that our proposed method is much more efficient than the other existing methods in terms of the ability of denoise and contributes to the accuracy of the Neural Network. In the future works, other Neural Network models can be developed to improve both training time and accuracy significantly for the motor fault diagnosis problem.

References

1. Noble, W.S.: What is a support vector machine? Nat. Biotechnol. **24**(12), 1565–1567 (2006)
2. Quinlan, J.R.: Simplifying decision trees. Int. J. Man Mach. Studies **27**(3), 221–234 (1987)
3. Sarkar, M., Leong, T.-Y.: Application of k-nearest neighbors algorithm on breast cancer diagnosis problem. In: Proceedings of the AMIA Symposium, p. 759. American Medical Informatics Association (2000)
4. Wold, S., Esbensen, K., Geladi, P.: Principal component analysis. Chemometr. Intell. Lab. Syst. **2**(1–3), 37–52 (1987)
5. Albawi, S., Mohammed, T.A., Al-Zawi, S.: Understanding of a convolutional neural network. In: 2017 International Conference on Engineering and Technology (ICET), pp. 1–6. IEEE (2017)
6. Zhang, D.: Wavelet transform. In: Fundamentals of Image Data Mining, pp. 35–44. Springer (2019)
7. Luo, G., Zhang, D., Baleanu, D.: Wavelet denoising. Adv. Wavelet Theory Appl. Eng. Phys. Technol. **634** (2012)
8. Donoho, D.L., Gavish, M.: The optimal hard threshold for singular values is $4/\sqrt{3}$. Citeseer (2013)
9. Savitzky, A., Golay, M.J.: Smoothing and differentiation of data by simplified least squares procedures. Anal. Chem. **36**(8), 1627–1639 (1964)
10. Peters, T.: Data-driven science and engineering: machine learning, dynamical systems, and control by SL Brunton and JN Kutz, pp. 472. Cambridge, Cambridge University Press, ISBN 9781108422093. Taylor & Francis (2019)
11. Cui, W., Liu, L., Yang, X., Wang, Y., Feng, L., Springel, V.: An ideal mass assignment scheme for measuring the power spectrum with fast fourier transforms. Astrophys. J. **687**(2), 738 (2008)
12. Zhang, S., Zhang, S., Wang, B., Habetler, T.G.: Deep learning algorithms for bearing fault diagnostics-a comprehensive review. IEEE Access **8**, 29857–29881 (2020)
13. Loparo, K.: Case western reserve university bearing data center. Bearings Vibration Data Sets, pp. 22–28. Case Western Reserve University (2012). https://csegroups.case.edu/bearingdatacenter/home
14. Lu, X., Liao, W., Huang, W., Xu, Y., Chen, X.: An improved linear quadratic regulator control method through convolutional neural network-based vibration identification. J. Vibr. Contr. **27**(7–8), 839–853 (2021)
15. Prasad, N.N., Rao, J.N.: The estimation of the mean squared error of small-area estimators. J. Am. Statist. Assoc. **85**(409), 163–171 (1990)

16. Sokolova, M., Japkowicz, N., Szpakowicz, S.: Beyond accuracy, f-score and roc: a family of discriminant measures for performance evaluation. In: Australasian Joint Conference on Artificial Intelligence, pp. 1015–1021. Springer (2006)
17. Alsberg, B.K., Woodward, A.M., Winson, M.K., Rowland, J., Kell, D.B.: Wavelet denoising of infrared spectra. Analyst **122**(7), 645–652 (1997)
18. Eren, L., Ince, T., Kiranyaz, S.: A generic intelligent bearing fault diagnosis system using compact adaptive 1d cnn classifier. J. Signal Process. Syst. **91**(2), 179–189 (2019)

Power Optimisation of DFIG Based WECS Using SMC and Metaheuristic Algorithms

Ouassima El Qouarti[1(✉)], Ahmed Essadki[1], Hammadi Laghridat[1], and Tamou Nasser[2]

[1] High National School of Arts and Crafts (ENSAM), Mohammed V University in Rabat, Rabat, Morocco
ouassima_elqouarti@um5.ac.ma

[2] High National School for Computer Science and Systems Analysis (ENSIAS), Mohammed V University in Rabat, Rabat, Morocco

Abstract. The renewable energy topic is nowadays, studied worldwide from different aspects, since it is a very promoting and reasonable issue to explore. In this paper we focus on studying the wind energy based on Doubly Fed Induction Generator (DFIG); We aim to extract the maximum power from the wind by proposing two approaches, first we present a combined control based on Maximum Power Point tracking (MPPT) and sliding mode control, then we present the super twisting control based on Particle Swarm Optimisation (PSO) and Grey Wolf Optimisation (GWO) techniques. Results show that super twisting control tuned by PSO algorithms does better contribute to the optimization of the power extraction compared to other analysed strategies. This is done by comparing the generator speed signals of different control scenarios, namely : MPPT, MPPT based SMC, Super Twisting (ST), ST based PSO, and ST based GWO.

Keywords: DFIG · WECS · Optimisation · Metaheuristic Algorithm · Power · Control · GWO · PSO · MPPT · Sliding Mode · Super Twisting

1 Introduction

Based on a global context dominated by a high electricity demand, the production based on Renewable Energies (RnEs) has been further solicited, since they present an abundant and unlimited source of energy. The main issue is that, this energy is unpredictable and intermittent. Therefore controlling the technology based on RnE presents a high and an urgent necessity to meet the countries' needs in electricity. Thus, Wind Turbines (WTs) based on Doubly Fed Induction Generators (DFIGs), subject of our paper, has been largely studied in literature in order to maximise the power extracted from the wind despite of the intermittency matter.

L. Azrar et al. (Eds.): CIP 2022, LNME, pp. 541–550, 2023.
https://doi.org/10.1007/978-3-031-23615-0_55

Getting the maximum efficiency can be gained using several techniques, authors in [1] have developed a new Adaptive Neuro Fuzzy Inference System (ANFIS) based method to estimate the effective wind velocity and thus the optimal tracked rotor speed, and thereby extracting the maximum efficiency from the variable speed WT. Also, three controls based on high order sliding mode controllers (1st, 2nd and 3rd SMC) were applied in order to reach the maximum power point tracking value. A non linear perturbation observer was added to SMC, in [2] as to optimize the extracted power with enhanced fault-ride through (FRT) capabilities. In this approach, only rotor speed and reactive power measurements are required without the need of the classical auxilary d-q axis current regulation loops. For the same purposes, works in [3] proposed a novel robust perturbation observer (RPO) based fractional order sliding mode controller combined with a Multi-Objective Grasshopper Optimization Algorithm (MOGOA), favorable comparison was done between RPO based fractional order sliding mode controller, perturbation observer-based second order SMC (SOSMC) and SOSMC under different scenarios. Similarly and in the same perspective, authors in [4] worked on a coordinated HOSMC using the Super Twisting (ST) algorithm and a novel reaching law. This approach aims to ensure power optimization and grid synchronization for the DFIG where the optimal rotor speed and rotor current are considered as the reference signals for power optimization. A comparative study was conducted, involving the first order sliding mode control (FOSMC) and Proportional Integral (PI) control. Results show the superiority of the studied approach over the two others. For the purposes of extracted power maximization, null stator reactive power regulation, with free undesirable chattering and optimal control performances, authors in [5], proposed the PSO and Gravitational Search Algorithm (GSA) based FOSMC (PSO-GSA based FOSMC). A comparison with PSO based FSMC and GSA based FSMC systems, highlights, through simulation, the high performances of the proposed technique in achieving the control objective.

Metaheuristic algorithms such as PSO and Grey Wolf Optimisation (GWO) techniques may serve in tuning systems parameter such as PID and super twisting controls and thus help reaching the optimal power; Authors in [6] have worked on these two algorithms and applied them to PI controllers of Rotor and Grid Side Converters, (respectively, RSC and GSC) in order to optimally tune the on-grid 9 MW DFIG system with a defined objective function. In this study, a 1.5 MW DFIG WT was studied, and the tuning control is used to tune the super twisting (ST) parameters in the speed control loop level. We note that the system we are studying is the wind turbine coupled to the DFIG; DFIGs are composed of two converters separated by a dc-bus; The RSC is linked to the rotor of the DFIG, and the GSC is attached to the grid via a filter. The mathematical model can be found at [7]. The RSC is controlled with ST algorithm as in [8], and for simplification purposes, the GSC has not been considered in the present work. A phase-locked loop (PLL) is used to get stator angular speed ws. Figure 1 schematizes the different components of the studied system. In this paper, we will focus on the turbine control; First, we are going to recall the well known Maximum Power Point tracking (MPPT) technique and apply it to the DFIG's

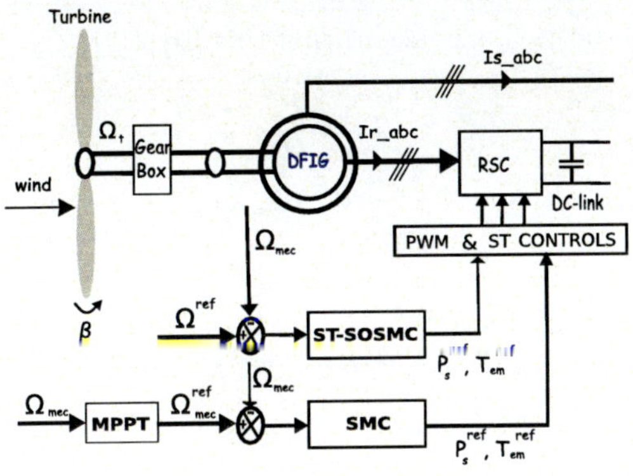

Fig. 1. DFIG model

turbine, then we will combine it to SMC as to show if there is a remarkable power optimisation, otherwise we will omit the whole MPPT technique and replace it by the ST algorithm since we hypothesise that it presents better performance. Our second hypothesis is that tuning super twisting parameters helps increase the wind power extraction and thus ensures a better efficiency of the system. We will then compare the performances of both GWO and PSO algorithms. A part of the adopted strategy in this work was investigated in [10] for a 149.2 kW Squirrel-Cage Induction Generator (SCIG). Authors applied ST algorithm for MPPT, and compared the method with conventional and fuzzy logic PI controllers. But authors did not explore the metaheuristic optimization algorithms to tune the parameters of the ST, as investigated in the present work. The results were evaluated in terms of chattering elimination, robustness and effectiveness, and have proven the superiority of the proposed ST-SMC over conventional and fuzzy logic PI controllers. Many other works have used ST to regulate DFIG's active power or torque in the entry of the ST bloc as in [8,9]. In this work, ST is used to regulate the speed of a grid connected DFIG, as investigated in [10] for a SCIG. The aim is to compare the generator speed responses for different scenarios, in terms of tracking capability and dynamic performances. The structure of this paper is as follow. A general introduction is presented in Sect. 1. An overview on the Turbine modeling is shown in Sect. 2. Control strategies are detailed in Sect. 3. Simulation results and discussion are listed in Sect. 4. Finally, a conclusion is drawn in Sect. 5.

2 Turbine Model

Based on [11], we have the following equations:

The mechanical power extracted from the wind and transferred to the rotor shaft is [11] :

$$P_a = \frac{1}{2} C_p (\lambda, \beta) \rho \pi R^2 V^3 \tag{1}$$

The corresponding torque is deduced using this Eq. [11]:

$$T_a = \frac{P_a}{\Omega_t} \tag{2}$$

Power coefficient Cp(λ,β) and tip-speed ratio λ of the WT are given by following relations [11].

$$C_p(\lambda, \beta) = c_1(c_2\frac{1}{\lambda^i} - c_3\beta - c_4)exp^{(\frac{-c_5}{\lambda^i})} + c_6\lambda \tag{3}$$

where:

$$\frac{1}{\lambda^i} = \frac{1}{\lambda + c_7\beta} - \frac{c_8}{\beta^3 + 1} \tag{4}$$

with the power coefficient constants: $c_1 = 0.5176$, $c_2 = 166$, $c_3 = 0.4$, $c_4 = 5$, $c_5 = 21$, $c_6 = 0.0068$, $c_7 = 0.08$, $c_8 = 0.035$.

$$\lambda = \frac{R\Omega_t}{V} \tag{5}$$

Where Ω_t is the turbine speed, and V is the wind speed.
Figure 2 shows, respectively, the Simulink blocks of the turbine, the gearbox and the shaft.

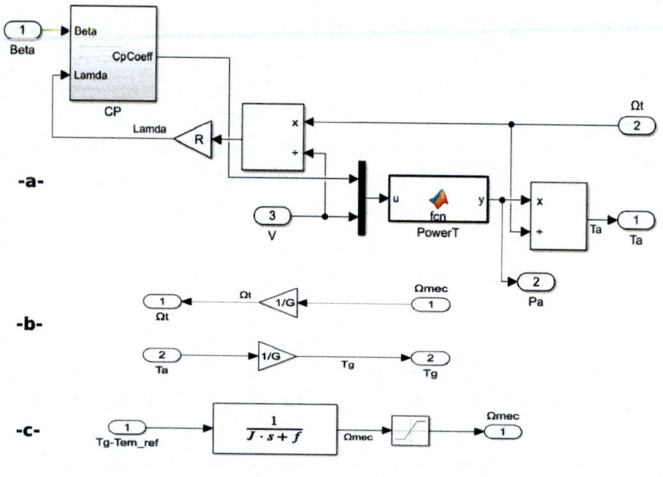

Fig. 2. a)Turbine model, b) gearbox model, c) shaft model

3 Turbine Control

3.1 MPPT

We use the Maximum Power Point Tracking (MPPT) technique to capture the maximum of power. In the literature we find different ways to explore the MPPT optimisation algorithm [12]. In this paper, we apply MPPT without speed control. Figure 3 shows the block diagram of the chosen technique. The estimated wind speed is described as in [9]:

$$V_{est} = \frac{R\Omega_t}{\lambda_{opt}} \tag{6}$$

We deduce the optimal power of the turbine and the reference torque as shown bellow :

$$P_{amax} = \frac{1}{2}\rho C_{pmax}(\lambda, \beta)SV^3 \tag{7}$$

$$T_{emref} = \frac{1}{2}C_{pmax}(\lambda, \beta)\frac{\rho\pi R^5}{(G\lambda_{opt})^3}\Omega_{mec}^2 \tag{8}$$

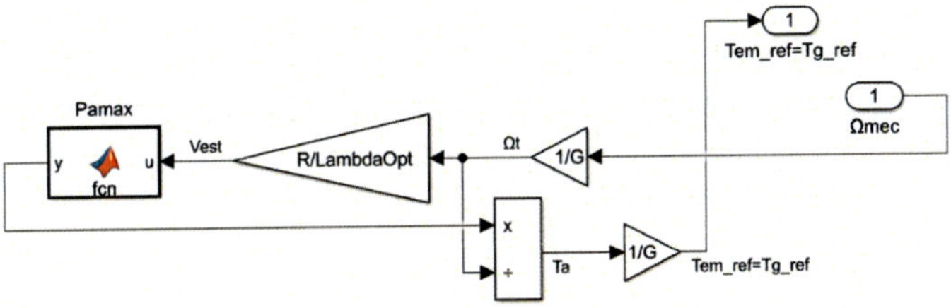

Fig. 3. MPPT control of wind turbine without speed control

3.1.1 MPPT Based on SMC Control We use SMC to control non-linear systems [13]. This control basics are presented in [8,9,14]. Figure 4 presents the simulated bloc diagram of SMC applied to MPPT technique.

$$S = \Omega_{mec\_ref} - \Omega_{mec} \tag{9}$$

$$\dot{S} = \dot{\Omega}_{mec\_ref} - \dot{\Omega}_{mec} \tag{10}$$

$$\dot{S} = \dot{\Omega}_{mec\_ref} - \frac{1}{J}(\dot{T}g - Tem - \Omega_{mec}f) \tag{11}$$

Based on SMC, we write:

$$Tem = Te\_eq + Te\_n \tag{12}$$

$$Te\_n = -Ksign(S) \tag{13}$$

$$Te\_eq = -J\dot{\Omega}_{mec\_ref} + Tg + f\Omega_{mec} \tag{14}$$

Constant K must be positive to verify the system stability.

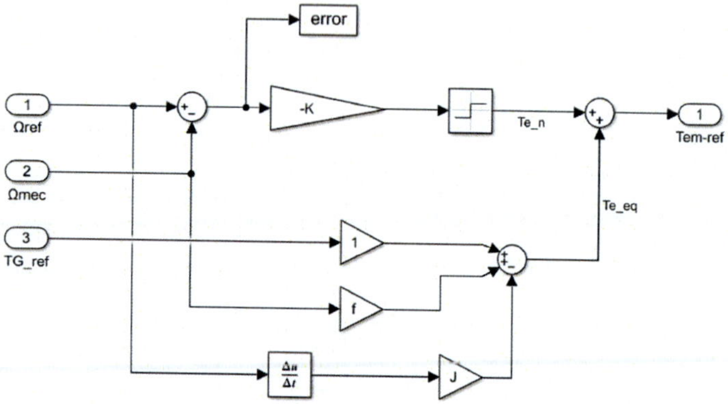

Fig. 4. Block diagram of MPPT based SMC

3.1.2 ST Control

ST control is one variant of SOSMC. The definition can be found in [8]. By applying this control on the turbine, we get the following reference torque equation. Figure 5 gives the block diagram of the implemented control.

$$Tem\_ref = -k2|S|^{0.5} sign(S_{\Omega_{mec}}) - \int \mathrm{k1sign}(\mathrm{S}_{\Omega_{\mathrm{mec}}}) \tag{15}$$

Where : $\mathrm{S}_{\Omega_{mec}}$ is the sliding surface. k1 and k2 satisfy the stability conditions.

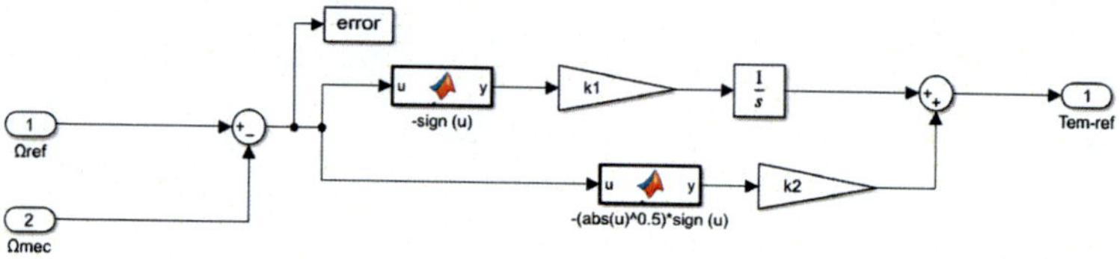

Fig. 5. Block diagram of ST control for the DFIG's turbine

3.2 Tuning Algorithms

The PSO algorithm we used was inspired from [15,16], and the GWO algorithm from [17]. The objective function we want to minimize is:

$$o = sum(abs(error. * error)) \tag{16}$$

PSO and GWO work to reduce the objective function and thus optimizing the ST parameters. The parameters we want to tune are k1 and k2.

4 Simulation Results and Discussion

The proposed control model was simulated under MATLAB/Simulink simulation tool. Data related to the DFIG are presented in Table 1. The wind has been set to 10 m/s and the simulation time is 1 s. Figure 6 shows the reference signals for the speed generator and the speed responses to different controls. Figures 7 and 8 expose, respectively, the behaviour of the PSO and GWO objective functions along the 20 iterations. The PSO and GWO tuned parameters are shown in Table 2. The swarm size for the PSO and the search agents for the GWO were set to 4. Constants k1 and k2 were bounded in between 5000 and 100000.

Table 1. System parameters

Parameters	Values	Parameters	Values
Rated power Pn	1.5 MW	Rotor resistance R_r	8.28e-3 $m\Omega$
Rated voltage Un	690 V	Magnetizing inductance M	26.96 mH
Nominal frequency	50 Hz	Stator inductance L_s	27.2 mH
Rated rotor speed	1750 tr/min	Rotor inductance L_r	27.1 mH
Number of pole pairs	2	Blade radius of the turbine	31 m
Stator resistance R_s	10.3e-3 $m\Omega$	The gearbox ratio G	62
Pitch angle β	0		

Table 2. Optimized parameters of ST controller

Control	k1	k2
ST without PSO and without GWO	5000	3000
ST with PSO	43244	5000
ST with GWO	9054.1775	5000

From the simulation results, we can tell that ST algorithm is more efficient for turbine control in terms of speed tracking objective. In addition, applying a tuning method did not bring much performances, but still they present a good solution to better track the reference speed and optimise the power extraction from the wind. Comparing our results with those of [10], we find that both works support the superiority of ST over the other studied strategies in each work; SMC and classical MPPT in our case, in terms of speed and tracking performances.

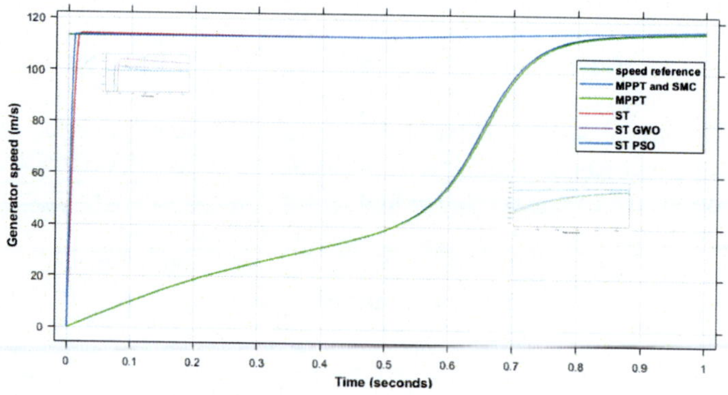

Fig. 6. Generator speed

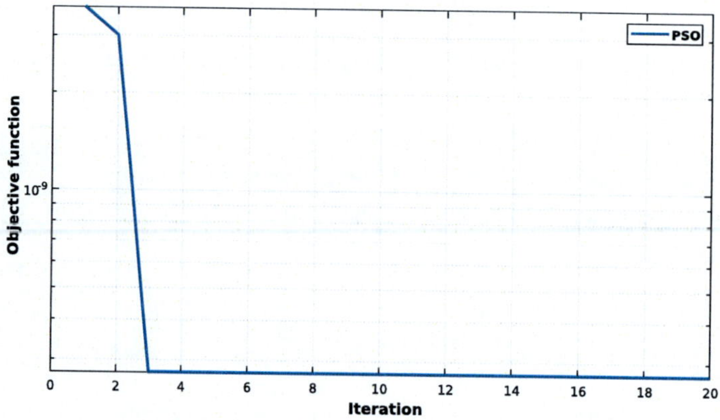

Fig. 7. PSO objective function

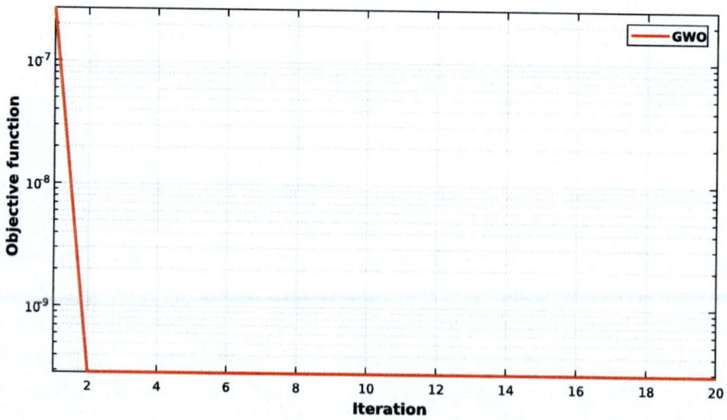

Fig. 8. GWO objective function

5 Conclusion

In this article, we studied the control of the turbine connected to a DFIG based WECSs in order to maximize wind power extraction by proposing two approaches; first we presented a combined control based on the classical MPPT without speed control combined to SMC, then we presented : the ST control, ST based on PSO and ST based on GWO techniques. Results have revealed that ST control tuned by PSO algorithm does better contribute to the optimization of the power extraction compared to other analysed strategies. The use of ST-SOSMC instead of MPPT-SMC is encouraged, and the combination of metaheuristic algorithms with ST algorithm did not bring a significant improvement. As perspectives, we might consider to apply more algorithms other than PSO and GWO, use more prominent objective functions, and conduct more simulations in order to excel the performances of the DFIG.

References

1. Golnary, F., Moradi, H.: Dynamic modelling and design of various robust sliding mode controls for the wind turbine with estimation of wind speed. Appl. Math. Model **65**, 566–585 (2019)
2. Yang, B., Yu, T., Shu, H., Dong, J., Jiang, L.: Robust sliding-mode control of wind energy conversion systems for optimal power extraction via nonlinear perturbation observers. Appl. Energy **210**, 711–723 (2018)
3. Falehi, A.D.: An innovative optimal RPO-FOSMC based on multi-objective grasshopper optimization algorithm for DFIG-based wind turbine to augment MPPT and FRT capabilities. Chaos, Solitons Fractals **130**, 109407 (2020)
4. Xiong, L., Li, P., Wu, F., Ma, M., Khan, M.W., Wang, J.: A coordinated high-order sliding mode control of DFIG wind turbine for power optimization and grid synchronization. Int. J. Electr. Power Energy Syst. **105**, 679–689 (2019)
5. Bounar, N., Labdai, S., Boulkroune, A.: Pso-Gsa based fuzzy sliding mode controller for DFIG-based wind turbine. ISA Trans. **85**, 177–188 (2019)
6. Soued, S., Ramadan, H.S., Becherif, M.: Dynamic behavior analysis for optimally tuned on-grid DFIG systems. Energy Procedia **162**, 339–348 (2019)
7. Laghridat, H., Essadki, A., Annoukoubi, M., Nasser, T.: A novel adaptive active disturbance rejection control strategy to improve the stability and robustness for a wind turbine using a doubly fed induction generator. J. Electr. Comput. Eng. **2020** (2020)
8. El Qouarti, O., Essadki, A., Nasser, T.: High order sliding mode control of active and reactive powers for DFIG based Wind turbine. In : E3S Web Conf 351, 01008 (2022)
9. Kelkoul, B., Boumediene, A.: Stability analysis and study between classical sliding mode control (SMC) and super twisting algorithm (STA) for doubly fed induction generator (DFIG) under wind turbine. Energy **214**, 118871 (2021)
10. Amina, B., Mohammed, K., Merabet, B.H.: Super-twisting SMC for MPPT and grid-connected WECS based on SCIG. Int. J. Power Electron. Drive Syst. **12**(1), 520 (2021)
11. Mensou, S., Essadki, A., Nasser, T., Idrissi, B.B., Tarla, L.B.: Dspace DS1104 implementation of a robust nonlinear controller applied for DFIG driven by wind turbine. Renew. Energ. **147**, 1759–1771 (2020)

12. Boukhriss, M. A.: Commande robuste par ADRC de l'éolienne à base de la machine asynchrone à double alimentation (2016)
13. Farida, M.: Contrôle Des Puissances Active et Réactive Dans Les Aérogénérateurs Doubles Alimentés. Thesis. Université Batna 2 - Mostefa Ben Boulaïd Faculté de Technologie, Département de l'Electrotechnique, Algérie
14. Amimeur, H., Aouzellag, D., Abdessemed, R., Ghedamsi, K.: Sliding mode control of a dual-stator induction generator for wind energy conversion systems. Int. J. Electr. Power Energy Syst. **42**(1), 60–70 (2012)
15. Mostapha Kalami Heris, Particle Swarm Optimization in MATLAB. Yarpiz (2015). URL: https://yarpiz.com/50/ypea102-particle-swarm-optimization
16. Miry, Abbas. (2020). Re: How can I connect a PSO or other Optimization Algorithms code to a PI controller in Simulink?. Retrieved from: https://researchgate.net/post/How_can_I_connect_a_PSO_or_other_Optimization_Algorithms_code_to_a_PI_controller_in_Simulink/5f70933164ce1b496c567907/citation/download
17. Mirjalili, S.: Grey Wolf Optimizer (GWO). https://mathworks.com/matlabcentral/fileexchange/44974-grey-wolf-optimizer-gwo, MATLAB Central File Exchange. Retrieved January 30, 2022

Workforce Assignment Problem Considering Versatility in a Collaborative Robot System

Taji Hajar[1(✉)], Ayad Ghassane[1], and Zaki Abdelhamid[2]

[1] Mechanical, Material and Thermal Laboratory, National School of Mines In Rabat (ENIM), Rabat, Morocco
hajar.tj@gmail.com, ghassane.ayad@enim.ac.ma

[2] Artificial Intelligence Complex Systems Engineering Laboratory (LAICSE), National School of Arts and Crafts (ENSAM), Hassan II University, Casablanca, Morocco

Abstract. With the upcoming concept of collaborative robots, definitions of competencies, workload performance, and efficiency become obsolete. In a changing context where the products are highly personalized, the diversity of activities intensifies the difficulties of a proper workforce assignment. This complexity becomes deeper when facing the variability and the dynamic evolution of each operator's competencies. Providing the proper versal training, the dynamic evolution of competencies can be related to human factors such as motivation, preference, and fatigue. This dynamic is also highly correlated to history assignment: repeating the same activity or activities highly similar for a multi-period can enhance the operator's workforce performance. This study aims to redefine the workload performance in a collaborative robot's environment, once this is done, we will be able to propose a proper workforce assignment to a set of activities to optimize total production time giving the operators real performance. The scope of this paper is to review the current state of the art regarding dynamic performance's definition to propose a mathematical model of this problem revisiting the particularities of a collaborative robot's structure.

Keywords: Workforce performance · Collaborative robot · Workforce assignment · Competencies · Versatility

1 Introduction

After the extensive progress of industrial evolution, Industrial organizations are facing a shift where clients become more aware of their precise needs, and concurrence reaches its highest level. For them to prevail, it became vital to produce personalized products of various sizes while ensuring high efficiency at a low cost. With the emergence of the collaborative robotics concept, the definition of performance evolved from the standards of the time, cost, and quality components. While the quality time and cost can be highly predicted when it comes to dealing with a robotic structure, the competencies requirement shift from standard job's sequence mastery into a set of new competencies such as flexibility, problem-solving, analytical skills, and more importantly the ability

L. Azrar et al. (Eds.): CIP 2022, LNME, pp. 551–564, 2023.
https://doi.org/10.1007/978-3-031-23615-0_56

to use digital technologies at its advantage. In a collaborative robotic system, operators are meant to be able to interpret the needed transformations and determine the proper programming to meet those particular needs, after validating the program, they have to proceed to a configuration of the robotic system and insert the proper setting, choose the right tools and monitor the system. When facing a scheduling problem, the chosen operator must exhibit the proper behavior according to a specifically changing environment (Fruggiero et al., 2020). While the operators considered for the study present a set of competencies, behavior, and knowledge that will ensure the rightful execution of the activities demanded, they remain human and can be subjected to several factors changing their performance level.

In this paper, I will start by presenting the properties of the collaborative robotic system. I will review the current state of the t and emphasize the competencies required for the use of the collaborative robot. This first step will lead to a proposition of a mathematical model of performance while taking into consideration the dynamic aspect of the latter. We will review several parameters that describe the evolution of the operator's performance and complete the mathematical model with dynamic scope. This paper will provide an illustrative example of performance calculation and its application scheduling solution to a case study.

2 Context of the Study

Since 2011, Industry 4.0 has become the main talk of academic researchers and industrial entities (Ciffolilli and Muscio, 2018, p. 4). The conceptual framework of industry 4.0 comprises two main aspects: on one hand, it needs a set of technologies that makes automation, communication, data storing, and analysis optimal. On the other hand, the manufacturing system evolved from its standard configuration into a SMART one, Smart manufacturing is based on a smart supply chain, smart working, and smart products, and is fuelled by smart operators working hand in hand with the latest advanced technologies (Frank et al., 2019).

2.1 Smart Manufacturing

Smart manufacturing is based on three important principles: collaborative environment, responsive structures, and real-time data exchange. These ingredients help adapt to constant demand changes in a shifting factory environment (Frank et al., 2019). It is based on context awareness, modularity, data-driven decision-making, and an easy self-organized internal structure (Lu et al., 2020). It shifts automated operational processes (robots) into collaborative robots designed to work with qualified humans to enhance the productivity and flexibility of personalized products in low quantities at a low cost (Frank et al., 2019).

Inside a smart factory, smart working is designed to facilitate several tasks by getting proper access to all types of information to ensure a better, safer, and easy, er work environment (Frank et al., 2019). Providing workers with technological tools will improve the visibility ty of the process, this global vision, added to the expertise and multi-skills of the workforce will orient a better decision making (Frank et al., 2019).

Smart factories are founded on interoperability principles (Fruggiero et al., 2020): machine tools, operators, robots, customers, and supply chain associates are working in collaboration and exhibit flexible behavior in a changing environment. They can be described as a flexible ecosystem of various resource types collaborator to meet the customers' needs in the bin best way possible (Mourtzis and Vlachou, 2018).

What's remarkable about this type of structure is that different entities with different capacities communicate, interpret and react to each other without any difficulties to come up with the best deepest-decentralized (Hofmann and Rüsch, 2017).

2.2 Collaborative Robot

Collaborative robots (or Cobots) can be defined as a structure where a human works hand in hand with a robot in a collaborative way to perform common tasks on various products Bendel (2018). Collaborative robots are also defined by El Zataari et al. (2019) as any structure composed of a human and a robot in interaction without any physical barriers.

Collaborative robots pillar is the versatility and adaptability aspects of the operators giving a large scope of possible courses of action, therefore a large variety of personalized products at a different batch size.

There are four collaboratives scenarios. The first one is the independent where humans and Cobots work on separate workpieces, independently, sharing a workspace without cages or fences. Simultaneous collaborative scenarios mean that two simultaneous different operations are taking place on the same workpiece: human operators operate on a manufacturing process different from the one the Cobots are operating on. Sequential collaborative scenarios suppose that the human operator executes one or several operations before the Cobots continue the manufacturing process, this sequence can happen several times. Although this scenario presents time dependencies challenges, it has the advantage of avoiding the human operators performing monotonous or highly complicated processes. The last collaborative scenario is the supportive one, where human operators and Cobots work in complete interaction on the same process on the same workpiece. This scenario promotes full dependencies between the human and the robot, El Zataari et al. (2019).

Knudsen et al. (2020) stress the impact of the social and technological dimensions, while the first one supposes that safety issues (Duffy, 2016), ergonomic concerns, and human factors must be integrated into the thinking of the collaborative robot (Villani et al., 2018). The technological one aims to produce more advanced understanding capabilities and AI-aided anticipation skills (Knudsen and KaiVo-Oja, 2020). Traditional robot programming becomes the role of operators of the shop floor rather than engineers. The enhancement of human-machine interaction comes from a full integration of human behavior leading to the capacity to differentiate between profitable and non-profitable operators' actions (Maurtua et al., 2017). On one hand Cobots might also have the awareness level that helps to predict their future actions (Ivaldi, 2018), on the other hand, trust between operators and robots is the key to performing several tasks in collaboration with Cobots without being separated by physical barriers (Maurtua et al., 2017).

2.3 Collaborative Robot in the Industry 4.0 Evolution

Industry 4.0 is based on using high-level technologies to ensure efficiency and cost reduction and provide a large catalog of personalized products at different batch sizes to satisfy the demand. Competitive manufacturing companies' main objective becomes increasing their flexibility to be able to reduce costs with high levels of quality and personalization. Robotic integration is the first step to realizing this aim, in some cases, this solution might lack the flexibility and versatility required to adapt to dynamic demand (Knudsen and KaiVo-Oja, 2020). The first comparison between traditional robots and collaborative robots in terms of productivity, profitability, and interaction with workers reveals that Cobots are expected to develop flexible structures to accommodate the ongoing changes in task specifics or order, whereas traditional robots have a standard implementation and are highly profitable due to the lack of changes and innovation in tasks. The profit in this circumstance is strongly linked to the repetitiveness of the activity as well as the medium or large size of the batch in a standard robotic structure. In contrast, the profit in a collaborative robot structure must be substantial with just one batch lot. Another distinction is the degree of human interaction, which is minimal in standard robotic structure and can be seen primarily during the programming process. While human involvement with collaborative robots is becoming more widespread and frequent, there is a greater concern for the operator's safety (El Zaatari et al. 2019).

3 Related Work

3.1 Theoretical Work

In a standard job shop manufacturing system, production is a series of successive activities that need to respect time, quality, and precise material consumption. On the organizational level, multi-competency operators are the core of an efficient production line.

The core of every scheduling resolution should focus on the definition of an operator's competencies. Every operator can master, to some degree, more than one activity. The workforce competencies can be characterized relying on three dimensions:

- Work performance
- Execution quality
- Consumption ratio

We are then interested in reducing the cost of execution of each activity, which means optimizing the use of the resources and ensuring the required quality services in a short period (Zaki et al., 2005).

This approach can not be adapted to a collaborative robot's context, as manual activities are replaced by automated ones (Rødseth et al., 2019). The shift-like task can be realized by making robots emulate the operator's actions to execute tasks that operators used to work on using their motor functions. To that end, two directions exist to program a Cobots: online and offline programming (Folscher and Kruger, 2016), each one presenting advantages and limitations regarding time programming or operational time. El Zaatari et al., (2019) proposed learning by demonstrations is a way to transfer manual activities into automatic actions. This solution has proved to consistently reduce the programming time (Folscher and Kruger, 2016).

In a Cobots structure, operators are highly solicited for their decision-making, the main added value is the input of the program and monitoring of the system's evolution using Cyber-Physical Technologies (CPS). As the operator's main function changes, the definition of performance and the scheduling proposed by (Zaki et al., 2021) must be revisited to adapt more to the technological evolution that comes with the collaborative robots.

3.2 New Context of Competencies Required in Collaborative Robot

A systematic review of studies involving Industry 4.0 stresses a noticeable gap regarding Human Resources Management. Many issues regarding human factors have been raised, the lack of skilled human operators (Liboni et al., 2019), and efficient training programs (Schallock et al., 2018) are one of the many areas of research. Many studies have been conducted to deal with the identification of job roles, proposing the requirements for education and training qualifications to embthe ody industry 4.0 concept. Knowledge management issues have been raised regarding tacit information that must be identified and passed to other team members to raise the versatility and flexibility of operators (Salvadorinho and Teixeira, 2021). To summarize, operators 4.0 must learn a wide range of abilities that go beyond the realm of normal operational technical expertise. Media skills, IT and security abilities, and coding skills can all be connected to technical competence. These skills must be complemented by methodological competencies such as decision-making and research skills, problem-solving, entrepreneurial thinking, conflict resolution, social competencies such as language, communication, and networking skills, leadership skills and problem-solving, and personal competencies such as flexibility, creativity, research orientation, and the ability to work under pressure (Hernandez-de-Menendez et al., 2020).

It becomes clear that Operators 4.0 represents few features and personal characteristics and skills far beyond the common ones that have been needed before the fourth evolution.

With the wild range of solutions CPS has brought, operators 4.0 are equipped with the necessary digital support to improve decision-making (Rødseth et al., 2019). The use of a proper interface can give enough real-time data (Mourtzis and Vlachou, 2018) cultivating the dynamic use of resources and giving a realistic input to take an informed decision.

Few frameworks have been proposed to get operators to the required competencies, knowledge, and habits. One way to do that is to start with defining the factory 4.0 relevant stakeholders, this step will help get enough perspective to concede the proper technology needed to help operators 4.0 increase process efficiency. Lastly, comes the learning phase, where operators should be enrolled in the proper training program according to their job requirements, this step should take into account the needs of the factory as well as the personal cognitive and personal characteristics of the operator (Rødseth et al., 2019).

The limitations of this approach can be linked to a few aspects: the large scope of technological evolution offers many possible solutions to meet the needs of the organization's stakeholders, this means the choice should be highly linked to the training offered as well as the pertinence of the solution since operators are meant to develop versatility and flexibility, otherwise, it becomes inevitable for a smart factory to develop their proper personalized training while avoiding any eruption in production activities. Since humans represent different cognitive and personal capabilities, the training must be adaptive while being highly personalized. Nowadays, training goes beyond the traditional scope of learning to propose a large variety of learning framework: a reward augmentation and a repair explanation framework has been introduced to operator's training via reinforcement learning (Tabrez et al., 2019).

Another framework of training came to challenge the reinforcement learning approach: cross-training presented the advantage of increasing human performance while enhancing their trust in the robots (Nikolaidis et al., 2015).

3.3 The Dynamic Evolution of the Individual Performance

In a more dynamic approach, the necessity of taking into consideration the learning curve as well as the forgetting curve refutes the static definition of performance. It goes under the assumption that previous experiences are key to defining the evolution of the workforce performance, therefore the workforce competencies level must always be re-evaluated.

When faced with the challenge of scheduling diverse activities, this understanding of the workforce performance is key to optimizing job shop productivity.

3.3.1 Learning Curve

The pillar of the learning curve theory must be Wright's learning curve is of the form:

$$Y_x = Y1._{x-b} \tag{1}$$

where Y_x is the cumulative average time to produce the xth unit and Y1 is the time to produce the first unit, and b is the learning exponent, which is computed with LR being the learning rate (Glock et al., 2019).

This formulation presents the evolution of the operator's production time depending on the cumulative production volume. It proves the instinctive observation that after several repetitions, operators can significantly reduce the standard time of execution depending on their learning rate (Wright 1936).

The proposed equation has two limitations: first, the learning curve fails the determine when and how should be reshaped, second in a flexible smart manufacturing system, the operation can be similar to some extent, so the repetitiveness is highly correlated to the degree the activities presented in the training stage are similar to the one the operators is executing daily.

Many other models have emerged, Jong (1957) a proposed model including the interaction with the machine have been proposed by Jong (1957).

$$Y_x = Y1.\left(M + (1 - M)x^{-b}\right) \tag{2}$$

where M has binary values assuming the presence or not of the automated part of the action or not (Glock et al., 2019).

Although the latter expression introduces the presence or the absence of a machine, it doesn't relate to the collaborative robot's configuration. As said before, the aforementioned configuration supposes that the human works hand in hand with the Cobots, meaning that almost every action involves human and machine intervention to orient the decision-making. When using the latest technologies, the robots can offer a large scope of immediate assistance that can reshape the learning curve of the operators' using cues, visual aids, or proposing analogic similarities to help with the programming phase.

3.3.2 Forgetting Curve

On the other hand, the phenomenon of forgetting is a human condition that can highly affect the performance of operators. It starts with the assumption that the interruption of any task performed for some time will disorient the operators, therefore more time for remembering can be needed to continue the action that was taking place. There is also another cause justifying the forgetting curve: the turnover and workload flexibility can highly affect not only the time of the execution of a certain action, but it can also negatively affect the performance in which the operators might execute the demanded task (Glock et al., 2019).

The equation of the forgetting curve has been developed by Globerson et al. 1989:

$$\breve{Y}_{\mathrm{x}} = \breve{Y}_1 \mathrm{x}^{\mathrm{f}} \tag{3}$$

where $\breve{Y}_{\mathrm{x}}$ is the time to produce correspondent to the Xth repetition and f presents the forgetting constraint also the opposite of the learning rate (Glock et al., 2019). The question regarding the learning curve and the forgetting curve is when the first stops and when the last begins.

3.3.3 The Implications of the Fatigue

Although repetitive execution of the same task could lead to an increase in the performance of the operators, excessive repetition can lead to a decrease in the concentration and their cognitive capacities to properly conduct a task previously mastered. Fatigue also is a synonym for impaired judgment and decisions can cause many accidents because of a lack of concentration and might increase the response time of the operators (Asadayoobi et al., 2021).

Fatigue can be highly correlated to the complexity of the actions to be executed in a condensed period. The performance of the operators might decrease as he moves through the day's work from one activity to another.

Mályusz, Pém and Digiesi et al. criticized the WLC learning curve and induced fatigue parameters to the equation developed a learning curve with the fatigue parameter, with the hypothesis that after a certain number of repetitions the learning rate decreases. This phenomenon has also treated by Asadayoobi et al. 2021. Although the solutions proposed are highly praised, this learning curve has been related only to a particular repetitive task, without keeping in mind the fatigue implication throughout the day to any activity at all. In this approach, we don't consider the similarities between tasks that can also generate a concentration decrease and fatigue symptoms.

4 Proposition

In this article, we aim to solve a scheduling issue in a collaborative robot's environment. Based on the assumption that the operators work hand in hand with the machine that facilitates the learning while executing the activity. This work applies to operators that have been given proper and eclectic training. The versatility of the workforce and the robot's assistance will raise the probability that the operators will be able to carry on a large number of actions while still learning. This study is based on the hypothesis that the learning curve is highly affected by the robots, as well as the similarities between the activities. The similarity assumes that the learning continues to occur when giving relatively similar tasks in a multi-period. It supposes also that the fatigue dimension can occur during the execution of relatively similar tasks. The dynamic aspect of performance can also be related to the operator's motivation level throughout the day. This proposition aims to evaluate the outcome of the flexibility of the operators in terms of performance and to propose the proper assignment to a scheduling problem. The dynamic evolution of their competencies is the key indicator to propose who does what.

4.1 Illustration of Human-Machine Interaction for One Activity

To have a better understanding of the collaborative robotic structure, Fig. 1 presents a framework that specifies the process of collaborating with the machine in a job shop configuration. In this framework, we suppose that the individual has enough training and understanding of any type of operations that can be conducted in this robotic system, we also suppose that operators start with a preliminary program and an evaluation of execution time for a single sample. Operators can proceed to evaluate different possibilities that can optimize execution time. Once the program is loaded, operators proceed to the verification of the program, different scenarios can be used for the verification: simulation, demonstration, or launching the program to verify a prototype. Once the program is validated, operators proceed to the final configuration and setting to include data such as batch sizes, details of the verification, or other details regarding packaging or transfer. During the execution, operators have to monitor the execution and validate the phases of the program. Once the execution is done, operators should proceed to final verification and the transfer. The total time of execution should be calculated by adding the duration of all the actions that have been conducted from order reception to order finalization.

The execution time depends on the degree of implication of the machine and the human operators, in other words, it is subject to the collaboration scenarios existent as presented in Sect. 2.2. The other times are mostly related to the human interference with a light help from the robots. The optimization of the total time execution is correlated to the optimization of the time of preliminary programming and demonstration and optimization, although the operators are fully engaged in monitoring the proper execution of the program, if the first two steps are not adequately realized, operators must review the program and generating more cost since the total time of execution has increased.

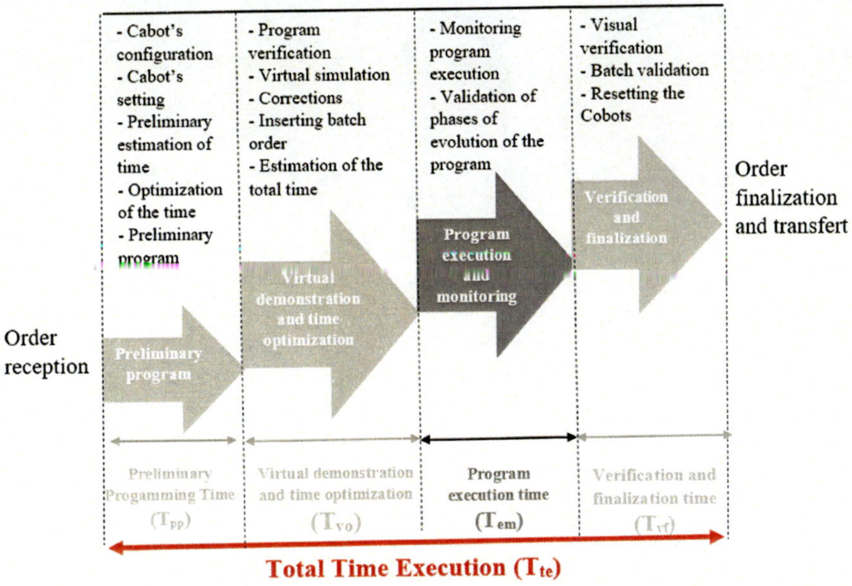

Fig. 1. Framework of execution times of one activity in a collaborative robot's context

4.2 Illustration of Human-Machine Interaction for Several Activities

In a scenario where the collaborative robot has to execute more than one activity, the constant change in the program and cobot's configuration can intensify the waste of time and resources. As presented in Fig. 2, to move from activity 1 to activity 2, several steps must be validated, the finalization step comprises resetting the cobot, and operators must move from program 1 for activity 1 to program 2 for activity 2, going through the same process again. Navigating from one program to another is the essence of the operator's competencies. If this intermediate phase is not properly conducted, it can be time-consuming and increase the total time of production. The illustration presents only two activities, when it comes to a large number of activities, the way this intermediate phase has been prepared and conducted is key to enhancing collaborative robots' performances.

The first step to reducing the total time of production is to act on the intermediate phases, if we give operators the chance to propose a schedule, the proposition might be highly subjective. This is the reason why we propose a scheduled gathering of semi-similar activities in one block and then moving into another one. This reflection can reduce significantly the effort of coming up with or recalling a new program in every intermediate phase.

The key is to look for similarities between activities that need to be realized in a period to reduce the changes performed from one activity to another, this phase can be highly affected by the operator's style and preferences, that's why the training should also focus on proposing a standard understanding of similarities and the duration of each setting since the operator should choose the shorter one.

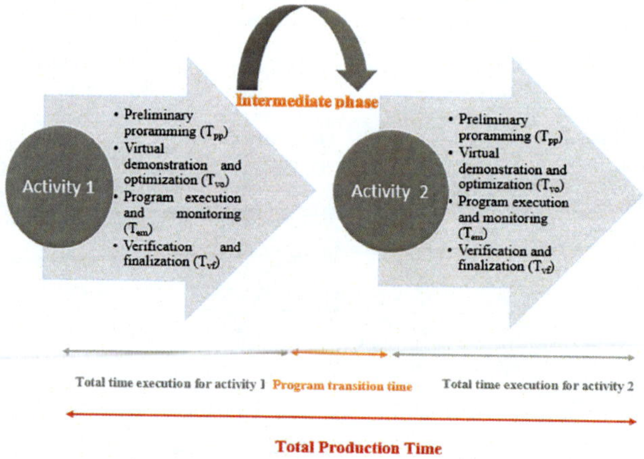

Fig. 2. Illustration of time consumption of two activities in a collaborative robot's context

4.3 Framework of the Solution

To give a better understanding of the proposed solution to the assignment problem in a collaborative robot's context, we develop the framework presented in Fig. 3.

Our vision starts from a preliminary phase that involves training operators to perform a large and diverse set of activities, the training must take into account the necessity for the operator to not only master activities but also maximize his performance level. The ideal scenario supposes an equitable assignment of operators to all the activities. While this scenario seems difficult to put in place, we must aim to attain a satisfying degree of versatility that gives us a sufficient number of operators that can execute the same activity with high performance. We must stress that this phase is highly important, as it may give us an insight into the personal characteristics, learning rate, preferences, fatigue curve, and the evolution of the level of motivation for each operator. At the end of this phase, we can also propose a standard time for each step of every activity, this baseline will be very useful to evaluate an operator's performance as well as the evolution of its level when repetition occurs. Another amazing output could be linked to the desired performance required for each activity; this is a simple result of the definition of the previous standards time for an activity to be executed, this definition can modify the modalities of job continuous training to make it more efficient.

The second phase of this framework will start once the training program matures. When we come across a set of different activities to perform in a period, the dynamic approach of performance evolution, suppose that each operator's competencies level has been changed from the moment he started the training until now. Many scenarios can happen, performance level can increase after repetition as it can decrease if the assignment didn't prepare the operator for that particular set of activities, it can also reach its the maximum personal performance level and stay faithful to it depending on the personal learning curve according to the personal learning curve.

In this paper, we rely on the assumption that given the proper training, operators develop a set of skills that can execute a large scope of activities. Once the operators have the proper tools the learning curve continues to develop when giving a relatively similar task. The repetitiveness doesn't assume the same course of action, which means that the operators should be lucid with the differences between activities as well as the program similarities to ensure a high level of performance, this level of awareness can decrease due to a fatigue factor. There is also a motivation factor that might be correlated to performance, the motivation in early stages can be at its highest level since the discovery of new concepts is exciting, but as time goes by, this factor can decrease the productivity of operators. To avoid an objective schedule proposal, we think it's preferable to come up with optimal scheduling while keeping in mind the similarities between activities to reduce total production time. The proposed schedule must take into consideration several constraints such as customer needs, maintenance problems, policy, and internal strategy; but most importantly it must reduce the intermediate phase between two activities and have an estimation of the required performance level of each activity as well as an estimation of a total time of production.

The last phase comes in combining the data at our disposal. With the real performance level of each operator and with the required level of performance of each activity we can come up with a proper assignment of operators to the optimal schedule previously proposed. Lastly, once the production is finished, we must update the operator's performance.

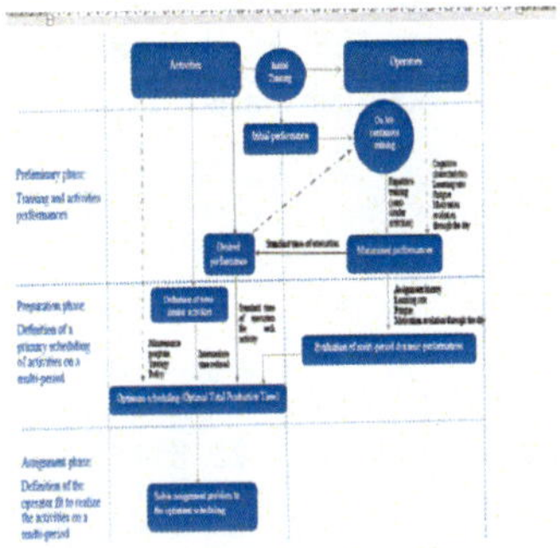

Fig. 3. Framework of workforce assignment solution in a collaborative robot's context

5 Discussion

In this paper, we consider the particularity of the collaborative robot's context. As we previously stated, this new environment requires a particular set of skills and flexible learning programs. While the operator's performances continue to evolve during on job continuous training. The versatility of activities challenges operators to adapt to several conditions using the training program. The abundance of activities presents thc difficulties of pinpointing certain similarities to reduce the configurations and the setting time of when we move from one activity to another. This phase can present a large number of subjective propositions. The particularity of this solution is to standardize scheduling to decrease Total Production Time and to train operators to navigate from one activity to another in a short period. The solution proposed, combines the activity's particularities as well as the dynamic workload performance level. The level of performance required is defined after maximizing all individual performance levels. It also takes into consideration several human factors such as motivation, fatigue, learning rate, and assignment history to make an informed decision about operators' assignments.

6 Conclusion

The proposed study aims to solve an assignment workload problem in the context of the collaborative robot. The dynamic approach comes from the evolution of the time needed to launch a program while being highly affected by the learning curve occurring during the execution and based on the relative similarities between the training program and a set of activities in a period. The dynamic evaluation of the performance is inspired by the learning curve, fatigue, motivation, and assignment history. This dynamic helps propose the proper assignment that ensures not only time optimization, but proper use of resources. The main limitations related to our proposition are linked to the development of tacit knowledge and competencies while dealing with a scheduling issue. The future contribution aims to raise the issue of the standard optimal scheduling to include in the training program. Another limitation can be related to the degree to which the human operator is implicated with the Cobots in workpiece transformation, simultaneous, sequential, and supportive collaborative scenarios present a various level of dependency that toughens the dissociation between human performance and machine performance.

References

Ciffolilli, A., Muscio, A.: Industry 4.0: national and regional comparative advantages in key enabling technologies. Eur. Plan. Stud. **26**, 2323–2343 (2018). https://doi.org/10.1080/09654313.2018.1529145

Zaatari, E. et al.:Cobot programming for collaborative industrial tas.pdf, n.d (2019)

El Zaatari, S., Marei, M., Li, W., Usman, Z.: Cobot programming for collaborative industrial tasks: an overview. Robot. Auton. Syst. **116**, 162–180 (2019). https://doi.org/10.1016/j.robot.2019.03.003

Folscher, D.J., Kruger, K.: Saving time on robot programming: programming by demonstration using stereoscopic motion capturing. In: 2016 Pattern Recognition Association of South Africa and Robotics and Mechatronics International Conference (PRASA-RobMech), pp. 1–6. Presented at the 2016 PRASA-RobMech International Conference, IEEE, Stellenbosch, South Africa (2016). https://doi.org/10.1109/RoboMech.2016.7813133

Frank, A.G., Dalenogare, L.S., Ayala, N.F.: Industry 4.0 technologies: Implementation patterns in manufacturing companies. Int. J. Prod. Econ. **210**, 15–26 (2019). https://doi.org/10.1016/j.ijpe.2019.01.004

Fruggiero, F., Lambiase, A., Panagou, S., Sabattini, L.: Cognitive human modeling in collaborative robotics. Proc. Manufact. **51**, 584–591 (2020). https://doi.org/10.1016/j.promfg.2020.10.082

Glock, C.H., Grosse, E.H., Jaber, M.Y., Smunt, T.L.: Applications of learning curves in production and operations management: a systematic literature review. Comput. Ind. Eng. **131**, 422–441 (2019). https://doi.org/10.1016/j.cie.2018.10.030

Hernandez-de-Menendez, M., Morales-Menendez, R., Escobar, C.A., McGovern, M.: Competencies for Industry 4.0. Int. J. Interact. Design Manuf. (IJIDeM) **14**(4), 1511–1524 (2020). https://doi.org/10.1007/s12008-020-00716-2

Hofmann, E., Rüsch, M.: Industry 4.0 and the current status as well as future prospects on logistics. Comput. Ind. **89**, 23–34 (2017). https://doi.org/10.1016/j.compind.2017.04.002

Knudsen, M., KaiVo-Oja, J.: Collaborative robots: frontiers of current literature. J. Intell. Syst. Theory Appl. 13–20 (2020). https://doi.org/10.38016/jista.682479

Liboni, L.B., Cezarino, L.O., Jabbour, C.J.C., Oliveira, B.G., Stefanelli, N.O.: Smart industry and the pathways to HRM 4.0: implications for SCM. Supp. Chain Manage. **24**, 124–146 (2019). https://doi.org/10.1108/SCM-03-2018-0150

Lu, Y., Xu, X., Wang, L.: Smart manufacturing process and system automation—a critical review of the standards and envisioned scenarios. J. Manuf. Syst. **56**, 312–325 (2020). https://doi.org/10.1016/j.jmsy.2020.06.010

Mourtzis, D., Vlachou, E.: A cloud-based cyber-physical system for adaptive shop-floor scheduling and condition-based maintenance. J. Manuf. Syst. **47**, 179–198 (2018). https://doi.org/10.1016/j.jmsy.2018.05.008

Rødseth, H., Eleftheriadis, R., Lodgaard, E., Fordal, J.M.: Operator 4.0 – Emerging Job Categories in Manufacturing. In: Wang, K., Wang, Y., Strandhagen, J.O., Yu, T. (eds.) IWAMA 2018. LNEE, vol. 484, pp. 114–121. Springer, Singapore (2019). https://doi.org/10.1007/978-981-13-2375-1_16

Salvadorinho, J., Teixeira, L.: Organizational knowledge in the I4.0 using BPMN: a case study. Proc. Comput. Sci. **181**, 981–988 (2021). https://doi.org/10.1016/j.procs.2021.01.266

Schallock, B., Rybski, C., Jochem, R., Kohl, H.: Learning factory for industry 4.0 to provide future skills beyond technical training. Proc. Manuf. **23**, 27–32 (2018). https://doi.org/10.1016/j.promfg.2018.03.156

Wright.: Factors Affecting the Cost of Airplanes.pdf, n.d (1936)

Zaki, A., Benbrahim, M., Benchekroun, B.: Proposition of a Model for Multi-Period Workforce Assignment Problem Considering Versatility, vol. 17 (2005)

Zaki, A., Benbrahim, M., Benchekroun, B., Ayad, G.: n.d. Using AHP and TOPSIS techniques for assessment of multi-skilled workforce in manufacturing industry 27

El Zataari, S., Marei, M., Li, W., Usman, Z.: Cobot programming for collaborative industrial tasks: an overview. Robot. Auton. Syst. **116**, 161–180 (2019)

Bendel, O.: Co-robots from an Ethical Perspective. In: Dornberger, R. (ed.) Business Information Systems and Technology 4.0. SSDC, vol. 141, pp. 275–288. Springer, Cham (2018). https://doi.org/10.1007/978-3-319-74322-6_18

Villani, V., Pini, F., Leali, F., Secchi, C.: Survey on human-robot collaboration in industrial settings: safety, intuitive interfaces, and applications. Mechatronics **55**, 248–266 (2018)

Duffy, V. G.: Modern human-robot interaction in smart services and value co-creation. In: Duffy, V. G. (ed.), Digital Human Modeling: Applications in Health, Safety, Ergonomics and Risk Management. Springer International Publishing (2016)

Schou, C., Andersen, R. S., Chrysostomou, D., Bøgh, S., Madsen, O.: Skill-based instruction of collaborative robots in industrial settings. Rob. Comput. Integr. Manuf. 72–80 (2018)

Maurtua, I., Fernandez, I., Tellaeche, A., Kildal, J., Susperregi, L., Ibarguren, A., Sierra, B.: Natural multimodal communication for human-robot collaboration. Int. J. Adv. Rob. Syst. 1–12 (2017)

Ivaldi, S.: Intelligent human-robot collaboration with prediction and anticipation. ERCIM News (2018)

Mályusz, L., Pém, A.: Model for—bath Tub|| effect in construction. Creat. Constr. Conf. (2014). http://2015.creative-construction-conference.com/wp-content/uploads/2015/01/CCC2014_L_Malyusz.pdf

Digest, S., Mossa, G., Mummolo, G.: Learning and tiredness phenomena in manual operation performed in lean automated manufacturing systems: a reference model. In: International IMS Forum, May, 2004, pp. 17–19. Erba, Italy (2004)

Wright, T.P.: Factors affecting the cost of engineering. J. Aeronaut. Sci. **3**, 122–128 (1936). https://doi.org/10.2514/8.155

Jong, J.R.: The effects of increasing skill on cycle time and its consequences for time standards. Ergonomics **1**(1), 51–60 (1957)

Globerson, S.: Introducing the repetition pattern of a task into its learning curve. Int. J. (1980)

Integrated Energy Production and Management

On the Parameters Identification of a Wind Turbine Emulation Problem

Amar Amour(✉), Yassine Ouakki, Mahmoud Ouhdan, and Abdelaziz Arbaoui

INSCM Team, LM2I, National School of Arts and Crafts (ENSAM), Moulay Ismail University, BP 4024, Meknes 50000, Morocco
amar.amour.2015@gmail.com

Abstract. Using emulators is an intelligent way to validate a control solution and to understand the energy conversion process and interactions between system components of a wind turbine. These emulators use generic models in the form of algebraic equations to simulate a wind turbine's aerodynamic performance. In order to be used effectively, there are important rotor parameters that need to be accurately extracted. This paper reports on some encountered issues related to the emulation of a stall constant speed control solution. Experimental data of the NREL Ames wind turbine are used in the parameters identification process of the well-cited parametric model [Eqs. (2) and (3)]. This latter is implemented in Scilab open source software using genetic algorithms. The results show that the proposed model, is not adequate for a stall-regulated rotor compared to pitch regulated model (see e.g. in Saint-Drenan et al. Renew Energy 157:754–768 [1]) and that the GA is not reliable for a such problem given its stochastic nature.

Keywords: Parameters identification · Genetic algorithm · Optimization · Wind turbine · Emulators

1 Introduction

During the design phase of wind turbines, the choice of control solution is of great practical interest, since it is used to size the wind turbine components which largely dictate the cost of the system and therefore the cost of wind energy [2]. The control solution choice is, therefore, an important key for the successful integration of wind turbines in existing and future energy systems [3].

Wind turbine emulators are used in industry as well as for educational purposes to validate a control solution and to understand the energy conversion process and interactions between system components. To emulate a wind turbine rotor, the generic model proposed by Saint-Drenan [1] is used to estimate the power coefficient of the turbine C_p (λ,β). This model is implanted in MATLAB toolboxes and is widely used by the researcher [2,4–7] and [8].

For control solution design and validation, there are important rotor parameters that need to be accurately extracted, before using this model. Since the

L. Azrar et al. (Eds.): CIP 2022, LNME, pp. 567–575, 2023.
https://doi.org/10.1007/978-3-031-23615-0_57

parameters identification problem is multi-variable and nonlinear, traditional extraction techniques are not reliable because many local optima may exist [3]. This paper presents a Genetic Algorithm (GA) based approach for estimating rotor emulator parameters. In Sects. 2 and 3 discusses the wind turbine rotor modeling and optimization problem along with the proposed approach. Section 3 presents obtained and simulation results. The conclusion is given in Sect. 4.

2 Wind Turbine Rotor Emulator Problem

In this work we focus on the stall-constant speed control solution. Experimental data of the NREL Ames wind turbine are used in parameters identification process. This wind turbine is a constant speed (72 rpm), two-bladed, upwind and stall-regulated rotor. His rotor blade has a linear chord taper with a nonlinear twist distribution, and operates with a 3-degree tip pitch. The radius from the center of rotation, which includes both blade and hub, is 5.03 m. The S809 airfoil is used from blade root to tip [9]. Before proceeding with the parameter's identification, it is essential to have a mathematical model that accurately represents the aerodynamics performance of the given wind turbine and then adjust the predicted power function to fit the experimentation points. For that reason, the output power of the turbine is given by the following equation [2] (Fig. 1):

Fig. 1. Wind turbine emulator under construction

$$P_{mp} = \frac{1}{2}\rho A C_p(\lambda, \beta)(V cos(\delta))^3 \quad (1)$$

where:

- P_{mp} : Predicted mechanical power of the turbine (W);
- C_p : Performance coefficient of the turbine;
- ρ : Air density (kg/m^3);
- A: Turbine swept area (m^2);
- V: Wind speed (m/s);
- δ: furling angle (rad).

A generic equation is used to model $C_p(\lambda_i, \beta)$. This equation, based on the work in Reference [1], as represented by the following equation:

$$C_p(\lambda, \delta) = C_1(\frac{C_2}{\lambda_i} - C_3\beta - C_4\beta\lambda_i - C_5\beta^{C_6} - C_7)\exp(\frac{-C_8}{\lambda_i}) + C_9\lambda \quad (2)$$

with:

$$\frac{1}{\lambda_i} = \frac{1}{\lambda + C_{10}\beta} - \frac{C_{11}}{\beta^3 + 1} \quad (3)$$

where:

- C_p: Performance coefficient of the turbine;
- λ: Tip speed ratio of the turbine($\lambda = \frac{\Omega R}{V}$);
- β: Blade pitch angle (deg);
- $C_i \ldots C_j$: Unknown parameters to be estimated.

Given the non-linearity of Eqs. (2) and (3) that involve 11 unknown parameters, the analytical solutions for C_p coefficient and therefore the predicted power P_m is quietly difficult. Instead, numerical methods, curve fitting methods, artificial intelligence (AI) techniques are often used to solve such problem [10]. In this paper GA algorithm technique is employed to estimate the given parameters by minimizing a pre-selected fitness function. The proposed fitness function to be minimized is defined by the following equation:

$$F(C_k) = \frac{1}{n}\sum_{i=1}^{i=n}\sqrt{(C_{pr}^i - C_{pe}^i)^2} \quad (4)$$

where:

- F: Fitness function to be minimized;
- C_{pr}: Predicted power coefficient of the turbine;
- C_{pe}: Experimental power coefficient of the turbine;
- n : Number of data points.

3 Solution Method Using GA

Least-square (LS), instrumental variable and maximum likelihood are some of the well-developed techniques commonly used to estimate such model's parameters. Whereas those techniques often fail in search for the global optimum, notably if the search space is nonlinear. Up to now, artificial intelligence (AI) techniques have become the most powerful tools that overcome the nonlinearity issues. The genetic algorithm GA) is one of those tools, which has been grossly used to accurately identify the parameters of some very complex models. For that reason, the GA has become the high ranked technique that widely applied for optimization problem with several local minima which standard algorithm fails ([11]).

The power of GA lies in its evaluation of many points at the same time in the parameters space and finds the best optimum rapidly without struggling in a local optimum. The search space doesn't need to be differentiable or continuous [12]. Many scientists have used this technique to best estimate non-linear systems parameters. The wind turbine rotor system is a sophisticated nonlinear system that traditional LS techniques can't give a best estimation of its parameters. The genetic algorithm was developed by John Holland, a colleague and student at the University of Michigan [12], is based on natural selection, the process that govern the biological evolution. The GA uses fitness functions (4) to be optimized while fixing some performance criterion. The powerful side of GA is that doesn't require a good initial estimate (Nolan 1994). The GA selects random individuals from the current population to be parents and produce the children form it to be used in the next generation. Through successive generations, the population "evolves" toward an optimal solution. GA is commonly used to solve many optimization problems that are not well suited to standard optimization algorithms, including problems in which the objective function is discontinuous, non-differentiable, stochastic, or highly nonlinear [11].

A simple genetic algorithm that produces good results in many practical problems is composed of three main types of rules:

- The selection rule selects individuals called parents who will contribute to the next generation of population;
- The crossover rule combines two parents to create the next generation of children;
- Mutation rules make each parent subject to random changes to form children.

The pseudo code of GA illustrates the procedure of the computation as follows (Fig. 2):

4 Results

In this paper we have focused on identifying some parameters necessary to emulate the behavior of a wind turbine rotor. Measured data of the stall-regulated NREL Ames wind turbine are considered in this work. The AG method was chosen because it performs better than standard algorithm especially when the number of experimental data is less than the number of unknown parameters.

The GA needs to specify, inter alia, the up and down bounds to start the algorithm. Those parameters were determined based on the literature data given in Table 1 [1]. Since the GA is a stochastic algorithm, it's necessary to test the

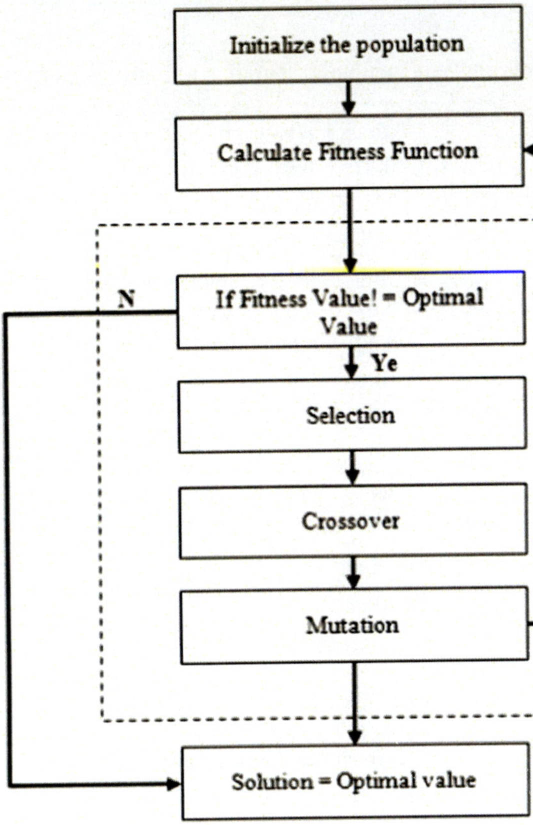

Fig. 2. GA procedure of the computation

model in a loop for large number of runs. In our case we test the model using 1000 runs. Table 2 shows the obtained best and worst estimate value.

Figure 3 shows that the GA algorithm gives good results in the pre-stall zone (wind speed less than 9 m/s). However, since the blades start to stall, the model can't well fit the experimentation, which leads to consider that the proposed model is not well suited for stall wind turbines.

Furthermore, we notice by analyzing the difference between the best and worst fit (Figs. 3 and 4) that the proposed algorithm is not reliable due to the stochastic nature of the GA.

Moreover, it should be noted that the error in predicted power for high wind speed is large due to fact that the power is proportional to the cube of wind speed as indicated by Eq. (1).

Table 1. Coefficients estimation found in the literature

	[4]	[2]	[5]	[6]	[7]	[8]	Max	Min
c1	0.73	0.5	0.5176	0.77	0.5	0.22	0.77	0.22
c2	151.00	116.00	116.00	151.00	116.00	120.00	151.00	116.00
c3	0.58	0.4	0.4	0.00	0.00	0.4	0.58	0.00
c4	0.00	0.00	0.00	0.00	0.4	0.00	0.4	0.00
c5	0.002	0.00	0.00	0.00	0.00	0.00	0.002	0.00
c6	2.14	0.00	0.00	0.00	0.00	0.00	2.14	0.00
c7	13.2	5.00	5.00	13.65	5.00	5.00	13.65	5.00
c8	18.4	21.00	21.00	18.4	21.00	12.5	21.00	12.5
c9	0.00	0.00	0.006795	0.00	0.00	0.00	0.006795	0.00
c10	−0.02	0.089	0.089	0.00	0.08	0.08	0.089	−0.02
c11	0.003	0.035	0.035	0.00	0.035	0.035	0.035	0.00

Table 2. Best and worst fit value obtained from the GA

	Worst values	Best values
Iterations	742	197
C1	0.26675	0.45375
C2	136.12500	130.17500
C3	0.12470	0.08410
C4	0.25000	0.16200
C5	0.00083	0.00117
C6	0.46010	1.23050
C7	5.04325	7.37875
C8	12.79750	15.77250
C9	0.00588	0.00554
C10	−0.00965	0.02524
C11	0.01698	0.03448
RMSE	0.03881	0.00388
Rsq	0.92778	0.99928
RsqAdj	0.85557	0.99856
NB_Gen	37	38
Time (s)	1.1550	1.1630

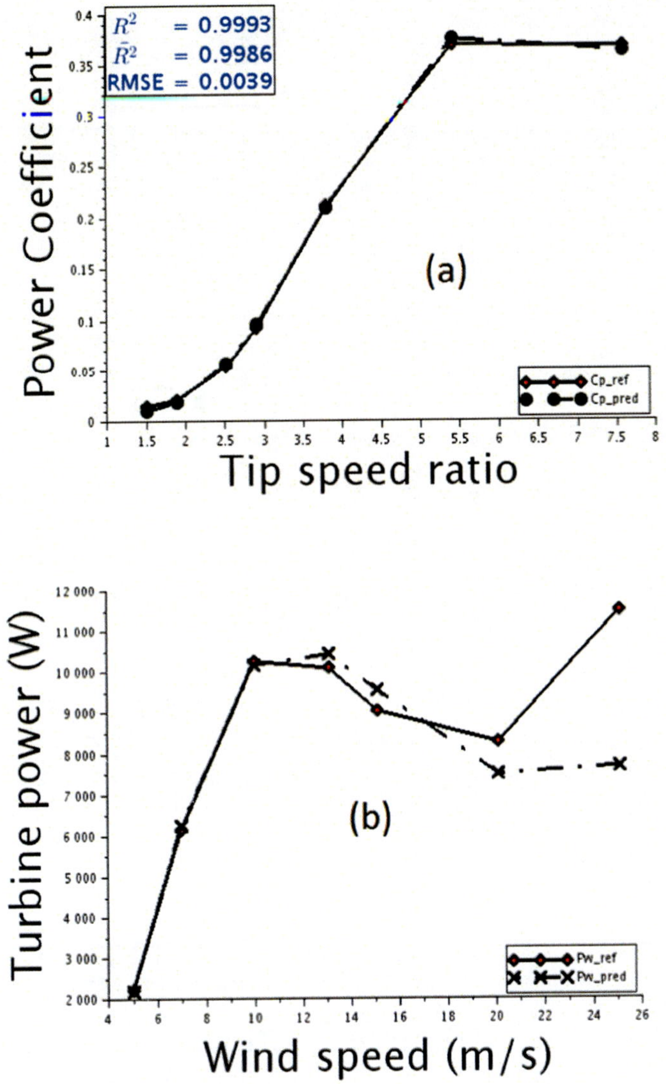

Fig. 3. C_p (a) and Power (b) curves for the best fit value

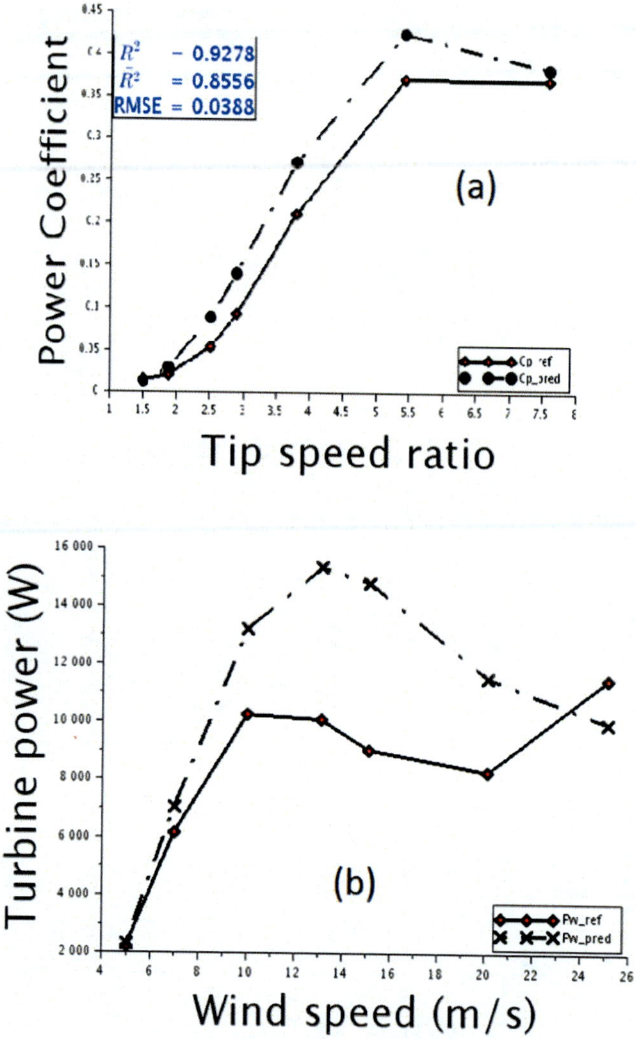

Fig. 4. C_p (a) and Power (b) curves for the worst fit value

5 Conclusion

This paper presents a GA algorithm for estimating the wind turbine power performances. The GA algorithm is used to identify the parameters of the well-cited parametric model [Eqs. (2) and (3)], to predict the power of the stall-regulated rotor. The GA algorithm showed its robustness and gives a good estimate. However, the results show that the proposed model [Eqs. (2) and (3)], is not suitable for a stall-regulated rotor compared to pitch regulated model (see e.g. [1]) and that the GA is not reliable for such a problem. In future work, a robust parametric model will be developed to predict the power of a stall constant speed control solution to be implemented in the next generation of stall-regulated control strategy; we will also develop a reliable algorithm to identify the unknown parameters of the future parametric model.

References

1. Saint-Drenan, Y.-M., Besseau, R., Jansen, M., Staffell, I., Troccoli, A., Dubus, L., Schmidt, J., Gruber, K., Simões, S.G., Heier, S.: A parametric model for wind turbine power curves incorporating environmental conditions. Renew. Energy **157**, 754–768 (2020)
2. Heier, S.: Grid Integration of Wind Energy: Onshore and Offshore Conversion Systems. John Wiley & Sons (2014)
3. Bianchi, F.D., De Battista, H., Mantz, R.J.: Wind Turbine Control Systems: Principles, Modelling and Gain Scheduling Design, vol. 19. Springer (2007)
4. Slootweg, J., Polinder, H., Kling, W.L.: Dynamic modelling of a wind turbine with doubly fed induction generator. In: 2001 Power Engineering Society Summer Meeting. Conference Proceedings (Cat. No. 01CH37262), vol. 1, pp. 644–649 (2001). IEEE
5. Thongam, J., Bouchard, P., Ezzaidi, H., Ouhrouche, M.: Wind speed sensorless maximum power point tracking control of variable speed wind energy conversion systems. In: 2009 IEEE International Electric Machines and Drives Conference, pp. 1832–1837 (2009). IEEE
6. De Kooning, J.D., Gevaert, L., Van de Vyver, J., Vandoorn, T.L., Vandevelde, L.: Online estimation of the power coefficient versus tip-speed ratio curve of wind turbines. In: IECON 2013-39th Annual Conference of the IEEE Industrial Electronics Society, pp. 1792–1797 (2013). IEEE
7. Ochieng, P., Manyonge, A., Oduor, A.O.: Mathematical analysis of tip speed ratio of a wind turbine and its effects on power coefficient (2014)
8. Dai, J., Liu, D., Wen, L., Long, X.: Research on power coefficient of wind turbines based on SCADA data. Renew. Energy **86**, 206–215 (2016)
9. Tangler, J., Kocurek, D.: Wind turbine post-stall airfoil performance characteristics guidelines for blade-element momentum methods. In: 43rd AIAA Aerospace Sciences Meeting and Exhibit, p. 591 (2005)
10. Yuan, Y.-X.: Recent advances in numerical methods for nonlinear equations and nonlinear least squares. Numer. Algebra, Control Optim. **1**(1), 15 (2011)
11. Katoch, S., Chauhan, S.S., Kumar, V.: A review on genetic algorithm: past, present, and future. Multim. Tools Appl. **80**(5), 8091–8126 (2021)
12. Mirjalili, S.: Genetic algorithm. In: Evolutionary Algorithms and Neural Networks, pp. 43–55. Springer (2019)

Harmonic Reduction Analysis of Generated Power by a Wind Energy Conversion System Using 7 Levels and 9 Levels Inverters

Maha Annoukoubi[1(✉)], Ahmed Essadki[1], Hammadi Laghridat[1], and Tamou Nasser[2]

[1] National School of Arts and Crafts in Rabat (ENSAM), Mohammed V University, Rabat, Morocco
m.annoukoubi@gmail.com
[2] National High School for Computer Science and Systems Analysis (ENSIAS), Mohammed V University, Rabat, Morocco

Abstract. Recently various sources of renewable energies are used to generate electrical power in order to respond to important growth of the electrical energy demand. This generation requires power inverters systems in order to inject and regulate the generated power to the electrical grid. The commonly used inverter is the two level one which generate a square output voltage instead of a sinusoidal one which impacts negatively the grid in term of harmonics. Hence, this work aims to ameliorate the Total Harmonic Distortion (THD) of power injected to grid in a way to be less than 5% as the IEEE condition limited and that by using multilevel inverters instead of the two level conventional one. A comparative analyze between the use of seven levels and nine levels in terms of harmonics emission and output voltage waveforms will be done using MATLAB/SIMULINK simulator and that in order to conclude about the positive and important impact of using multilevel inverter in a wind energy conversion system for a better integration to the elctrical grid.

Keywords: Wind energy · Renewable energy · Harmonic distortion · 7 levels inverter · 9 levels inverters · DFIG

1 Introduction

In the last years, we have assisted to an increasing rate in the demand of the electrical power in order to please the industrial and daily demands. This augmentation is accompanied by the risks of shortage of fossil energy and their effects on the climate change and the environment, which leads to look for other less risky source of energy such renewable energies which represent a clean solution.

L. Azrar et al. (Eds.): CIP 2022, LNME, pp. 576–586, 2023.
https://doi.org/10.1007/978-3-031-23615-0_58

Among the new renewable sources of energy there is wind energy that is the subject of our work. Wind is directly created from the heating of the atmosphere caused by sun, rotation of the earth and the fluctuation in earth's surface. Mountains, bodies of water, vegetation and another parameters impact the rate of wind presence in earth. The kinetic energy of the wind is transformed into electrical energy through a system called wind energy conversion system. This system is composed of 3 main devices starting from the mechanism that transform the kinetic energy into an adaptive mechanical one that is called the wind turbine, then the conversion of it to an electrical one is done by a generator, finally the electrical power is injected to the electrical grid through a power inverters system that must adapt the electrical energy generated to the limits and configuration on the grid. The control strategies of these machines and their possible grid connection interfaces must make it possible to capture as much energy as possible over the widest possible range of wind speed variation, with the aim of improving the profitability of wind power.

The increasing use of electronic power devices in electrical systems has resulted many problems related to disturbances or harmonic distortions of electrical grid as presented in [1]. Harmonic distortion is generated by non-linear components connected to the grid and which absorb non-sinusoidal currents. These harmonic currents also generate harmonic voltages at various points of connection to the grid. For other electrical equipment connected to these points, this harmonic pollution has negative effects as presented in work [2]. Among these effects, we can in particular cite the distortion of the grid voltage at the connection point between the generator of energy and the grid. This pollution can also lead to the heating of cables and electrical equipment or even the result a sudden shutdown of machines, or even the total destruction of all this equipment [9].

That's why the aim of our work is to reduce the harmonics distortion of the electrical current generated by the WECS in order to be injected to the grid and that's by respecting the IEEE condition in terms of total harmonics which should be less than 5 %. For that we will study the use of multilevel inverters instead of conventional inverters, for this work we will compare the use of 7 level inverter and 9 level inverter in term of harmonics, electromagnetic interference, filter size and costs. The choice of those 7 and 9 level inverters is a sequel of our past research work on three and five level invereter and that in order to compare their THD ameliorations results with result that can be presented by using 7 or 9 level inverters. Our article is organized as follow, the introduction is covered in Sect. 1. Our WECS model is presented in Sect. 2. Then, the study, model and of the the 7 and 9 levels inverter is presented in Sect. 3. In the Sect. 4, we will present and discuss simulation result in MATLAB / SIMULINK and then conclude in the last section.

2 Modeling of WECS Subsystems

The modeling of any system is an essential part for the study and control of its functioning especially for applying a particular command to it. On the other hand, it is an essential phase in the development of the design of the systems. Advances in computer science and software engineering make it possible to perform efficient modeling, for our work we are using MATLAB/SIMULINK in order to model and simulate the functioning of our systems. In this section we will present the model of the each subsystem of our WECS separately as shown in the Fig. 1 . The kinetic energy of wind is transformed into electrical one through the wind turbine that will be described firstly after that comes the drive train that adapt the speed and torque to the electrical generator that transform the mechanical power into an electrical one for this work we are using the doubly fed induction generator then this power is injected to the grid using power inverters that will be described in the next section.

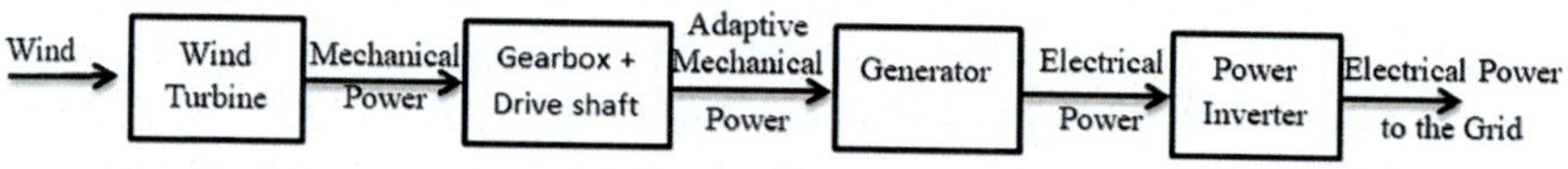

Fig. 1. WECS subsystems scheme

2.1 Wind Turbine and Gearbox Model

In this part, we gonna present the aerodynamic model of the wind turbine system and the gearbox we have already sutdied and presented in our last works [[2]-[3]]. The wind of speed v, applied to the blades of the wind turbine, causes it to rotation and creates a mechanical power on the turbine shaft, Pt that is expressed as follow:

$$P_t = C_p(\lambda, \beta)\ \frac{1}{2}\ \rho\ A\ v^3 \tag{1}$$

C_p is the power coefficient which represents the aerodynamic efficiency of the wind turbine, it depends on the turbine parameters : the blades pitch angle β and the turbine tip speed ratio λ; A, is the turbine area; ρ is air density and v is the wind speed .

Knowing the speed of rotation of the turbine, the mechanical torque T_t available on the shaft of the turbine can be expressed as :

$$T_t = \frac{P_t}{\Omega} = C_p(\lambda, \beta)\ \frac{1}{2\Omega}\ \rho\ A\ v^3 \tag{2}$$

The mechanical part of the turbine is composed of three orientable blades of length R, fixed at a drive shaft rotating at a rotational speed Ω_t , connected to a gain multiplier G.This multiplier drives the electric generator. Mechanical angular speed Ω_m and torque T_m are expressed as follow, where G is the gearbox coefficient.

$$T_m = \frac{T_t}{G}; \Omega_m = G\ \Omega_t \tag{3}$$

The mechanical equation of our system is:

$$J\,\frac{\mathrm{d}\Omega_{mec}}{\mathrm{d}t} = T_m - T_{em} - f\,\Omega_m \tag{4}$$

T_{em} and f are the DFIG electromagnetic torque and the friction coefficient . J represents the total inertia.

2.2 Modeling of the DFIG

For our WECS we are using as generator the doubly fed induction generator because of the many advantages it presents in tems of the quality of power generated and its control. The detailed model of the DFIG has been presented in [4].

The DFIG model is represented by the following equations :

$$\begin{bmatrix} v_{sd} \\ v_{sq} \\ v_{rd} \\ v_{rq} \end{bmatrix} = - \begin{bmatrix} R_s & 0 & 0 & 0 \\ 0 & R_s & 0 & 0 \\ 0 & 0 & R_r & 0 \\ 0 & 0 & 0 & R_r \end{bmatrix} \begin{bmatrix} i_{sd} \\ i_{sq} \\ i_{rd} \\ i_{rq} \end{bmatrix} + \frac{\mathrm{d}}{\mathrm{d}t} \begin{bmatrix} \Psi_{sd} \\ \Psi_{sq} \\ \Psi_{rd} \\ \Psi_{rq} \end{bmatrix} + \begin{bmatrix} 0 & -\Omega_s & 0 & 0 \\ \omega_s & 0 & 0 & 0 \\ 0 & 0 & 0 & -s\Omega_s \\ 0 & 0 & s\Omega_s & 0 \end{bmatrix} \begin{bmatrix} \Psi_{sd} \\ \Psi_{sq} \\ \Psi_{rd} \\ \Psi_{rq} \end{bmatrix} \tag{5}$$

i, v and Ψ are the current, the voltage and the DFIG field. L and R are the inductance and the resistance; s and r indicate the side stator and the side rotor of generator and Ω_s is the angular speed of the DFIG. Stator and rotor magnetic fields equations are:

$$\begin{bmatrix} \Psi_{sd} \\ \Psi_{sq} \\ \Psi_{rd} \\ \Psi_{rq} \end{bmatrix} = \begin{bmatrix} -L_s & 0 & L_m & 0 \\ 0 & -L_s & 0 & L_m \\ L_m & 0 & -L_r & 0 \\ 0 & L_m & 0 & -L_r \end{bmatrix} \begin{bmatrix} i_{sd} \\ i_{sq} \\ i_{rd} \\ i_{rq} \end{bmatrix} \tag{6}$$

L_s; L_r and L_m are the stator inductance; the rotor inductance and the mutual inductance .

Equation of electrical active and reactive powers delivered by the DFIG isexpressed as follow:

$$\begin{bmatrix} P \\ Q \end{bmatrix} = \frac{3}{2} \begin{bmatrix} v_d & v_q \\ v_q & -v_d \end{bmatrix} \begin{bmatrix} i_d \\ i_q \end{bmatrix} \tag{7}$$

2.3 WECS Control Strategie

2.3.1 MPPT Controller In order to extract maximum power possible from the kinetic energy of wind, a maximum power point tracking algorithm has been developped. For our WECS we used MPPT controllerin order to track a specific speed for extracting maximum power from the wind despite the wind speed variation. The principle of this controller is to keep the tip speed ratio equal to an optimum value in order to maximaze the power coefficient ($\lambda = \lambda^{opt}$;

$C_p = C_p^{max}$).
The maximum aerodynamic power can expressed as follow:

$$P_{aero}^{max} = \frac{1}{2}\,\rho\, A_t\, \frac{C_p^{max}}{\lambda^{opt3}}\, \frac{R^3}{G^3}\, \Omega_{mec}^3 \tag{8}$$

2.3.2 PI Control Strategie of Rotor and Grid Sides Inverter To control our inverters , we are using a simple PI controller for a separate control of the injected electromechanical torque and flux of the DFIG. PI controller is also used to ensure the tracking of active and reactive power, secure a good disturbance rejection and secure a structred robutness. The general PI controller expression is:

$$C(s) = K_p + \frac{K_i}{s} \quad ; \tag{9}$$

Kp is the proportional coefficient, Ki is the integral one and tr is the settling time for the rotor currents.

$$K_p = \frac{3\,\rho\, L_r}{t_r} \quad ; K_i = \frac{3\, R_r}{t_r} \quad ; \tag{10}$$

In Fig. 2 we present the block diagram of the our WECS .

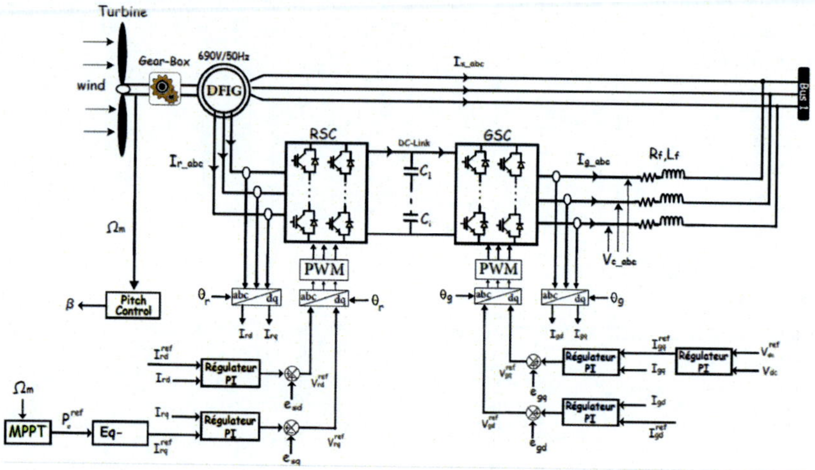

Fig. 2. PI Controller loop

3 Model of 7 Levels and 9 Levels Multilevel Inverters

After generating the electrical power by the DFIG and before injecting it to the grid, this power passes through a power inverter AC/DC/DC/AC because the frequency and voltage of the power produced by the electric generator vary with wind speed, so the power converter is used to convert the variable AC to the fixed frequency and voltage required by the electrical grid. The commonly used inverter is the two level one that produce an output voltage that varies between two values, the fact is that this output voltage is not smooth and present a high level of harmonics distortion that is up to 68% working alone and the THD of generated current is about 7% whereas the THD of the injected current to grid should be less than 5% as presented in our last work [5]. Indeed in order to ameliorate the THD factor of current injected to the grid we are going to study the use of developed technology of inverters that is called multilevel inverters [6].

Multilevel inverters are a developed approach used to have a high quality output voltage with a reduced harmonics without using transformers. Multilevel inverters use several power semiconductors and capacitor voltage sources in order to generate a multilevel smoother output voltage. The commutation of the switches result the addition of the voltages provided by the capacitor. In Fig. 3 we present a scheme of one phase Multilevel inverters that can give from 2 to n level depending on numbers of capacitors and switching frequency. 3 different topologies of multilevel inverters have been proposed: diode-clamped by [6]; capacitor-clamped by [7] and cascaded H-bridge [8] with separate dc sources that we will be using for this work.

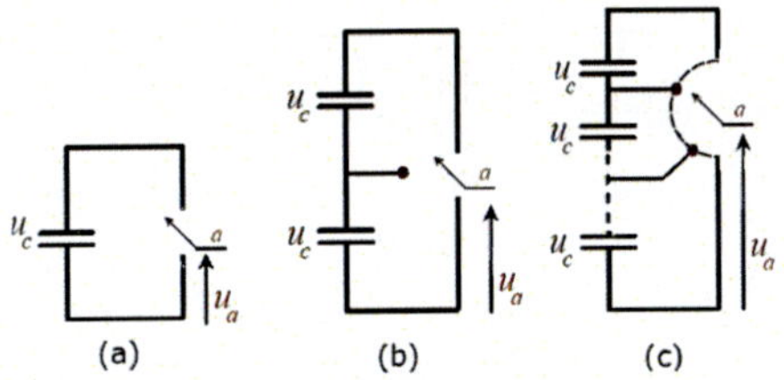

Fig. 3. Multilevel inverter scheme: (a) Two level (b) three level (c) n level

3.1 Model of Seven Level Inverter

For this work, we are using a Cascaded H-Bridge multilevel inverters topologies. It is obtained by connecting two or more single phase H-Bridge inverters in series each one is composed of 4 switches for our topologies we are using an IGBT [9]. Generally when we connect k H-Bridges in series, we obtain 2k+ 1 levels for the output voltage and the maximum output voltage is k Vdc. So, to release a 7 level inverter we are using 3 H-Bridges connected in series , so that a total of 12 IGBT and the maximum output voltage is 3 Vdc, the scheme of our 7 level inverter is represented in Fig. 4.

The switching sequences of the switches must respect the fact that two switches in the same leg don't have to conduct at same time and respect the switching state presented in the following table .

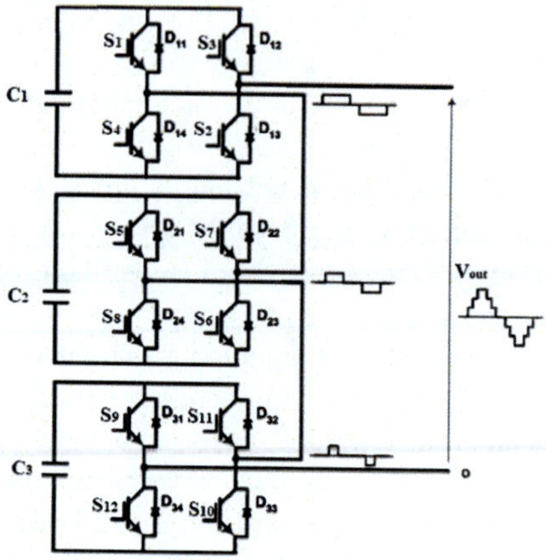

Fig. 4. One phase leg of a 7 level inveretr

Vout	SWITCHING STATES (7 level)											
	S_1	S_2	S_3	S_4	S_5	S_6	S_7	S_8	S_9	S_{10}	S_{11}	S_{12}
+Vdc	1	1	0	0	0	1	0	1	0	1	0	1
+Vdc/2	1	1	0	0	1	1	0	0	0	1	0	1
+Vdc/2	1	1	0	0	1	1	0	0	1	1	0	0
0	0	1	0	1	0	1	0	1	0	1	0	1
−Vdc	0	0	1	1	0	1	0	1	0	1	0	1
−Vdc/2	0	0	1	1	0	0	1	1	0	1	0	1
−Vdc/3	0	0	1	1	0	0	1	1	0	0	1	1

We realsed the the model of 7 sevel using power electronic component in MATLAB/SIMULINK, the output voltage obtained of one phase and a Vdc = 1300V is presented in Fig. 5.

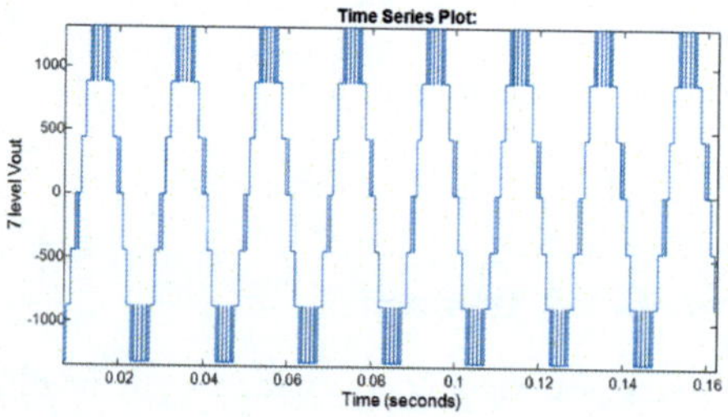

Fig. 5. One phase output voltage of a 7 level inverter

3.2 Model of Nine Level Inverter

To release a 9 level inverter we are using 4 H-Bridges connected in series , so that a total of 16 MOSFET and the maximum output voltage is 4 Vdc, the scheme adapted is presented in Fig. 6.

The switching state of 9 level inverter is presented in the following table .

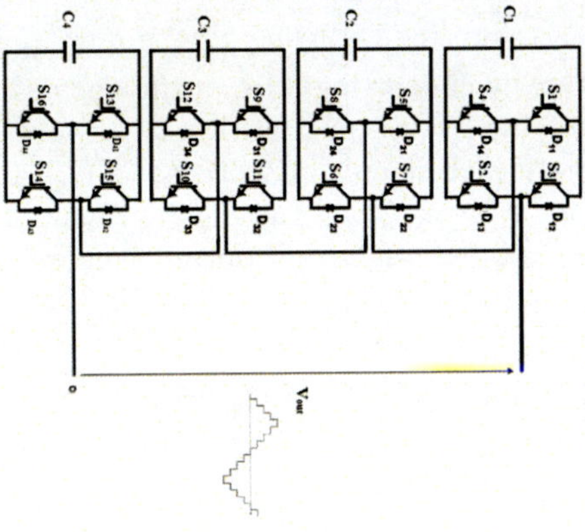

Fig. 6. One phase leg of a 9 level inveretr

Vout	SWITCHING STATES (9 level)															
	S_1	S_2	S_3	S_4	S_5	S_6	S_7	S_8	S_9	S_{10}	S_{11}	S_{12}	S_{13}	S_{14}	S_{15}	S_{16}
+Vdc	1	1	0	0	0	1	0	1	0	1	0	1	0	1	0	1
+Vdc/2	1	1	0	0	1	1	0	0	0	1	0	1	0	1	0	1
+Vdc/3	1	1	0	0	1	1	0	0	1	1	0	0	0	1	0	1
+Vdc/4	1	1	0	0	1	1	0	0	1	1	0	0	1	1	0	0
0	0	1	0	1	0	1	0	1	0	1	0	1	0	1	0	1
−Vdc	0	0	1	1	0	1	0	1	0	1	0	1	0	1	0	1
−Vdc/	0	0	1	1	0	0	1	1	0	1	0	1	0	1	0	1
−Vdc/3	0	0	1	1	0	0	1	1	0	0	1	1	0	1	0	1
−Vdc/4	0	0	1	1	0	0	1	1	0	0	1	1	0	0	1	1

We realsed the the model of 9 sevel using power electronic component in MATLAB/SIMULINK, the output voltage obtained of one phase and a Vdc = 1300V is presented in Fig. 7.

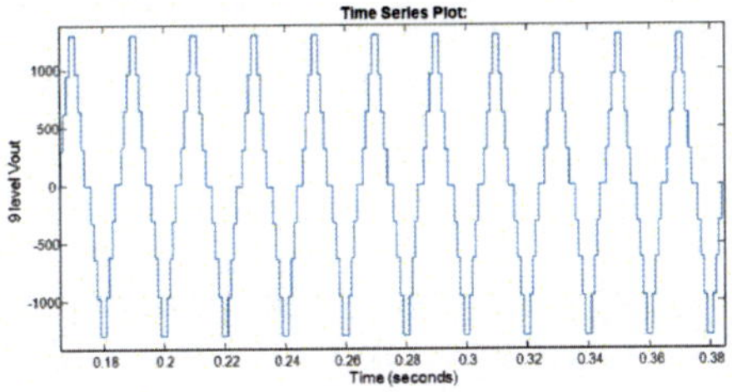

Fig. 7. One phase output voltage of a 9 level inverter

4 Simulation Result and Discussion

In MATLAB/SIMULINK we realized the model of all the described part of our WECS using the equations presented in Sect. 2, for 7 level and 9 level inverters we used the power electronic component of SIMULINK and realized 3 phases 7 level and 9 level inverter by connecting 3 leg of the one described on Fig. 4 for 7 level inverter and Fig. 6 for 9 level inverter. First, we simulated the 7 level and 9 level inverters separately their output voltages form was presented in Fig. 5 for 7 level and Fig. 7 for 9 level . Then the Fig. 8 present the result of the FFT analyze used to obtain the THD factor of output voltage, so as we can see the THD factor of the 9 level inverter is less by about 10% than the 7 level inverter.

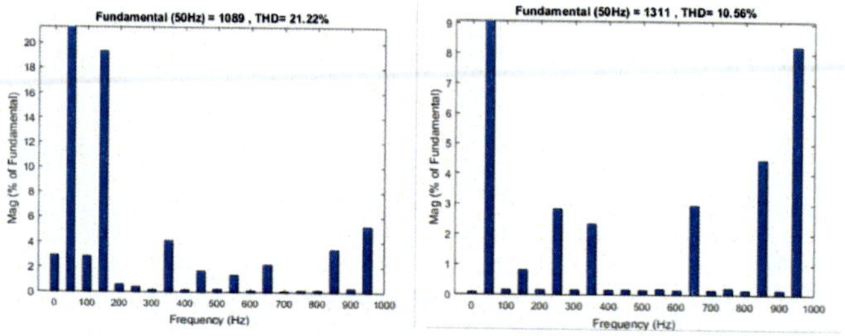

Fig. 8. Simulation results of FFT Analyze: (a) THD of 7 level inverter; (b) THD of 9 level inevrter

Figure 9 present the the current and voltage that will be injected to the grid.

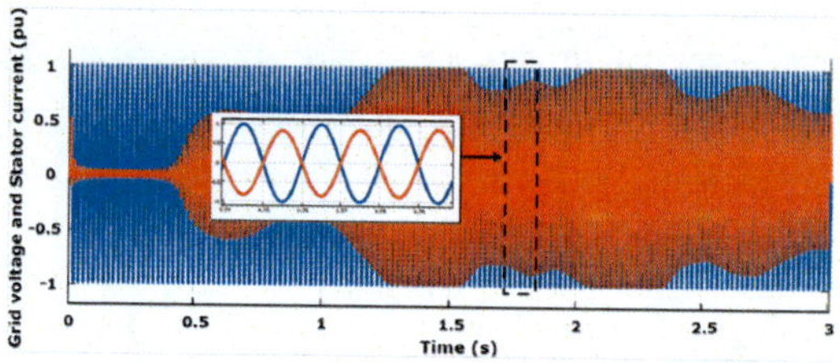

Fig. 9. Voltage and current injected to the grid

Figure 10 present the FFT analyze of the current injected to the grid ig, produced by the WECS using 7 level inverter at first and the using 9 level.

Finally, we can conclude that the THD of ig was reduced by about 2% using the nine level inverter but the complexity of 9 level controler is more important that 7 level which lead us to study a compromise between Harmonic reduction and complexity of controller systems.

5 Conclusion

This paper has confirmed by results of simulation using MATLAB/SIMULINK that the use of multilevel inverter present many advantages: lower dv/dt, the output voltage waveform is smoother using more voltage levels so it is less distorted,

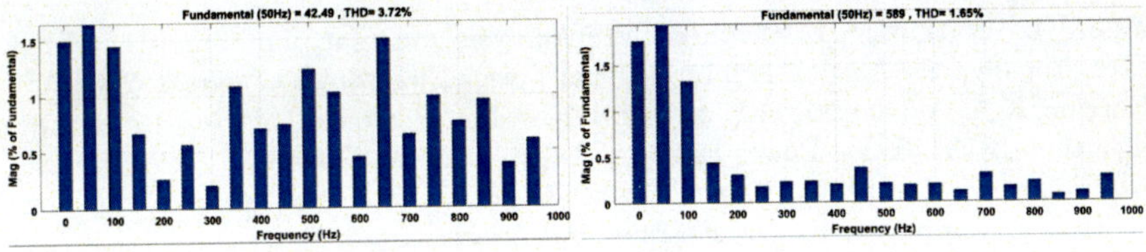

Fig. 10. THD of ig: (a) Using 7 level inverter; (b) Using 9 level inevrter

also we used lower switching frequency. The use of 2 different level multilevel inverter for our WECS instead of the conventional one confirmed that the THD of injected current to the grid is reduced using the 9 level compared to the 7 level one but the control command of the nine level one is more complex than the seven level inverter. Indeed, we conclude that for optimizing our WECS in order to generate the maximum quality power respecting the grid condition in term of voltage, frequency and component cost we should develop a compromise between control complexity and higher level inverter.

Appendix A

The parameters of the our WECS are given in the following Table.

Parameters	Values	Parameters	Values
Rated power	1.5 MW	Nominal Frequency	50 Hz
Rated wind speed	12 m/s	Rated Rotor Speed	1750 tr/min
Density of Air	1.225 kg/m^3	Number of Pole Pairs	2
Blade radius	31 m	Stator Resistance R_s	2.65 $m\Omega$
Gearbox ratio G	59	Rotor Resistance R_r	2.63 $m\Omega$
Rated Power Pn	1.5 MW	Stator Leakage Inductance $L_{s\sigma}$	0.1687 mH
Rated Voltage Un	690 V	Rotor Leakage Inductance $L_{r\sigma}$	0.1337 mH
		Magnetizing Inductance L_m	5.4749 mH

References

1. Rashid, M.H.: Power Electronics, Circuit, Devices, and Application, 3rd edn, pp. 253–255 (2009)
2. Annoukoubi, M., Essadki, A., Laghridat, H.: Comparative study between the performances of a three-level and two-level converter for a wind energy conversion system. In: WITS, 2018
3. Laghridat, H., Essadki, A., Annoukoubi, M.: A novel adaptive active disturbance rejection control strategy to improve the stability and robustness for a wind turbine using a doubly fed induction generator. J. Electr. Comput. Eng. (2020)
4. Deicke, M., De Doncker, P.W.: Doubly-fed induction generators systems for wind turbines. IEEE Ind. Appl. Mag. (2000)
5. Annoukoubi, M., Essadki, A., Nasser, T.: Cascade H-bridge multilevel inverter for a wind energy conversion system applications. In: IRSEC, 2020
6. Nabae, A., Takahashi, I., Akagi, H.: A new neutral-point-clamped PWM inverter. IEEE Trans. Ind. Appl. (1981)

7. Zhang, L., Watkins, S.J., Shepherd, W.: Analysis and control of a multi-level flying capacitor inverter. In: IEEE International Power Electronics Congress, 2002
8. Corzine, K.A., Wielebski, M.W., Peng, F.l., Wang, J.: Control of cascaded multilevel inverters. IEEE Trans. Power Electron. 19(3), 732–738 (2004)
9. Rodriguez, J., Lai, J.S., Peng, F.Z.: Multilevel inverters: a survey of topologies, controls and applications. IEEE Trans. Ind. Electron (2002)

Eco-Innovation for Recycling/Remanufacturing Electric Vehicle Engines

P. Jacquet[1,2](✉), A. Vaucheret[1,2], G. Grimaud[3], and T. Gallone[4]

[1] Ecole Catholique des Arts et Métiers, LaBECAM, 40, Montée Saint Barthélemy, 69005 Lyon, France
{philippe.jacquet,alexis.vaucheret}@ecam.eu

[2] Arts et Métiers Institute of Technology, LaBoMaP, HESAM Université, 71250 Cluny, France

[3] MTB RECYCLING, Quartier de la gare, 38460 Trept, France
g.grimaud@mtb.fr

[4] ICRM «Impulse Circula'r Raw Materials», 6 rue de la Richarderie, 78760 Jouars-Pontchartrain, France

Abstract. Because of climate changes and the increasing cost of raw materials, the mobility will become more and more electric in the future years. It is already the case for automotive and the first electric car from Renault company, "Renault Zoé" appeared in 2009. So the first models are now reaching their End-Of-Life and it is time to consider their dismantling. The traditional recycling industry used for Internal Combustion Engine vehicles is not adapted to electric vehicle powertrains. This study deals with the disassembly of 2 electric powertrains of Renault Zoé vehicle. All the components have been separated, identified, their chemical composition has been determined as well as their potential recyclability. We have highlighted that sometimes, some materials should be substituted by more recyclable others and that some subsets should be redesigned in order to facilitate the disassembly. A strategy to maximize the gain, reduce wastes has been built and some recommendations for the future design of these electric powertrains has been proposed.

Keywords: Recycling · Electric vehicle powertrain · Eco-design

1 Introduction

The goal of e-REVE project (Eco-Innovation for Recycling/Remanufacturing Electric Vehicle Engines) is to find solutions for recycling and add value to the different materials contained inside electric vehicle powertrains. Up to now, because of the specific manufacturing of electric vehicle powertrains it is not possible to use the traditional recycling methods which are currently used for Internal Combustion Engines Vehicles. Some people are worrying about this for some years (Hermine 2014; Rapport RSE Renault 2015) and a team project has been created to bring answers to this challenge. The team is made up of.

MTB Recycling, a world leader in industrial waste management which focuses its activities on three major areas: *recycling*, equipment *manufacturing* and *engineering*.

L. Azrar et al. (Eds.): CIP 2022, LNME, pp. 587–594, 2023.
https://doi.org/10.1007/978-3-031-23615-0_59

RENAULT for its skills in electric vehicles design.
INDRA, automotive recycler which provides a global offer in the End-Of-Life Vehicle (ELV) sector.
ECAM Lyon, an engineering school for its skills in materials characterization.
ICRM, a circular economy consulting company.

At the beginning of the 2000's, several events have made electric engines for vehicles more and more present on the market: atmospheric pollution, depletion of natural resources,… According to European Union proposals "Fit for 55", greenhouse gas emissions must be reduced by 55% by 2030 and 100% by 2035 (compared to 1990), so that Electrical Engines should progressively substitute Internal Combustion Engines for vehicles (see Fig. 1).

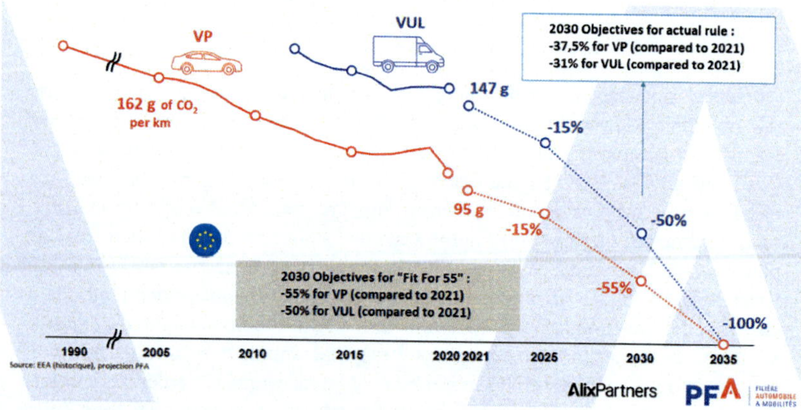

Fig. 1. Evolution perspective for Internal Combustion Engines. (Plate-forme automobile, filière automobile & mobilité 2021)

In 2019, D'Adamo and Rosa published a large literature review (129 references) about the recycling of Electric Vehicles. Most of the references deal with the recycling of batteries which seems to be a key point for numerous laboratories but very few concern the recycling of the different components of electric engines (Bdiwi et al. 2016; Li et al. 2014, 2018; Elwert 2015; Soo et al. 2018; Kibira and Jain 2011). It seems that this global approach for electric Vehicles is still few developed for the moment.

In the first part, this paper will present the Ecodesign approach which is to our point of view a key point to have in mind if a manufactured product is designed for future recycling. Then experimental results are presented and divided in two parts: from a technical point of view with the dismantling of the different subsets and their material analysis but also from an economic point of view because recycling will be developed only if a financial gain is obtained.

2 Ecodesign

The description of the term "eco-design" is given in the standard NF X 30-264 (Management environmental – Aide à la mise en place d'une demarche d'éco-conception 2013). The Mechanical Engineering, Design and Manufacture Department from the Manchester Metropolitan University also published a few years ago (Simon and Dowie 1993) some recommandations about this subject. Before thinking about the different recycling technics for materials from End-Of-Life Vehicles (electric or not), the first point to take into consideration is the disassembly of a manufactured product. That is why the sooner this step is considered, the easier the recycling operations will be and if possible from the design. For a manufactured product, three subsets have been identified: materials, fasteners and connectors and finally the product structure.

Concerning the materials, the requirements are:

Minimize the number of different materials.
In case of subsets difficult to disassembly, use the same material.
In case of polymer materials, make the symbols visible.
Favor recyclable materials.
Clearly identify dangerous parts if present to put aside quickly.

Concerning fasteners and connectors, the requirements are:

Minimize the number of fasteners.
Easy access fasteners.
Fasteners should be disassembly with standard tools, if not it should be easily breakable.
Avoid adhesive junction between 2 parts which are incompatible with recycling.
Minimize length and number of connection cables.

Concerning the product structure, the requirements are:

Minimize the number of parts.
Realize a design as modular as possible with separated functions.
Easy access for parts which have a high value regarding recycling.
Aggregate non-recyclable parts at the same place, easy to remove.
Avoid filled polymers.

The way of dismantling and recycling End-Of-Life Vehicles has been described by several authors (Hao et al. 2017; Pan and Li 2020; Zhou et al. 2019). The first step consist on separating the different subsets and salvage all the fluids and toxic or dangerous products (oil, batteries, brake fluid, engine coolant,…). All the components in good condition which can easily be sold are put aside (tires, mirrors, body parts,…). Non-metallic parts which have their own recycling channel are sorted out (glass, rubber, plastics,…). The remainder is then milled and form what is called "Automotive Shredded Residue" (ASR). These ASR are either landfilled or sorted for recycling.

The benefit of a manual disassembly prior an automatic grinding is the recovery rate of materials for recycling. According to a Swedish study (Tasala Gradin et al.

2013), an automatic grinding lead to a result for recycling materials of only 73% but 85% if a manual disassembly is realized before. Anyway, a manual disassembly is time consuming: as an example, it took more than 6h for three operators to dismantle a "Volkswagen Vista" vehicle (Tian and Chan 2016). The ratio cost/benefit of recycled materials should be taken into account.

3 Experimental Results

Our work focused on both understanding the technical and economic elements. The challenge was to enable short loop reuse of different materials from electric vehicle powertrains.

3.1 Technical Point of View

The different subsets of an electric powertrain are presented on Fig. 2. The main components are: wire harnesses (1), electrical gear motor, electric power unit and junction box (2). Each subset has been completely disassembled but to maximize the recovery of the different materials (especially strategic materials) and respond to a short loop challenge, a manual disassembly is necessary.

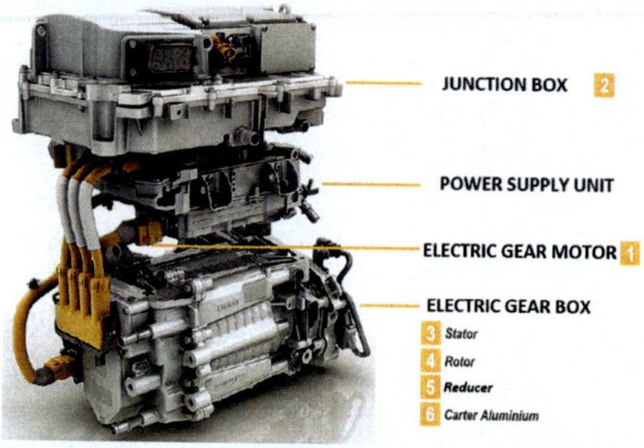

Fig. 2. Description of the different subsets of an electric powertrain (Renault) (Grimaud 2018).

As an example a complete disassembly of a junction box is presented on Fig. 3.

It was also fundamental to involve an important sorting during this disassembly in order to think precisely about the rest of the process for each organ (see Fig. 4).

Then, for each component, its composition has been determined by different chemical analysis (IR Spectroscopy, Glow-Discharge Optical Emission Spectrometry, Energy Dispersive Spectroscopy) depending on its shape, size, materials family,… As an example, the composition of some containing copper parts is given in Table 1; most of them are almost pure copper or copper alloys with some tin (bronze).

Fig. 3. Junction box after complete disassembly.

Fig. 4. Wire harnesses (left) and screws and bolts (right)

Table 1. Composition of containing copper parts (GDOES analysis).

	Cu	Sn	Sb	As	Co
Part identification	Mass percentage				
1	99.93	0.01	0.02	0.03	0.01
2	99.94	0.01	0.01	0.03	0.01
3	99.66	0.34	–	–	–
4	99.69	0.31	–	–	–
5	98.51	1.49	–	–	–
6	98.75	1.25	–	–	–
7	99.42	0.58	–	–	–
8	99.34	0.66	–	–	–
9	97.23	2.77	–	–	–

3.2 Economic Point of View

The constitution of the three main families of powertrains from a "Renault Fluence" vehicle is presented in Table 2. The major difference concerns the content of copper which is very much higher in case of electric powertrain.

As the mineral resources concerning this metal are limited, its price is continuously increasing that is why it is more and more interesting to salvage it.

Furthermore, as shown on Fig. 5, economic gains are exponential when quality approaches 100%, that is why the chemical composition of each part is important to be known in order to avoid mixture of different alloys.

Table 2. Constitution of different powertrains (Grimaud 2018).

	Petrol engine	Diesel engine	Electric engine
Powertrain (kg)	141	141	113
Aluminum (kg)	24.6	25.5	32.8
Copper (kg)	0.6	0.7	15.9
Steel (kg)	102.9	106.6	48
Others (kg) (polymers. ...)	12.8	14.2	15.7

The possible gains are a balance between the costs (essentially workforce + energy) and the material recovery. The dismantling of the first powertrain was a complete discovery for our team so that the necessary time to find the right tool (screwdriver, spanner, grinder,..), the right procedure was quite long. However, the duration to reach each component has been measured.

After this first experience, a second electric powertrain (Renault Zoé Vehicle) has been disassembled with an optimized duration and it took approximately 95 min for 2 operators.

Then a complete table was realized. This table contains for each component, its weight, its composition, the price per kg of this material if recycled and the necessary duration to collect it. With the help of this table, it has been possible to draw the diagram presented on Fig. 6.

We can notice that after only 32% of the total disassembly duration (95 min), already 65% of the possible gain is obtained. Of course, it is possible to obtain a higher gain but a balance between the gain and the necessary time to get it should be deeper studied.

4 Conclusion

After this work, the exact composition of each component of an electric powertrain and the necessary time to collect it are known. Some of them can be reuse (screw, connectors, wires,...), some others can be sold for recycling (steel, copper, aluminum, polymers,...). This study allowed us to realize that the partial dismantling then the treatment of different fractions of GMPe makes it to obtain at the same time a suitable

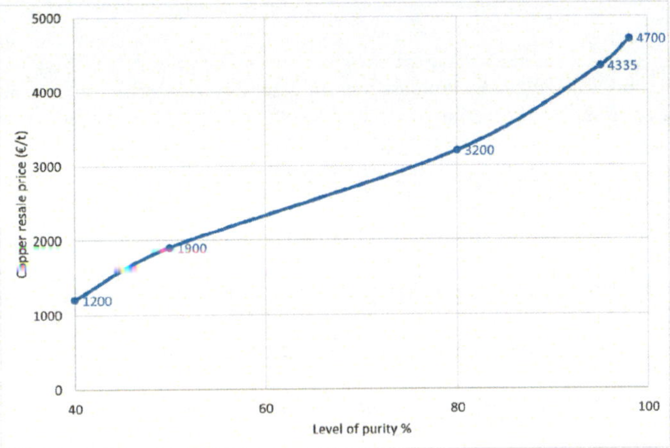

Fig. 5. Data from the London Metal Exchange (October 2018)

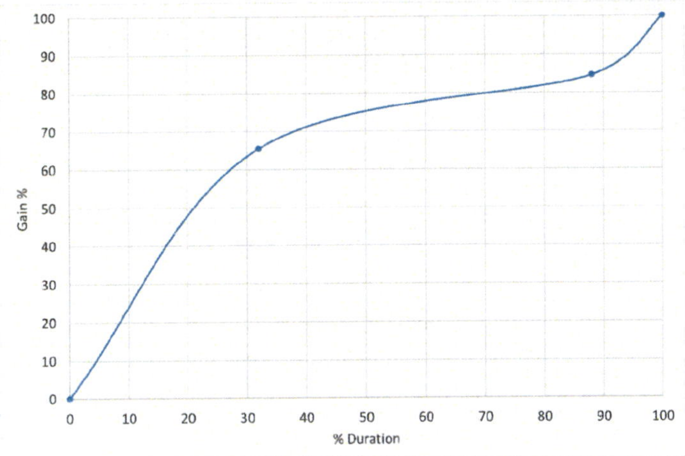

Fig. 6. Evolution of the possible gain versus time.

economic profitability with a final machine valuation over 700 €/t. At this material value, it is necessary to subtract all the processing costs. We have highlighted that sometimes, some materials should be substituted by more recyclable others and that some subsets should be redesigned in order to facilitate the disassembly. In the next years, electric vehicles will be more and more present (the first ones are already reaching their End-Of-Life) so that it is necessary to follow these recommendations, as soon as the design for cost saving and reduce future wastes.

Acknowledgements. The authors wish to thank BPI France which has partially supported this work.

References

Bdiwi, M., Rashid, A., Putz, M.: Autonomous disassembly of electric vehicle motors based on robot cognition. In: Proceedings of the ICRA 2016—IEEE International Conference on Robotics and Automation, pp. 2500–2505. IEEE: Stockholm, Sweden

Elwert, T., et al.: Current developments and challenges in the recycling of key components of (hybrid) electric vehicles. Recycling **1**, 25–60 (2015)

Grimaud, G.: Recyclage des groupes motopropulseurs électriques. Internal Report (2018)

Hao, H., Qiao, Q., Liu, Z., Zhao, F.: Resources, conservation and recycling impact of recycling on energy consumption and greenhouse gas emissions from electric vehicle production: The China 2025 case. **122**, 114–125 (2017)

Herminc, J.: La mise en oeuvre de l'économie circulaire au sein du groupe Renault. Responsab. Environ. **76**, 45–49 (2014)

Kibira, D., Jain. S.: Impact of hybrid and electric vehicles on automobile recycling infrastructure. In: Proceedings of the WSC'11—2011 IEEE Winter Simulation Conference, Phoenix, AZ, USA, 11–14 Dec 2011; pp. 1072–1083

Li, J., Barwood, M., Rahimifard, S.: An automated approach for disassembly and recycling of Electric Vehicle components. In: Proceedings of the IEVC 2014—IEEE International Electric Vehicle Conference, Florence, Italy, 17–19 Dec 2014, pp. 1–6

Li, J., Barwood, M., Rahimifard, S.: Robotic disassembly for increased recovery of strategically important materials from electrical vehicles. Robot. Comput. Integr. Manuf. **50**, 203–212 (2018)

Pan, Y., Li, H.: Sustainability evaluation of end-of-life vehicle recycling based on energy analysis : a case study of an end-of-life vehicle recycling enterprise in China. J. Clean. Prod. **131**(2016), 219–227 (2020)

Plate-forme automobile, filière automobile & mobilité (2021). https://pfa-auto.fr/wp-content/uploads/2021/10/Feuille-de-route-filie%CC%80re-auto-a%CC%80-2030-vF.pdf. Accessed 10 Jan 2022

Rapport RSE 2015: Groupe Renault. https://www.renaultgroup.com/wp-content/uploads/2016/07/rapport-rse-2015_vf_.pdf. Accessed 26 Jan 2022

Simon, M., Dowie, T.: Disassembly process planning. In: Proceedings of the Thirtieth International MATADOR Conference. Palgrave, London, 1993

Soo, V.K., Peeters, J.R., Paraskevas, D., Compston, P., Doolan, M., Duflou, J.R.: Sustainable aluminium recycling of end-of-life products: a joining techniques perspective. J. Clean. Prod. **178**, 119–132 (2018)

Tasala Gradin, K., Luttropp, C., Björklund, A.: Investigating improved vehicle dismantling and fragmentation technology **54**, 23–29 (2013)

Tian, J., Chen, M.: Assessing the economics of processing end-of-life vehicles through manual dismantling. Waste Manag. **56**, 384–395 (2016)

Zhou, F., Lim, M.K., He, Y., Lin, Y., Chen, S.: End-of-life vehicle (ELV) recycling management: improving performance using an ISM approach. J. Clean. Prod. **228**, 231–243 (2019)

Modeling and Control of Wind Turbine System Based on PMSG in Grid-Connected AC Microgrid

Youssef Akarne[1(✉)], Ahmed Essadki[1], Tamou Nasser[2], and Hammadi Laghridat[1]

[1] National School of Arts and Crafts (ENSAM), Mohammed V University, Rabat, Morocco
youssef_akarne@um5.ac.ma

[2] National High School for Computer Science and Systems (ENSIAS), Mohammed V University, Rabat, Morocco

Abstract. The development of microgrid (MG) concept is the key for assuring high reliability, good performance optimization and management requirements in electric power distribution systems. This paper presents a modeling and control of wind turbine system (WTs) in AC microgrid. Our system comprehends of permanent magnet synchronous generator (PMSG) driven by two wind turbines and interfaced by back to back control topology. The aim of this paper is to inject the active and reactive power into the AC bus and also exchanging these powers with the main grid with ensuring the unity power factor. The system is capable to operate at the maximum power during wind speed changes. In term of participating to the grid regulation, the MG provides these ancillary services like a frequency stabilization. The proposed work is implemented in Matlab/Simulink environment. The results demonstrate that our proposed system guarantee excellent performance in terms of stability, tracking and power quality injected.

Keywords: Microgrid · Wind turbine · PMSG · Power electronic converters · Back-to-back converter · Power controllers

1 Introduction

The limitation of the world's nuclear and fossil fuel resources has necessitated an immediate search for alternative sources of energy. A new way of balancing supply and demand without the use of coal- and gas-fired generators must therefore be sought [1]. Today, Morocco's renewable electricity system is extremely diversified and comprises a mix of solar, wind, and hydroelectric power plants. Morocco is actually considered as one of the leaders in the global energy transition, particularly in Africa, with different programs to generate electricity from renewable sources [2–10].

L. Azrar et al. (Eds.): CIP 2022, LNME, pp. 595–604, 2023.
https://doi.org/10.1007/978-3-031-23615-0_60

Renewable energy sources integration with the power systems like solar system (PV), wind turbine, fuel cell is one of the main concepts of microgrids. A microgrid comprises the generation sources and loads, it is considered a flexible system which can operate in two modes: grid-connected mode, at the point when it is connected to the main grid that can provides many ancillary services like frequency stabilization and exchanging the power with the grid. In addition, the MG can be disconnected from the main grid (islanded mode).

Wind energy system is classified one of the most suitable distributed generation and widely utilized in microgrids. This system is based on the variation of the wind speed. WTS provide a high penetration of renewable energy sources (RES) without affecting the MG stability. In addition the WTS improves the reliability of MG and supporting the grid management system.

There are numerous types of variable speed generators used for wind turbines. Currently, permanent magnet synchronous generator (PMSG) are becoming very popular in wind power systems, it is direct drive that can be used without gearbox. In addition to the high efficiency and low maintenance, the PMSG has high power conversion efficiency and simple structure. The power converters devices are essential to adapt the power produced from the wind turbine to the microgrid requirements. Currently, the most adopted power converter control topology is the back-to-back converter.

The architecture of the AC Microgrid connected to the main grid is depicted in Fig. 1. Recall that the microgrid is composed of PVs, energy storage systems (ESS) and wind-turbines (WT) and Electrical vehicle (EV) with loads comsumption.

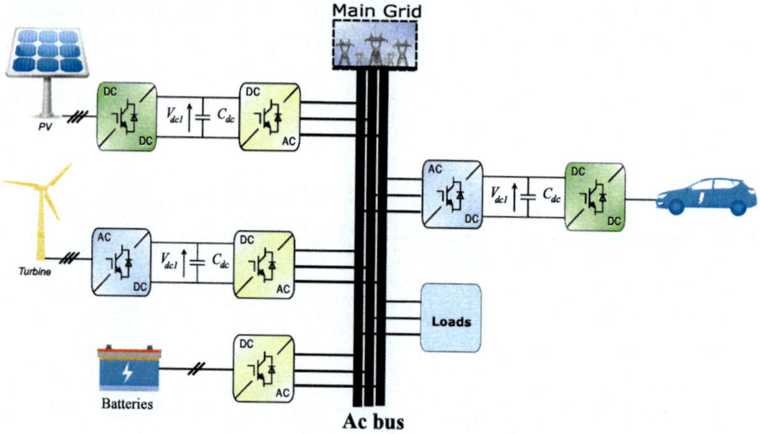

Fig. 1. Concept of a AC microgrid

In the literature, several contributions exist which deal with the modeling and control of PMSG in AC or DC MG. In [3], author designs WTS in DC MG, which this systems contains the PMSG based WECS (wind energy conversion system) connected to DC MG. PMSG system is modeled in [4], the PMSG is driven by micro-WT and connected into the DC Microgrid by a rectifier converter and a boost converter as depicted in Fig. 2. In addition, in term of AC MG, a modeling of wind turbine based PMSG and its control in AC MG presented in [5], Back to back control is adopted in [5].

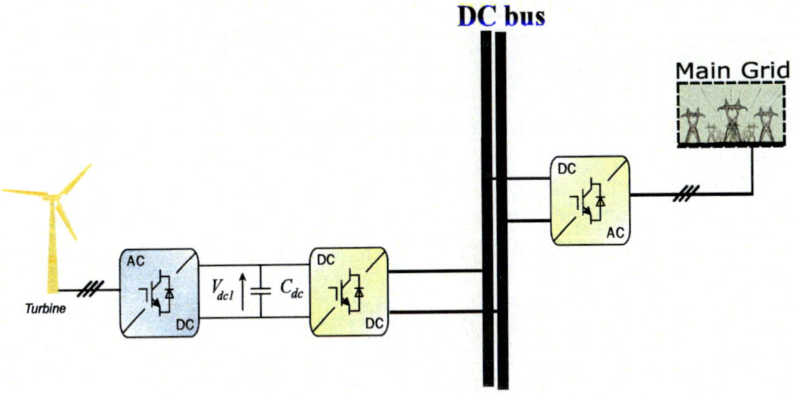

Fig. 2. MWT connected to a DC microgrid

The main contributions in this work are modeling and control of two wind turbine based PMSG in AC MG connected to the main grid. The WTS is capable of exchanging active power with main grid at unity power factor. The paper is organized as follows: Sect. 2 presents the modeling of wind turbine system based on PMSG. Section 3 presents Control of PMSG Wind Turbine. Finally, in Sect. 4 we presented the simulation results and discussed them.

2 Modeling of WTS Based PMSG in Microgrid

Our proposed architecture of two wind turbines based PMSG in AC Microgrid connected to the main grid is shown in Fig. 3.

2.1 Wind Turbine Model

The wind turbine is used in order to convert the wind kinetic energy to the mechanical energy. The aerodynamic power P_{aer} available from turbine rotor blades can be expressed as follow [6]

$$P_{aer} = C_p(\lambda, \beta)P_w = C_p(\lambda, \beta)\ \frac{1}{2}\ \rho\ A_t\ w^3 \tag{1}$$

where P_w is the wind's kinetic power; C_p is the energy coefficients and it depends on the blades pitch angle β and the turbine tip speed ratio λ; A_t, ρ and w_t are

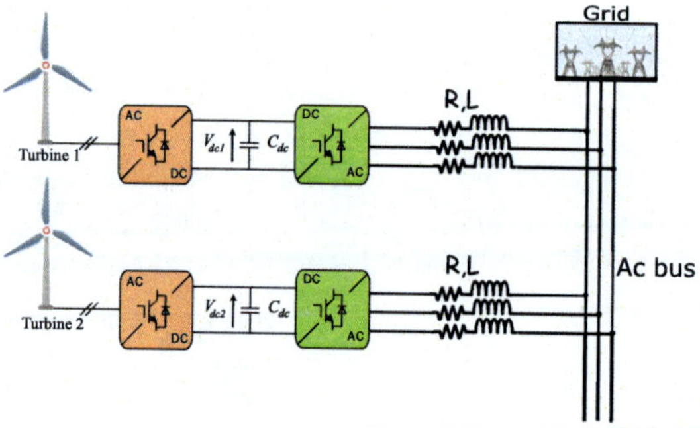

Fig. 3. Proposed architecture of AC microgrid connected to the main grid based PMSG

the turbine blades area, the air density and the wind speed of the wind turbine respectively. The expression of turbine tip speed ratio is given by [6]:

$$\lambda = \frac{R\,\Omega_{tur}}{w} \tag{2}$$

The aerodynamic torque T_{aer} is given by the following expression [6]:

$$T_{\text{aer}} = \frac{1}{2\omega_{\text{tur}}}\rho A_t C_p(\lambda, \beta) w^3 \tag{3}$$

The expression of the power coefficient is expressed as follow:

$$C_p(\lambda, \beta) = k_1 \left(k_2 \frac{1}{\lambda^j} - k_3\beta - k_4 \right) \exp^{(-k_5/\lambda^j)} + k_6\lambda \tag{4}$$

with $k_1 = 0.5176, k_2 = 166, k_3 = 0.4, k_4 = 5, k_5 = 21, k_6 = 0.0068,\ k_7 = 0.08,$ and $k_8 = 0.035$.

2.2 Dynamic Model of PMSG

The mathematical model of PMSG can be described in synchronous reference frame as follows [7]:

$$V_d = R_s i_d + L_d \frac{di_d}{dt} - \omega_e L_q i_q \tag{5}$$

$$V_q = R_s i_q + L_q \frac{di_q}{dt} + \omega_e L_d i_d + \omega_e \psi_f \tag{6}$$

$$\omega_e = p\omega_m \tag{7}$$

where R_s is the stator winding resistance (Ohm). L_d and L_q are the direct and the quadrature inductances (H). $V_d\,(V_q)$ are the direct (quadrature) generator voltages (V). $I_d\,(I_q)$ are the direct (quadrature) generator currents (A). ω_e and p are the electrical angular speed of the generator and the number of pole pairs of the PMSM and ψ_f is the permanent magnet flux(Wb).

The electromagnetic torque of PMSG is given by [7]:

$$T_e = \frac{3}{2}p\left[\left(L_d - L_q\right)i_q i_d - \psi_f i_q\right] \tag{8}$$

The mechanical equation is expressed as follows:

$$T_e - T_m = J\frac{d\omega_m}{dt} + f\omega_m \tag{9}$$

where T_e is the electromagnetic torque (N.m)

3 Control of Microgrid Based PMSG WTS

3.1 Maximum Power Extraction and Pitch Control

3.1.1 MPPT Control The output power of wind turbine system is intermittent and highly dependent on atmospheric conditions. For this aim, the MPPT algorithms are required to improve the efficiency and the reliability and extract the maximum power from wind turbine.

As mentioned previously, the MPPT should be used in order to extracting the maximum power from wind turbine during wind speed changes. This method is based on keeping the tip speed ratio at its optimal value with the aim of maximizing the power coefficient.

$$\Omega_{opt} = \frac{w}{R}\lambda_{opt} \tag{10}$$

The maximum mechanical output power of the wind turbine is expressed as follows:

$$P_{Tmax} = \frac{1}{2}\rho A_t C_{p\,\max}\left(\frac{R\Omega_{opt}}{\lambda_{opt}}\right)^3 \tag{11}$$

3.1.2 Pitch Angle Control In term of limiting the power of wind energy conversion system. A pitch control is used in order to avoid the exceeding of power its rated value in case of strong wind speeds and protecting the generator before overloading at high wind speeds by acting on the orientation of the turbine blades area β [9].

The equation between β and β^{ref} is given as follow [9]:

$$\beta = \frac{1}{1+T_\beta s}\beta^{\text{ref}} \tag{12}$$

with T_β is the time constant of the pitch servo and s is a Laplace operator.

3.2 Generator-Side Controller

The generator currents are expressed in following equations [6–8]:

$$\begin{aligned}\frac{\mathrm{d}i_d}{\mathrm{d}t} &= -k_1 i_d + \omega i_q + k_2 v_d \\ \frac{\mathrm{d}i_q}{\mathrm{d}t} &= -k_1 i_q - \omega i_d - k_2 \omega \psi_f + k_2 v_q\end{aligned} \tag{13}$$

where: $k_1 = \frac{R_s}{L_s}$; $k_2 = \frac{1}{L_s}$.
The compensation and the coupling terms are:

$$\begin{aligned}e_d &= \frac{1}{k_2}\omega i_q, \\ e_q &= \frac{1}{k_2}\omega i_d + \frac{k_2}{\omega}\Psi_p.\end{aligned} \tag{14}$$

The electromagnetic torque is controlled by the quadrature current:

$$T_{e-ref} = \frac{3}{2}P\psi_f i_{qref} \tag{15}$$

Consequently, the quadrature reference current i_q^{ref} is [7]:

$$i_{\mathrm{qref}} = \frac{2}{3P\psi_f}T_{e-ref} \tag{16}$$

3.3 Grid-Side Controller

3.3.1 Grid Synchronization The main element of the control structure is the grid synchronization. A synchronization control performed by phase-locked-loop (PLL) in order to obtain the phase angle and frequency from the grid voltage. The block of phase-locked-loop is shown in Fig. 4.

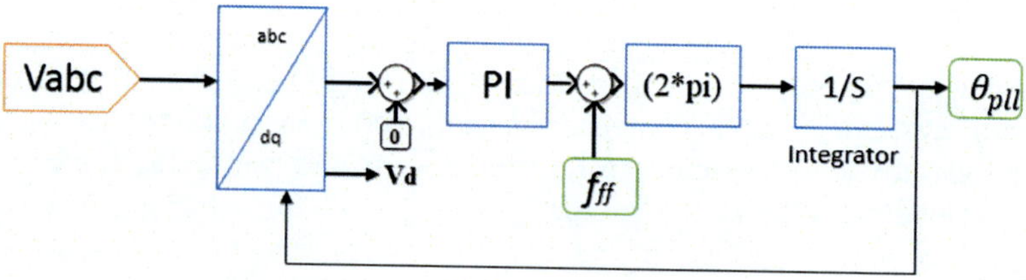

Fig. 4. Synchronous reference frame PLL

3.3.2 DC Bus Voltage Control The modeling of DC-link bus is given by [6–8]:

$$\frac{\mathrm{d}U_{\mathrm{dc}}}{\mathrm{d}t} = \frac{1}{C}\left(i_{\mathrm{w}} - i_{\mathrm{inv}}\right) \tag{17}$$

where C is the DC-link capacitor

3.3.3 Grid Current Control In term of grid current control, the wind turbine system converter currents control loops are based on synchronous reference frame, proportional-integral (PI) controllers [6].

The grid currents are expressed in following equations [6–8]:

$$\frac{di_{gd}}{dt} = -k_3 i_{gd} + \omega_g i_{gq} + k_4 v_{id} \tag{18}$$

$$\frac{di_{gq}}{dt} = -k_3 i_{gq} - \omega_g i_{gd} + k_4 (v_{iq} - v_g) \tag{19}$$

where: $k_3 = \frac{R_f}{L_f}$; $k_4 = \frac{1}{L_f}$.
The compensation and the coupling terms are:

$$\begin{aligned} e_{gd} &= \frac{1}{k_4}\omega_g i_{gq}, \\ e_{gq} &= v_g + \frac{1}{k_4}\omega_g i_{gd}. \end{aligned} \tag{20}$$

The direct reference current i_{gd}^{ref} is:

$$i_{gd}^{ref} = \frac{2}{3v_g} Q_g^{ref} \tag{21}$$

with Q_g^{ref} is the reference reactive power, it is given zero to obtain a unit power factor in grid side.

The active P_g and reactive power Q_g injected into the grid based on voltage oriented control are expressed in following equations [6–8]:

$$P_g = \frac{3}{2} v_g i_{gd} \tag{22}$$

$$Q_g = -\frac{3}{2} v_g i_{gq} \tag{23}$$

The control strategy of wind turbine system based on PMSG is shown in Fig. 5.

4 Simulation Results

The proposed model and control of WTS in MG is simulated in MATLAB/Simulink environment. The simulation parameters are provided in Table 1. In order to evaluate the performance of our system, different values of wind speed profile are adopted. Figure 6 shows the variation of wind speed for the first and second wind turbine system.

Figure 7 illustrates the simulation of the active and reactive power injected into the grid form the first and second wind turbine system respectively. We can observe that the MPPT algorithm track accurately the maximum power. As we can see the reactive power was set to zero to ensure a unit power factor.

Figure 8 depicts the total of active and reactive power injected into the AC bus.

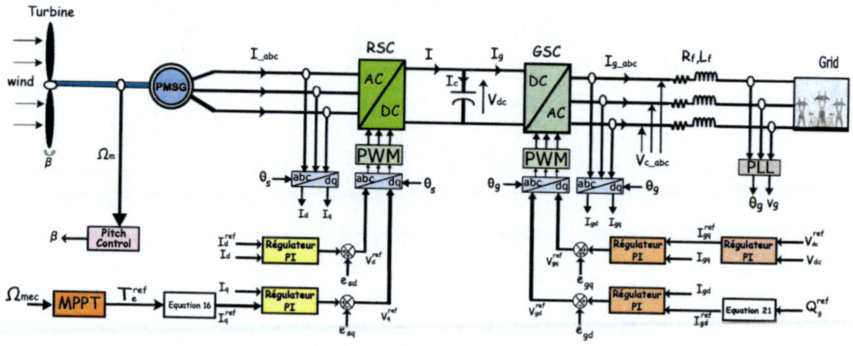

Fig. 5. Architecture control of wind turbine based PMSG

Table 1. Wind turbine and PMSG parameters

PMSG parameters	Values	WT parameters	Values
Rated power	2 MW	Air density	1.225 kg/m^3
Stator resistance	$7.30 \times 10^{-4}\Omega$	Rotor radius	R = 35 m
Stator inductance	$Ld = 1.21 \times 10^{-3}$H; $Lq = 2.31 \times 10^{-3}$H	Rated power	2.0093 MW
Number of pole pairs	30	Rated wind speed	12 m/s

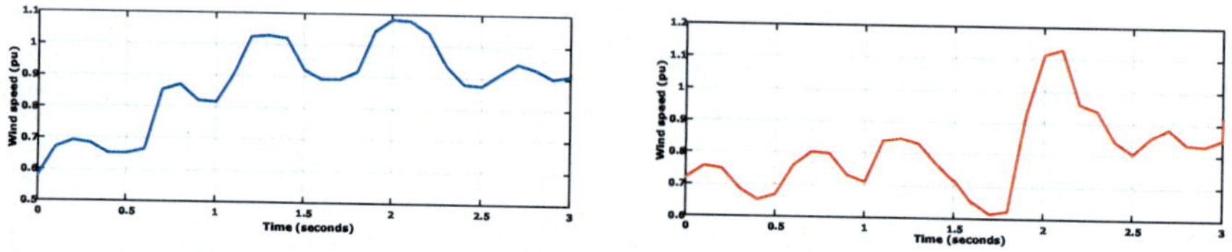

(a) (b)

Fig. 6. The wind speed profile for (a): first turbine and (b): second turbine

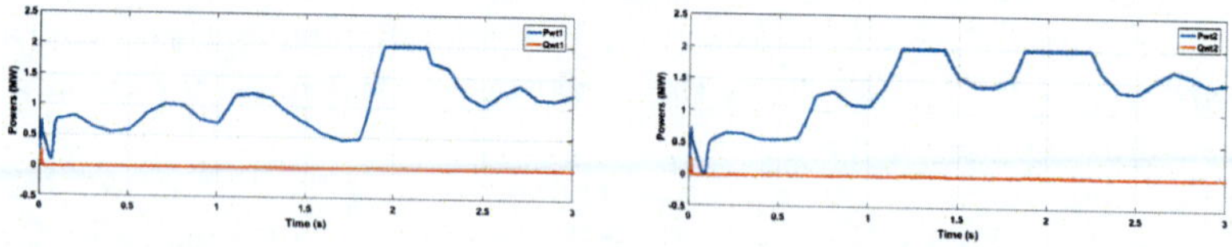

(a) (b)

Fig. 7. The active and reactive power injected into the grid of (a): WTS1 and (b): WTS2

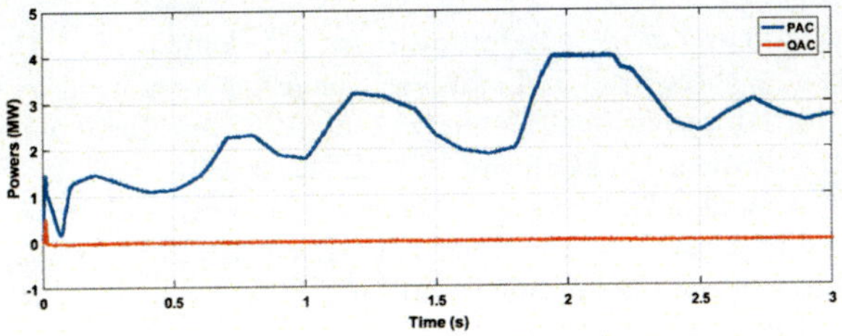

Fig. 8. Total active and reactive power injected into AC bus

5 Conclusion

The aims of our work was confirmed by the Simulations. The modeling and control of two wind turbines based PMSG in AC MG connected to the main grid and their operation are presented and discussed in this article. A presentation of some contributions of wind turbine based PMSG in AC or DC MG is presented. The system is capable to operate at the maximum power during wind speed changes. The control system based on back to back topology control is validate through dynamic simulation. This work is implemented using MATLAB/Simulink environment. The simulation gives considerable results. It's shown that the design and control system guarantee excellent performance in terms of stability, tracking and power quality injected.

References

1. Meliani, M., El Barkany, A., El Abbassi, I., et al.: Energy management in the smart grid: state-of-the-art and future trends. Int. J. Eng. Bus. Manag. **13**, 18479790211032920 (2021)
2. Choukri, K., Naddami, A, Hayani, S.: Renewable energy in emergent countries: lessons from energy transition in Morocco. Energy, Sustain. Soc. **7**(1), 1–11 (2017)
3. Oussama, H., Othmane, A., Abdeselem, C., et al.: Wind turbine generator based on PMSG connected to DC microgrid system. Int. J. Control, Energy Electr. Eng. **7**, 40–43 (2019)
4. Zammit, D., Staines, C.S., Micallef, A., et al.: Incremental current based MPPT for a PMSG micro wind turbine in a grid-connected DC microgrid. Energy Procedia **142**, 2284–2294 (2017)
5. El Hassane Margoum, Krami, N., Harmouch, F.Z., et al.: Control design and operation of photovoltaic systems in low voltage AC MicroGrid. In: 2016 17th International Conference on Sciences and Techniques of Automatic Control and Computer Engineering (STA), pp. 569–574. IEEE (2016)
6. Laghridat, H., Essadki, A., Annoukoubi, M., et al.: A novel adaptive active disturbance rejection control strategy to improve the stability and robustness for a wind turbine using a doubly fed induction generator. J. Electr. Comput. Eng. **2020** (2020)
7. Aboudrar, I., El Hani, S., Mediouni, H., et al.: Modeling and robust control of a grid connected direct driven PMSG wind turbine by ADRC. Adv. Electr. Electron. Eng. **16**(4), 402–413 (2018)

8. Annoukoubi, M., Essadki, A., Laghridat, H., et al.: Comparative study between the performances of a three-level and two-level converter for a wind energy conversion system. In: 2019 International Conference on Wireless Technologies, Embedded and Intelligent Systems (WITS), pp. 1–6. IEEE (2019)
9. Laghridat, H., Essadki, A., Nasser, T.: Comparative analysis between PI and linear-ADRC control of a grid connected variable speed wind energy conversion system based on a squirrel cage induction generator. Math. Prob. Eng. **2019** (2019)
10. Meliani, M., El Barkany, A., El Abbassi, I., et al.: Control system in the smart grid: state of the art and opportunities. In: 2020 IEEE 13th International Colloquium of Logistics and Supply Chain Management (LOGISTIQUA), pp. 1–6. IEEE (2020)

Analysis of the Magnetic Field Effect on Thermosolutal Convection Heat and Mass Transfer in a Square Cavity Filled with Nanofluid

Maryam Bernatchou(✉), Kamal Gueraoui, Mohammed Cherraj, Ahmed Rtibi, and Mustapha El Hamma

Faculty of Sciences, MSME Team, Mohammed V University in Rabat, Rabat, Morocco
maryam_bernatchou@um5.ac.ma

Abstract. The problem of combined heat and mass transfer of thermosolutal natural convection in a rectangular cavity filled with a metallic nanofluid $Cu - H_2O$, under the influence of magnetic field and considering the Soret effect was investigated. The longer walls of the cavity are maintained at constant temperatures and concentrations while the short ones are adiabatic and impermeable except in the left lower region, a constant temperature is applied. The finite volume method is used to solve the governing equations in order to predict the effect of varying Hartmann number, Thermal Rayleigh number, Buoyancy ratio, and cavity inclination angle on heat and mass transfer, and on the variation of temperature, concentration, velocities, and stream function inside the cavity. The main results obtained from this study show that increasing Hartmann number improves mass transfer and decreases heat transfer, increasing Buoyancy ratio improves heat transfer and decreases mass transfer, reducing thermal Rayleigh number reduces heat transfer and increases mass transfer.

Keywords: Thermosolutal convection · Heat and mass transfer · Magnetic field · Nanofluid · Soret

1 Introduction

The phenomenon of natural convection heat and mass transfer in enclosed spaces is of great importance especially in fields such as chemistry, biology, astrophysics, and geosciences. In this sense, many geometries have been used to study the natural convection in cavities, but the majority focused on simple geometries like square or rectangular ones. The Soret effect, which results in a concentration gradient when the fluid is exposed to a temperature gradient, has also been subject to numerous studies.

The need to improve natural convection heat transfer has been of great interest to researchers. Adding a small concentration of nanoparticles to the base fluids such as water and oils can significantly enhance thermal properties of these fluids [1–4]. The resulted suspensions are called nanofluids. Nanofluids are good thermal and electrical

L. Azrar et al. (Eds.): CIP 2022, LNME, pp. 605–615, 2023.
https://doi.org/10.1007/978-3-031-23615-0_61

conductors [5]. In the presence of a magnetic field, the nanofluid is subject to two forces of buoyancy and Lorentz. The Lorentz force induces the magnetohydrodynamics (MHD). A sub-discipline of MHD, the magneto-convection, studies the natural convection of electrically conducting fluids in the presence of magnetic field. Magneto-convection is present in many processes, mainly in metals processing, power plants, petrochemicals, and nuclear reactors [5–9]. Applying magnetic field to electrically conductive fluids has potential applications for controlling convection [10].

Significant research studying the effects of magnetic field on convection in nanofluids has been done. Dutta et al. [11] analyzed heat transfer and entropy generation of MHD natural convection in an inclined cavity filled with $Cu - H_2O$ nanofluid. Hussam et al. [12] studied the natural convection of $Cu - water$ nanofluid in a rectangular cavity subject to sinusoidal boundary temperature, in the presence of magnetic field. Mliki et al. [13] discussed the influence of magnetic field on natural convection of $Cu - water$ nanofluid in a cavity subject to sinusoidal boundary conditions. Ghasemi et al.[14] performed a numerical study to analyze MHD convection in a square cavity filled with $Al_2O_3 - water$ nanofluid. Ouyahia et al. [15] studied the convection of $Al_2O_3 - H_2O$ nanofluid in an inclined square cavity subject to a magnetic field. A number of studies discussed MHD double diffusive natural convection. Moolya et al. [8] studied the effect of applying constant magnetic field on thermosolutal convection in an inclined rectangular cavity. The MHD thermosolutal convection in rectangular cavity, with constant heat and mass fluxes imposed has been treated by Chamkha et al. [16]. Kefayati [17] discussed the double diffusive convection of a non-Newtonian fluid in open cavity, considering the Soret and Dufour effects. Mojumder et al.[18] analyzed the entropy generation for double diffusive MHD convection in rectangular cavity. Teamah et al. [19] studied the influence of magnetic field on heat, mass, and flow transfer by thermosolutal convec tion. For thermosolutal convection in square cavities filled with nanofluids, in the presence of magnetic field, it has been discussed by Kushawaha et al. [6]. They studied thermosolutal convection of nanofluids in cavity subjected to sinusoidal boundary conditions. Mahapatra et al. [7] studied MHD double diffusive convection heat, mass, and flow transfer of nanofluid in trapezoidal cavity. Parveen et al. [9] analyzed heat and mass transfer of MHD double diffusive convection in $Al_2O_3 - H_2O$ nanofluid. Aly [20] studied the influence of Soret and Dufour effects on convection of $Cu - H_2O$ nanofluid inside a square annulus. In this article, we investigate the thermosolutal convection in rectangular cavity filled with $Cu - H_2O$ nanofluid containing Copper nanoparticles suspended in a binary base fluid in the presence of uniform magnetic field, and taking into consideration Soret effect. After presenting some articles dealing with this topic, we presented the geometrical configuration followed by the mathematical model. The numerical method is presented. We finish by presenting the obtained results and their interpretation.

2 Mathematical Formulation

The geometry of the present work is a rectangular cavity with the height h, length L, and inclination angle θ saturated by $Cu - H_2O$ nanofluid, as shown in Fig. 1. The cavity aspect ratio is $A_r = \frac{L}{h} = 2$. An external magnetic field of flux density B_0 is applied in an arbitrary direction in the $X - Z$ plane. The longer walls are fixed at uniform temperatures and concentrations (T_c, T_f) and (C_{inf}, C_{sup}) respectively with $T_c > T_f$ and $C_{inf} > C_{sup}$, while the short walls are adiabatic and impermeable except in the left wall for $z < \frac{h}{2}$, a constant temperature T_c is applied.

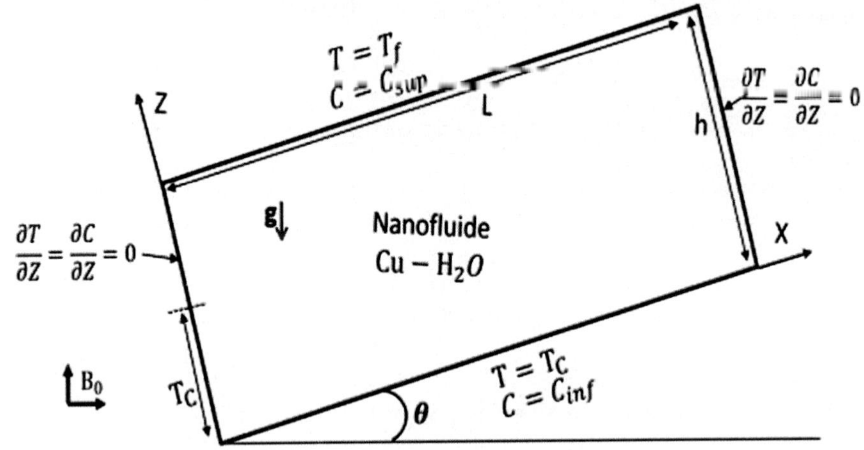

Fig. 1. Schematic diagram of the problem

The dimensionless equations for conservation of energy, concentration, and momentum are as follows:

$$U\frac{\partial T}{\partial X} + W\frac{\partial T}{\partial Z} = \frac{K_e}{\alpha_f}\left(\frac{\partial^2 T}{\partial X^2} + \frac{\partial^2 T}{\partial Z^2}\right) \quad (1)$$

$$U\frac{\partial C}{\partial X} + W\frac{\partial C}{\partial Z} = \frac{1}{L_e}\left(\frac{\partial^2 C}{\partial X^2} + \frac{\partial^2 C}{\partial Z^2}\right) + \frac{S_r}{L_e}\left(\frac{\partial^2 T}{\partial X^2} + \frac{\partial^2 T}{\partial Z^2}\right) \quad (2)$$

$$\begin{aligned} &\Omega\frac{\partial U}{\partial x} + \Omega\frac{\partial W}{\partial z} - U\frac{\partial \Omega}{\partial x} - W\frac{\partial \Omega}{\partial z} \\ &= P_r R_T\left[\left(\frac{\partial T}{\partial x}\cos\theta - \frac{\partial T}{\partial z}\sin\theta\right)\right. \\ &\left. - N\left(\frac{\partial C}{\partial x}\cos\theta - \frac{\partial C}{\partial z}\sin\theta\right)\right] \\ &+ P_r\left(\frac{\partial^2 \Omega}{\partial x^2} + \frac{\partial^2 \Omega}{\partial z^2}\right) - P_r H_a^2 \Omega \end{aligned} \quad (3)$$

$$\Omega = \frac{\partial^2 \Psi}{\partial X^2} + \frac{\partial^2 \Psi}{\partial Z^2} \quad (4)$$

where: $K_e = \frac{K_{nf}}{(\rho C_p)_{nf}}$.

The thermal conductivity and specific heat capacity of the nanofluid are calculated using the expressions in Table 1.

Table 1. Thermophysical properties of the nanofluid

Thermophysical properties	Expression
Thermal conductivity	$K_{nf} = K_f \frac{K_s+2K_f+2\varphi(K_s-K_f)}{K_s+2K_f-\varphi(K_s-K_f)}$
Specific heat capacity	$(\rho C_p)_{nf} = \varphi(\rho C_p)_s + (1-\varphi)(\rho C_p)_f$

R_T, L_e, S_r, H_a are, respectively, numbers of Rayleigh, Prandtl, Lewis, Soret, and Hartmann.

N is the Buoyancy ratio.

$$\mathrm{R_T} = \frac{g\beta_T\Delta T^* h^3}{\nu_f\alpha_T}\ \mathrm{P}_r = \frac{\nu_f}{\alpha_T}\ \mathrm{L_e} = \frac{\mathrm{S_c}}{\mathrm{P}_r}$$

$$\mathrm{N} = \frac{\Delta C^*\beta_S}{\Delta T^*\beta_T}\ \mathrm{S_r} = \frac{\Delta T^* K_{CT}}{D\Delta C^*}\ \mathrm{H_a} = \mathrm{L.B_0}\sqrt{\frac{\sigma_{nf}}{\rho_{nf}\nu_f}}$$

The boundary conditions in non-dimensional form associated to the equations are given below.

$$Z = h, 0 < X < L: \quad T = 0, \quad C = 0, \quad \Omega_{(Z=h)} = \frac{1}{X}\frac{2}{\Delta Z}\Psi_{(Z=h-\Delta Z)} \tag{5}$$

$$Z = 0, 0 < X < L: \quad T = 1, \quad C = 1, \quad \Omega_{(Z=0)} = \frac{1}{X}\frac{2}{\Delta Z}\Psi_{(Z=\Delta Z)}$$

$$z < \frac{h}{2}, X = 0: \quad T = 1, \quad \frac{\partial C}{\partial Z} = 0, \quad \Omega_{(X=O)} = \frac{2}{\Delta X}\Psi_{(X=\Delta X)}$$

$$z > \frac{h}{2}, X = 0: \quad \frac{\partial T}{\partial Z} = 0, \quad \frac{\partial C}{\partial Z} = 0, \quad \Omega_{(X=O)} = \frac{2}{\Delta X}\Psi_{(X=\Delta X)}$$

$$0 < z < L, X = L: \quad \frac{\partial T}{\partial Z} = 0, \quad \frac{\partial C}{\partial Z} = 0, \quad \Omega_{(X=1)} = \frac{2}{\Delta X}\Psi_{(X=1-\Delta X)}$$

The local and average Nusselt numbers are expressed by:

$$\mathrm{Nu} = -(\frac{\partial T}{\partial Z})_{Z=0}\quad \overline{Nu} = -\frac{1}{A_r}\int_0^{A_r}\frac{\partial T}{\partial Z}dX$$

The local and average Sherwood numbers are expressed by:

$$\mathrm{Sh} = -(\frac{\partial C}{\partial Z})_{Z=0}\quad \overline{Sh} = -\frac{1}{A_r}\int_0^{A_r}\frac{\partial C}{\partial Z}dX$$

3 Numerical Method

The finite volume method is used to solve the governing equations of energy, concentration, and momentum (1)–(4). The discretized equations are solved using double sweep method [24]. The stream function equation is discretized using the method of Successive Over Relaxation (S.O.R.). For each time step, we use an iterative procedure to calculate velocities, vorticity, stream function, temperature, and concentration. We repeat the procedure until convergence. The convergence criteria is $\frac{\sum_i \sum_j \left| f_{i,j}^{t+\Delta t} - f_{i,j}^{t} \right|}{\sum_i \sum_j \left| f_{i,j}^{t+\Delta t} \right|} \leq 10^{-5}$ where f represents the temperature, concentration, and vorticity.

The calculation code we developed in Fortran-95 has been validated with the results found by other authors The presented results are obtained for a mesh size of 41×41.

4 Results and Discussion

4.1 Influence of Hartmann Number

The variation of temperature and concentration by varying H_a at $X = \frac{1}{2}$ are presented in Figs. 2 and 3. From Fig. 2, temperature reduces by increasing H_a from 0 to 100 for $R_T = 10^4$ and 10^6. The rate of this decrease is 6.008×10^{-6} for $R_T = 10^4$, and 6.04×10^{-4} for $R_T = 10^6$. We notice also that temperature rises with R_T. For concentration, we can observe in Fig. 3 that it decreases with H_a. The rate of this reduction is 5.16×10^{-7} for $R_T = 10^4$ and 5.05×10^{-5} for $R_T = 10^6$. In addition, we note that concentration increases with R_T. Thus, we can conclude that temperature and concentration reduce with H_a. We conclude also that concentration and temperature decrease by decreasing R_T number.

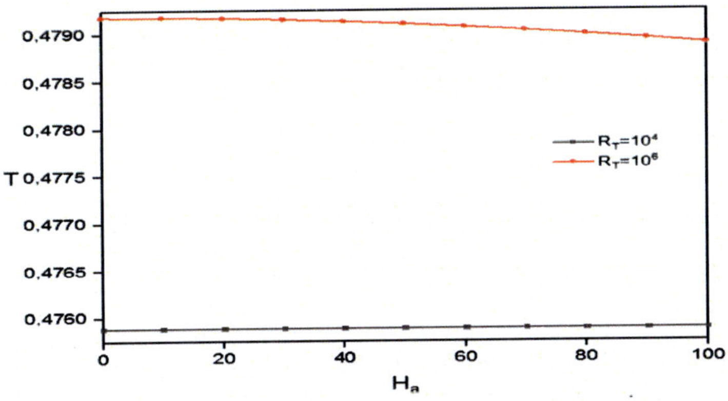

Fig. 2. Effect of H_a on temperature for $N = 1$ and $\theta = 30^{\circ}$, at $X = \frac{1}{2}$ and $Z = \frac{1}{2}$

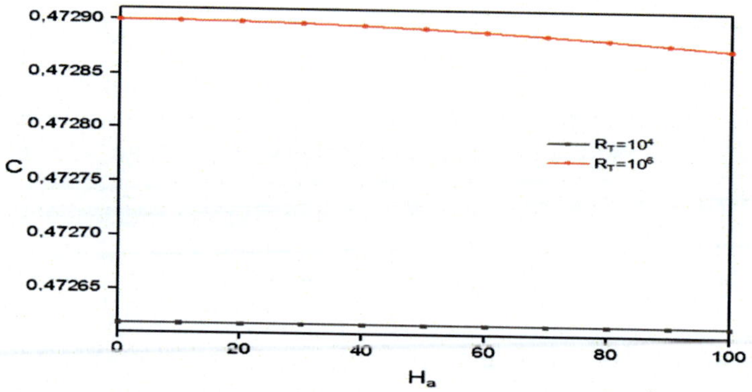

Fig. 3. Effect of H_a on concentration for $N = 1$ and $\theta = 30^\circ$, at $X = \frac{1}{2}$ and $Z = \frac{1}{2}$

4.2 Velocity, Vorticity, and Stream Function Distributions

The distribution of velocity, vorticity and stream function along the Z axis at $X = 0.51$ are presented in Figs. 4, 5, 6 and 7. In Fig. 4, the U velocity declines from 0 to -0.082, -0.08, -0.075 at $Z = 0.047$, then increases to 0.041, 0.04, 0.037 at $Z = 0.073$ to approach the value 0 at $Z = 0.1$, for H_a equal to 0, 50, and 100 respectively. In the interval [0.1, 1], velocity is constant. The three curves are almost identical so that velocity increases with increasing H_a. For the velocity W (see Fig. 5), it declines from 0 to 9.65×10^{-4}, 9.42×10^{-4} and 8.81×10^{-4} at $Z = 0.047$, and then increases to approaches 0 at $Z = 0.1$. Above the value 0.1, the velocity is nearly constant in the three cases, and the curves of velocity are practically identical. In this case, effect of magnetic field on velocity can be considered as negligible. From Fig. 6, for Z between 0 and 0.414, vorticity Ω is nearly constant and equal to 0 in the three cases. It takes the values -783.46, -765.72, -717.02 at $Z = 0.463$, and the values 751.49, 734.49, 687.83 at $Z = 0.487$ for $H_a = 0, 50, 100$. For Z between 0.5 and 1, variation of Ω is negligible. It takes the value 0 for the three cases of H_a. In addition, we note that Ω takes largest values with increasing H_a for $Z \in [0, 0.463]$, while it decreases with increasing H_a for $Z \in [0.463, 1]$. Distribution of ψ is given in Fig. 7. We can see that it decreases from 0 to -0.001 at $Z = 0.047$, then it increases to approach 0 at $Z = 0.073$. In the interval $Z \in [0.097, 1]$, ψ diminishes by increasing H_a.

4.3 Influence of the Thermal Rayleigh Number

Effect of Hartmann number on the average Nusselt and Sherwood numbers ($\overline{Nu}$ and $\overline{Sh}$) by varying the cavity inclination angle and the thermal Rayleigh number, is shown in Figs. 8 and 9. $\overline{Nu}$ reduces with the growth of H_a for all values of θ and R_T (see Fig. 8). This decrease is due to the increase of Lorentz force by increasing magnetic field intensity, which affects the thermosolutal convection. From Fig. 8, we see that $\overline{Nu}$ grows with reducing R_T from 10^4 to 10^6. This variation is a consequence of the increase in thermal volume forces, due to the increase of R_T. We can also see from Fig. 8 that $\overline{Nu}$ takes highest values at $\theta = 45°$. We conclude that 45° is the optimum inclination angle

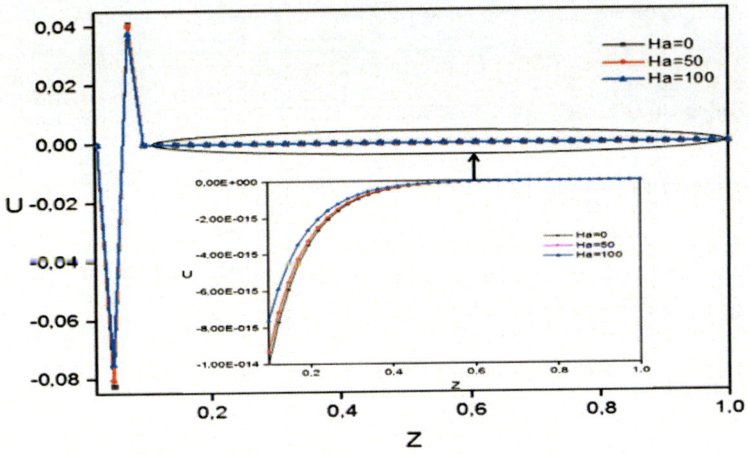

Fig. 4. Distribution of velocity U along Z axis at $X = 0.51$ for $R_T = 10^6$, $N = 1$ and $\theta = 30°$

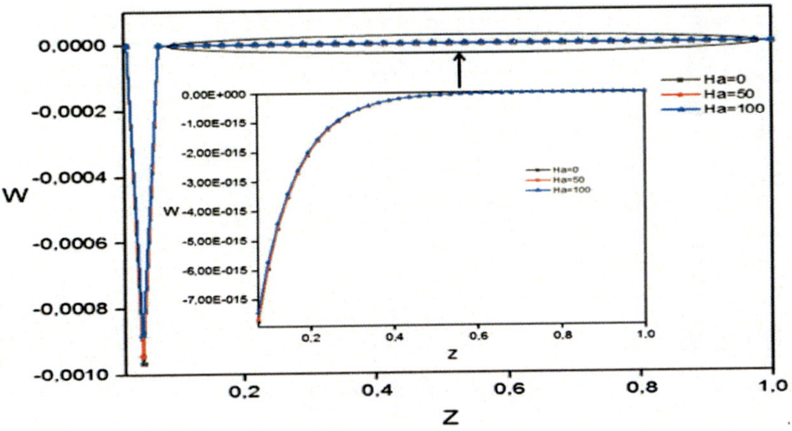

Fig. 5. Distribution of velocity W along Z axis at $X = 0.51$ for $R_T = 10^6$, $N = 1$ and $\theta = 30°$

of the cavity for better heat transfer. In Fig. 9, the mass transfer increases with decreasing R_T from 10^6 to 10^4, due to the increase in solutal volume forces. $\overline{Sh}$ increases also by reducing θ from 90° to 0°, for both cases of R_T. Then, the optimal cavity inclination angle for better mass transfer is 0°. We conclude that the mass transfer is reduced with the increase of H_a. It is reduced also by diminishing R_T. However, the heat transfer is improved with increasing H_a for all values of R_T and θ, and also with decreasing R_T and θ. Heat transfer takes the highest value for $\theta = 45°$, while the mass transfer is highest at $\theta = 0°$.

4.4 Influence of the Buoyancy Ratio

The influence of varying H_a on heat and mass transfer for three different values of Buoyancy ratio $N(1, 5, 10)$ is illustrated in Figs. 10 and 11. From Fig. 10, $\overline{Nu}$ decreases

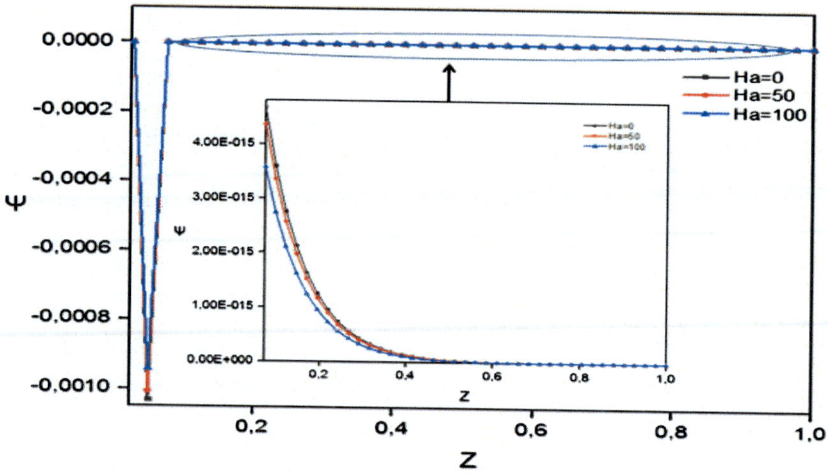

Fig. 6. Distribution of ψ along Z axis at $X = 0.51$ for $R_T = 10^6$, $N = 1$ and $\theta = 30°$

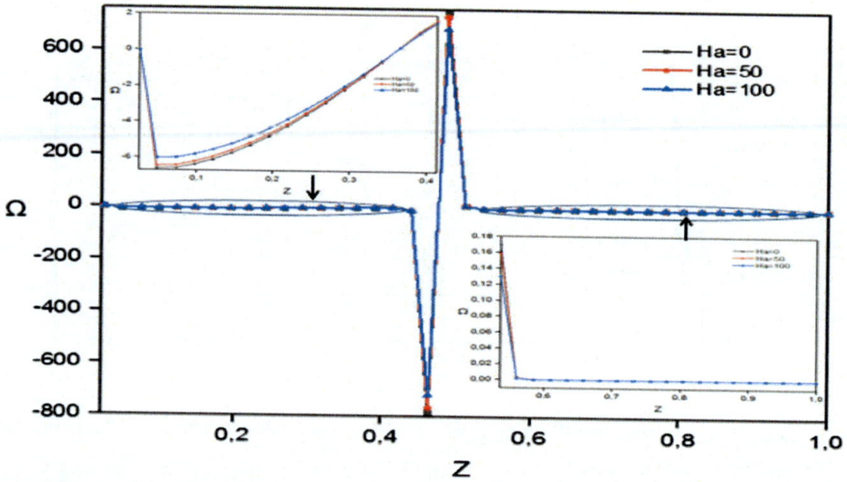

Fig. 7. Distribution of Ω along Z axis at $X = 0.51$ for $R_T = 10^6$, $N = 1$ and $\theta = 30°$

with increasing H_a for all the values of N. $\overline{Nu}$ is improved with increasing N, because increasing the thermal buoyancy rate within the enclosure increases the value of $\overline{Nu}$. Contrary to mass transfer, and from Fig. 11, we see that $\overline{Sh}$ increases with H_a for the three values of N, and decreases with increasing N. Thus, we can conclude that increasing N improves the heat transfer on the one hand, and on the other hand, it decreases the mass transfer, and that increasing H_a decreases heat transfer and improves mass transfer, mainly for low values of N.

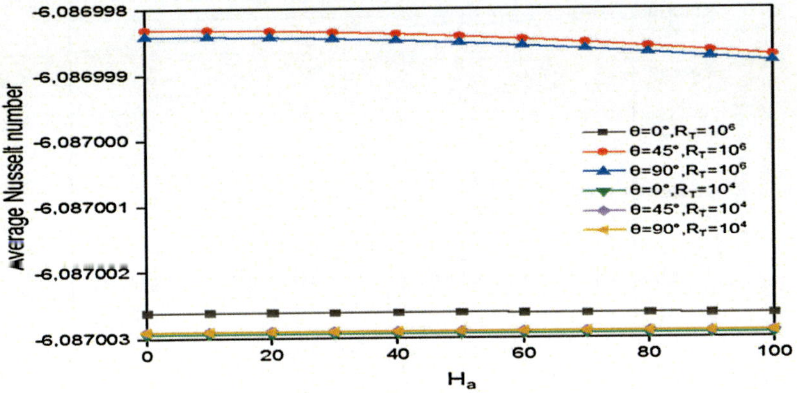

Fig. 8. Effect of $\mathrm{H_a}$ on $\overline{Nu}$ for $N = 1$

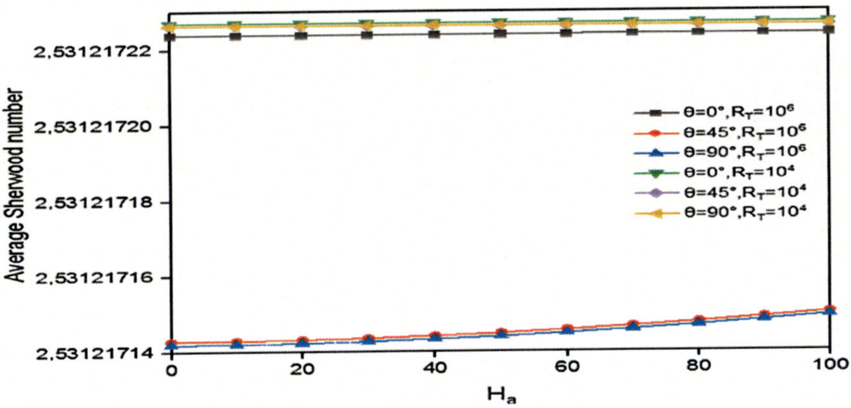

Fig. 9. Effect of $\mathrm{H_a}$ on $\overline{Sh}$ for $N = 1$

5 Conclusion

Double diffusive magneto-convection in an inclined rectangular enclosure filled with $Cu - H_2O$ nanofluid has been studied. It can be concluded from this study that:

- Increasing Hartmann number improves mass transfer and decreases heat transfer.
- Along the Z axis, increasing $\mathrm{H_a}$ increases the velocity U and decreases the stream function. The vorticity increases with increasing $\mathrm{H_a}$ for $Z \in [0, 0.463]$, and decreases in the rest of the interval.
- Reducing the thermal Rayleigh number reduces heat transfer and increases mass transfer.
- The mass transfer increases with decreasing cavity inclination angle and takes the highest value at $\theta = 0^\circ$, while the heat transfer takes the highest value at $\theta = 45^\circ$.
- Temperature and concentration reduce with rising the Hartmann number and decreasing thermal Rayleigh number.

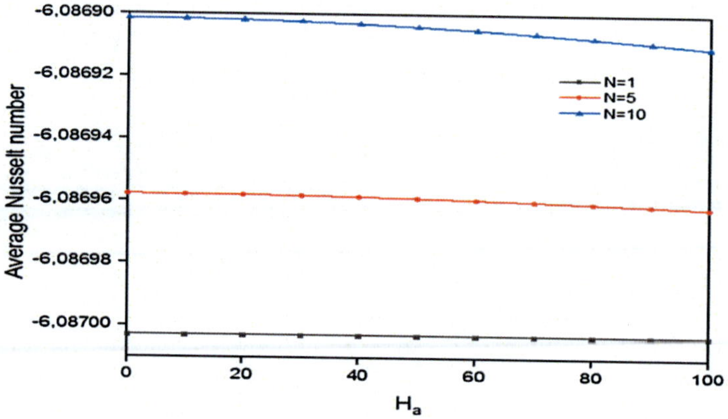

Fig. 10. Effect of H_a on $\overline{Nu}$ for $R_T = 10^4$ and $\theta = 30°$

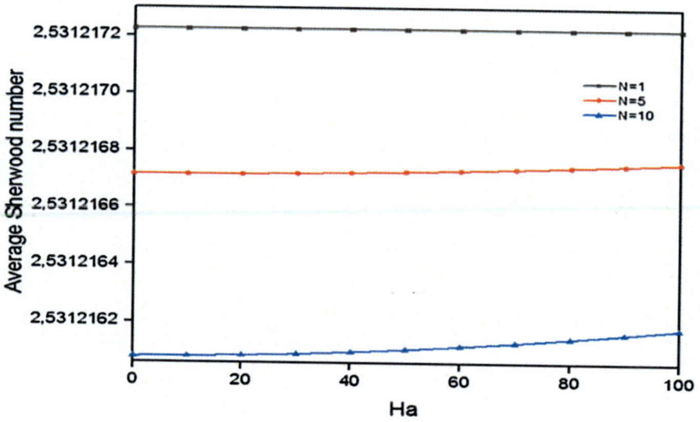

Fig. 11. Effect of H_a on $\overline{Sh}$ for $R_T = 10^4$ and $\theta = 30°$

- Increasing the Buoyancy ratio improves heat transfer and decreases mass transfer.

References

1. Bernatchou, M., Gueraoui, K., Rtibi, A., Cherraj, M., El Hamma, M.: Effect of magnetic field on double diffusive natural convection in an inclined cavity filled with nanofluid considering the Soret effect. JP J. Heat Mass Transf. **25**, 1–25 (2022)
2. Bernatchou, M., Rtibi, A., Gueraoui, K., El Hamma, M., Cherraj, M.: Study of thermosolutal natural convection of Cu-water nanofluid in an inclined cavity. JP J. Heat Mass Transf. **27**, 37–56 (2022)
3. Ganvir, R.B., Walke, P.V., Kriplani, V.M.: Heat transfer characteristics in nanofluid—a review. Renew. Sustain. Energy Rev. **75**, 451–460 (2017)
4. Gupta, M., Singh, V., Kumar, R., Said, Z.: A review on thermophysical properties of nanofluids and heat transfert applications. Renew. Sustain. Energy Rev. **74**, 638–670 (2017)

5. Giwa, S.O., Sharifpur, M., Ahmadi, M.H., Meyer, J.P.: A review of magnetic field influence on natural convection heat. J. Therm. Anal. Calorim. 2581–2623 (2020)
6. Kushawaha, D., Yadav, S., Singh, D.K.: Magnetic field effect on double-diffusion with magnetic and non-magnetic nanofluids. Int. J. Mech. Sci. (2021)
7. Mahapatra, T.R., Saha, B.C., Pal, D.: Magnetohydrodynamic double-diffusive natural convection for nanofluid within a trapezoidal enclosure. Comput. Appl. Math. **37**(5), 6132–6151 (2018). https://doi.org/10.1007/s40314-018-0676-5
8. Moolya, S., Satheesh, A.: Role of magnetic field and cavity inclination on double diffusive mixed convection in rectangular enclosed domain. Int. Commun. Heat Mass Transf. **118** (2020)
9. Parveen, R., Mahapatra, T.R.: Numerical simulation of MHD double diffusive natural convection and entropy generation in a wavy enclosure filled with nanofluid with discrete heating. Heliyon **5** (2019)
10. Kaddeche, S., Henry, D., Putelat, T., Hadid, H.B.: Instabilities in liquid metals controlled by constant magnetic field—part I: vertical magnetic field. J. Cryst. Growth **242**, 491–500 (2002)
11. Dutta, S., Goswami, N., Biswas, A.K., Pati, S.: Numerical investigation of magnetohydrodynamic natural convection heat transfer and entropy generation in a rhombic enclosure filled with Cu-water nanofluid. Inter. J. Heat Mass Transf. **136**, 777–798 (2019)
12. Hussam, W.K., Khanafer, K., Salem, H.J., Sheard, G.J.: Natural convection heat transfer utilizing nanofluid in a cavity with a periodic side-wall temperature in the presence of a magnetic field. Int. Commun. Heat Mass Transf. **104**, 127–135 (2019)
13. Mliki, B., Ali, C., Abbassi, M.A.: Lattice Boltzmann simulation of MHD natural convection heat transfer of Cu-water nanofluid in a linearly/sinusoidally heated cavity. Int. J. Phys. Math. Sci. **14** (2020)
14. Ghasemi, B., Aminossadati, S.M., Raisi, A.: Magnetic field effect on natural convection in a nanofluid-filled square enclosure. Int. J. Therm. Sci. **50**, 1748–1756 (2011)
15. Ouyahia, S., Benkahla, Y.K., Benzema, M., Berabou, W.: Effet du champ magnétique sur la convection naturelle dans une cavité inclinée remplie d'un nanofluide. Congrès de la Société Française de Thermique (2015)
16. Chamkha, A.J., Al-Naser, H.: Hydromagnetic double-diffusive convection in a rectangular enclosure with uniform side heat and mass fluxes and opposing temperature and concentration gradients. Int. J. Therm. Sci. **41**, 936–948 (2002)
17. Kefayati, G.H.R.: Simulation of double diffusive MHD (magnetohydrodynamic) natural convection and entropy generation in an open cavity filled with power-law fluids in the presence of Soret and Dufour effects (part I: study of fluid flow, heat and mass transfer). Energy **107**, 889–916 (2016)
18. Mojumder, S.: Analysis of entropy generation for double diffusive MHD convection in a square cavity with isothermal hollow cylinder. In: AIP Conference Proceedings (2016)
19. Teamah MA: Numerical simulation of double diffusive natural convection in rectangular enclosure in the presence of magnetic field and heat source. Int. J. Therm. Sci. **47**, 237–248 (2008)
20. Aly, A.M.: ISPH method for MHD convective flow from grooves inside a nanofluid-filled cavity under the effects of Soret and Dufour numbers. Physica **546** (2020)
21. Mrabti, A.: Simulation numérique d'écoulements de convection naturelle dans une géométric cylindrique à axe vertical soumise à l'effet d'un champ magnétique ou d'un gradient solutal. Dissertation, University Mohamed 5 of Rabat (1999)

Author Index

L. Azrar et al. (Eds.): CIP 2022, LNME, 2023.
https://doi.org/10.1007/978-3-031-23615-0